To Brad
With Warmest
Best Wishes
Stan

STANFORD R. OVSHINSKY

THE SCIENCE AND TECHNOLOGY OF AN AMERICAN GENIUS

STANFORD R. OVSHINSKY
THE SCIENCE AND TECHNOLOGY OF AN AMERICAN GENIUS

HELLMUT FRITZSCHE
University of Chicago, USA

BRIAN SCHWARTZ
The Graduate Center of the
City University of New York, USA

World Scientific

NEW JERSEY · LONDON · SINGAPORE · BEIJING · SHANGHAI · HONG KONG · TAIPEI · CHENNAI

Published by

World Scientific Publishing Co. Pte. Ltd.
5 Toh Tuck Link, Singapore 596224
USA office: 27 Warren Street, Suite 401-402, Hackensack, NJ 07601
UK office: 57 Shelton Street, Covent Garden, London WC2H 9HE

British Library Cataloguing-in-Publication Data
A catalogue record for this book is available from the British Library.

THE SCIENCE AND TECHNOLOGY OF AN AMERICAN GENIUS — **Stanford R. Ovshinsky**

ISBN-13 978-981-281-839-3
ISBN-10 981-281-839-1

Desk Editor: Tjan Kwang Wei

Printed in Singapore by World Scientific Printers

Presented to

Stanford R. Ovshinsky

on the occasion of his

85th Birthday

November 24, 2007

CONTENTS

Chapter I: Stan Ovshinsky

The last century has been the century of unrivaled expansion of our understanding and exploitation of Nature. For the first time we learned how and when the Universe was formed, the nature of its most elementary and smallest constituents, and the origin and nature of our species. Our incredibly advanced information technology has made our world accessible and small.

During the past century an ever increasing rate of specialization produced new records of accomplishment and innovation. These developments grew out of the collective efforts of creative minds at our centers of learning, and the inventive and curious spirits that were supported by technologically advanced industries.

Enter Stan Ovshinsky. Born in 1922 in Akron, Ohio, he came from a humble background. He obtained his education by taking an armful of books each week from the Akron Public Library. His endlessly active mind needed no teacher. His academy and his college were, and continue to be throughout his life, books feeding his intellectual hunger. Working as a tool maker and lathe operator in machine shops, he questioned why different materials, such as cast iron or the different steels, vary in ductility or strength; what happens during annealing, and, why do cutting tools lose their edge? Can one stop the shaking and the vibrations of the lathe at high speeds; why is the whole machine out of balance? These simple questions going through Ovshinsky's mind opened for him a deep insight into the functions of the elements, their combinations and the different atomic arrangements in materials, defining their properties. These probing questions did not let go of him until he had invented a well balanced lathe without loose tolerances which ran at ten times higher speed and had cutting tools on both sides. His first important invention was soon followed by automated machine tools. Important lessons he learned were first that problems can be solved by thoroughly analyzing and understanding them and second that one has to build and demonstrate their solutions in order to convince others. Words are not enough. Ovshinsky followed these lessons throughout his life inviting his friends, his doubters as well as his opponents, to come to his company and let the results speak for themselves. No one left unconvinced after such a visit.

It is astonishing that this one remarkable individual, Stanford R. Ovshinsky, self taught and without special training, could compete with the well funded establishments of learning and industry in the second half of this past century and leave us an incredible legacy of brilliant innovations with a lasting impact on our lives. He has navigated the world of science and technology without formal academic training, nor was he funded by high tech industry. He has taken an individual path that places him more in line with Thomas Edison or Charles Darwin, those self taught geniuses of the 19th Century, who laid the ground work for the advances of the 20th Century.

Ovshinsky is the inventor of the nickel-metal hydride battery, which powers a large fraction of our electronic tools and which is indispensable for electric and hybrid automobiles. He invented our rewritable CD and DVD optical disks, as well as new forms of non-volatile computer memories which are being commercialized through Intel, STMicroelectronics and Samsung among others. He holds crucial patents relating to flat panel displays, non-silver photography, hydrogen storage materials, and thin-film solar cells. Moreover, this large range of apparently disparate inventions did not grow from the solid base of accepted knowledge of materials science. They evolved from a new paradigm of materials discovered and created by Ovshinsky, which at that time contradicted the established teachings of what constitutes useful and scientifically interesting materials. These path-breaking new ideas and inventions were based on his new paradigm of compositional and structural disorder in materials. These ideas broke with the reliance on the crystalline ordered structures, which dominated the conventional thinking of the time and form the contents of this book of selected publications of Ovshinsky and the list of his important patents.

The number of inventions and patents of Ovshinsky rank with those of the master inventor Edison. Yet, this is only a fraction of his accomplishments. He and his wife Iris founded Energy Conversion Devices, Inc. in 1960 and followed a vision "using creative science to solve societal problems": energy conversion devices for freeing our society from our dependence on fossil fuels, using instead the sun as well as hydrogen, the primary element that fuels the universe. According to Ovshinsky, information is encoded energy, therefore, information and energy technologies overlap in his work, they are two sides of the same coin. Working outside the accepted materials technologies, Ovshinsky had to do more than invent devices he needed to realize his vision. He had to create and nurture the scientific foundation for understanding his disordered materials, develop the equipment for making them, invent the machines for their manufacture, and manage his ever growing company: a scientist, inventor, product developer, machine builder, manufacturer, and entrepreneur all in one person.

Ovshinsky's strong and unwavering belief in himself and his awareness of his superior intelligence as well as his unusually dogged mind guided him through times when the establishment just could not accept that he was right. Ovshinsky knew that pure reason and clear logic as well as the laws of nature were on his side, he had thought through and understood each problem and saw the results of his ideas in his laboratory. He concluded that his opponents just needed more time to understand, to be able to overcome the boundaries of their academic thinking. The scientific and industrial communities are often astonishingly conservative and adverse to new concepts and ideas. They are also averse to listening seriously to people outside their league who are not part of their world, like the self taught Ovshinsky. Academic education not only conditions and confines thinking in specialized disciplines, it also narrows the choice of problems that are considered worth solving and sets up blinders to other areas. Lacking this formal education and prejudice, Ovshinsky was able to see all fields of science and engineering broadly as one intellectual unity. His ship of imagination did not stay close to the safe shores of known territory.

Materials science has become ever more important for opening new areas of technology. It is therefore astonishing that a large new family of materials lay dormant, waiting to be discovered and used. The sparks of Ovshinsky's ideas caught fire in Japan, China, and Europe. The Japanese were particularly inspired by the fact that one individual could be so creative and yet outside the establishment. The founders of Sony, Sharp and Cannon asked Ovshinsky to show them how to change their traditional ways and how to encourage individual thought and imagination in their country. Some seeds must have germinated because the PBS science program, Nova, called Ovshinsky "Japan's American Genius". Ovshinsky's new field became known as "Ovonics" for <u>Ov</u>shinsky+electr<u>onics.</u> Soon, international scientific conferences and journals started to focus on this new materials science. Some adopted names like Ovonic science and Ovonic materials. The Ovshinsky Award honors major contributors to the field.

There is a mysterious quality in Ovshinsky's persona that attracts people into his sphere, builds life long friendships and awakens deep respect and devotion. Meeting him leaves each person with a deep impression of his superior intellect, his self confidence, his compassion to improve society combined with his certainty that his vision can be realized. His enthusiasm is contagious. In his presence you feel how exciting it would be to join him in his endeavors. As a result he and his wife Iris attracted many great minds with diverse expertise and broad talents as friends, supporters and collaborators.

Energy Conversion Devices, the company Stan and Iris founded in 1960, always had a different working atmosphere from that of other companies. People felt less as employees than as collaborators participating in Ovshinsky's thought process and sharing the excitement of success and the process of inventing. Most discussions and brainstorming happened at a big round table where its head was anyone who offered an interesting idea. There was no hierarchy and advancement just meant larger responsibilities. This book does not name the large number of collaborators who contributed to the success of Ovshinsky's work. They are found as coauthors of his publications and patents and are acknowledged at the end of his published papers.

The publications selected for this book, though overwhelming in their breadth and scope, portray only the scientific and technical accomplishments. It is impossible to fully describe or bring to life for the reader the richness of Ovshinsky's mind and the greatness of his personality. The very soul of who we are at our very best is expressed in our curiosity. Ovshinsky's curious and inquiring mind is boundless. He explores and admires excellence in any field, be it music, art, painting, poetry, theatre, history ... his over 20,000 volume library, are all beautiful expressions of the best in humans.

Ovshinsky never stopped being fascinated by the functions of the brain, beginning with his first publications on neurophysiology. His original work and life long fascination spawned his, as yet, unrealized idea of a new generation of computers, a cognitive computer based on his lifetime inventions of devices. But these ideas will reach far into the future. This book deals with the past and the present. It tries to bring to life the multifaceted ideas and accomplishments of Ovshinsky during the past 20 years. His earlier publications have been presented and discussed in a prior volume. [1]

[1] Disordered Materials, Science and Technology, Ed. D. Adler, B.B. Schwartz and M. Silver (Plenum Press, New York and London. 1991)

Chapter II: New Science

A fundamentally new science of amorphous and disordered materials has been pioneered by Ovshinsky. How was it possible that this large family of materials remained virtually unexplored; materials, which subsequently opened up new technologies with far-reaching applications? Amorphous materials, i.e. glasses, have been known for thousands of years. The art of making stained glass windows, art glass, and the glass lenses of telescopes that opened our view of the universe is much older than the refined use of crystalline materials which came to dominate the attention of scientists in the past century. The first issue of the Encyclopedia Britannica devotes seven full pages to glass, whereas crystals are discussed on just half a page.

A science of amorphous and disordered materials did not exist before Ovshinsky announced in the 1950s his discovery of extraordinary electronic effects in glassy semiconductors because of two main reasons. First, solid materials are agglomerates of trillion of trillions of atoms, and an understanding of their detailed properties seemed possible only when these atoms are arranged in ordered periodic structures, when the solids are crystals. One of the great successes of quantum mechanics in the 1930s was the theoretical understanding the basic properties of these ordered structures of crystalline materials. Because of this success, generations of scientists were taught by physics textbooks that solids are synonymous with crystals. There was no word about amorphous and disordered solids. The second reason was the invention of the transistor in 1947 which revolutionized electronics and elevated the study of highly purified and nearly perfect crystals to a major success story of science.

Ovshinsky's discovery therefore caught the solid state community by surprise. How could his non-crystalline materials which were mixtures of three, four or more elements be of any importance? This contradicted their teaching and scientists don't easily change their mode of thinking. A very different mindset indeed was necessary to explore these novel semiconducting glasses. None of the scientific concepts taught in the textbooks could explain electronic phenomena in glassy or amorphous solids. One did not even know where the atoms of such non-crystalline materials were located, and why should one even care. The conventional crystals worked miracles in the electronic industry.

Ovshinsky's discoveries made scientists aware that a huge gap existed in our understanding of materials in general. Scientists were soon drawn to the challenge of exploring this virgin territory. Ovshinsky's laboratory became the leading research institution for many years and a sort of Mecca for scientists working on these problems. The early rejection by scientists of the old school was soon replaced by an outburst of scientific conferences, journals and books solely devoted to non-crystalline semiconductors.

A major force driving the scientific exploration was the prospect of new optoelectronic devices reaching beyond the practical boundaries of the electronic devices of that time. The glassy, non-crystalline semiconductors could be prepared inexpensively over large areas, well beyond the limits of the sizes of crystals. One could envision electronic circuits leading to flat panel displays and flat television screens. Many of the early expectations have over the years become reality. Ovshinsky's early switching and phase change memory devices lead to rewritable CDs, DVDs and optical disc memories. Laptop computers are inconceivable without flat panel displays and large area electronics has become reality in form of huge solar panels manufactured by the mile. These developments are topics of several of the following chapters.

In the first two of his papers following these notes, Ovshinsky describes what he calls the Rosetta Stone for deciphering the enigma of disorder. His important first discoveries of the threshold and memory action, which started the field of amorphous semiconductors, are discussed there as well as his experiments which ushered in flat screen displays and the end of the bulky conventional cathode-ray-tube displays. Shown also in these papers is the first rewritable optical memory disk player built by Panasonic and based on Ovshinsky's invention. More on this important development can be found in the section on Phase Change Memory.

Ovshinsky was drawn to the mystery of non-crystalline materials through his work and interest in neurophysiology, attempting to understand the disorder of the surfaces of nerve cells. Since most of the nerve's actions occur at their surfaces, he paid special attention to the surfaces of his materials where the atoms have unusual and novel bonding activities. That is the location of surface catalysis which plays an essential role in enabling or in speeding up chemical reactions. At a Gordon Conference in 1978, Ovshinsky explained how compositional and structural disorder can activate catalytic sites at the nano- and micro-crystalline surface regions obviating the need for expensive catalysts such as platinum and palladium. That idea again was contrary to accepted academic teaching which held that catalyst must have clean and perfect surfaces. Ovshinsky's idea of disorder promoting catalysis, called "completely untenable" by the experts, proved to be correct. It enabled him to invent the now indispensable Nickel-Metal Hydride battery, his regenerative fuel cell as well as hydride alloys in which large amounts of hydrogen can be safely stored and transported. These important inventions are explained in the later sections on "Batteries" and "Hydrogen Storage and Fuel Cells".

A common theme in Ovshinsky's work is his ability to garner desired properties and functions out of synthesized materials by what one can call atomic engineering. A key concept advanced by Ovshinsky is the application of the new and varied degrees of freedom afforded by the disordered and amorphous states of matter. He recognized that the ordered crystalline lattice imposed many constraints on the structure and properties of materials due to a rigid adherence of atoms to a prescribed structural lattice. He exploited the enormous flexibility in chemical bonding, intermolecular interactions, and structural configurations allowed by disordered and amorphous states of matter. He viewed his atomically engineered materials in terms of constituent local structures, each of which has unique properties according to the chemical elements and topology present, which collectively interact to produce macroscopic materials having novel properties.

Some of the many awards and honors bestowed on Ovshinsky are mentioned in the introductions to the various chapters in this book. We close this section by noting that in 2007, the Scientific Research Society Sigma Xi honored Ovshinsky by awarding him the Walston Chubb Award specifically "for pioneering the fundamentally new science of amorphous and disordered materials which has opened up new technologies with far-reaching benefits."

FUNDAMENTALS OF AMORPHOUS MATERIALS

Stanford R. Ovshinsky

Energy Conversion Devices, Inc., 1675 West Maple Road, Troy, Michigan 48084

I. INTRODUCTION

When I first began to study amorphous materials in the mid 1950's, the field appeared to be as mysterious as hieroglyphics had been to renaissance scholars. While it was taken for granted that amorphous materials had no real significance scientifically or technologically, it was clear to me, even then, that this was a rich, unexplored, and important area of science [for early references see 1-5]. Until then its major thrust was in the ancient art of glass making, and glass meetings devoted inordinate amounts of time to discussing "What is glass?"

Just as there was a Rosetta Stone which allowed the deciphering of the hieroglyphics of ancient civilizations, the following is the key I provided to make the nature of amorphous materials clear and to understand their physical properties.

It is the purpose of this paper to discuss how we broke the code, and how we have applied this insight to the development of an array of new devices, several of which will be described in detail. Such an understanding of our field is not yet widespread. For example, I recently received a book from Professors Yonezawa and Ninomiya,[6] both fine scientists. In discussing topologically disordered systems, i.e., amorphous materials, they state, "In this kind of disordered system, long-range order in the atomic distribution is completely broken while the short-range order (...referred to as SRO), is maintained in the sense that the coordination number of each atom remains the same as in the case of a corresponding ordered crystal, although bond lengths and angles in a disordered system fluctuate."

That statement is insufficient and can be misleading since the characteristics of amorphous materials are controlled not only by the fluctuations of bond lengths and bond angles with the consequent loss of periodicity and the establishment of chemical short-range order,[7] but also by the following interrelated factors which make up the Rosetta Stone for understanding amorphous materials. First, there is an average coordination number which defines the structural integrity of the material and its gap and is determined only by the chemistry of the constituent atoms; I have called this its normal structural bonding (NSB). Second, it is the deviations from the optimal coordination number, the deviant electronic configurations (DECs), that are essential to the understanding of the important phenomena in amorphous materials.[8,9] It is these DECs which determine the transport properties of amorphous materials and are responsible for the states in the gap. Third, there need not be "corresponding crystal structures," the central dogma of many working in the amorphous field, a leftover from crystalline physics with its inherent dependence on a lattice structure. The ability to design and synthesize a great variety of amorphous materials depends on the fact that many do not have corresponding crystal structures.

There is a subtle but important insight which should be kept in mind. It is that while short-range order and deviant electronic bonding represent distinct configurations whose total energy can be calculated, there is another distinction that reflects a localized region, the total interactive environment (TIE). This TIE depends on a number of factors of which the nearest-neighbor bonding is but one; others include the effects of nearby chemical forces and of electrical charge distribution which are reflected in the overall three-dimensional topology and in the character of the states in the gap. Perturbations of the TIE can occur by excitational processes.[10]

It is difficult to understand now, but the absolutist belief of physicists in the dogma of the crystalline lattice as the basis of semiconductor science can be appreciated by tracing the attitude of Ziman, one of the leading figures in solid-state theory. In 1965 he wrote in his well-known introductory book on solid-state physics,[11] "A theory of the physical properties of solids would be practically impossible if the most stable structure for most solids were not a regular crystal lattice." Later, in 1969, at the Third International Conference on Amorphous and Liquid Semiconductors, he delivered a paper [12] entitled "How It It Possible To Have An Amorphous Semiconductor?" In this talk he proved that, since there is no regular lattice in amorphous

Reprinted by permission from D. Adler, B.B. Schwartz, and M.C. Steele, eds., *Physical Properties of Amorphous Materials,* Plenum Press, New York, 1985, pp. 105–155.

materials, there can be no band gap, and therefore these materials cannot be semiconductors. Of course, this misses the whole point of the CFO model with its concept of a mobility gap,[13] illustrated in Fig. 1. Finally, indicating how science progresses, or better, how scientists progress, Ziman later published another book, Models of Disorder,[14] in which he states, "Condensed-matter physics has expanded in recent years and shifted its centre of interest to encompass a whole new range of materials and phenomena. Fundamental investigations on the molecular structure of liquids, on amorphous semiconductors, on polymer solutions, on magnetic phase transitions, on the electrical and optical properties of liquid metals, on the glassy state, on metal ammonia solutions, on disordered alloys, on metallic vapours--and many other interesting systems--now constitute a significant proportion of the activity of innumerable physical and chemical laboratories around the world." He continues, "This research is not purely academic: disordered phases of condensed matter--steel and glass, earth and water, if not fire and air--are far more abundant, and of no less technological value, than the idealized single crystals that used to be the sole object of study of 'solid state physics.'" These contradictory quotes [15] suggest the climate in which we were living when I first discussed amorphous materials at scientific meetings. While the situation is much better these days, there are still remaining misconceptions which this paper will attempt to clarify. In so doing, we will address the fundamental principles of amorphous materials.

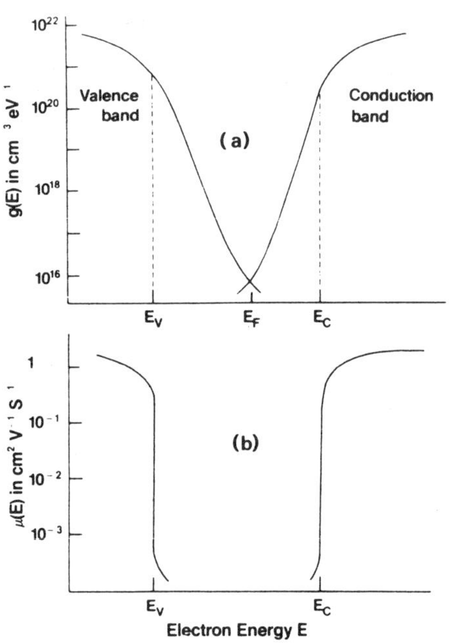

Fig. 1. Sketch of the densities of states of the valence and conduction bands and corresponding electron and hole mobilities. The magnitude of the mobilities should be regarded as approximate because no quantitative calculations have been made. States which are neutral when occupied are associated with the valence band, those neutral when empty with the conduction band; they overlap in the mobility gap. (Ref. 13.)

Kuhn's book [16] should be required reading for the historical and philosophical understanding of how scientific paradigms are developed. He discusses how anomalies appear in physical understanding, how the new solutions based upon original thinking unacceptable to the status quo physicists emerge, and then after a period of struggle, how a new mindset is generated and a new field is created. "More clearly than most other episodes in the history of at least the physical sciences, these display what all scientific revolutions are about. Each of them necessitated the community's rejection of one time-honored scientific theory in favor of another incompatible with it. Each produced a consequent shift in the problems available for scientific scrutiny and in the standards by which the profession determined what should count as an admissible problem or as a legitimate problem-solution. And each transformed the scientific imagination in ways that we shall ultimately need to describe as a transformation of the world within which scientific work was done. Such changes, together with controversies that almost always accompany them, are the defining characteristics of scientific revolutions." It is amazing how Kuhn's description applies to the development of amorphous materials in contradiction to the accepted and dogmatically defended crystalline approach. I was particularly struck by his description of how scientists attempt to explain new phenomena by seeking to extend their conventional approaches to the point of irrationality. This can have unfortunate consequences. For example, what a waste of time it was for us to have to prove over and over that threshold switching was really electronic rather than thermal![17-32] Shaw [33], who for a time embraced the thermal theory, has recently put the final nail in its coffin.

II. CHEMICAL CONSIDERATIONS

It is important to discuss some of the still-remaining misconceptions. It appears puzzling to many that materials composed of exactly the same elements can have completely different structural and electronic properties, depending upon how they are processed. The reason that many amorphous materials are preparation-dependent is that the same elements can combine with each other in a number of different and distinct configurations. The local order actually chosen depends on the nature of the chemical bonding, which in turn is predicated on several factors, including dynamic considerations, for ours is not a chemistry of equilibrium states. The possibility of steric isomerism results in the same elements in different configurations displaying very different chemical reactivities and electronic properties. The internal freedom for placement of atoms in three-dimensional space without long-range order allows for new design possibilities not found in crystals. Indeed, it was stereo- and polymer chemistry that was my guide from the beginning.

Instead of a lattice of repetitive atoms, amorphous solids form a matrix where bonding and nonbonding orbitals with different energies interact in three-dimensional space, sometimes yielding charged centers and thus internal electric fields. The particular bonding option chosen by an atom as it seeks out an equilibrium position on the surface of a growing film is dictated by the kinetics, the orbital directionality, the state of excitation of the relevant atoms, and the temperature distribution during the deposition process. With the constraints of crystalline symmetry and lattice specificity lifted, new internal configurations can and do develop. Rheology plays a role, disclosing important differences between amorphous and crystalline materials, with the former exhibiting unique electron-phonon relaxation processes and pseudo-equilibria.

The chemical foundation of amorphous materials can be clarified by considering, e.g., why and how the carbon atom forms the basis of organic chemistry. For just as the varied bonding possibilities of carbon can generate many different configurations,[9,34] even though the same elements are involved, the multi-orbital choices of the elements in an amorphous material can lead to differing configurations. This is the basis for my broadly classifying our materials as synthetic. The difference between amorphous inorganic materials and synthetic organic materials is qualitatively important since we can make not only high temperature, chemically stable, passive materials which in themselves outperform plastics, but we can also make amorphous solids that are electronically active and can be used as switches, memories, transducers, photovoltaic cells, batteries, catalysts, superconductors, etc. Be reminded that there was no such list in 1960.

Directionality of bonding, multi-valences, and varied coordination possibilities, all of which are involved in the offering of multi-orbital choices, become the building blocks for amorphicity. It should not be a surprise that the temperature of the substrate, the state of excitation of the atoms, and the reequilibration kinetics all affect how orbital relationships are formed and how atoms select one another to make up a desired material. Therefore, substrate temperature, orbital directionality, multi-atomic interactions, sticking coefficients, free radical chemistry, and diffusion coefficients are important considerations in how atoms in amorphous materials relate to other atoms and build up their local geometries. These controllable parameters are important assets for they permit us to engineer many new and useful materials as well as being of great scientific value.

It is often asked if there is a fundamental difference between glasses and amorphous materials. The difference is simply that scientists who prepare materials by quenching from the melt make use of a longer time scale than those who deposit atoms on surfaces directly from vapor or plasma phases. Therefore, more equilibrium structures can be expected in glasses. The time and energy required for two atoms to bond to one another can be considered to be design parameters. For example, if there are four outer p electrons, as in a chalcogenide material, but only two bond in the NSB configuration, it is easier to prepare an amorphous material than if one has to bond four outer sp^3 electrons, as in elemental amorphous silicon. We can chemically aid the process by making it easier for atoms to bond to each other. How do we accomplish this? It is exceptionally difficult, if not impossible, to form amorphous silicon from the melt (except under laser energization), but it is easy to form amorphous selenium in this way. In the former case, the liquid is not tetrahedrally coordinated and quickly crystallizes upon quenching. One has to add an interfering additive to prevent this crystallization and, more importantly, one has to bond all four outer electrons to obtain the tetrahedral structure. The rigidity of that structure can be understood by anyone who has tried to fit four surfaces together. In mechanics, one must insert a shim or a gib to do so. In stereo- and polymer chemistry, a "fitting link," either a crosslink or a bridge, is needed. In elemental amorphous silicon, it costs too much strain energy to try to bond all four orbitals [35-38] when all must be distorted to fit the local geometry. The result is that there are many strained bonds, dangling bonds are prevalent, and voids are formed in the solid. One would expect this from free energy considerations.

In contrast, in chalcogen elements, only two of the outer electrons need to be utilized for structural bonding. The remaining lone pair can assume a spectrum of nonbonding or bonding relationships.[39] Consequently, more flexible chain and ring structures result in the chalcogenides, more rigid structures in the tetrahedral materials. In both cases, I utilized stereo- and polymer chemistry concepts to control rigidity. In the tetrahedral materials, additional alloying elements are needed to reduce the strain and <u>lower</u> the average coordination of the structure. They can also act in a bridging manner, like oxygen in fused silica. In the chalcogenide materials, alloying elements should be preferentially those that effectively crosslink the material, thus increasing the average coordination,[17,35,36,38-40] and making for more stable structures; i.e., they should <u>add</u> rigidity. If these alloying and bridging rules are not followed, then the rapid quench rate that one achieves, e.g., by sputtering, only leads to the freezing in of local atomic mismatches and strains. Wherever there are strains or, more importantly, wherever there are <u>bonding options</u>, DECs are ordinarily created yielding large densities of localized states whose origin and significance need to be understood, especially if one wants to control or eliminate them. For example, the DECs in elemental silicon are generated by undercoordination; the DECs in chalcogenide materials arise from the various lone-pair configurations.[35,36,39] We can control them in the former by compensating the dangling bonds, e.g., with fluorine and hydrogen,[41] and in the latter by interacting the lone pairs with modifying elements.[8,38,42,43]

Chemical understanding must be translated into specific topological configurations, since the local geometries reflect the appropriate chemistry in amorphous materials and structure and function are indivisible. Grigorovici [44] was early interested in the structural configurations and internal topology of amorphous materials. Our work emphasizes the correlation of internal geometries with electronic properties. Surface topology in periodic materials is related to the lifting of restrictions in the free space above the surface. Therefore, the study of crystalline surfaces can be a useful first step in the study of bulk amorphous materials.[45,46] In fact, the unusual back-bondings at crystalline surfaces can provide clues of internal bulk configurations of amorphous materials. The TIE is different on the surface than in the bulk, for the third dimen-

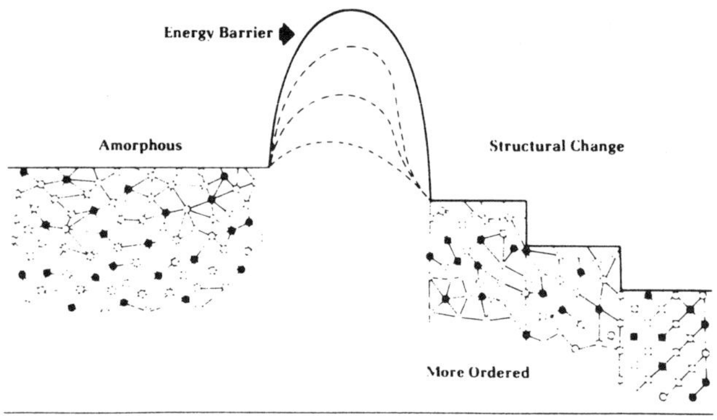

Energy barrier can be reduced by any of the following-applied singly or in combination:
- Light
- Heat
- Electric field
- Chemical catalyst
- Stress-tension pressure

Transformations in amorphous materials produce changes in:
- Resistance
- Capacitance
- Dielectric constant
- Charge retention
- Index of refraction
- Surface reflection
- Light absorption, transmission and scattering
- Differential wetting and sorption
- Others, including Magnetic Susceptibility

Fig. 2. Information storage/retrieval and display by structural transformation. (Ref. 50.)

sion in the bulk sets up its own chemical and electrical constraints.

Understanding that the types of defects available in amorphous materials are intimately related to the internal degrees of freedom unique to noncrystalline solids, one can appreciate that the defects are really part of the total interactive environment and part of the energy considerations therein. Defects need not be only dangling bonds, but can be very similar to the unusual bonding configurations that occur in amorphous chalcogenides or variations of the back-bonding that occur at surfaces. In the same amorphous material, there can be a whole spectrum of bonds including metallic, covalent, ionic and coordinate.[2] Whether they appear as defects or not depends upon the particular design of the material.

III. THERMODYNAMIC CONSIDERATIONS

Thermodynamically, if we have a system that has several possible configurations with essentially equal bulk energies open to it, depending, e.g., on the temperature distribution, we may well ask how the atoms developing into a solid choose between them. I would like to briefly discuss the meaning of metastability in amorphous solids. How often have we heard that amorphous materials are metastable? We should bear in mind that so is diamond! Should we consider tectites as unstable? Amorphous materials can be very stable indeed. When we want to utilize their metastability, we do so by design. The understanding of energy barriers on an atomic scale, as well as on a more macroscopic scale, is a crucial point. As we have shown, the barrier between the amorphous and crystalline phases can be controlled, as is sketched in Fig. 2. It is adjustable by altering the bond strengths of the atoms involved, and it can be lowered or overcome by external energy sources. For crystallization to occur, there must be a cooperative action of a large cluster of atoms, but many subtle changes can occur first. Far more subtle barriers exist than the one between the amorphous and crystalline phases. Slight differences in energy can have important influences on the various conformations and configurations that are inherent in amorphous materials and the transformations available to them. One internal structure can be converted into another without affecting important properties of the material or, for that matter, without even breaking bonds, as indicated in Fig. 3. (However, the TIEs would be affected.) The closeness of energy of the various conformations and configurations can be masked by thermal vibrations (phonons) down to very low temperatures.[35] I interpret the so-called universal, low-temperature, two-level atomic tunneling systems seen in glasses and many amorphous materials as direct evidence of the multi-equilibrium possibilities that I have been describing. Other such evidence includes the photostructural changes that characterize both chalcogenide and tetrahedral alloys.

Rather than postulate that only bond switching is the source of the specific heat anomalies which have been viewed as atomic tunneling phenomena, larger-scale relaxations unique to the disordered and amorphous state could be the most accurate explanation, especially since these represent the conformational changes discussed in this paper. Such changes are directly related to variations in the TIE which reposition atoms, ions, and charged as well as neutral defects, to new positions related to the rest of their environment; i.e., there can be a new TIE as one changes the phonon concentration. This is reflected in the character and number of states in the gap. From the very beginning of my work in this field, I have been emphasizing that the coupling between electrons and phonons is, and must be, basically different in amorphous and crystalline materials. It is in pursuit of a direct demonstration of this concept that I have been actively working on superconductors since the early 1970's.[47,48] I am certain that investigations of the phonon spectra of amorphous material will some day be one of the most exciting new areas of scientific research.

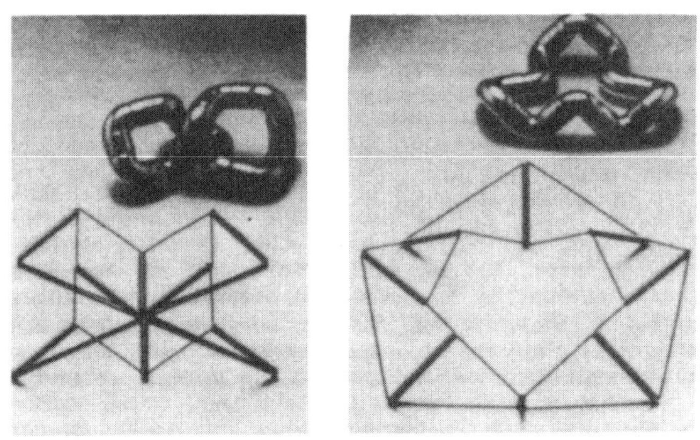

Fig. 3. Models illustrating conformational changes without bond breaking--the interconversion of one structural configuration into another.

Our concept of metastability begins on an atomic level, or, because atoms are not isolated in amorphous materials, rather on a molecular one. Let us assume that a local atomic cluster has been excited by inducing a transition from a low-energy molecular orbital to a higher-energy one, and ask what happens to the TIE? It must change, but how? It will change transiently if the local environment absorbs the excitation energy as it does in the Ovonic Threshold Switch; but if the added energy is dissipated through structural interactions that cannot contain the local conformational changes, as in the Ovonic Memory Switch, then the surrounding structure will disperse the energy in a manner which not only reshapes the conformation with an attendant redistribution of charge but also results in a configurational change, i.e., a breaking of bonds. These configurational changes can be designed to be reversible. There can be a whole spectrum of such changes, including the formation of crystallites. Whether the process is reversible or irreversible is basically a matter of the bond strengths, the size of the crystallites, and the topological and chemical environment. There are not only energy barriers in amorphous materials inhibiting crystallization but also many more subtle barriers involved with atomic and molecular scale changes which are part of the relaxation process unique in amorphous materials.

Reversible amorphization can be pictured as the dissolving of the periodic structure into the surrounding matrix.[49,50] This solute-solvent concept is an apt analogy since it conjures up the picture of precipitating under certain sets of conditions and dissolving under others. Unlike the absorption of energy in a crystal which then propagates throughout the entire lattice, such events in amorphous materials can be very localized. That is why recombination of carriers has more important consequences in amorphous materials than in crystalline. A knowledge of the principles and processes of relaxation, nucleation, and of catalytic effects is necessary for the understanding of crystallization mechanisms in amorphous materials.

Not satisfied with the conventional wisdom that one had to have melting, i.e., a transition to the liquid phase, in order to reach the amorphous state, I proposed the concept of "amorphization" to describe the process of going from an ordered to a disordered system, and placed emphasis on this process occurring from chemical interactions without the temperature having to exceed the melting point although, of course, it may.[51] My theory, which has now been vindicated by many experiments, was that there is a dynamic chemical force tending to bring about the amorphous state, which can represent a configuration equally as attractive as the crystalline one under certain conditions. As an example of what we might call an "anticrystalline" configuration, let us consider a tellurium atom which is initially part of a chain as in crystalline tellurium but is also near an arsenic atom. If energy is supplied to the vicinity of this local area, the resultant displacement of the tellurium atom under consideration can cause it to align its orbitals within the chemical field of the arsenic atom and form a tellurium-arsenic bond. Since this is more stable than the tellurium-tellurium bond which it replaced, absorption of the energy in this case has led to a destruction of the crystal structure. The tellurium-arsenic configuration and, even more, the selenium-arsenic configuration are cross-linked, disordered ones, and can be thought of as anticrystalline. We have shown that there is an analog of the amorphization process in the mechanism of crystallization. If many free carriers are generated by light or electric field, then the relaxation processes favor ordering without the need for melting.

IV. CHEMISTRY AS A DESIGN TOOL

Right from the beginning of my work in amorphous materials, I have used a chemical approach as a basic design tool. The Periodic Chart of the Elements shown in Fig. 4 [52] has been for me primarily a means of deciding which elements could bond to each other in such a way as to control not only the shape and magnitude of the mobility gap but also the density of localized states in the gap. Since many of our materials are multi-component alloys, this concept can be illustrated by examples which will be detailed subsequently, but whose simple premises follow.

I have emphasized that it is not only the bond strengths but the type and number of crosslinks which control the barrier to crystallization. One can frustrate crystallization by <u>steric hindrances</u>. For a <u>unistable</u> material, we utilize maximum numbers of strongly bonded atoms and crosslinks, e.g., silicon, germanium, arsenic, and oxygen, as the crosslinks for a tellurium-based alloy. For a <u>bistable</u> material, we reduce the bond strengths of the alloying elements and also reduce their concentration, e.g., some or all of the arsenic can be replaced by antimony, which forms weaker bonds, and some or all of the silicon can be replaced by germanium, or even by tin or lead, for the same reasons. A glance at Fig. 4 shows the chemical logic in this method. It also follows that as we reduce the bond strengths, we concomitantly reduce the band gap of the material. The lone pairs in chalcogenides and the various configurations that they enter into control the transport properties of these materials.[39] We utilize small amounts of additional elements in our multi-component materials not only for their spatial, structural and chemical

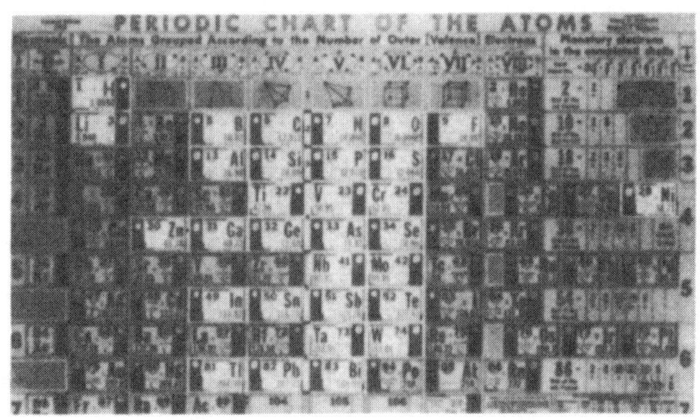

Fig. 4. Periodic chart of the elements with examples of the various elements that can be utilized to fabricate amorphous materials or to modify or dope them.

effects, but also for the influence they have on the electronic activity.

From the above, we can see that mere atomic displacements or simple distortions of a crystalline structure do not do the concept of amorphization justice. The two phases of our bistable materials can co-exist at room temperature. The balance can be shifted from one state to the other. For example, excitation in a memory material can result not only in crystallization but also in a tendency toward the amorphous state because of the chemical forces discussed above. We showed, e.g., that excitation could either inhibit crystallization or expedite it.[53] The outer electron lone pairs of the chalcogens are analogous to the double bonds of carbon in that their possible conversion into bonding-antibonding pairs opens up a host of different configurational structures with nearly the same energy. These new configurations fall into the category of DECs.

Fundamental to my way of thinking relative to the Periodic Chart has been the fact that low average coordination ordinarily favors the amorphous state. In multi-component alloys, the additional elements aid in assuring optimal coordination.(9,35] The balance between adding constraints, completing structures, and assuring rigidity becomes a chemical design parameter as will be seen from our subsequent discussion of tetrahedral materials.

Chemistry and structure are related through the concept of connectivity,[38] for lattice constraints limit the ways in which atoms connect to each other, but the different possibilities in three-dimensional space in amorphous materials allow many new geometric configurations. The consequences of this concept, together with free energy considerations, is that there is not just one equilibrium but various equilibria. Structure and function are connected. If one wants to design and define a local order, then the entire local environment, the TIE, must be taken into account. Selective excitation can, in fact, add an important dimension in designing new configurations that would otherwise not be available through the usual thermodynamic considerations. We can also use such electronic pathways and their recombination events in amorphous materials for various memory and photographic applications.[2,45,46,49-51,55,56]

By perturbing the TIE, one also perturbs the density of states. It is no wonder that the Staebler-Wronski effect [54] can be understood as an example of a photostructural change [41] instead of appearing as some new esoteric phenomenon. If one excites carriers and recombination events occur in a material which has several different structural relaxations available to it, one can forget the conventional picture of a well-defined density of states;[55,57] one can readily see that there would be a redistribution of the localized states as a consequence of the redistribution of atomic configurations in three-dimensional space. From the beginning of our work we have used electro- and photostructural effects constructively for device applications. As has been shown in our laboratory by Guha et al.,[58] recombination is the mechanism which explains the worrisome Staebler-Wronski effect in hydrogenated amorphous silicon alloys.

Fritzsche's "hills and valleys" model [59] puts into perspective some of the consequences of the atomic fluctuations related to the density of states. The charge density of fluctuations in such situations is connected with positional relationships. Therefore, while local chemical bonding is of great importance because it has calculable bond strengths, and therefore short-range order, it does not adequately reflect the true spatial state of affairs of an amorphous solid. The overall positional charge-density fluctuations in three-dimensional space as well as the nature of the chemical bonds are an integral part of the total interactive environment.

It is important to reemphasize that in amorphous materials we do not use the concept of lattice but of matrix.[60] The normal structural bonding (NSB) makes up the great majority of bonding configurations, and therefore is responsible for the cohesive energy, the structural integrity, and the optical energy gap of the material. This gap, as we have discussed, can be adjusted by alloying, and is related to the bond strengths of the elements involved. Compositional, positional, and translational disorder inherent in amorphous materials are reflected in the shape and sharpness of the mobility edge and the density of states in its vicinity. The origin of the density of states in the gap as well as its control are also now quite clear. More subtle effects related to states near the mobility edge itself are still interesting areas of investigation, for these can act as traps and thus can have important device consequences. I am sure that as research progresses, we will be finding fine structure near and in the edge itself. We have already been successful in affecting the sharpness and steepness of the mobility edge by the choice of materials, the control of impurities and the generation of intermediate order.

The use of the term "disorder" is unfortunate since it ordinarily means deviations from periodic reference points, but if periodicity is not dominant, then we must substitute our own basic and specific noncrystalline principles. As pointed out previously, one can tailor the optical gap by the use of different covalently bonding elements which also affect the cohesive energy of the material. The alloying elements can further act as structural crosslinks, assuring amorphicity. Following our rules, one can very specifically design materials: e.g., to increase the band gap of a tellurium alloy, add germanium; to increase it further, add stronger-bonding silicon. Similar increase of the band gap occurs if one substitutes arsenic for antimony, or adds selenium, sulfur, or oxygen. It is not unusual for amorphous materials to be multi-component alloys, with four or more elements. The bond strengths of all the elements affect and determine the overall gap. In terms of defects, DECs are generated by the three-dimensional spatial freedom of individual atoms counterbalanced by the chemical and electrical forces surrounding them, i.e., their environment. Therefore, a silicon alloy is primarily tetrahedral but its electronic properties, i.e., its transport properties, are controlled by the deviations from the NSB. These DECs are primarily responsible for the deep states in the gap of amorphous materials, and, depending upon their position in energy, can also play a role in the aforementioned shape of the mobility edge. The matrix that we are discussing not only has relaxation modes which are different from a lattice structure, but has a degree of elasticity which becomes an exceedingly important parameter in material design.[35,36,38,39,61]

V. MECHANICAL PROPERTIES

The concept of elasticity is a common theme throughout this paper. For simplicity, consider the fact that as one changes the average coordination by replacing divalent materials in Group VI by tetrahedral materials in Group IV, e.g., silicon, the elasticity decreases. In order to attain necessary elasticity to make useful materials, atoms of lower valence are utilized. In contrast, if we start with divalent materials, we must add crosslinks to assure and control rigidity and stability. If we begin with tetrahedral materials, we add monovalent atoms such as hydrogen and fluorine to decrease the rigidity and to control and assure the tetrahedral structure of the Group IV atoms.

In order to understand how one goes from a flexible to a rigid structure, I proposed that the controlling influence was the network connectivity, which is characterized by a single parameter, the average coordination number C.[38] This average coordination number is related, of course, to the NSB. As one goes from primarily divalent materials, which have the greatest tendency for flexibility and the formation of glass, to tetrahedral materials with the greatest rigidity, the alloying and cross-linking elements that are added accomplish two purposes. They not only play a structural role, e.g., as can be seen in Fig. 5 [39] where the nonchalcogenide elements add rigidity to the solid, but they also provide an increase of the average coordination number. As the average coordination number is increased, the freedom in three-dimensional space is limited by placing a greater number of constraints on each atom; however, just as important, the freedom of chain and ring folding and twisting is also controlled and inhibited. We need not go to more tetrahedral materials to increase the coordination of tellurium; oxygen and/or arsenic can increase both the average coordination and the size of the gap. In tetrahedral materials, there is much strain added as the bonding orbitals seek to complete their configurations. To relieve the strain, one alloys with atoms of a lower valency or with those which tend to form ionic bonds. As coordination is increased, e.g., in elemental amorphous silicon, the alloying atoms play the role of permitting completion of the tetrahedral structure by providing flexibility and electronic compensation. If they did not, DECs would be induced by virtue of the resulting under-coordination, and dangling bonds would be formed. Therefore, elasticity is intimately connected with coordination number: in chalcogenides, crosslinks and bridges play an important role; in a material such as amorphous silicon, alloying reduces the coordination. However, the average coordination of the silicon atoms themselves is increased, e.g., by the addition of fluorine, carbon, oxygen, nitrogen, etc. due to the reduction of the concentration of dangling bonds.

Fig. 5. Model of an Ovonic Threshold Switch illustrating a large amount of strongly-bonded crosslinks assuring stability. The dark balls are Ge, Si, and As atoms. The light balls are Te atoms. (Ref. 39.)

VI. PHONONS

I wish to emphasize that phonon activity in amorphous materials differs basically from that in crystalline materials, although there is a wide spectrum and in some materials similarities can exist. One should start with the simple premise that although crystalline solid only exhibit extended phonons, both localized and extended phonons characterize amorphous solids. There is a tendency for strong localized coupling in the flexible materials and weaker coupling in the tetrahedral materials. The matrix mediates the orbital energies in the divalent materials. The resulting spin pairing usually produces completely diamagnetic material. When the matrix is not deformable enough, such as in an as-deposited elemental amorphous silicon material, the electron-phonon interactions cannot provide the necessary pairing. The relaxations that are inherent in amorphous materials are, therefore, different from those of crystalline materials.

VII. MATERIALS SYNTHESIS FOR DEVICE APPLICATIONS

Let us see how these principles actually work in synthesizing materials for device purposes. We will start with the chalcogens and end with tetrahedral materials. As was pointed out earlier, amorphous devices fall into two categories.[17] The first are unistable materials, whose bond strengths and steric hindrances act to prevent crystallization; this class is illustrated in Fig. 5 by an Ovonic Threshold Switch. The crosslinks are numerous and the bonding is strong, and therefore structural changes such as crystallization do not occur within the device operating range. The second are bistable materials, in which there are fewer crosslinks and the bond strengths are weaker so that the barrier to crystallization can be overcome. An example is the Ovonic Memory Switch, shown in Fig. 6. Note how flexible and elastic the chalcogenide bistable memory material is compared to the unistable threshold switch. The average coordination for each is significantly different, C = 2.3 in the memory material while C = 2.9 in the threshold material.

Fig. 6. Model of an Ovonic Memory Switch showing fewer and weaker crosslinks and inherent flexibility which permit the reversible bistability. The light balls are Te atoms. The dark balls are Ge atoms. The darkest balls are Sb and S atoms. (Ref. 39.)

Figure 2 shows how the unique structural changes in amorphous materials, ranging from subtle relaxations to changes of phase including crystallization, become the basis of a whole new field of information and encoding devices, including new types of photography. Especially interesting is the fact that structural reversibility characterizes the more flexible materials so that one can cycle from, e.g., the crystalline state back to the amorphous. These changes of phase can be driven reversibly for more than hundreds of billions of cycles without degradation.

In all materials, we know the origin of the normal structural bonds. We already pointed out that the primary origin of the DECs in chalcogenides are the lone-pair electrons, either nonbonded or forced by the internal chemical and topological environment to assume a spectrum of bonding states, including one- and three-electron states.[39] In an important paper, Kastner, Adler, and Fritzsche [62] further developed this theme to explain the nature of the charged defects that result from these lone-pair interactions. They called the low energy one- and three-fold coordinated defect states valence alternation pairs (VAPs). It is interesting that these VAPs have the property suggested by the original CFO model,[13] large and equal concentration of positively and negatively charged centers which can act as efficient traps for excess electrons and holes. The elucidation of the lone-pair nature of the chalcogenides by Kastner [63] allowed us [35,39,64] to explain why there is no ESR signal in most chalcogenides despite the typical presence of a high density of states in the gap.[65] The fact that lone pairs are spin-compensated in all their variety of free or bonded conditions explains the above as well as how one can have a negative correlation energy. The difference between my explanation [35-37] and those of Street and Mott [66] and of Anderson [67] is that theirs are based upon disorder as sufficient for the negative correlation energy and allow unduly for dangling bonds, and therefore fail to distinguish between the chalcogenide and the tetrahedrally-bonded materials. My position was, and is, that it is the inherent flexibility of the divalent state which permits the lone pairs to have the strong electron-phonon interactions that are the basis of the induced spin pairing.

Since switching and memory are such basic functions in our information-oriented society, it is important to note the special characteristics of the chalcogenide-based Ovonic devices and correlate them with the explanations given above. In the Ovonic Threshold Switch (see Fig. 7), we see a unique reversible transition between a high impedance and a low impedance state in less than 120 picoseconds at room temperature. (I have never understood the interest in Josephson Effect switches for computers, since they require liquid helium temperatures to achieve comparable switching times.) Such a device is completely independent of polarity and is made preferably in thin-film form, from less than 0.5 μm to many μm in thickness, depending upon the threshold voltage required.

When I first invented these devices, I called them Quantrols,[68,69] since I believed that the switching mechanism was electronic in nature and that such speeds could be observed only if there were a quantum basis for the electronic change of state. From 1960 on, I described the electrical characteristics of these devices.[70,71] In my 1968 paper,[17] I emphasized the electronic nature of the switching process, and explained the basis of the mechanisms of both threshold and memory switching phenomena. The application of a high electric field to specifically designed chalcogenide glasses induces a rapid switching process to a nonequilibrium conducting state followed by injection. The electronic basis of the process has been proven,[17-33] and it has many implications both to solid-state theory and to device potential. In the early 1960's, I performed a simple experiment to prove the electronic nature of the phenomenon by adding some selenium to the threshold materials (preserving the high-impedance state even in the liquid phase) and demonstrating switching above the melting point. Obviously, switching therefore was not based on a solid-to-liquid transition.

Now to discuss chemical-topological correlations. As pointed out, the Ovonic Threshold Switch is a heavily crosslinked material with strong bonds, and is therefore unistable; i.e., the electronic excitation does not change the basic structure. The Ovonic Memory Switch is deliberately made with fewer crosslinks and weaker bonds. Referring to Figs. 4-6, it can be seen that, e.g., if germanium is

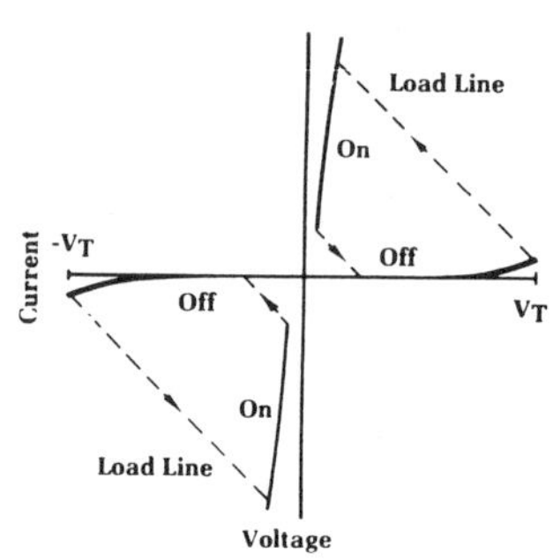

Fig. 7. Current-voltage characteristics of an Ovonic Threshold Switch. (Refs. 17 and 68.)

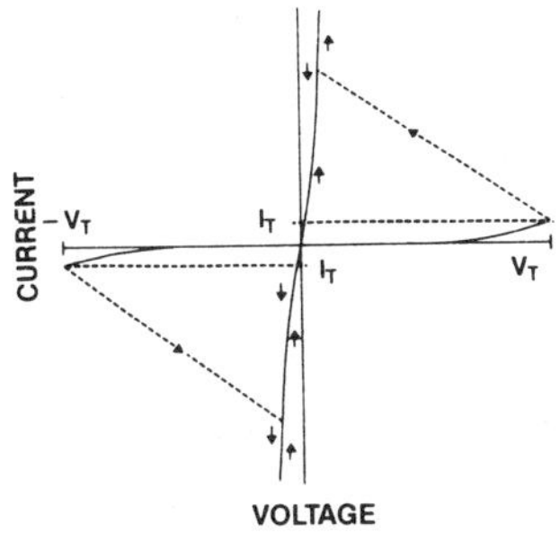

Fig. 8. Current-voltage characteristics of an Ovonic Memory Switch. (Refs. 17, 68 and 72.)

substituted for silicon, or antimony for arsenic, and the nonequilibrium threshold electronic switching effect is used to make the material reactive to electronic excitation, thus weakening or breaking the bonds, the subsequent thermal action which permits cooperative movement of atoms helps induce the memory state. Memory changes can be generated as well as accelerated by diffusion processes.[3] I have called materials which are based upon changes of local order bistable or phasechange materials. A typical current-voltage characteristic of an Ovonic Memory Switch is shown in Fig. 8.

Nowadays, the topic of artificial intelligence is of great interest. I feel that we are completing the grand circle that my wife and collaborator, Iris Ovshinsky, and I originally started in 1955 when we set out to understand the physical basis of intelligence, i.e., information, how it is encoded, switched and transmitted, and the energy transformations connected with it.[3,72] I proposed that this little-understood area of neurophysiology could be illuminated by considering that "disorder," i.e., local order, could play a crucial role. I felt that the energy transformations, excitations, and structural changes associated with amorphous materials would be valid models for nerve-cell action, and I built my first nerve-cell switching model and memory to prove the analogy.[72] I was particularly interested in the adaptive memory aspect of my model [73] and have continued with the adaptive memory concept as a learning "machine" in micron thicknesses ever since. To illustrate my work in this area since 1960, consider Fig. 9. We are also pursuing the three-dimensional circuit potential of amorphous materials.

We predicted and observed memory effects in amorphous semiconductors in response to electrical or optical pulses, and associated them with either irreversible changes if the materials were strongly bonded and did not have inherent flexibility, e.g.,

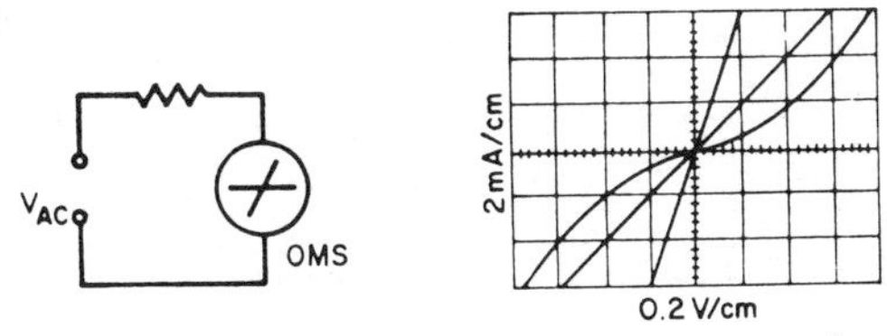

Fig. 9. Ovonic adaptive memory. (Ref. 72.)

in amorphous silicon, or reversible changes in cases where more flexible structures were generated by utilizing, e.g., more weakly-bonded divalent alloys such as the chalcogenides. The adaptive memory, therefore, reflects the ratio of the amount of order to the energy input. Before 1960, I showed switching, memory, and adaptive memory action in transition metal oxides.[72]

I find it stimulating and fascinating to connect the new cosmological theories with the work being described here since they deal with the same types of problems, i.e., phase changes, supercooling, freezing-in of defects, nucleation, broken symmetry, etc., except that the time scale is a bit different when we are dealing with the origin of our universe (or universes)! Guth [74,75] assumes a liquid-to-crystal analogy whereas I would suggest that asymmetries of the amorphous state and the changes that can occur in it as described herein are more to the point; i.e., in the early transitional phase of the evolution of the universe, the theories of the amorphous phases discussed in this paper are more relevant than those of a crystalline phase. In fact, I believe that my multi-equilibria concept may have some connection and applicability to Guth's general theory. The unity of science is a marvel indeed!

As discussed previously, chalcogenide glasses have low values of the average coordination number. The network is not overconstrained and intermediate-range order is often observed. For nearly pure chalcogens, such as glasses in the Te-Se system, chemical crosslinking is very low. However, "mechanical" entanglements, especially as longer chains and rings are formed, serve as an energy barrier to crystallization, albeit a low one. Recall our rule that as one generates stronger chemical bonds, the band gap goes up as well; e.g., sulfur and oxygen both have stronger bonds to tellurium than does selenium, and thus the gap increases progressively as one replaces selenium by either sulfur or oxygen. If one combines several elements, then the bond strengths are averaged and the gap changes accordingly. Depending on the design of the material, especially the use of crosslinks involving particular bond strengths, crystallization can proceed at relatively rapid rates, especially in the presence of some activation such as increased temperature, incident light, or applied electric field. Similar energy input can be used in conjunction with rapid quench rates to return the material to the amorphous phase. Since the two phases are very distinct electrically and optically, and both phases are essentially completely stable at ambient conditions, such materials can be used as the basis for reversible nonvolatile memory systems. Either the crystalline or the amorphous phase can be used as the "zero" memory state. If the amorphous phase is considered the zero, writing can be accomplished by, e.g., applying a voltage pulse to crystallize a filament between two electrodes. As noted previously, chalcogenides ordinarily possess equal concentrations of positively and negatively charged defect centers which act as effective traps for injected free carriers. When an electric field is applied so that double injection takes place, the traps fill. Under these nonequilibrium conditions in an amorphous memory material, the large concentration of carriers weakens the structure, bond reconstructions take place, many covalent bonds are broken, and the rate of crystallization is enhanced (electrocrystallization). Typically, filaments of the order of 1 μm can be grown in times of less than 1 ms. The same results can be induced by optical excitation. It is important to point out that in the Ovonic Threshold Switch this filament is composed of carriers originating from nonbonding configurations, and therefore there is no structural change, i.e., no crystallization; in the Ovonic Memory Switch, the electronic threshold switching effect leads to desired structural changes.

A wide array of materials can be utilized for Ovonic memories. These include, but are not restricted to, chalcogenide alloys. Having worked on a particularly attractive chalcogenide system, i.e., tellurium-based materials, since 1960, we have been reporting on it in the scientific literature for many years. While multi-component alloys are ordinarily used, the memory mechanism can be understood by considering a simple example. For an alloy such as $Te_{83}Ge_{17}$, the eutectic composition for the Te-Ge system, application of a voltage pulse leads to a phase separation into Te-rich and GeTe-rich regions. Since tellurium crystallizes even at room temperatures, the Te-rich regions quickly form crystallites. Both Te and GeTe under somewhat nonstoichiometric conditions are semi-metals with conductivities over 10 billion times larger than the Te-Ge glass at room temperature. We have found that Te crystallites always grow when sufficient energy is coupled to virtually any Te-based memory glass, which can include O, As, Sb, Pb, etc. The differences in physical properties which serve as the memory mechanism are due to the properties of the Te crystallites on the one hand and the amorphous matrix on the other. In the case of electronic memories, the written filament is highly conductive whereas the unwritten glass is highly resistive. One can also write, and it is often preferable, by amorphizing an originally crystalline film. This has been accomplished in the nanosecond range.[76,77] The memory thus can be easily read by applying a small voltage across the contacts. In the Ovonic memory, we utilize other parameters such as large changes in reflectivity.[4,78] The amorphous-crystalline transition is a completely reversible one, a very important attribute.

To electrically erase, application of a sharp current pulse with a rapid trailing edge is all that is necessary. The electronic effects plus the consequent Joule heating are localized within the conducting filament, while the surrounding medium remains at room temperature, thus quenching the active material and reforming the nonconducting glass. I have found it helpful to consider both the precipitation of the crystallites from the matrix and their dissolving back in from a chemical point of view.[45,49] As I pointed out in 1973,[2] "This is not only a 'melt' condition, but one in which the chemical affinities of the crosslinking atoms aid in the establishment of the amorphous state."

A system designed on this principle acts as a nonvolatile electrically erasable programmable read-only memory (EEPROM), an important link between volatile random access memories (RAMs) and unalterable read-only memories (ROMs). Ours were the first EEPROMs made, and were commercially available in the 1960's and 1970's. Their characteristics have been continually improved since then.

Fig. 10. Ovonic high-speed, high-density Programmable Read-Only Memory (PROM) manufactured by Raytheon.

Because both the crystalline and amorphous phases of the material are completely stable at operating temperatures, it is evident that Ovonic memory switching can be used as the basis for ordinary ROMs and for archival applications by using a write-once mode. If one wants to assure irreversibility, it takes a change of chemistry, following the rules we have outlined, by utilizing stronger bonds than those in the reversible material. Figure 10 shows an Ovonic amorphous silicon-based, high-speed, high-density electrical PROM manufactured by Raytheon. Figure 11 is a commercially available (Panasonic) optical Ovonic memory based upon the chalcogenide crystalline-to-amorphous transition utilized as an optical PROM.

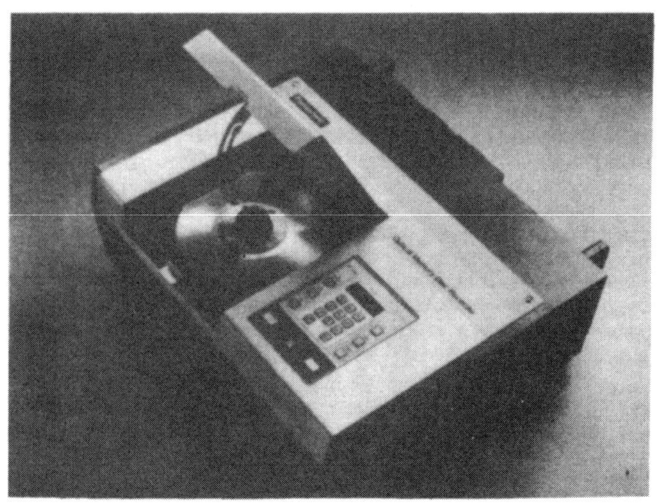

Fig. 11. Panasonic Optical Memory Disc Recorder.

The use of light to induce structural and phase changes has been very rewarding for us. Not wanting to hear again the dreary litany of thermal versus electronic models as the mechanism for changes in amorphous materials, I decided to utilize light to produce new types of optical recording and photographic imaging with unique properties. In addition, I showed that one could use lasers, electron beams, and ordinary light to create information encoding systems, both series and parallel. We were the first to accomplish the laser crystallization of amorphous materials, i.e., to cause crystallization to occur by utilizing a laser interacting with the materials.[4,79,80] I took the position that the simple explanation of melting and recrystallization was not adequate to describe fast laser crystallization. Melting in amorphous materials is the first refuge of ignorance. I proposed that in a material that is unstable to a large amount of excited carriers (initiated by light or electric field), changes of conformation occur and can result in exceedingly fast configurational changes such as crystallization.[3,80] In elemental tetrahedral materials, such relaxations are minor since crystallization requires very little more than eliminating the distortions, which are primarily bond angle changes. In the more flexible chalcogenide alloy materials, bond switching can also take place. This view was supported by much experimental evidence and in a 1971 paper in Applied Physics Letters,[81] we stated that "We have observed a high-speed crystallization of amorphous semiconductor films and the reversal of this crystallization back to the amorphous state using short pulses of laser light and evidenced by a sharp change in optical transmission and reflection. This optical switching behavior is analogous to the memory-type electrical switching effect in these materials which has received wide attention since the observation by S.R. Ovshinsky of both threshold and memory switching in amorphous semiconductors. ...we propose a model which closely relates the optical and electrical switching behavior, and shows that the phase change from amorphous to crystalline state is not only a thermal phenomenon but is directly influenced by the creation of excess electron-hole carriers by either the light, or, for the electrical device, by the electric field. The reversibility of the phenomenon in this model is obtained through the large difference in crystallization rates with the light on or off."

The idea of optical mass memory systems, for example, with the entire contents of a library stored on several disks, is a very appealing one, but was not seriously considered prior to our work on reversible phase changes, such as amorphous-to-crystalline or crystalline-to-amorphous transitions. That work opened up the possibility of optically writing, erasing, and reading via, e.g., the use of a laser which could be focused down to a 1-μm spot size. Resolution of this order of magnitude could provide information storage capacities of 10^8 bits/cm^2 and dramatically higher densities with the use of electron beams. Present-day technology allows the storage of about 250,000 pages of information on a single video disk, an even greater bit density. We showed that a number of multi-component alloys, such as glasses in the Te-Ge system, had very different reflectivities from the same material in its crystallized form, irrespective of the particular components of the alloy. Similar properties can be attained with a multitude of other amorphous alloys as well. One can optimize specific properties by varying the composition. With the use of antireflection coatings tuned to either the crystalline or amorphous phase, the phase transition then provides many orders-of-magnitude changes in transmission upon writing and erasing.

My collaborators and I showed [81] that an ordinary laser pulse could both crystallize and amorphize a spot less than 1μm in diameter in under 1 μs, and that the same or another laser could be used to read in either a reflection or a transmission mode. We were operating optical disk systems based on these ideas in the 1960's and early 1970's. Exposure of the glass to a laser beam with characteristic frequency greater than the energy gap excites large concentrations of electron-hole pairs. These can have several effects including recombination and trapping by the charged defect centers, resulting in large densities of bond switching and broken bonds. Under these conditions, crystallization proceeds at an extremely enhanced rate (photocrystallization). This process is similar to the electric field-induced crystallization discussed previously. In the crystallized form, the material is more light absorbent than the glass. Exposure to the same laser beam thus can transfer an increased amount of energy to the written spot, returning it to the disordered state. Since the surrounding matrix is unaffected by the focused laser beam, it serves to provide the proper thermal as well as chemical environment, quenching in the anticrystalline configuration and thus reforming the glass. Consequently, either writing or erasing can be accomplished by the same laser pulse.

We have continually improved the parameters of reversible optical data storage disks, obtaining sharp increases in resolution, contract, and lifetime. In addition, optical memory techniques other than amorphous-to-crystalline transitions have been developed. One such technique [81] uses the self-focusing property of many glasses to rapidly nucleate a vapor bubble at the interface between the chalcogenide glass and an inert transparent layer. Self-focusing occurs whenever the energy gap of the glass decreases with increasing temperature, a common phenomenon. If the laser has a characteristic frequency very near the energy gap at room temperature, a small amount of laser-induced heating just below the interface will cause ever-increasing absorption in the same region, rapidly nucleating a small bubble. The bubble scatters light effectively, enabling the spot to be read easily. The entire memory can not only be laser-erased but can be block-erased by gentle heating with an infrared lamp. The advantages of such a system are smaller spot sizes (and thus higher resolution), faster write times, and lower energy cost per bit.

Other ideas conceived by us for optical memory applications include photostructural changes such as photodispersion (utilized in our MicrOvonic File,[60] photodoping, photodarkening, and holographic storage.[82] There is now no question that the much-needed mass memories of the near future will be optically written, erased, and read. Finally, we note the present-day importance of laser crystalliz-

Fig. 12. Electrostatic printout obtained with ECD photostructural film printer. Each of the typewriter-size characters were generated on a 5x7 matrix by computer tape. (Ref. 4.)

Fig. 14. Ovonic Continuous Tone Imaging Film exposed through a high resolution test mask, exhibiting a resolution in excess of 1200 line pairs per millimeter.

ation of amorphous materials as an example of how new areas of technology can spring from basic scientific investigations. This use of amorphous solids as a vital step in preparing improved crystallites was not a subject of general scientific investigation until we demonstrated such phase changes. Laser printout is now accepted as a matter of course. We were the first to utilize lasers for such applications.[2,4,46,51,53] Figure 12 shows a printout of an early laser copying and printing demonstration. As the old Chinese proverb teaches, one should always leave a golden bridge of retreat so as not to humiliate one's opposition. In this vein, it is relevant to emphasize how the use of amorphous materials has proved to be crucial in the understanding, control, and operation of crystalline MOS devices. More and more crystalline scientists and technologists are appreciating the value of amorphous materials. To me, it has been a needless controversy since the understanding of disorder illuminates the inherent deviations from order in crystalline materials.

Looking farther in the future, still higher capacity memories will be essential. For such purposes, only x-rays, electron beams, or ion beams can yield the necessary resolution. The most promising technique at present involves the use of electron beams. Recent advances in electron optics suggest that 1000Å beams will soon be available, and even 100Å beams are a possibility. In the early 1960's, we showed that electron beams can be used to either crystallize or amorphize alloys. Furthermore, the crystalline and amorphous phases are quite distinct with regard to secondary-electron emission, so that the memory can be easily read by the electron beam. If 100Å resolution can be achieved, about 10^{15} bits of information can be sorted on a 30-cm disk, more information than is contained in the books in all the libraries of, e.g., a highly literate country such as Japan.

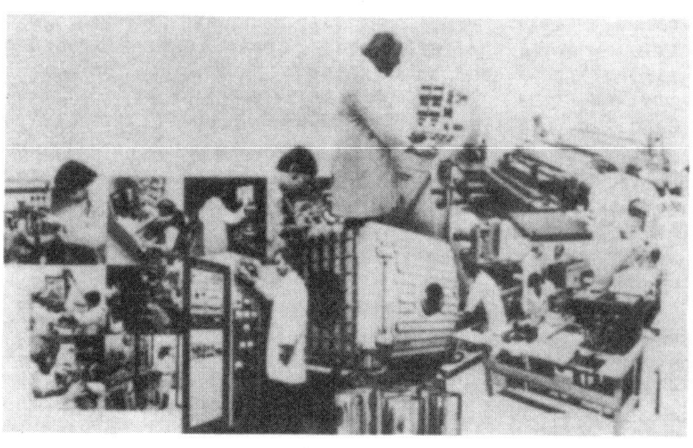

Fig. 13. Ovonic nonsilver photo-duplication film. (Ref. 61.)

The Ovonic memory concept forms the basis for preparation of many types of instant, dry, stable, photographic films with unique amplification, high resolution, and gradation of tones. This is accomplished by varying the fraction of the glass which has been crystallized and the grain size of the crystallites.[45,46,55] ECD has produced an array of films with either ultra-high contrast or exceptional continuous tones for imaging applications. Additional flexibility arises from the fact that the image can be obtained either directly after exposure, as discussed previously, or in latent form, to be developed subsequently when desired. One mechanism for the latter approach is to use our proprietary organo-tellurides as the film material.[83] In this case, exposure to light induces nucleation centers which form the latent image. Subsequent annealing above the glass transition temperature then induces crystallization of the latent region which produces the desired image (see Fig. 13). Excitation also permits the diffusion of tellurium. Using these procedures, we have been able to attain significant amplification factors.

In addition to using the crystalline-to-amorphous/amorphous-to-crystalline transitions, we have developed materials in which local structural changes can be induced and detected optically. These have proven useful in updating or correcting images well after exposure. While the materials described here are of the instant dry development type, an exciting feature by itself, we have also designed materials which have excellent etching properties. These have been used for high resolution masks (see Fig. 14) and other photographic applications.[2]

VIII. CHEMICAL MODIFICATION

It was taken for granted that in amorphous materials certain important parameters were in lock step with each other, e.g., if one had a large band gap material, low electrical conductivity would necessarily result. I decided to challenge this dogma by showing that amorphous materials could be chemically modified, and that by controlling the states in the gap one could for the first time independently control the conductivity changes over many orders of magnitude (see Fig. 15). What was so exciting about these results was that we could

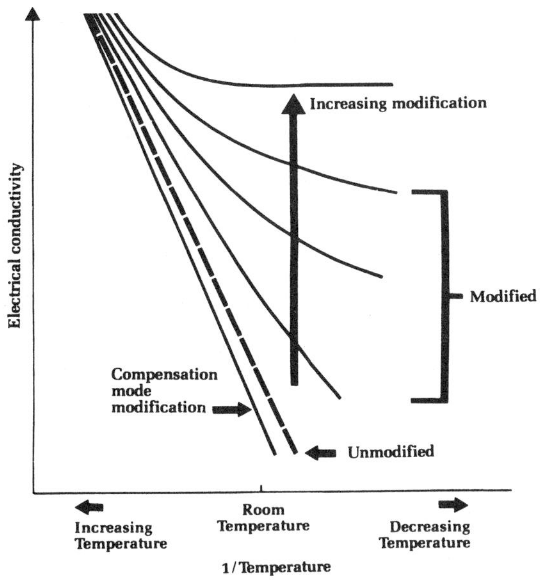

Fig. 15. Effect of chemical modification on the electrical properties of amorphous films. (Refs. 8 and 43.)

obtain large conductivity changes in elemental materials and in alloys containing elements from Group III through Group VI, including materials with drastically different band gaps.[8,9,42,43,84] (In the Periodic Chart of the Elements, Fig. 4, various atoms are darkened to show most of those used in the modification process.) In many cases, a small amount of modifier could increase electrical resistance, while larger amounts decrease it.

I had previously shown that lithium could achieve the same effect in chalcogenide glasses.[38] This was during the same period of time that, following the work of Chittick et al.,[85] Spear and LeComber [86] were demonstrating the possibility of substitutional doping in "amorphous silicon." (There still is a question about the effectiveness of p-doping in amorphous Si-H alloys.[8]) The fact that we could alter the conductivity of such a large variety of materials showed that we could outwit equilibrium and design a whole new family of materials with characteristics heretofore considered impossible. To put this in historical perspective, note the paper of Hamakawa,[87] which states, "the electrical properties of chalcogenide glasses could not be controlled so widely before the sensational appearance of 'chemical modification' proposed by Ovshinsky." While I appreciate the statement, I wish to reiterate that my paper on modification covered elements and alloys from columns III-VI in the Periodic Chart and was not limited to chalcogenides.[8,84] I was very pleased that Davis and Mytilineou corroborated chemical modification in amorphous arsenic with nickel as the modifier.[88]

Figure 16 shows typical materials that were modified, and it can be seen that the various active modifiers are either d-orbital or multi-orbital elements. The d-orbitals act as "pin cushions" when co-deposited so that they interact with the primary elements being deposited in a manner so as to create new TIEs. These TIEs would not exist if the modifying elements were deposited conventionally.[8,42] I believe the achievement of modification proves my point about multi-equilibria, since the normal structural bonds need not be affected at all by the modifying element (although they can be, if desired), i.e., the optical gap remains the same while the electrical conductivity can increase by over 10 orders of magnitude. It should be quite clear that the three-dimensional freedom of the amorphous state permits unusual and stable orbital interactions of a highly nonequilibrium nature. As can be seen from Fig. 16, various multi-orbital elements can be used and new nonequilibrium TIEs can be generated even without cosputtering, since the very fact that they are multi-orbital permits several different configurations. We have also utilized excitation as a means of having an atom or molecule enter into and interact with the matrix in such a manner as to effect modification. It is of interest that we have accomplished modification through dual nozzle melt spinning as well.[89] It should be kept in mind that the quenching process itself is a method of achieving nonequilibrium configurations.

In our technique, substantial concentrations of an appropriately chosen modifier are introduced into the amorphous network in a nonequilibrium manner so that it need not enter in its "optimal" chemical configuration. The modifier in small amounts can decrease electrical conductivity, but in larger amounts ordinarily increases it. When the concentration of the modifier exceeds that of the intrinsic defect centers, the Fermi level begins to move. In other words, in small concentrations, the modifiers can compensate and convert positively charged DECs to negatively charged ones, or vice versa; however, in larger concentrations, the modifiers yield many more DECs than would have been present in an equilibrium material. Therefore, the chemical modifier alters the localized states in the gap that control transport, while alloying alters the optical gap without changing the transport properties significantly. We therefore can independently separate the electrical activation energy from the optical gap and control them individually.[90] In a sense, an alloying element is also a modifier since it modifies the overall band gap, but I have used the term chemical modifier to describe situations in which transport or active chemical sites are the properties of interest, for in such cases the purpose of modification is to

Host Material	Active Modifier
Ge Te Se As	Ni, Fe, Mo
As	Ni, W
B_4C	W
Si C	W
Si	Ni, B, C
Si_3N_4	W
BN	W
TeO_2	Ni
Ge	Ni
SiO_2	W
$GeSe_2$	Ni
$Se_{95}As_5$	Ni

Fig. 16. Chemical modification of amorphous semiconductors. (refs. 8 and 43.)

alter the localized states within rather than the positions of the mobility edges.

I utilized surface chemical modification during the 1950's when I was mostly working with oxides, particularly those of the transition metals; I used amphoteric atoms and ions to change the conductivity by over 14 orders of magnitude,[70] utilizing such interactions to design switches and memories, both digital and adaptive.[3,72]

During the early 1960's, I investigated many amorphous and disordered phases, combining primary atoms with many types of alloying elements, and was the first to make amorphous gallium arsenide films. Our laboratory also made the first amorphous silicon carbide films.[91]

Another method of modification in amorphous materials is doping. Following Chittick et al.,[85] Spear and LeComber [86] reported doping experiments on what they considered to be amorphous silicon.[92] As we have pointed out, elemental amorphous silicon is not useful as an electronic material because free-energy considerations lead to an immense density of defects, including dangling bonds and voids. How is one then to utilize the silicon atom in amorphous materials for worthwhile electronic purposes? A means must be found to allow silicon atoms to be connected so that a completed tetrahedral structure ensues. The atoms that achieve such connections must play two roles. First, they must saturate the dangling bonds, i.e., they must be chemical compensators. Equally as important, they must also fulfill the role of structural links which act to provide flexibility to the matrix, relieving the stresses and strains of the pure silicon matrix and compensating it structurally so that the local order retains the electronic properties of the completed silicon configuration. I was therefore dubious about the usefulness of "amorphous silicon" since I thought that in its elemental form it held little electronic interest. When Fritzsche and his colleagues [93] showed that the dopable "amorphous silicon" really contained a large percentage of hydrogen and was therefore an alloy, I was pleased since it meant that my point of view and understanding were justified and correct, and, since alloys were where our talents lay, that we could make superior alloys based upon our chemical and structural concepts. This led me to suggest fluorine as a more suitable element since elements such as hydrogen and fluorine both terminate dangling bonds and at the same time enter into the structural network. I postulated that fluorine, due to its superhalogen qualities, i.e., its extreme electronegativity, small size, specificity and reactivity, not only terminates dangling bonds and can become a bridge, but also induces new local order and affects the TIE by several means, including controlling the way hydrogen bonds in the material since it can bond with silicon in several different ways, some of which produce defects.[60] Fluorinated materials are intrinsically different and fluorine is responsible for new TIEs. Lee, deNeufville and I [94] showed that fluorine does induce a new configuration when combined with silicon and hydrogen. Therefore, in amorphous silicon alloys, the addition of fluorine [41,56,95-97] minimizes defects, including dangling bonds, by generating new beneficial short-range order and TIEs.

I have utilized this concept for other materials such as germanium [97] and was able to solve the problem of "anomalous" density of states of amorphous germanium alloys, which most physicists consider to be tetrahedral materials. They, indeed, are tetrahedral in terms of their NSB; however, they are not in their DECs. There are various divalent and other configurations due to the "inert" lone pairs found in crystalline germanium compounds as well as in those of other elements in Group IV such as tin and lead.[98] Applying our chemical approach, I was able to show that this tendency away from tetrahedralness is even more prevalent in amorphous materials. The lack of tetrahedralness leads to increased DECs and unless compensated for can make germanium-containing alloys inferior as low density-of-states electronic materials. I consider that a very important attribute of fluorine is its tendency to expand the valence of many atoms by making use of the orbitals that are within its strong chemical attraction, and it is therefore particularly valuable where defects are involved with undercoordination.

The above is particularly relevant since in previous work I had considered that germanium could be two-fold coordinated,[39] and that silicon under certain conditions in the amorphous state could also have more than one orbital available that could result in additional defects, and therefore that fluorine could terminate and compensate as well the defect states that were not available to hydrogen. Adler,[99] using thermodynamic considerations, has proposed that two-fold coordination plays a role in the defect centers of amorphous silicon-hydrogen alloys. I felt that in germanium-containing materials, fluorine would interact with the "Sedgwick" lone pairs to force germanium into a more tetrahedral structure, thereby making an intrinsic material with an inherently low number of DECs. Fluorine also introduces an ionic character to the bonding, helpful in relieving strains. It decreases the fluctuations in potential on an atomic scale caused by the disorder of the amorphous state. The results are that we now make silicon-based alloys with a concentration of localized states in the low $10^{15}cm^{-3}$ range, and achieving this quality with our germanium-based materials. By lowering the DEC noise level, we can more effectively substitutionally dope these materials, and through the use of Raman spectroscopy we have been able to show that they also have more intermediate range order. The use of fluorine assures far more stable amorphous materials and is crucial in making these and similar materials into superior microcrystalline films.[100] Free radical chemistry, the leitmotif of my work since the very beginning in the 1950's, is involved in these processes. It is a subject that cannot be covered comprehensively here. Suffice it to say that it plays a very important role in the plasma decomposition processes which lead to many of the condensed materials that are discussed here.[56] We have performed experiments which clearly show the important role that free radicals play in producing better tetrahedral materials.[101]

While amorphous silicon-hydrogen alloys can be substitutionally doped n-type, just as crystalline materials, and boron doping yields p-type material, the boron doping is not very efficient. Following the chemical arguments of this paper, one can see why. Instead of being constrained to enter the

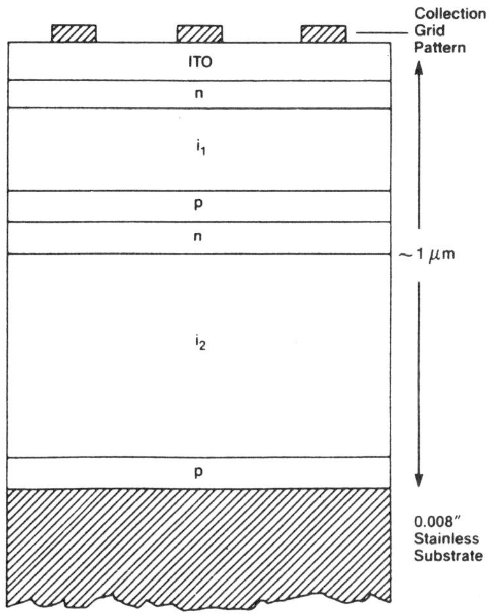

Fig. 17. Schematic cross section of 1.8eV/1.8eV ECD-Sharp Ovonic Tandem Solar Cell. (Refs. 106 and 107.)

matrix in a tetrahedral position, as in crystalline silicon, boron can form three-center bonds especially with bridging hydrogen atoms, and therefore can generate nontetrahedral structure.[8,9] Having discovered the unusual properties of boron many years ago upon reading Lipscomb's brilliant and profound work,[102] I was prepared to apply my understanding of it to amorphous materials, not only to explain boron's difficulty to adequately become a substitutional dopant in materials made from silane but also to utilize its various configurations as one of the elements to achieve chemical modification. Using this approach, we can see why it can be not only a chemical modifier even without co-deposition, because of the many configurations it can assume (this is one of the reasons it is a good glass former), but can also generate unneeded DECs, particularly in an alloy containing hydrogen. In fluorinated materials, due to the increase of intermediate-range order, both n and p substitutional doping is greatly improved. For tetrahedral materials, boron's empty orbital can also be used for coordinate bonding, making use of ordinarily non-bonded lone pairs with very interesting results. As I have pointed out, coordinate bonding can play an important role in amorphous materials.[2,8,39] Boron's natural glass-forming tendencies can be used to good avail since in proper amounts it is an excellent structural element and acts in its own way to affect coordination in a manner that can be as important as hydrogen and fluorine. These attributes give a structural stability to, for example, boron-containing tetrahedral amorphous materials. Boron and fluorine, therefore, are important elements in generating a superior photovoltaic material.

IX. CONCLUSION

We might well ask, where are we now? We have come a long way in 30 years. Photovoltaic devices based upon the superb electronic properties of amorphous silicon alloys are now in production. Figure 17 shows a commercial tandem cell which has the highest energy-to-weight ratio of any amorphous photovoltaic device. New ultra-light devices have been developed.[103] Our new generation multi-cell devices are very stable and have solar conversion efficiencies similar to those of our single band gap cells [104] which are over 10%,[105] and, because of the principles which I have outlined above, we are now making small band gap materials which are approaching the low defect concentrations of our amorphous silicon alloys. This means that we can expect efficiencies as high as 30% when we optimize the different alloys in a three-layer cell.

The information industry, the semiconductor industry, and increasingly the telecommunication industry, all are presently tied to the crystalline structure of silicon. The battle of the electronic giants is taking place on wafers that are now close to their maximum size of about six inches. More and more investment is being made to achieve higher chip densities by photolithographic means. Gordon Moore of Intel, whose expertise in crystalline materials is well known, facetiously proposed that what is needed for the circuits of the mid-1980's and 1990's is an impossible chip the size of the cardboard "wafer" shown in Fig. 18. Since wafers larger than about six inches are not to be expected from the melt-pulling techniques used to grow crystalline silicon, it seemed obvious that this could never be realized, and the industry would be limited to the approximate size of the hand-held chip in the inset. However, the discussion of amorphous silicon alloys in this paper shows that we can make materials that are not only analogous in their circuit functions to crystalline materials but have unique attributes as well. The 1000-foot long, 16-inch wide tandem cell that is shown on the right is a complete photovoltaic cell (electrical contacts are later printed on). We are making an amorphous analog of an infinitely long "crystal" of any desired width. (The width is a matter only of machine design.) We are, therefore,

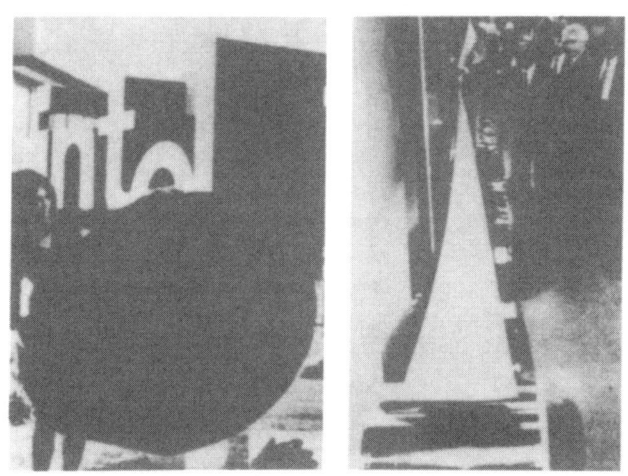

Fig. 18. Cardboard model of an imaginary "crystal wafer" on the left; a typical real wafer is shown in the inset; Ovonic thin-film semiconducting material, an "infinite crystal," on the right.

taking up the challenge and are designing large-area, totally integrated thin-film circuits which we believe will soon transform the information and telecommunication industries.

There are many other areas where I have utilized our synthetic materials approach. Our laboratory has worked successfully for years in fields as diverse as coatings, batteries, catalysts, electrochemistry, hydrogen storage, thermoelectrics and superconductors. We are already in production in a number of these areas. These fields are very important not only commercially and technically but have significant theoretical implications. Obviously, these cannot all be covered here and will be the subject of another paper; some of these areas have been described elsewhere.[108]

My intention here has been to discuss the fundamental concepts involved in amorphous materials and to show how our basic understanding can be directly related to the devices that have been and are being developed. Ours is a synthetic materials approach. When one frees oneself from the restrictions of crystalline symmetry, then not only excellent crystalline analogs such as transistors can be made,[109] but many new nonequilibrium phenomena and materials can be developed. In fact, the whole dogma of bulk homogeneity can be re-evaluated. I described exceedingly thin multi-layer and compositionally modulated devices in our patent literature years ago. Based on these concepts, we have been successfully developing devices, some of which are being utilized commercially.[110] We have reported in the scientific literature on the unusual effects seen in such materials.[48,111-114] This subject has now been rediscovered,[115] and most likely will grow into a new field.

As can be seen, even though we have arbitrarily divided this paper into electronic, chemical, and mechanical sections, one cannot really speak of one parameter divorced from another. The <u>electronic density of states</u> can be the equivalent of <u>active chemical sites</u>, etc. Amorphous materials science is a synthesis of many different disciplines, and therefore has through this process been transmuted into a discipline of its own. Amorphous materials indeed are characterized by the total interactive environment.

The voyage into the amorphous field has been one of discovery and delight, and I take great pleasure in being represented in this institute and in this volume.

ACKNOWLEDGEMENTS

Since it has been a personal odyssey and began at a time when there were few in the field, I have written this paper reflecting my own travels and travails. However, this work could not have been accomplished without my collaborators and colleagues of many years. As always, this work could not have been done without the partnership of my wife, Iris.

It is impossible to give adequate thanks to all the people with whom I have worked in the past 30 years. In the neurophysiological area where I first started, I wish to express my appreciation for their encouragement to both the late Ernest Gardner, Dean of Wayne State University Medical School, and the late Fernando Morin, Chairman of the Department of Anatomy there. I wish to also express my thanks to I.I. Rabi, Nevill Mott, and Kenichi Fukui for their encouragement through the years; also to Boris Kolomiets, an early and major figure in Russian amorphous chalcogenide work, for his statement at the IV Symposium on Vitreous Chalcogenide Semiconductors in Leningrad in 1967, when I first scientifically discussed switching. He said that this work would transform the amorphous field whose progress to that date he denoted as a very slowly-rising, essentially horizontal line to an almost vertical one, a prediction that fortunately came true. It was at that same meeting that I met Radu Grigorovici, a friend and important contributor to our field.

From the 1960's on, I benefitted immensely from the colleagual collaboration of Hellmut Fritzsche, Morrel Cohen, and David Adler. I owe David a special debt of gratitude. Our close association has been of great help to me. I acknowledge with appreciation the collaboration of Arthur Bienenstock, John deNeufville, Heinz Henisch, Marc Kastner, and Krishna Sapru, among others. The work described in this paper was made possible through the years by my colleagues, especially Wally Czubatyj, Steve Hudgens, Masat Izu, and the rest of the superb group of people that make up ECD.

REFERENCES

1. D. Adler, ed., <u>Disordered Materials: Science and Technology</u>, Selected Papers by S.R. Ovshinsky, (Amorphous Institute Press, Bloomfield Hills, 1-296 (1982).

2. S.R. Ovshinsky and H. Fritzsche, "Amorphous Semiconductors for Switching, Memory, and Imaging Applications," IEEE Trans. on Electron Devices <u>ED-20</u>, 91 (1973).

3. S.R. Ovshinsky and I.M. Ovshinsky, "Analog Models for Information Storage and Transmission in Physiological Systems," Mat. Res. Bull. 5, 681 (1970). (Mott Festschrift.)

4. S.R. Ovshinsky, "The Ovshinsky Switch," <u>Proc. 5th Annual National Conference on Industrial Research</u>, Chicago, 86-90 (1969). The Ovographic work was done in collaboration with P. Klose.

5. S.R. Ovshinsky, "An Introduction to Ovonic Research," J. Noncryst. Solids <u>2</u>, 99-106 (1970).

6. F. Yonezawa and T. Ninomiya, eds., "Topological Disorder in Condensed Matter," <u>Proc. 5th Taniguchi Int. Symp.</u>, Japan, 2 (1982).

7. For a discussion of short-range order, see: A.F. Ioffe and A.R. Regel, "Non-Crystalline, Amorphous, and Liquid Electronic Semiconductors" in <u>Progress in Semiconductors</u>, vol. 4, John Wiley, New York, 237-291 (1960).

8. S.R. Ovshinsky, "Chemical Modification of Amorphous Chalcogenides," <u>Proc. of the 7th</u>

Int. Conf. on Amorphous and Liquid Semiconductors, Edinburgh, Scotland, 519-523 (1977).

9. S.R. Ovshinsky and D. Adler, "Local Structure, Bonding, and Electronic Properties of Covalent Amorphous Semiconductors," Contemp. Phys. 19, 109 (1978).

10. S.R. Ovshinsky, "The Chemical Basis of Amorphicity: Structure and Function," Rev. Roum. Phys. 26, 893-903 (1981). (Grigorovici Festschrift.)

11. J.M. Ziman, Principles of the Theory of Solids, Cambridge University Press, 1 (1965).

12. J.M. Ziman, "How Is It Possible To Have An Amorphous Semiconductor?" J. Noncryst. Solids 4, 426-427 (1970).

13. M.H. Cohen, H. Fritzsche and S.R. Ovshinsky, "Simple Band Model for Amorphous Semiconducting Alloys," Phys. Rev. Lett. 22, 1065-1068 (1969). See also N.F. Mott, Adv. Phys. 16, 49 (1967).

14. J.M. Ziman, Models of Disorder, Cambridge University Press, ix (1979).

15. I thank Dennis Weaire for bringing these contradictory remarks to my attention.

16. Thomas S. Kuhn, The Structure of Scientific Revolutions, University of Chicago Press, 6 (1962).

17. S.R. Ovshinsky, "Reversible Electrical Switching Phenomena in Disordered Structures," Phys. Rev. Lett. 21, 1450-1453 (1968).

18. H.K. Henisch, S.R. Ovshinsky and R.W. Pryor, "Switching Effects in Amorphous Semiconductor Thin Films," in Proc. of the Intl. Congress on Thin Films, Cannes, October 5-10, 1970 (published by Societe Francaise des Ingenieurs et Techniciens du Vide).

19. R.W. Pryor and H.K. Henisch, "Mechanism of Threshold Switching," Appl. Phys. Lett. 18, 324 (1971).

20. H.K. Henisch and R.W. Pryor, "Mechanism of Ovonic Threshold Switching," Solid State Elec. 14, 765 (1971).

21. H.K. Henisch, R.W. Pryor and G.J. Vendura, "Characteristics and Mechanisms of Threshold Switching," J. Noncryst. Solids 8-10, 415 (1972).

22. R.W. Pryor and H.K. Henisch, "Nature of the On-State in Chalcogenide Glass Threshold Switches," J. Noncryst. Solids 7, 181 (1972).

23. W. Smith and H.K. Henisch, "Threshold Switching in the Presence of Photo-Excited Charge Carriers," Phys. Stat. Sol. A 17, K81 (1973).

24. M.P. Shaw, S.H. Holmberg and S.A. Kostylev, "Reversible Switching in Thin Amorphous Chalcogenide Films--Electronic Effects," Phys. Rev. Lett. 31, 542 (1973).

25. H.K. Henisch, W.R. Smith and M. Wihl, "Field-Dependent Photo-response of Threshold Switching Systems," Proc. of the 5th Intl. Conf. on Amorphous and Liquid Semiconductors, Garmisch-Partenkirchen, Germany, September 1973, J. Stuke and W. Brenig, eds., Taylor and Francis, London, 567 (1974).

26. W.D. Buckley and S.H. Holmberg, "Nanosecond Pulse Study of Memory Material of Different Thicknesses," Sol. State Elec. 18, 127 (1975).

27. K.E. Petersen and D. Adler, "Probe of the Properties of the On-State Filament," J. Appl. Phys. 47, 256 (1976).

28. K.E. Petersen, D. Adler and M.P. Shaw, "Amorphous-Crystalline Heterojunction Transistor," IEEE Trans. 23, 471 (1976).

29. D.K. Reinhard, "Response of the OTS to Pulse Burst Waveforms (Critical Power Density)," Appl. Phys. Lett. 31, 527 (1977).

30. D. Adler, H.K. Henisch and N. Mott, "The Mechanism of Threshold Switching in Amorphous Alloys," Rev. Mod. Phys. 50, 209 (1978).

31. D. Adler, M.S. Shur, M. Silver and S.R. Ovshinsky, "Threshold Switching in Chalcogenide-glass Thin Films," J. Appl. Phys. 51, 3289 (1980).

32. M.P. Shaw and N. Yildirim, "Thermal and Electrothermal Instabilities in Semiconductors," Adv. in Elec. and Electron Phys. 60, 307-385 (1983).

33. J. Kotz and M.P. Shaw, "Thermophonic Investigation of Switching and Memory Phenomena in Thick Amorphous Chalcogenide Films," Appl. Phys. Lett. 42, 199 (1983).

34. S.R. Ovshinsky and D. Adler, to be published.

35. S.R. Ovshinsky, "Localized States in the Gap of Amorphous Semiconductors," Phys. Rev. Lett. 36, 1469-1472 (1976).

36. S.R. Ovshinsky, "Amorphous Materials as Interactive Systems," in Proc. of the Sixth Int. Conf. on Amorphous and Liquid Semiconductors, Leningrad, USSR (1975) 426-436: Structure and Properties of Non-Crystalline Semiconductors, B.T. Kolomiets, ed., Nauka, Leningrad 426-436 (1976).

37. H. Fritzsche, "Summary Remarks," in Proc. of the Sixth Int. Conf. on Amorphous and Liquid Semiconductors, Leningrad, USSR, 1975: Electronic Phenomena in Non-Crystalline Semiconductors, B.T. Kolomiets, ed., Nauka, Leningrad, 65-68 (1976).

38. S.R. Ovshinsky, "Lone-Pair Relationships and the Origin of Excited States in Amorphous Chalcogenides," AIP Conf. Proc. 31, 31-36 (1976).

39. S.R. Ovshinsky and K. Sapru, "Three Dimensional Model of Structure and Electronic Properties of Chalcogenide Glasses," in Proc. of the Fifth Int. Conf. on Amorphous and Liquid Semiconductors, Garmisch-Partenkirchen, Germany, 447-452 (1974).

40. A. Bienenstock, F. Betts and S.R. Ovshinsky, "Structural Studies of Amorphous Semiconductors," J. Noncryst. Solids 2, 347 (1970).

41. S.R. Ovshinsky and A. Madan, "A New Amorphous Silicon-Based Alloy for Electronic Applications," Nature 276, 482-484 (1978).

42. R.A. Flasck, M. Izu, K. Sapru, T. Anderson, S.R. Ovshinsky and H. Fritzsche, "Optical and Electronic Properties of Modified Amorphous Materials," in Proc. 7th Intl. Conf. on Amorphous and Liquid Semiconductors, Edinburgh, Scotland 524-528 (1977).

43. S.R. Ovshinsky, "The Chemistry of Glassy Materials and Their Relevance to Energy Conversion," Proc. Intl. Conf. on Frontiers of Glass Science, Los Angeles, California; J. Noncryst. Solids 42, 335-344 (1980).

44. For references to his work see Revue Roumaine de Physique 26, No. 809 (1981). (Grigorovici Festschrift.)

45. S.R. Ovshinsky, "Electronic and Structural Changes in Amorphous Materials as a Means of Information Storage and Imaging," in Proc. of the Fourth Int. Congress for Reprography and Information, Hanover, Germany 109-114 (1975).

46. S.R. Ovshinsky, "Amorphous Materials as Optical Information Media," J. Applied Photographic Engineering 3, 35-39 (1977).

47. S.R. Ovshinsky, unpublished data, 1975; S.R. Ovshinsky and K. Sapru, 1977.

48. The Francis Bitter National Magnet Lab. Annual Report for July 1982 to June 1983, 118.

49. S.R. Ovshinsky and H. Fritzsche, "Reversible Structural Transformations in Amorphous Semiconductors for Memory and Logic," Met. Trans. 2, 641 (1971).

50. S.R. Ovshinsky, "Optical Information Encoding in Amorphous Semiconductors," Topical Meeting on Optical Storage of Digital Data, Aspen, Colorado, MB5-1-MB5-4, 1973.

51. S.R. Ovshinsky and P.H. Klose, "Reversible High-Speed High-Resolution Imaging in Amorphous Semiconductors," Proc. SID 13, 188 (1972).

52. The reason for the darkening of some of the elements in this figure will be given when we discuss chemical modification.

53. S.R. Ovshinsky and P.H. Klose, "Imaging in Amorphous Materials by Structural Alteration," J. Noncryst. Solids 8-10, 892-898 (1972).

54. D.L. Staebler and C.R. Wronski, "Reversible Conductivity Changes in Discharge Produced Amorphous Si," Appl. Phys. Lett. 31, 292 (1977).

55. S.R. Ovshinsky and P.H. Klose, "Imaging by PhotostructuraL Changes," Proc. Symp. on Nonsilver Photographic Processes, New College, Oxford, 1973; R.J. Cox, ed., Academic Press, London, 61-70 (1975).

56. S.R. Ovshinsky, "The Role of Free Radicals in the Formation of Amorphous Thin Films," in Proc. Int. Ion Engineering Congress (ISIAT '83 & IPAT '83), Kyoto, Japan, 817-828 (1983).

57. D. Adler, "Origin of the Photo-Induced Changes in Hydrogenated Amorphous Silicon," Solar Cells 9, 133 (1983).

58. S. Guha, J. Yang, W. Czubatyj, S.J. Hudgens and M. Hack, "On the Mechanism of Light-Induced Effects in Hydrogenated Amorphous Silicon Alloys," Appl. Phys. Lett. 42, 588 (1983).

59. H. Fritzsche, "Optical and Electrical Energy Gaps in Amorphous Semiconductors," J. Noncryst. Solids 6, 49 (1971).

60. S.R. Ovshinsky, The Shape of Disorder," J. Noncryst. Solids 32, 17 (1979). (Mott Festschrift.)

61. S.R. Ovshinsky, "Principles and Applications of Amorphicity, Structural Change, and Optical Information Encoding," in: Proc. 8th Intl. Conf. on Amorphous and Liquid Semiconductors, Grenoble, France (1981): J. de Physique, Colloque C4, supplement au no. 10, 42, C4-1095-1104 (1981).

62. M. Kastner, D. Adler and H. Fritzsche, "Valence-Alternation Model for Localized Gap States in Lone-Pair Semiconductors," Phys. Rev. Lett. 37, 1504 (1976).

63. M. Kastner, "Bonding Bands, Lone-Pair Bands, and Impurity States in Chalcogenide Semiconductors," Phys. Rev. Lett. 28, 355 (1972).

64. S.R. Ovshinsky, "Electronic-Structural Transformations in Amorphous Materials--A Conceptual Model," July 13, 1972, unpublished.

65. S.C. Agarwal, "Nature of Localized States in Amorphous Semiconductors--A Study by Electron Spin Resonance," Phys. Rev. B 7, 685 (1973).

66. R.A. Street and N.F. Mott, "States in the Gap in Glassy Semiconductors," Phys. Rev. Lett. 35, 1293 (1975).

67. P.W. Anderson, "Model for the Electronic Structure of Amorphous Semiconductors," Phys. Rev. Lett. 34, 953 (1973).

68. M.P. Southworth, "The Threshold Switch: New Component for Ac Control," Control Engineering 11, 69 (1964).

69. J.R. Bosnell, "Amorphous Semiconducting Films," in Active and Passive Thin Film Devices, T.J. Coutts, ed., Academic Press, 288 (1978).

70. J.D. Cooney, "A Remarkable New Switching Form," Control Engineering 6, 121 (1959).

71. "How Liquid State Switch Controls A-C," Electronics 32, 76 (1959).

72. S.R. Ovshinsky, "The Physical Base of Intelligence-Model Studies," presented at the Detroit Physiological Society, 1959.

73. E.J. Evans, J.H. Helbers and S.R. Ovshinsky, "Reversible Conductivity Transformations in Chalcogenide Alloy Films," J. Noncryst. Solids 2, 339 (1970).

74. A.H. Guth, "Inflationary Universe: A Possible Solution to the Horizon and Flatness Problems," Phys. Rev. D 23, 347-356 (1981).

75. A.H. Guth and P.J. Steinhardt, "The Inflationary Universe," Sci. Am. 250, 116-128 (1984).

76. R.J. von Gutfeld and P. Chaudhari, "Laser Writing and Erasing on Chalcogenide Films," J. Appl. Phys. 43, 4688-4693 (1972).

77. A.W. Smith, "Injection Laser Writing on Chalcogenide Films," Appl. Optics. 13, 795 (1974).

78. J. Feinleib and S.R. Ovshinsky, "Reflectivity Studies of the Te(GeAs)-Based Amorphous Semiconductor in the Conducting and Insulating States," J. Noncryst. Solids 4, 564 (1970).

79. S.R. Ovshinsky, presented at the Gordon Conf. on Chemistry and Metallurgy of Semiconductors, Andover, N.H., 1969.

80. S.R. Ovshinsky, "Method and Apparatus for Storing and Retrieving Information," U.S. Patent No. 3,530,441.

81. J. Feinleib, J.P. deNeufville, S.C. Moss and S.R. Ovshinsky, "Rapid Reversible Light-Induced Crystallization of Amorphous Semiconductors," Appl. Phys. Lett. 18, 254 (1971). Earlier, Laurence Pellier and Peter Klose worked with me in this area.

82. J.P. deNeufville, "Optical Information Storage," Proc. 5th Intl. Conf. on Amorphous and Liquid Semiconductors, Garmisch-Partenkirchen, Germany 1973; J. Stuke and W. Brenig, eds., Taylor and Francis, London, 1351-1360 (1974).

83. Y.C. Chang and S.R. Ovshinsky, "Organo-Tellurium Imaging Materials," U.S. Patent No. 4,142,896.

84. "Amorphous Materials Modified to Form Photovoltaics," New Scientist 76, 491 (1977).

85. R.C. Chittick, J.H. Alexander and H.F. Sterling, "Preparation and Properties of Amorphous Silicon," J. Electrochem. Soc. 116, 77-81 (1969).

86. W.E. Spear and P.G. LeComber, "Electronic Properties of Substitutionally Doped Amorphous Si and Ge," Phil. Mag. 33, 935 (1976).

87. H. Okamoto and Y. Hamakawa, "Statistical Considerations on Electronic Behavior of the Gap States in Amorphous Semiconductors," J. Noncryst. Solids 33, 230 (1979).

88. E.A. Davis and E. Mytilineou, "Chemical Modification of Amorphous Arsenic," Solar Energy Mats. 8, 341-348 (1982). (Ovshinsky Festschrift.)

89. S.R. Ovshinsky and R.A. Flasck, "Method and Apparatus for Making a Modified Amorphous Glass Material," U.S. Patent No. 4,339,255.

90. We know that there is controversy as to what constitutes an optical gap in amorphous materials, but we feel that our chemical examples are quite clear.

91. E.A. Fagen, "Optical Properties of Amorphous Silicon Carbide Films," Silicon Carbide-1973, Proc. 3rd Intl. Conf. on Silicon Carbide, Miami Beach, Florida, R.C. Marshall, J.W. Faust, Jr. and C.E. Ryan, eds., University of Southern Carolina Press, 542-549 (1973).

92. For the work of others in the amorphous area, see, for example, Science and Technology of Noncrystalline Semiconductors, H. Fritzsche and D. Adler, eds., Solar Energy Materials 8, Nos. 1-3, 1-348 (1982).

93. H. Fritzsche, M. Tanielian, C.C. Tsai and P.J. Gaczi, "Hydrogen Content and Density of Plasma-Deposited Amorphous Hydrogen," J. Appl. Phys. 50, 3366 (1979).

94. H.U. Lee, J.P. deNeufville and S.R. Ovshinsky, "Laser-Induced Fluorescence Detection of Reactive Intermediates in Diffusion Flames and in Glow-Discharge Deposition Reactors," J. Noncryst. Solids 59-60, 671 (1983).

95. S.R. Ovshinsky and A. Madan, "Properties of Amorphous Si:F:H Alloys," Proc. 1978 Meeting of the American Section of the Intl. Solar Energy Soc., K.W. Boer and A.F. Jenkins, eds., AS of ISES, University of Delaware, 69 (1978).

96. A. Madan and S.R. Ovshinsky, "Properties of Amorphous Si:F:H," Proc. 8th Intl. Conf. on Amorphous and Liquid Semiconductors, Cambridge, Massachusetts 1979; J. Noncryst. Solids 35-36, 171-181 (1980).

97. S.R. Ovshinsky and M. Izu, "Amorphous Semiconductors Equivalent to Crystalline Semiconductors," U.S. Patent No. 4,217,374; S.R. Ovshinsky and A. Madan, "Amorphous Semiconductors Equivalent to Crystalline Semiconductors Produced by Glow-Discharge Process," U.S. Patent No. 4,226,898; S.R. Ovshinsky and M. Izu, "Method for Optimizing Photoresponsive Amorphous Alloys and Devices,"

U.S. Patent No. 4,342,044; S.R. Ovshinsky and A. Madan, "Amorphous Semiconductors Equivalent to Crystalline Semiconductors," U.S. Patent No. 4,409,605; S.R. Ovshinsky and M. Izu, "Amorphous Semiconductors Equivalent to Crystalline Semiconductors," U.S. Patent No. 4,485,389.

98. E. Cartmell and G.W.A. Fowles, Valency and Molecular Structure, Van Nostrand Reinhold, New York, 1970.

99. D. Adler, "Density of States in the Gap of Tetrahedrally Bonded Amorphous Semiconductors," Phys. Rev. Lett. 41, 1755 (1978).

100. R. Tsu, S.S. Chao, M. Izu, S.R. Ovshinsky, G.J. Jan and F.H. Pollak, "The Nature of Intermediate Range Order in Si:F:H:(P) Alloy Systems," Proc. 9th Intl. Conf. on Amorphous and Liquid Semiconductors, Grenoble, France, 1981; J. de Physique, Colloque C4, supplement au no. 010, 42, C4-269 (1981).

101. R. Tsu, D. Martin, J. Gonzales-Hernandez and S.R. Ovshinsky, to be published.

102. W.N. Lipscomb, Boron Hydrides, W.A. Benjamin, New York 1963. We are honored to be working with him on some of these important problems today.

103. J. Hanak, to be published.

104. J. Yang, R. Mohr and R. Ross, "High Efficiency Amorphous Silicon and Amorphous Silicon-Germanium Tandem Solar Cells," to be presented at the First International Photovoltaic Science and Engineering conference, Kobe, Japan, November 13-16, 1984. Our devices are the only ones that have both high efficiency and great stability.

105. W. Czubatyj, M. Hack and M.S. Shur, to be published.

106. S.R. Ovshinsky, "Commercial Development of Ovonic Thin Film Solar Cells," Proc. of SPIE Symposium on Photovoltaics for Solar Energy Applications II, Arlington, Virginia, vol. 407, 5-8 (1983).

107. M. Izu and S.R. Ovshinsky, "Production of Tandem Amorphous Silicon Alloy Solar Cells in a Continuous Roll-to-Roll Process," Proc. of SPIE Symposium on Photovoltaics for Solar Energy Applications II, Arlington, Virginia, vol. 407, 42-46 (1983).

108. S.R. Ovshinsky, Problems and Prospects for 2004, Symp. Glass Science and Technology, Vienna, 1984. To be published in J. Noncryst. Solids. (Kreidl Festschrift.)

109. Z. Yaniv, G. Hansell, M. Vijan and V. Cannella, "A Novel One-Micrometer Channel Length a-Si TFT," 1984 Materials Research Society Symposium, Albuquerque, New Mexico, 1984.

110. Ovonyx™ multi-layer x-ray dispersive mirrors.

111. S.R. Ovshinsky, unpublished.

112. L. Contardi, S.S. Chao, J. Keem and J. Tyler, "Detection of Nitrogen with a Layered Structure Analyzer in a Wavelength Dispersive X-ray Microanalyzer," Scann. Electron Microscopy II, 577 (1984).

113. J. Kakalios, H. Fritzsche, N. Ibaraki and S.R. Ovshinsky, "Properties of Amorphous Semiconducting Multi-layer Films," Proc. Intl. Topical Conf. on Transport and Defects in Amorphous Semiconductors, Institute for Amorphous Studies, Bloomfield Hills, Michigan; J. Noncryst. Solids 66, 339-344, H. Fritzsche and M.A. Kastner, eds., 1984.

114. S.R. Ovshinsky and M. Izu, "Method of Optimizing Photoresponsive Amorphous Alloys and Devices," U.S. Patent No. 4,342,044.

115. Proc. Intl. Topical Conf. on Transport and Defects in Amorphous Semiconductors, Institute for Amorphous Studies, Bloomfield Hills, Michigan; J. Noncryst. Solids 66, 1-392, H. Fritzsche and M.A. Kastner, eds., 1984.

AMORPHOUS MATERIALS—PAST, PRESENT, AND FUTURE

Stanford R. Ovshinsky

Energy Conversion Devices, Inc., Troy, Michigan 48084

It is fitting that we honor Norbert Kreidl by considering the future of glass for there is a dramatic "phase" transformation presently taking place in the area of glass science and technology. The glass meetings of old were centered on the much debated subject, "What is glass?" In a sense, this reflected not only scientific uneasiness but a search for identity, for the very basis of glassy materials, their inherent disorder, was the cause of much insecurity. The crystalline field had as its bedrock the crystal lattice, the order, if not the boredom, of repetitive atoms which looking in all directions saw sharply definable and identifiable neighbors over relatively long distances. It is no wonder that glassy materials from time immemorial through the 1950's appeared to be associated only with wide band gap materials whose optical and mechanical properties were of paramount importance. Here disorder could be considered as a positive attribute yielding, for example, isotropy, and short-range order could provide some feeling of comfort as far as reproducibility was concerned. Defects, if they could be identified at all, were identified when they could be localized in large band gap materials such as oxide glasses. The study of glass, therefore, had a deceptive transparency which tended to limit the potential applications. However, after much recent work, the question "Whither glass?" can now be answered with a great deal of assurance.

We can pay no better tribute to Norbert Kreidl than by showing that today we know what glass is and, more importantly, that this knowledge provides us firm ideas about the future of its science and technology. Kreidl's ever-inquiring creative mind, his intelligence, commitment and dedication to our field are examples of what science is about at its best and how it can be utilized to serve our world society.

My tribute to Norbert Kreidl will be to show how our present understanding has led to products undreamed of during the past when glass was only "glass." Of course, there are still many important uses for glass as a purely passive material, but it is its applications as active elements which will lead us into the 21st century.

Glass, which was previously considered primarily an optical, dielectric or passivating material, can now be used to create active devices--switches, memories, solar cells, catalysts, etc. We are not interested so much in its "old" properties, but rather want to exploit new properties, its exciting electronic phenomena, chemical reactivity or inertness, unusual phonon properties, exceptional superconductivity characteristics, important magnetic properties, etc. Instead of considering only wide band gap materials as the natural area of glassy materials, we are now able to design glassy metals, an immense range of glassy semiconductors from degenerate to wide band gap materials, and new types of information encoding devices where unusual structural changes play the decisive role.

Based upon my experience I make the bold prophecy that by the year 2004, when we celebrate Norbert's 100th birthday, most of the alternative energy sources involving materials which convert light, heat, or chemical energy into electricity will be made of amorphous or disordered materials. It is generally accepted that the two largest industries in the 1990's will be energy conversion and information processing, the latter involving computers, telecommunications, etc. As I will show here, I believe that in addition to the energy conversion devices, almost all of the informational devices will also be made of amorphous materials. The third leg of the stool representing the basic industries of the world is, of course, the materials from which everything is made. Here I believe that the powerful ability to engineer at the molecular level, that is, to synthesize countless new materials, is a unique feature of the amorphous state that will make amorphous materials technology the answer for those who wish to have products that can withstand high temperatures, corrosion, abrasion, etc. Furthermore, amorphous materials will also be utilized as catalysts, membranes for water desalination, and other basic chemical processing activities.

As I have pointed out,(1,2) the ages of humankind have always been defined by the predominant materials: the Stone Age, Iron Age, Bronze Age, etc. In this paper I wish to support

Reprinted by permission from *Journal of Non-Crystalline Solids,* Vol. 73, pp. 395–408 (1985).

with hard evidence my contention that the next century will be the Glass Age. In fact, we are already in the transition period between the age of order and the age of disorder. But unlike the emotive meaning of disorder and the literary meaning of amorphicity, both of which presently have negative connotations, I wish to address in this paper the meaning and value of disorder or, put in another way, the value of freedom associated with lifting the restrictions of crystalline symmetry. This freedom is reflected in the many new local bonding and nonbonding configurations that give noncrystalline solids unique attributes.(2-13)

The term "glass," commonly described as a supercooled liquid is too narrow since it excludes materials that are not formed from the melt. Such a definition creates an unnatural division and diversion that confuses the important underlying concept of non-periodic materials that atoms can be placed in three-dimensional configurations with the only restrictions being those that are imposed by the chemical and electrical force fields of the constituent atoms and their local environment.(2-13) This means that atoms can be placed into the solid by many means such as sputtering, vacuum deposition, ion implantation, electrochemical deposition, plasma decomposition, chemical vapor deposition, etc.

With the new synthetic materials that can be prepared using these methods, we can make devices in which the familiar benchmarks of equilibrium, stoichiometry and homogeneity do not have the same basic importance as they do in crystalline or in old fashioned glassy materials. In fact, the only rules that atoms deposited by our various techniques necessarily follow are those imposed by chemical reactivity, steric hindrances and local environmental forces. There are in fact <u>many</u> local equilibrium conditions available to atoms depending upon the dynamic conditions of their assembly (see, for example, refs. 1, 12 and 13). Therefore, the structure and properties of these materials are process dependent and are affected by temperature, excitation, sticking coefficients, etc. Each of this multitude of phases is metastable rather than globally stable, but can maintain its properties for millenia at room temperature, and, of course, be stable at actual operating conditions. There are chemical and topological barriers designed into the material which prevent structural changes. I remind you that diamond is a metastable material. Moreover, concepts which apply to the ordered world have different meaning when applied to noncrystalline solids. For example, I am in the process of redefining equilibria (14) where I show that the conventional metallurgical wisdom concerned is not relevant to many of the materials that we develop. I believe that in 2004 there will be many non-equilibrium, non-homogeneous materials designed for specific uses. Our science will be stimulated and benefited by the theories involved.

Crystallinity with its very limited specific choices of configuration pales when compared to the richness of possibilities of structures allowed in amorphous materials. Instead of building up materials where atoms have to fit in and match the lattice constraints inherent in a particular crystal structure, we can now design amorphous materials where we literally place the atoms one at a time in three-dimensional space to achieve properties undreamed of in the past. We develop layered and modulated structures where we develop and utilize new solid state phenomena and chemistry.(15)

Before we describe the devices which will be the basis for many future generations in the coming age of energy, information and synthetic materials, I will discuss what has been the Rosetta Stone for deciphering the enigma of disorder.

The Rosetta Stone for understanding amorphous materials is the concept that it is the deviations from normal structural bonding (NSB) that control the transport properties of amorphous materials just as perturbations from perfect long-range order do in crystalline materials, and that in order to control the electronic properties of amorphous materials, one must understand how to remove or add states in the gap which arise from the deviant electronic configurations (DECs).(7,8,10,11,16) The ability to tailor the band gap itself is directly related to the strength of the chemical bonds between the constituent elements which represent the NSB. By suitably choosing elements that make up the amorphous alloy, we design crosslinking structures which can retard crystallization. One can design these structures to be either unistable or bistable.(4,17-19) There is a vast range of materials that can be designed and therefore the relaxation processes which are involved can differ, e.g., elements ordinarily bonding with low coordinations can tend to remove strains in fully three-dimensional networks.(4,5,7) As I have said, "Physicists have taken for granted that there is a thermodynamic drive toward crystallinity. We emphasize that there is an equally important energetic process that leads to amorphicity, that is, the preferred chemical bonding of atoms and the charge field produced by the nonbonding electrons can alter a molecular structure so that it has an anticrystalline state. Crystals by definition have geometries that allow for repetition of the basic cell structure. The shapes that I am discussing are not rigid spherical balls but complex distorted shapes formed by localized pressures, repulsions and attractions of surrounding forces, compressed here, elongated there, twisted along another axis, the very antithesis of a crystal cell model. These tangled networks are further inhibited from crystallinity by crosslinks and bridging atoms."(2) The spectrum of configurations ranging from this description to more rigid tetrahedral type amorphous materials that can have intermediate order show the power of our field for the development of new phenomena.(20-22)

Not only can amorphous materials be designed to be very stable, but they can also be tailored to exhibit structural changes ranging from subtle bond switching to reversible crystallization. The latter include materials which are the basis of information encoding devices such as electrical or optical memories that will be described subsequently. Their design depends upon an understanding of how the energy barriers to crystallization can be overcome by optical, electrical or thermal excitation.(17,18,23,24)

I will now describe three areas where amorphous, disordered and glassy materials have been made into products which are the basis for developing a

Fig. 1 Mass production Ovonic Photovoltaic Processor.

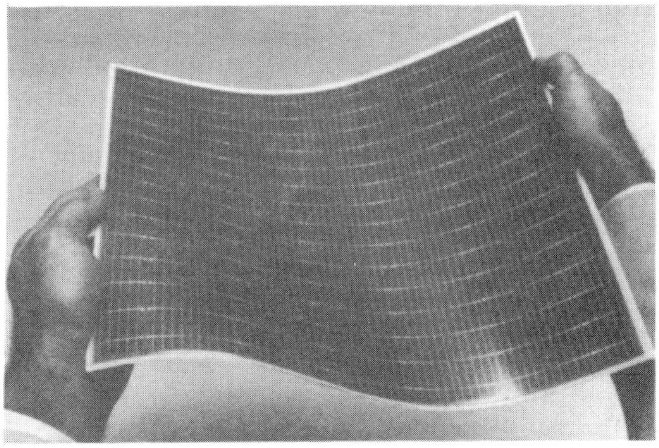

Fig. 2 Flexible one-square-foot Ovonic solar cell.

technology that will alter the way we generate electricity, store energy, and encode, transfer and distribute information. Since all of these depend upon the ability to synthesize new materials, I will also describe how synthetic coatings can advance industrial and commercial development.

In terms of energy, I will start with photovoltaics which has long been a great favorite of Kreidl in his activities to help solve the problems of providing electrical energy to meet the needs of developing countries. The ability to prepare amorphous solids inexpensively and in thin-film form over arbitrarily large areas is obviously of great value in the field of photovoltaics.(25,26)

At Energy Conversion Devices, Inc. (ECD), we have designed and built a mass production Ovonic roll-to-roll photovoltaic processor which has clearly demonstrated that the basic barrier to low-cost solar-cell production has already been broken and that one can now speak realistically of delivering power directly from the sun for under a dollar per peak watt merely by making larger versions of this continuous web, large-area thin-film machine (see Fig. 1).

We have made one-square-foot amorphous silicon alloy PIN devices (see Fig. 2) with conversion efficiencies in the 7% range and have reported smaller area laboratory PIN devices in the 10% conversion efficiency range. In addition, much higher efficiencies can be obtained with the same process by using multi-cell layered or tandem thin-film solar cell structures. These devices exhibit enhanced efficiency by utilizing a wider range of the solar spectrum.(25,26) Since the theoretical maximum efficiency for multi-cell structures is over 60%, more than twice that of a single-cell device, one can certainly realistically anticipate the production of thin-film amorphous photovoltaic devices with efficiencies as high as 30%. To accomplish this we will be utilizing narrow band gap materials which have the same low density of states as do our present silicon alloys. We already have firm experimental evidence based upon sound theoretical understanding of the material needs that make this achievable. Our production device is already a two-cell tandem, as we have solved not only the problems of interfacing the individual cell components but also the difficulties associated with a one-foot-square format deposited on a continuous web.

Figure 3 shows a continuous roll of Ovonic solar cells. Realistic calculations for a multi-layered tandem thin-film device using our amorphous semiconductor alloys with varying optical band gaps indicate that solar energy conversion efficiencies of 20-30% can be achieved.(27) Based upon production experience and laboratory results, I can predict with a great deal of certainty that we will be in the 30% range in the year 2004. Certainly, solar power will be cheaper at that time than electricity derived from coal, gas, oil or uranium.

Figure 4 shows an interior view of a portion of the Ovonic amorphous cell production plant in

Fig. 3 Continuous roll of Ovonic photovoltaic cells.

Fig. 4 Interior view of a portion of Sharp-ECD Solar, Inc.'s Ovonic Amorphous Solar Cell Production Plant in Japan.

Japan. Note the absence of operators. This plant is highly automated, another indication of how factories producing amorphous devices will look in the next century.

Another important source of alternative energy is the utilization of the age-old dream of turning waste heat directly into electricity without moving parts, that is, the heat analog of the photovoltaic device. Again I predict that by the end of this century, thermoelectric devices will be as ubiquitous as photovoltaic devices. The problems that held back this important form of energy conversion were entirely material-based and have been solved by ongoing work at ECD. Figure 5 shows an Ovonic Thermoelectric Generator which utilizes disordered materials as the active elements. In addition, materials of the same nature have been used for Peltier Effect devices for refrigeration.

Devices such as these are already in production in the United States as well as in Japan and represent the first realistic mass production approach to an area which combines the attributes of cogeneration, conservation and pollution control. Tapping waste heat, a necessary by-product of our industrial civilization, is in a sense tapping a nondepletable form of energy, for just as there will be photovoltaics as long as there is sunlight, there will be waste heat so long as we have civilization.

The need for batteries to store energy grows apace with every new development involving either portability or variable energy sources such as sunlight. Even without this dynamic, it has been a sad fact that batteries have not basically changed for over 100 years. Utilizing our synthetic materials approach, we have designed and built Ovonic batteries to have at least twice the energy density of conventional nickel-cadmium rechargeable batteries, as well as other improved features, while maintaining the same size and weight as conventional batteries. Initial production is being planned for 1984.

Figure 6 shows a typical battery. It is important to keep in mind that we can build these from the size of a pinhead to the size of a room, the latter for use in utility load levelling. My present prediction is that based upon laboratory studies, we will be seeing electric automobiles utilizing batteries based on these principles in widespread use by 2004. They also will fill the need for energy storage as the use of photovoltaic and thermoelectric devices grows.

Other interesting energy applications of amorphous materials include superconducting wires and magnets and the storage of hydrogen so as to make practical its use as a fuel. I predict that by 2004 we will see the widespread use of superconductivity for the storage and transmission of electricity and many vehicles will be on the road fueled by hydrogen stored in amorphous hydrides.

We have also developed a basic new approach to information technology which we believe can transform this field which until now has been dominated by the crystalline silicon industry. Our approach, utilizing proprietary amorphous materials and technology, enables us to manufacture large-area, single substrate, high density information systems which were not considered possible several years ago. For example, Ovonic technology lends itself to the development of large, flat-screen displays to replace bulky conventional cathode-ray-tube information displays. The display shown below can be made large enough to be a wall-size television set (see Fig. 7) This technology can also be applied to the fabrication of small, single substrate, completely integrated, thin-film main-frame computers which would replace room-sized ones or give small personal computers the processing capability of large mainframe units.

Fig. 5 Ovonic Thermoelectric Generator.

Fig. 6 Rechargeable Ovonic batteries.

Fig. 7 Large-area Ovonic flat screen display.

Crystalline silicon is limited to wafers no larger than six inches in diameter. By utilizing thin-film technology we can make devices many feet wide and of arbitrary length with characteristics analogous to crystalline materials at a fraction of the cost. This is the unexpected quantum leap that will transform the information and telecommunications industry by the year 2004. Instead of a circuit on a chip, we can have high density circuits vertically integrated in three-dimensional stacks of thin films on a single substrate. The various information functions we have already built are exemplified by the Ovonic write-once and reversible memory switches, PROMs, optical memories and thin-film transistors. In the 1960's I described optical memories that we were developing. In fact, we were the first to utilize lasers for what is now called "laser annealing."(28,29) The use of such optical memories is growing rapidly. Figure 8 shows an optical memory operating on the crystalline-amorphous phase-change principle. We have developed all of the other basic ingredients for an optical computer, and I believe that by the end of the century we will be seeing such new computers. I predict, based upon actual amorphous devices--both electrical and optical--already being developed and produced, a revolution sweeping the information fields as fundamental as the transition from the tube to the transistor.

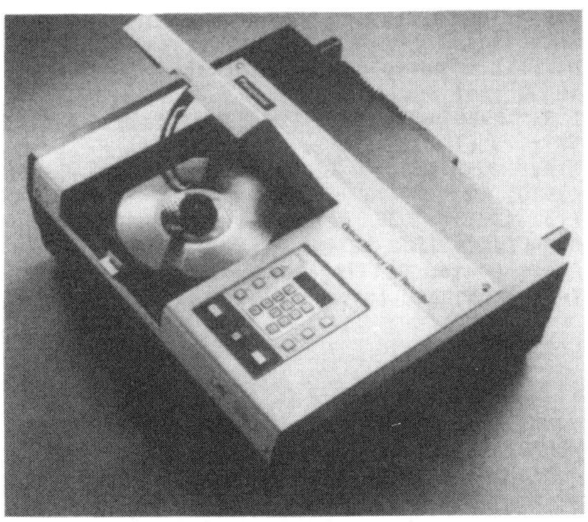

Fig. 8 Optical memory disc recorder.

Other information uses for amorphous materials include new types of photographic media that have amplification, continuous tone and very high resolution.(17,18)

Iris and I started in amorphous materials in 1955 through our interest in what is now called artificial intelligence.(30,31) Since the many products we are describing here have all developed from this approach, I would like to make another prediction. The adaptive materials (32,33) which I have used as micron-sized learning machines and the three-dimensional circuits which we are developing related to our thin-film work will, I believe, be the basis for much artificial intelligence work by the year 2004.

Another area of great interest to us is the design of passive synthetic materials. We specifically engineer coatings for a wide spectrum of other technologies, including the areas of turbines, data processing equipment, household fixtures, automotive and aviation equipment, etc.

One of the most unique coatings that we have developed is for X-ray optics for which exceedingly thin multiple layers of amorphous materials have been made into X-ray mirrors focussing elements and optical devices (15) (see Fig. 9).

There is increasing recognition that a new materials revolution is occurring. The need to overcome the limitations of existing materials technology can only be met by the ability to design, that is, synthesize, new materials with characteristics not found in either naturally occurring materials or plastics.(2)

Amorphous solids can be made by many different procedures--from powders to slabs. Of immediate technological importance is our unique ability to design and tailor-make films, coatings, wires and ribbons which can be extremely hard, resist corrosion, have increased lubricity, and/or have exceptional catalytic properties. The makers and users of industrial tools have been among the first to benefit from this new materials technology.(34)

I have permitted myself to predict what will be happening in our field in the year 2004 because the predictions I made in 1955 when there were virtually no amorphous products have now materialized into the

Fig. 9 Ovonyx X-ray mirrors.

many and varied applications I have described here. In 30 years we have transformed the amorphous field scientifically and technologically. I fully expect the fruition of this work to be seen in full in 20 years from now.

There are many other applications that cannot be covered in this paper. What I have sought to do is show that non-crystalline solids will be the vehicle to carry our industrial society into the new century. We all wish to see Professor Kreidl riding that vehicle urging us on.

References

1. S.R. Ovshinsky, in: Proc. 33rd Nat. Conf. on the Advancement of Research, Pennsylvania State University, State College, PA, Oct. 7-10, 1979 (Denver Res. Inst., Denver, USA, 1980) p. 15.
2. S.R. Ovshinsky, presented at 9th Int. Conf. on Amorphous and Liquid Semiconductors, Grenoble, France, July 2-8, 1981: also published in J. de Phys., Coll. C4, suppl. au no. 10, Tome 42 (October 1981) p. C4-1095.
3. S.R. Ovshinsky and H. Fritzsche, IEEE Trans. Electron Dev. ED-20 (1973) 91.
4. S.R. Ovshinsky and K. Sapru, in: Proc. 5th Int. Conf. on Amorphous and Liquid Semiconductors, Garmisch-Partenkirchen, Germany, September 1973, eds. J. Stuke and W. Brenig (Taylor and Francis, London 1974) p. 447.
5. S.R. Ovshinsky, Phys. Rev. Lett. 36 (1976) 1469.
6. S.R. Ovshinsky, in: Proc. 6th Int. Conf. on Amorphous and Liquid Semiconductors, Leningrad, USSR, Nov. 18-24, 1975 (Nauka, Leningrad,, 1979) p. 426.
7. S.R. Ovshinsky, in: AIP Conf. Proc. No. 31, Williamsburg, Virginia, March 25-27, 1976 (AIP, New York, 1976) p. 31.
8. S.R. Ovshinsky, in: Proc. 7th Int. conf. on Amorphous and Liquid Semiconductors, Edinburgh, Scotland, (June 27-July 1, 1977) p. 519.
9. R.A. Flasck, M. Izu, K. Sapru, T. Anderson, S.R. Ovshinsky and H. Fritzsche, in: Proc. 7th Int. Conf. on Amorphous and Liquid Semiconductors, Edinburgh, Scotland, (June 27-July 1, 1977).
10. S.R. Ovshinsky and D. Adler, Contemp. Phys. 19 (1978) 109; also presented at the APS March Meeting, March 27-30, 1978, Washington, D.C.
11. S.R. Ovshinsky, in: Proc. Int. Conf. on Frontiers of Glass Science, Los Angeles, California, 16-18 July 1980, J. Non-Crystalline solid 42 (1980) 3335.
12. S.R. Ovshinsky, Rev. Roum. Phys. 26 (1981) 893 (Grigorovici Festschrift).
13. D. Adler, ed., Disordered Materials: Science and Technology, Selected Papers by S.R. Ovshinsky (Amorphous Institute Press, Bloomfield Hills, Michigan, 1982).
14. S.R. Ovshinsky, to be published.
15. J. Kakalios, H. Fritzsche, N. Ibaraki and S.R. Ovshinsky, in: Proc. Int. Topical Conf. on Transport and Defects in Amorphous Semiconductors, Bloomfield Hills, Michigan, March 22-24, 1984; J. Non-Crystalline Solids 66 (1984) 339.
16. S.R. Ovshinsky, J. Non-Crystalline Solids 32 (1979) 17 (Mott Festschrift).
17. S.R. Ovshinsky and P.H. Klose, in: Proc. Soc. Inf. Display Int. Symp., May 4-6, 1971, Philadelphia, PA (Winner, New York, 1971) p. 58.
18. S.R. Ovshinsky, in: SPIE/SPSE Technical Symposium East, March 22-25, 1976, Reston, Virginia; J. Appl. Photo. Eng. 3 (1977) 35.
19. S.R. Ovshinsky, in: Proc. 4th Int. Congress for Reprography and Information, April 13-17, 1975, Hannover, Germany (Aussch. Wirtschaft, Frankfurt, 1975) p. 109.
20. S.R. Ovshinsky and A. Madan, Nature 276 (1978) 482.
21. R. Tsu, M. Izu, V. Cannella and S.R. Ovshinsky, Solid State Comm. 36 (1980) 817.
22. R. Tsu, S.S. Chao, M. Izu, S.R. Ovshinsky, G.J. Jan and F.H. Pollak, in: Proc. 9th Int. Conf. on Amorphous and Liquid Semiconductors, Grenoble, France (July 2-8, 1981); also published in J. de Phys. Coll. C4, suppl. au no. 10, Tome 42 (October 1981) C4-269.
23. S.R. Ovshinsky and H. Fritzsche, Metal. Trans. 2 (1971) 641.
24. S.R. Ovshinsky, Phys. Rev. Lett. 21 (1968) 1450.
25. S.R. Ovshinsky, presented at SPIE Symp. on Photovoltaics for Solar Energy Applications II, April 1983, Arlington, Virginia; published in Proc. SPIE 407 (1983) 5.
26. M. Izu and S.R. Ovshinsky, presented at SPIE symp. on Photovoltaics for Solar Energy Applications II, April 1983, Arlington, Virginia; published in Proc. SPIE 407 (1984) 42.
27. S.R. Ovshinsky, Invited presentation at the Int. Ion Engineering Congress ISIAT '83 & IPAT '83, September 12-16, 1983, Kyoto, Japan (Int. Ion Eng. Congr., Kyoto, 1983) p. 817.
28. S.R. Ovshinsky, in: Proc. 5th Annual National Conf. on Industrial Research, Chicago, September 19, 1969 (Ind. Res. Inc., Beverley Shores, Indiana, 1970) p. 86.
29. S.R. Ovshinsky, U.S. Patent No. 3,530,441. Applied August 22, 1968. Issued September 22, 1970.
30. S.R. Ovshinsky, The Physical Base of Intelligence--Model Studies, presented at Detroit Physiological Society (December 17, 1959) (not published).
31. S.R. Ovshinsky and I.M. Ovshinsky, Mat. Res. Bull. 5 (1970) 681.
32. S.R. Ovshinsky, J. Non-Crystalline Solids 2 (1970) 99.
33. E.J. Evans, J.H. Helbers and S.R. Ovshinsky, J. Non-Crystalline Solids 2 (1970) 334.
34. Glass Coated Drills, Ford Worldwide Productivity Idea Exchange, Issue 4 (1st quarter, 1983) p. 3.

AMORPHOUS AND DISORDERED MATERIALS – THE BASIS OF NEW INDUSTRIES

S.R. OVSHINSKY
Energy Conversion Devices, Inc., Troy, Michigan 48084

ABSTRACT

As in the past, materials will shape the new century. Dramatic changes are taking place in the fields of energy and information based on new synthetic materials. In energy, the generation of electricity by amorphous silicon alloy thin film photovoltaics; the storage of electricity in nickel metal hydride batteries which are the batteries of choice for electric and hybrid vehicles. In the information field, phase change memories based on a reversible amorphous to crystalline transformation are widely used as optical memories and are the choice for the new rewritable CDs and DVDs. The scientific and technological bases for these three fields that have become the enabling technologies are amorphous and disordered materials. We will discuss how disordered, multielemental, multiphase materials can throw new light upon metallic conductivity in both bulk and thin film materials. We will demonstrate new types of amorphous devices that have the ability to learn and adapt, making possible new concepts for computers.

INTRODUCTION

The great advances in civilization have been based on materials – the Stone Age, the Bronze Age, the Iron Age. The interaction between materials and the industries that have transformed society was the driving force for the Industrial Revolution. The twin pillars of our society today are energy and information. In a deep sense they are opposite sides of the same coin. They both must be generated, stored and transmitted. Information is structured energy that contains intelligence.

I will show how a new scientific approach to materials based upon disorder and local order can enable the development of new pollution free technology which will answer society's urgent needs to reduce its dependence upon uranium and fossil fuels, particularly oil; the latter is a causative factor not only in climate change but a root cause of war. Science and technology which can change the world's dependence on it can create new huge industries so necessary for economic growth. $30 trillion in 30 years for new electricity alone [1]. Furthermore, over 2 billion people in developing countries are without electricity. Electricity is the fundamental requirement of modern life and the common link between energy as an undifferentiated source of power and energy which can be encoded, switched and stored as information.

Devices made of amorphous and disordered materials have become the enabling technology for generating electricity through thin film photovoltaics which can be cost competitive to fossil fuels, for storage of electricity in batteries for electric and hybrid vehicles, ushering in a new, much needed transportation revolution, and high density switching and storage media based on phase change optical and electrical memories, so needed for our information society. Computers which have adaptability, can learn from experience and provide neuronal and synaptic type intelligence, are being made possible by devices described here.

How is it possible that multi-elemental disorder can be the basis for such "revolutionary" possibilities [2] when it is well-known that the great success of the 20th century, the transistor is based upon the periodicity of materials with particular emphasis on one element, silicon? Indeed, with the great success of the transistor based upon the crystal structure of germanium and silicon, we entered the historical era where achieving crystalline perfection over a very large distance became the sine qua non of materials science.

From a materials point of view, the physics that made the transistor possible was based upon the ability to utilize periodicity mathematically which permitted parts per million

Mat. Res. Soc. Symp. Proc. Vol. 554

perturbations of the crystalline lattice by substitutional doping. But right from the beginning, the plague of their disordered surfaces prevented for a decade the fulfillment of the field effect transistor. The disorder of the surface states swamped out the transistor action. Emphasizing again Pauli's statement "God created the solids, the devil their surfaces" [3]. The irony was that the solution that made not only the field effect transistor possible but also the integrated circuit which became the basis for the information age was the utilization of amorphous silicon oxide for photolithography and for the gate oxide.

When I introduced the idea that there was a new world of interesting physics and chemistry in minimizing and removing the constraints of periodicity [4,5], one can understand the resulting consternation of the solid state physicists who had received their Ph.Ds by accepting the dogma of periodicity as being the basis of condensed matter and of the theoretical physicists to whom the control of many elements was as incomprehensible as the conundrum of many-body theory. The change from periodicity to local order permitted atomic engineering of materials in a synthetic manner by opening up new degrees of design freedom. Literally many scores of new materials could be developed and new physical phenomena could be displayed, new products made and new process dependent production technology invented.

Disorder is the common theme in the minimization and lifting of lattice constraints, (what I call the tyranny of the crystalline lattice) which permits the placing of elements in three-dimensional space where they interact in ways that were not previously available. This allowed the use of multi-elements and complex materials including metals where positional, translational and compositional disorder removed the restrictions so that new local order environments [6,7] could be generated which controlled the physical, electronic and chemical properties of the material. Just as the control of conductivity through doping was the Rosetta Stone of understanding and utilizing the transistor, the unusual bonding, orbital configurations and interactions affecting carriers, including ions, are the controlling factors in disordered materials [8-11].

The tools that I utilize for generating these configurations are hydrogen, fluorine, f- and particularly d-orbitals, and nonbonding lone-pair orbitals of the chalcogens. The latter, like the d-orbitals, can be distinguished from their cohesive bonding electrons, freeing them for varied interactions. Even in a sea there are channels and currents affected by topology and climate; in a metallic sea of electrons, we can design paths, control flow and make hospitable environments for incoming ions/protons. Since we are designing new local environments, we also utilize rapid quench technology to make for non-equilibrium configurations and offer a new degree of freedom for the production of unusual local order. Of course, rapid quench and non-equilibrium are associated with vacuum deposition, sputtering and plasma generated materials, in brief, such materials are process dependent.

The understanding of these basic premises became a design tool which we applied universally across the periodic table to build new types of semiconductors, dielectrics and metals. We showed that we can control the density of states in a band/mobility gap affecting conductivity in several ways including chemical modification and the generation of chemically reactive sites [12,13] so as to design, for example, complex, disordered, metallic electrodes of our nickel metal hydride batteries. What we mean by complex is not just that there are many elements, but that we build into a material a chemical, electronic and topological system which performs in the same material various functions such as catalysis, hydrogen diffusion paths, varying density of electrons, acceptor sites for hydrogen, etc. Such a material system in a battery must provide high energy density, power, long life and robustness [14-17].

To put into perspective the principles of disorder, what is required is a metaphor. It is helpful to continue the analogy of the sea. In ancient times, the earliest explorers stayed as close to the shore as possible; there were many things to discover that way – new people, animals, physical environment, things could be strange but understandable. However, to explore the great unknown ocean, they were filled with anxiety for out there was the end of the world and where the dragons lay. Navigational skills were needed to avoid dangerous shoals, utilize favorable currents, etc.

As I pointed out in the 1970s [18-19], amorphous tetrahedral materials to be useful would have to be as close to the four-fold coordination of their crystalline cousins as possible, otherwise, the huge density of states of dangling bonds would prohibit their use. Hydrogen and fluorine were able to act as organizers to assure sufficient four-fold coordination resulting in a low density of states so that the materials could be electronically useful, a necessity for successful photovoltaic products [20,21]. Hydrogen not only capped dangling bonds, but was also valuable as a bridging element providing the connectivity between the silicon atoms and under the proper circumstances assuring enough four-fold coordination. I chose fluorine since it is the superhalogen and provides a much stronger bond and, most importantly, expands the undercoordinated bonds of silicon and germanium so that they can have their full tetrahedral structure. Fluorine also provides useful functions in the plasma and on the surface in the growth of the film. It does its job so well not only in intermediate order but as a preferred element to make microcrystalline tetrahedral materials [22] as well as thin-film diamond-like carbon [23]. It also played a role in making superior superconducting films [24].

The most exciting physics lay in the unexplored ocean that I have been working in since 1955 with our only nautical chart the periodic table and physical intuition our compass. Even though disordered materials could not be easily categorized mathematically, one can constructively design nonequilibrium, nonstoichiometric graded and mixed phase materials to discover new phenomena [8].

In order to follow the exploration process, we will intermingle the relevant scientific and technological approaches with the materials, products, and technologies made possible by utilizing the freedom permitted by disorder to design and atomically engineer local environments.

ENERGY GENERATION –PHOTOVOLTAICS

In photovoltaics, we started our exploration relatively close to the silicon shoreline but had to push further out because contrary to conventional thinking, elemental amorphous silicon had no possible electronic use and therefore required the alloying described above so as to eliminate dangling bonds and yet to retain its four-fold coordination. I felt that it also required new technology befitting its thin film form. Rather than choosing heavy glass as a substrate, we used flexible, thin stainless steel as well as other flexible materials such as kapton.

A historical perspective is needed to show the interaction of science, technology and product. Starting in the late 1970s, we invented the materials, the products and designed and built six generations of production machines utilizing our multi-junction, continuous web technology, the most recent, our 5 MW machine shown in Fig 1. These machines, designed and built at Energy Conversion Devices (ECD), manufacture much needed, simple products using our advanced science and technology and, most importantly, this continuous web approach has shown that it is possible in larger machines to make solar energy cost competitive to conventional fuels. This, of course, would set off an enormous positive change in the use and economics of energy. The plasma physics involved in such a machine makes it possible to discard the power consuming crystal growing methods of crystalline semiconductors and the billion dollar costs now involved in building crystal wafer plants. In crystalline materials, the investment and throughput are linearly coupled, in our amorphous thin film technology, a 4 times increase of capital investment (in the millions of dollars) would yield a 20 times increase in throughput.

Fig. 1

Fig. 2

Fig. 3

Fig. 2 shows a 3 inch crystalline wafer (at bottom, center), the state of the art of semiconductors at the time, and what Gordon Moore and Bob Noyes, founders of Intel, utilized as opening humor for their talks, a huge cardboard "wafer", representing what they felt would be the electronic requirements of the 90s. On the right, made in our second generation continuous web machine in 1982 for our joint venture with Sharp Corporation, is the first half mile long, over a foot wide roll of a sophisticated multi-junction thin film semiconductor solar cell continuously deposited on thin, flexible stainless steel. This revolutionary new process was the answer to the fantasy of the cardboard wafer. Obviously the roll could be made much wider and longer. I have called our photovoltaic roll an infinite "crystal".

In order to show the extreme light weight, high energy density potential of our approach, we deposited our films on kapton and demonstrated in 1984 the highest energy density per weight of any kind of solar cell. This is shown in Fig.3 in a water pumping application at that time. These days the same approach is beginning to be utilized for telecommunication, space and satellite applications by Guha, Yang and colleagues[25]. The power density in thin film stainless steel and especially kapton which can store almost 3000 watts per kilogram is so exceptional as to become the solar cell of choice for telecommunication. Just last month, in November 1998, an amorphous silicon solar array was installed on the MIR space station. It was fabricated in Troy, Michigan by United Solar Systems Corp (United Solar), (ECD's joint venture with Canon), and assembled in Russia by Sovlux, ECD's Russian joint venture with Kvant, the developer of the original photovoltaics on MIR, and the Russian Ministry of Atomic Energy.

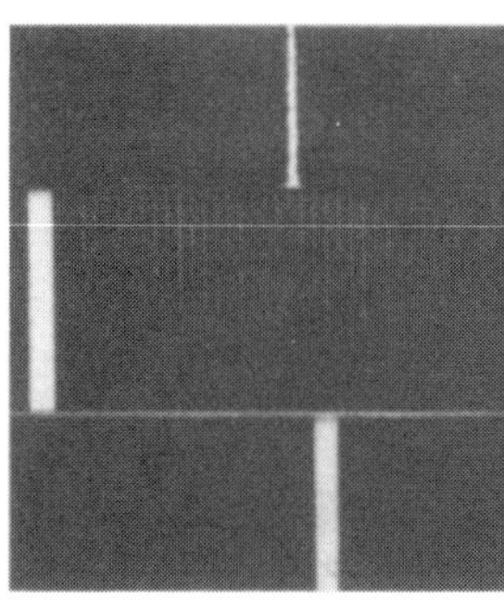

Fig. 4

Fig. 5

Fig. 6

Fig. 4 shows shingles made in 1980 [26] providing a new paradigm for energy generation. Paradigm shifting takes time. Figs 5 and 6 are current installations of the shingles and a standing seam roof. These products are gaining widespread approval. Fig 7 shows former Secretary of Energy Pena displaying our solar shingles in 1998.

Fig. 7

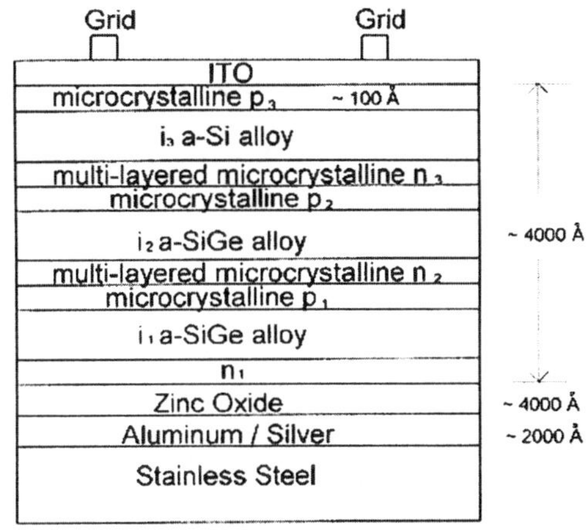

Fig. 8

Fig 8 is a schematic of our triple-junction solar cell showing its multi-layered configuration wherein light can be absorbed in sub-cells with different bandgaps so as to utilize as much of the sun's spectrum as possible. The blue, green and red light is absorbed in layered thin films of amorphous silicon and germanium alloys containing hydrogen/fluorine. ECD and United Solar have all the world's records for efficiency, culminating in the latest world record of 10.5% stable efficiency on a one square foot module and 13% stable efficiency on a .25 square cm cell [27].

Amorphicity is crucial for several reasons. Unlike crystalline tetrahedral materials in which quantum mechanical selection rules make for indirect bandgaps and require layers of 50 to 100 microns in order to absorb the light energy, amorphous materials act as direct band gap materials and therefore the entire triple layer system is less than 1 micron in total thickness. It is important to note that in crystalline materials of different bandgaps, lattice mismatch is a serious problem and therefore such multi-layered structures could not be made in a production manner. In this case, we can see that amorphicity and the physics make possible continuous web production. Amorphous photovoltaics illustrate that when one removes the lattice constraints, atomic engineering can be merged with machine engineering to provide a new, much needed approach to energy generation.

There has been much ongoing work in photovoltaics at ECD since 1977 [28], advancing the science and technology of materials, production processes and new products. At the MRS 1998 Spring Meeting, my collaborators, Guha, Yang and coworkers at United Solar, who have made very significant contributions to our work, gave an excellent review of our recent commercialization progress [29]. The ECD-United Solar team, which also includes Masat Izu, Prem Nath, Steve Hudgens, Joe Doehler, Scott

Jones and Herb Ovshinsky among others (the latter the head of ECD's Machine Division which has designed and built our continuous web processors) has through the years made important contributions to our field and has been working on products, production and plasma technology that increase throughput.

From a materials science point of view, we note several points of importance. While amorphous tetrahedral materials are close to the crystalline shoreline in their need for low density of states and substitutional doping, being direct bandgap materials, one can make large area, thin film multi-junction devices by the decomposition of plasmas in a continuous manner. Accomplishing this, we were able to basically alter the way materials could be laid down in a continuous manner, showing the tight coupling in amorphous materials between basic science and advanced technology.

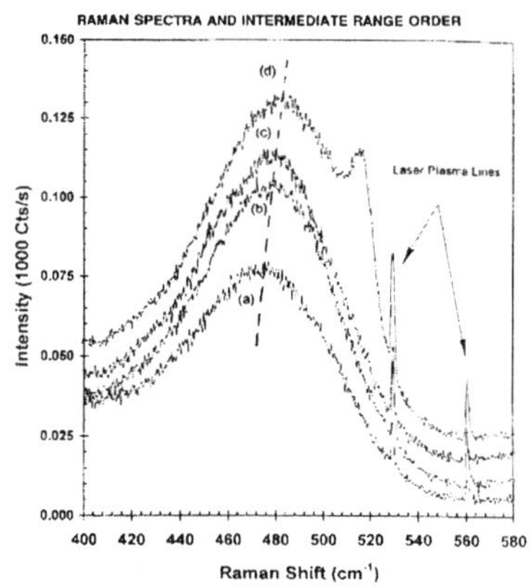

Fig. 9

Illustrating the scientific and technological richness of amorphous materials is the ability to develop intermediate range order in an amorphous matrix as we do in the intrinsic silicon alloy layer of our photovoltaic product [11,30]. (Fig. 9) It is of great interest that quasi one-dimensional ordering is accomplished without introducing grain boundaries that would interfere with electron and hole conductivity. This intermediate state is the signature of the best material. We have shown that fluorine is a great facilitator of intermediate order [31-34] and leads to crystallization. The intermediate order has important implications for the future. It is possible that the carriers have increased mobility due to the intermediate structures and could affect the important parameter of hole mobility in these materials. Microcrystalline materials can also have unique properties that bridge the gap between crystalline and disordered solids. We have incorporated them in our continuous web process by utilizing fluorine [11,21] to make under 120 angstrom micro-crystalline silicon p-layers. When one considers that this is accomplished in a continuous web, very high yield production process, it can be appreciated that atomic engineering and manufacturing are a reality.

ENERGY STORAGE – NICKEL METAL HYDRIDE BATTERIES

Nickel metal hydride batteries are in a real sense misnamed, for while nickel, by virtue of its filled d-orbital, plays an important catalytic role, there are usually seven or eight other elements that make up the alloy used in the negative electrode; certainly hydrogen is the key component. It is the smallest and simplest atom in the universe (which, by the way, in terms of actual matter is composed of over 90% hydrogen). Therefore, to attain the highest energy storage density not only for batteries but for hydrogen as a fuel is a matter of designing the highest density of reversible hydrogen sites. This can be accomplished by our principles of disorder and local order [11,14-17,35]. From 1960 on, we have demonstrated that hydrogen is not just a future source of energy but a here and now solution that, together with solar energy, offers the ultimate answer to society's need for clean, virtually inexhaustible energy.

Energy technology should be regarded as a system. We have described energy generation. Equally as important is energy storage. Now we must go much further from shore in order to discuss the basis for the materials used for energy storage and later for information which is structured energy and must also be stored.

In the energy storage area, we cannot discern the shoreline, we are in the ocean of many-body theory which has never been adequately understood. I have had a long interest and worked for many years with d-orbital materials [36]. A recent publication sums up present day advanced thinking regarding d-electrons: "d-electron-based systems, in particular, present the combined intrigue of a range of dramatic phenomena, including superconductivity and itinerant ferromagnetism, and the tendency toward inscrutability associated with the fact that key electronic states are often intermediate between the ideals of localization and itineracy which provide the starting points for most theory " [37]. By making multi-elemental, multi-phase d-orbital materials for hydrogen storage in a completely reproducible manner, it is clear that we have been able to understand how to take the mystery out of them and utilize these orbitals in new and unique ways.

We are literally at sea when we discuss metals for they are always described as being composed of a sea of electrons. Why does the sea not swamp out the background provided by local atomic environments? Paradoxically, this is because when we use many, for example, ten, different atoms, particularly those with directional d-orbitals, to make up a material such as the negative electrode in our nickel metal hydride battery, we provide through the disordered state regions of lower electronic density which have a larger probability to overlap with negative hydrogen ions in interstitial sites. To simplify, one can say that it is the s-electrons that provide the sea, the d-electrons sculpt out the channels and receptors. Obviously there is some hybridization possible in many of these materials.

It is in the multielemental f- and particularly the d-orbital material that we introduce a means of delocalizing electrons and still have them represent their parentage. This is where internal topology begins to play an important role in metallic conduction for, as noted, we need channels in the sea of electrons just as we take into account sea level/Fermi level. The different types of atoms provide the interatomic spacings for the hydrogen ions to operate in and the varying electronic density is the steering means for the ions to reach the preferred sites of low electron density. Disordered materials are therefore necessary to provide the spectrum of binding sites. In summary, large interatomic spacing and low energy density make for the optimal binding/storage of negative hydrogen ions. The electron environment surrounding the hydrogen provides the degree of negativity and coulombic repulsion is the steering means. The binding energy provided by the local environment is of such a nature as to assure reversibility so important for the rechargeable battery. While we list characteristics of individual atoms in Fig.10, it is how they act and interact in the alloy that makes for the mechanism that we have described above.

The acceptance of nickel metal hydride batteries has been very rapid. All significant manufacturers of nickel metal hydride batteries are under agreement with ECD and our Ovonic Battery Company (OBC) and over 600 million consumer batteries were sold last year with a 30% per year predicted growth rate. These high production volumes have made for a very low cost battery.

The problems of pollution, climate change and our strategic dependence on oil have provided a global impetus for the use of electric and hybrid electric vehicles [38,39]. The automotive industry has made our nickel metal hydride battery the battery of choice for these vehicles. GM, Toyota, Honda, Hyundai, Ford and Chrysler all have chosen nickel metal hydride batteries. We have a joint venture with GM, GM Ovonic, which manufactures and sells electric vehicle and hybrid electric vehicle batteries to all companies.

Battery Properties Are Changed Because of Compositional Disorder Introduced by <u>Specific Elemental Modifiers:</u>

- Hydrogen Binding Energy
- Density of Storage Sites
- Fast Electrode Kinetics
- Low Temperature Operation
- Charge Retention
- Chemical and Physical Stability (Cycle Life)

Element	Alloy Property Principally Affected
Ti	Hydrogen Absorption
Zr	Hydrogen Absorption
V	Hydrogen Absorption
Ni	Hydrogen Catalysis
Cr	Corrosion Resistant
Co	Surface Activity
Fe	Corrosion Resistant
Mn	Surface Activity

- Structural disorder on several scales of length in electrode alloys also affects battery properties, and these phenomena are not considered in this chart.

Fig.10

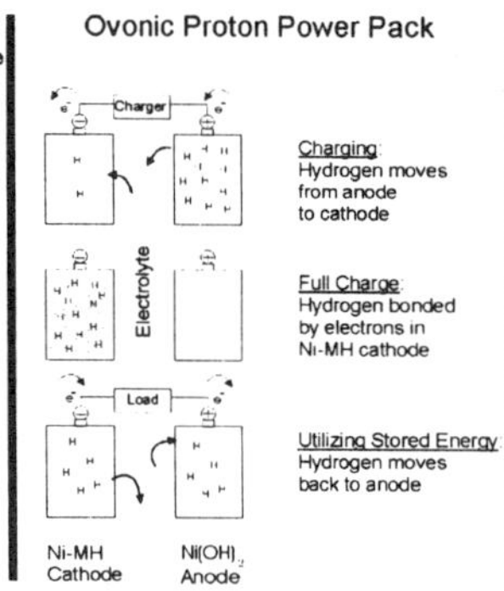

Fig. 11

Fig. 12

Fig. 11 shows a very simplified schematic of the nickel metal hydride battery. The battery is based on hydrogen transfer in which hydrogen is shuttled back and forth between the nickel hydroxide and metal hydride without soluble intermediates or complex phase changes. Fig. 12 shows an ingot of our materials as well as our green battery.

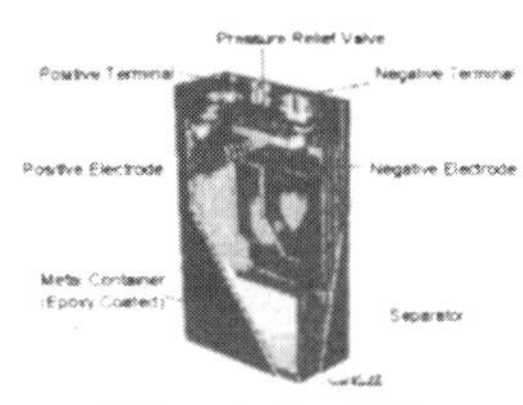

Fig. 13

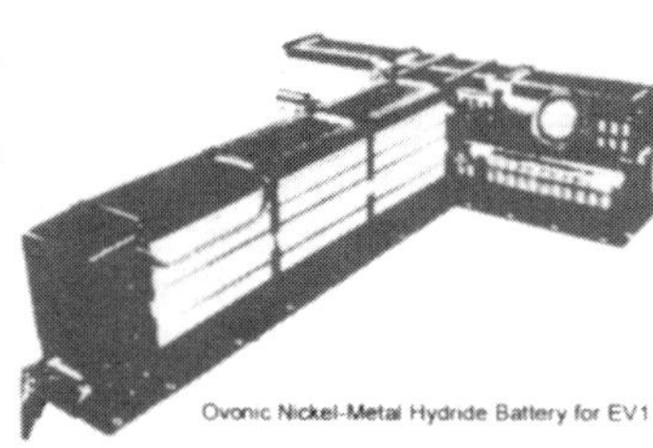

Fig. 14

Fig. 15

Fig.13 is a cutaway drawing of our NiMH automotive battery illustrating its simplicity. Fig. 14 is a GM EV1 battery pack; Fig. 15 is a photo of the EV1 car whose range is between 160 and over 200 miles, its acceleration is 0 to 60 mph in less than 8 seconds, less than 15 minutes for an over 60% recharge, very robust, environmentally benign, lifetime of the car battery.

Fig. 16 shows James Worden, cofounder of Solectria, having driven his 4 passenger Solectria Sunrise from Boston to New York on the equivalent BTU energy of less than 1 gallon of gas and he had 15% energy remaining [40]. Fig. 17 gives world record ranges of EVs using

Ovonic NiMH batteries,and Fig. 18 shows the capability of nickel metal hydride to be continuously improved by atomic engineering of the materials.

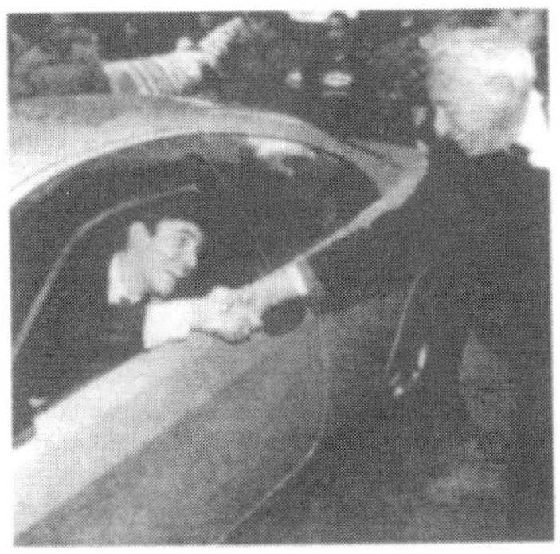

Fig. 16

World Record Ranges of Electric Vehicles Using Ovonic NiMH Batteries

Boston to New York One charge (15% unused)	217 miles
Ground-Up Tour de Sol 1996 (Solectria Sunrise)	375 miles
GM Impact (EV1)	245 miles
Conversion EV T. de Sol 1996 (Solectria Force) 1997 1998	214 miles 249 miles 241 miles
Pikes Peak Hill Climb (30% unused) GM S10	14,100 ft
Ovonic Prototype Scooter near hurricane conditions: normal weather conditions:	 69 miles 93 miles

Fig. 17

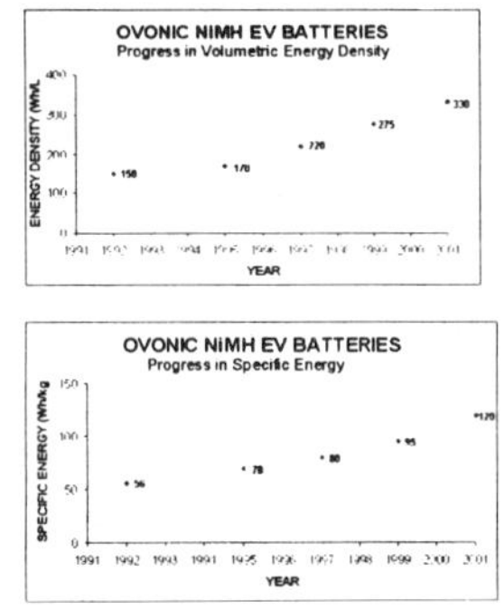

Progress in Gravimetric and Volumetric Energy Density

Fig. 18

A gasoline powered GM Geo Metro was tested against the same make and model car rebuilt by Solectria to be an electric vehicle with our Ovonic batteries. In New York City driving, the gasoline car provided an equivalent range of 120 miles compared to the 220 mile equivalent range of the electric vehicle with our batteries [41]. It is quite clear that scientific and technological issues are not what is holding back electric vehicles. Hybrid electric vehicles with our batteries will offer at least 80 miles per gallon in the charge sustaining mode and over 100 miles per gallon in the charge depletion mode.

INFORMATION

Disordered materials depend upon optional bonding configurations of atoms generating various kinds of new orbital relations. While boron and carbon are helpful in this regard, the elements of choice as I have shown for NiMH batteries are f- and d-orbitals. For the information side of our work, we prefer the chalcogenides characterized by Kastner as lone pair materials [42]. We utilize lone pair p-orbitals since they are not only nonbonding but have a spectrum of lone pair interactions that include various new bonding configurations [43-46]. In some respects, they have similarity to the directional d-orbitals, in other respects, they are different, for example, the empty or filled d-orbitals are very localized and do not reach out as far as the lone pair p-orbitals of the chalcogens, where two of the p-orbitals are deep in energy and serve as strong structural bonds responsible for the cohesiveness of the material. The many lone pair interactions are spread in energy throughout the mobility gap. Their similarity to d-orbitals is that neither the lone pairs nor the d-orbitals play a strong role in cohesive bonding but both are available for interesting electronic, optical and chemical interactions. However, their dissimilarity is that the lone pairs being the outer electrons (they are as far out as any valence electrons), can remain free or form weak or strong bonds, covalent or coordinate, depending on the environment. They are as far out as any valence electrons. The d-orbitals, on the other hand, form a narrow but designable band of high density states at the Fermi level and they too can act as receptors in the coordinate bond configuration. A more profound difference is that the divalency of the lone pair chalcogens allows a flexibility of structure which can be controlled by crosslinking so that we can make either an electronic Ovonic threshold switch (OTS) in which the excitation process does not affect bonding (see Fig. 19) or an Ovonic memory switch (OMS)

[4] in which the electronic processes initiate structural, that is, reversible phase change. We will concentrate here on the OMS which is now universally utilized in its optical phase change form (see Figs. 20 and 21) [47].

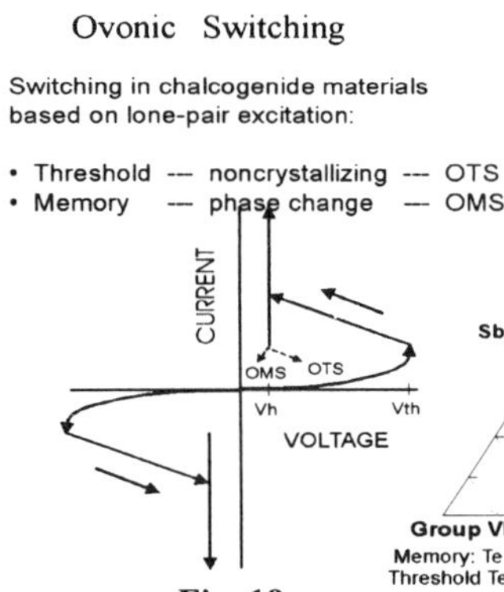

Fig. 19

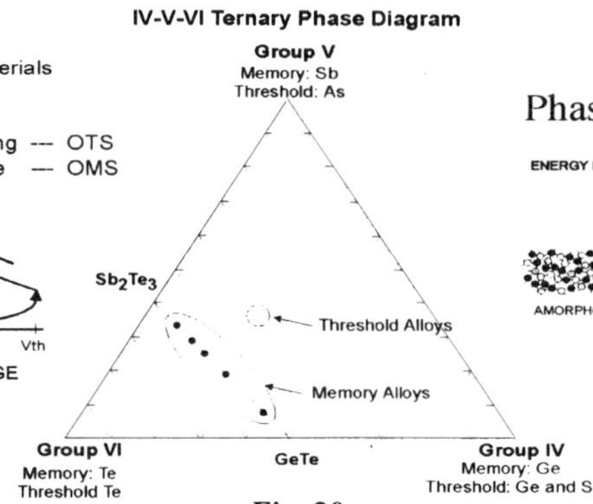

Fig. 20

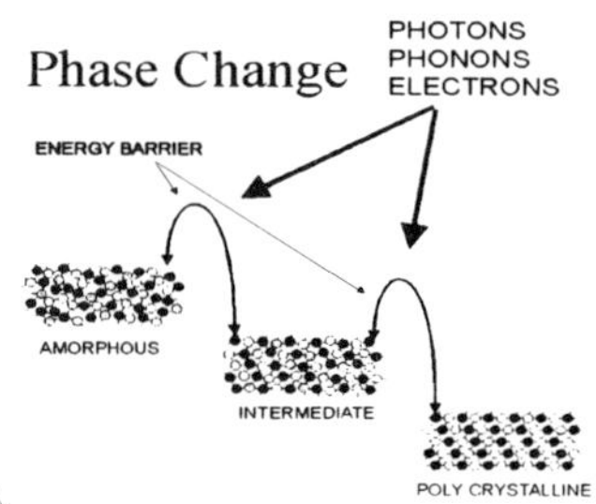

Fig. 21

Phase change rewritable memories have become the basis for the rapidly growing DVD rewritable market which holds so much promise for the future since it is replacing VCRs and CD-ROMs.

The energy necessary for an Ovonic optical memory material to change its atomic configuration is provided by a laser beam which couples to the non-bonding lone pairs so that the electronic energy exceeds a threshold value, causing a high atomic mobility state to occur and a change from the amorphous to the crystalline phase to take place. The same laser, but at different power, is used for recording, erasing and rewriting since the amorphous material can become crystalline again by rapid rearrangement through slight movements of atoms. The different structural phases of the material have different optical constants, so information is stored in the form of regions with different reflectivity. It is particularly favorable to use a phase congruent material for these applications and as we will show, the cycle life for phase change memories is exceptionally long. Electrical phase change memories have gone over 10^{13} cycles when testing was stopped [48,49].

Since Ovonic phase change optical memories are having great commercial success and the markets are growing rapidly, we are now entering the semiconductor memory market with devices using these materials. Conventional semiconductor memories are the basic building blocks of the information age. The data shown in Fig. 22 clearly show that conventional memories can be replaced by Ovonic semiconductor memories since a single plane of our memory can replace DRAM, SRAM and Flash. To indicate the great advantages that our multi-state memory offers in this highly competitive industry, we show a comparison with Intel's multi-level flash memory which was announced as "the Holy Grail" and which they said would have a "revolutionary" impact on the Flash market. (Figs. 23 and 24).

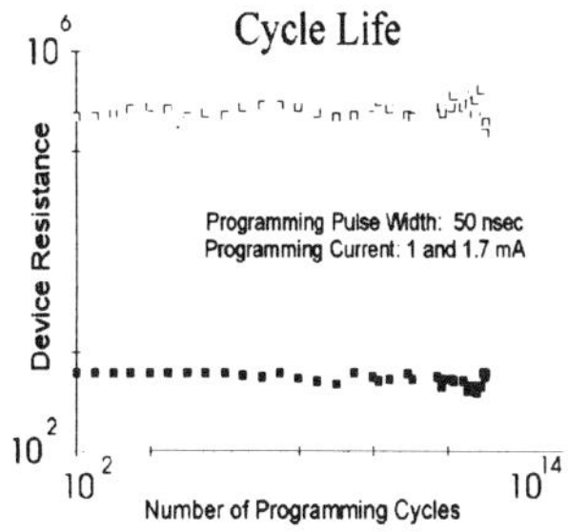

Multilevel Memory Comparison

Property	Intel Strataflash	Ovonic Multistate Memory
Density (bits)	2 bit (4 levels)	>4 bit (16 levels)
Write Cycles	10.000	$>10^{13}$ (no limitation)
Write Voltage	5V (with large 12V Charge pump)	< 4V
Write Time	10 microseconds (times n iterations)	0.6 microseconds (times n iterations)
Erase Time	1 second per block	No erase required
Minimum Erasure Block	must erase 128 Kbytes (before writing)	direct overwrite (byte-by-byte)
Read Access Time	~20 microseconds	1.0 microsecond (estimated)

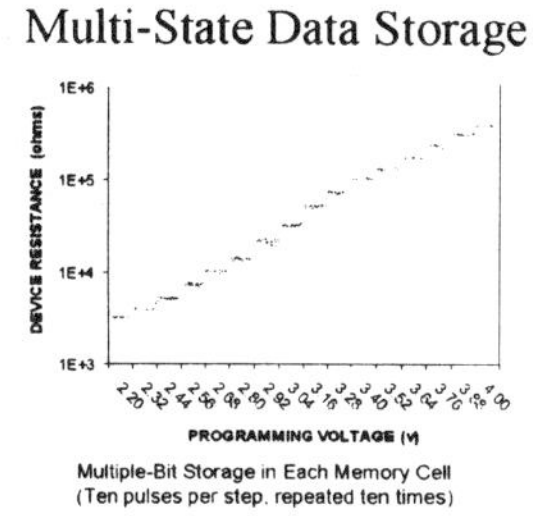

Fig. 22 **Fig. 23** **Fig. 24**

It can be appreciated that the multi-state memory is also a learning device since it adapts its electrical conductivity to the amount of information it receives. In other words, it displays what neurophysiologists refer to as plasticity as the basis for intelligence. We have also developed another version of this thin-film memory which is truly neuronal and synaptic in that it accumulates a number of sub-threshold pulses before it changes state. This device takes advantage of the fact that a small portion of the active volume of the memory will change phase upon application of every sub-threshold energy pulse. After a specified amount of energy has been deposited, a percolation path is established among the crystallized regions, and a large change in electrical resistance results. The device is reset into its virgin state in the same manner as our other semiconductor memory. This accumulation mode memory device will have important near-term applications in secure, tamper-proof information storage in smart cards and other devices.

What I have in mind for the future is an all thin-film intelligent computer. We have designed all thin-film circuits which can have logic, memory, and adaptive or intelligent memory integrated into a circuit. We have already proven that these devices can be made in three-dimensional and multi-layered circuits, can receive information from various sources, integrate it, remember it and learn from it. This is the basis of a truly cognitive machine, not artificial intelligence as we now know it, nor just a large number of parallel circuits, but a huge density of switching points, receiving and integrating information, in other words, many neurons of different thresholds and frequencies, receiving huge amounts of synaptic information, responding to it and utilizing it [50]. This is what I have wanted to build since 1955; this is what we can build now. I feel that this kind of computer is the computer of the next millenium. Combined with amorphous sensors and displays which we introduced many years ago and which have found widespread use, such a computer could perform tasks which are now beyond the ken of present "dumb" computers. All the various parts have been shown to work and they are all based on amorphous and disordered materials.

CONCLUSION

I believe that I have shown that science and technology can be utilized to build new industries that are responsive to societal problems and needs, providing jobs, educational opportunities and the chance to express the creative urge that has driven humankind since time immemorial. Fig. 25 shows a young woman climbing a mountain barefooted with her future on her back, our photovoltaics, and her future in front of her, her child, bringing our photovoltaics to a village that does not have electricity. There can be no civilization without energy and without knowledge (information). We take this picture as inspiration to continue our work.

Fig. 25

ACKNOWLEDGEMENTS

This work could not have been accomplished without the immense help and loving and fruitful collaboration of Iris Ovshinsky. Amorphous and disordered materials could not have reached its state in science without Hellmut Fritzsche. I am honored to have him as my longest scientific collaborator about whom it can be truly said that he is a giant in our field. Dave Adler's untimely death cut short our warm, close and productive friendship and collaboration. We and the amorphous field owe so much to him. I wish to thank the talented and creative teams of ECD, United Solar, and the Ovonic Battery Company. In the battery group, Subhash Dhar is an inspired multi-talented leader; Mike Fetcenko has made and is making many invaluable, critical contributions, his collaboration with me has been essential to the success of the battery company. Dennis Corrigan is owed thanks by me and the entire EV industry for his essential contributions to EV batteries; Srini Venkatesan has been a mainstay, his finger- and brain- prints are on everything that happens in batteries; Benny Reichman's great creativity has always been deeply appreciated; Art Holland's mechanical contributions have been of great importance in our battery work; John deNeufville's contributions through the years have not only been of great value to us but his latest activities contribute to the development of the amorphous silicon and germanium alloy field; we thank Paul Gifford not only for his battery talents, but for representing us in developing the new industry; particular thanks to our "young tigers" who are literally building this industry from the ground up. Our computer authority, Guy Wicker fits into that category very well. We also wish to thank Krishna Sapru for her early and current valuable work with me on hydrogen storage, batteries and atomic modeling. Our optical memory work would not be possible without the talent, commitment and hard work of Dave Strand; the electrical memory activity is dependent on the innovative talents and motivation of Wally Czubatyj and his group including Sergey Kostylev, Pat Klersy, and Boil Pashmakov; Ben Chao, the head of our analytical laboratory, is a resource beyond comparison to the entire company;

Rosa Young is a unique, extraordinarily innovative talent whose important contributions have spanned all of our areas in a remarkable manner; Steve Hudgens has contributed greatly to all areas of our work, his collaboration and advice have been extremely helpful to me and essential to the company. We are indeed fortunate to be working with Bob Stempel, the pioneer and leading figure in electric vehicles, a great engineer, leader and most appreciated partner. I am prouder of the organization and working climate that we have built than of any of my inventions.

REFERENCES

1. The Economist, p, 98, October 2, 1993.
2. U.S. Department of Energy announcement, January 18, 1994.
3. U. Hofer, Science **279**, 190 (Jan. 9, 1998).
4. S.R. Ovshinsky, Phys. Rev. Lett. **21**, 1450 (1968).
5. M.H. Cohen, H. Fritzsche and S.R. Ovshinsky, Phys. Rev. Lett. **22**, 1065 (1969).
6. S.R. Ovshinsky, Rev. Roum. Phys. **26**, 893 (1981).
7. S.R. Ovshinsky and D. Adler, Contemp. Phys. **19**, 109 (1978).
8. S.R. Ovshinsky, in *Physics of Disordered Materials*, edited by D. Adler, H. Fritzsche and S.R. Ovshinsky, (Inst. for Amorphous Studies Series, Plenum Press, New York, 1985) p. 37.
9. S.R. Ovshinsky, J. Non-Cryst. Solids **32**, 17 (1979).
10. S.R. Ovshinsky, in *Insulating and Semiconducting Glasses*, edited by P. Boolchand (World Scientific Press, Singapore, 1999).
11. For further references, see *Disordered Materials: Science and Technology, Selected papers by Stanford R. Ovshinsky*, edited by D. Adler, B.B. Schwartz and M. Silver (Institute for Amorphous Studies Series, Plenum Press, New York, 1991).
12. S.R. Ovshinsky, in *Proc. of the Seventh International Conference on Amorphous and Liquid Semiconductors*, Edinburgh, Scotland, 27 June-1 July, 1977, p. 519.
13. S.R. Ovshinsky, J. of Non-Cryst. Solids **42**, 335 (1980).
14. S.R. Ovshinsky, Presented at 1978 Gordon Research Conference on Catalysis (unpublished).
15. S.R. Ovshinsky, presented in May 1980 at Lake Angelus, MI (unpublished).
16. K. Sapru, B. Reichman, A. Reger and S.R. Ovshinsky, U.S. Pat. No. 4 633 597 (18 Nov. 1986).
17. S.R. Ovshinsky, M.A. Fetcenko and J. Ross, Science **260**, 176 (9 April 1993).
18. S.R. Ovshinsky, in *Proc. of the Sixth International Conference on Amorphous and Liquid Semiconductors*, Leningrad, USSR, 18-24 November 1975, p. 426.
19. S.R. Ovshinsky, in *Proc. of the International Topical Conference on Structure and Excitation of Amorphous Solids*, Williamsburg, Virginia, 24-27 March 1976, p. 31.
20. S.R. Ovshinsky, New Scientist **80** (1131), 674-677 (1978).
21. S.R. Ovshinsky, Solar Energy Mats. and Solar Cells **32**, 443-449 (1994).
22. S.R. Ovshinsky, in *Proc. of the International Ion Engineering Congress, ISIAT '83 & IPAT '83*, Kyoto, 12-16 September 1983, p. 817.
23. S.R. Ovshinsky and J. Flasck, U.S. Patent No. 4 770 940 (13 September 1988).
24. S.R. Ovshinsky and R.T. Young, in *Proc. of the SPIE Symposium on Modeling of Optical Thin Films II*, **1324**, San Diego, California, 12-13 July 1990, p. 32.
25. S. Guha, J. Yang, A. Banerjee, T. Glatfelter, G.J. Vendura, Jr., A. Garcia and M. Kruer, presented at the 2nd World Conference and Exhibition on Photovoltaic Solar Energy Conversion, Vienna, Austria, 6-10 July 1998.
26. J. Glorioso, Energy Management, June/July 1980, p. 45. Shown in 1980 by Stan and Iris Ovshinsky to Domtar, a Canadian roofing company and to Allside, an aluminum siding company of Akron, Ohio.

27. J. Yang, A. Banerjee, K. Lord and S. Guha, presented at the 2nd World Conference and Exhibition on Photovoltaic Solar Energy Conversion, Vienna, Austria, 6-10 July 1998.
28. S.R. Ovshinsky, presentation at the British House of Commons, July 1977.
29. S. Guha, J. Yang, A. Banerjee and S. Sugiyama, presented at the 1998 MRS Spring Meeting, San Francisco, CA, 1998 (invited).
30. D.V. Tsu, B.S. Chao, S.R. Ovshinsky, S. Guha and J. Yang, App. Phys. Lett. **71**, 1317 (1997).
31. R. Tsu, M. Izu, V. Cannella, S.R. Ovshinsky, G-J. Jan and F.H. Pollak, J. Phys. Soc. Japan Suppl. **A49**, 1249 (1980).
32. R. Tsu, M. Izu, V. Cannella, S.R. Ovshinsky and F.H. Pollak, Solid State Comm. **36**, 817 (1981).
33. R. Tsu, S.S. Chao, S.R. Ovshinsky, G-J. Jan and F.H. Pollak, J. de Physique **42**, 269 (1981).
34. R. Tsu, J. Gonzalez-Hernandez, J. Doehler and S.R. Ovshinsky, Solid State Comm. **46**, 79 (1983).
35. M.A. Fetcenko, S.J. Hudgens and S.R. Ovshinsky, Daido Journal (Denki-Seiko) **66** (2) 123-136 (April 1995). (Special Issue Electronics & Functional Materials.)
36. See for ex. 1959 Control Engineering on the Ovitron.
37. P. Kostic, Y. Okada, N.C. Collins, Z. Schlesinger, J.W. Reiner, L. Klein, A. Kapitulnik, T. H. Geballe, and M. R. Beasley, Phys. Rev. Lett. **81**, 2498, (1998).
38. S.R. Ovshinsky and R.C. Stempel, invited presentation at the 13th Electric Vehicle Symposium (EVS-13), Osaka, Japan, 13-16 October 1996.
39. R.C. Stempel, S.R. Ovshinsky, P.R. Gifford and D.A. Corrigan, IEEE Spectrum, November 1998, p. 29.
40. IEEE Spectrum, December 1997, p. 68.
41. 10th Anniversary American Tour de Sol Competition, run in New York City by Northeast Sustainable Energy Association (NESEA), (1998).
42. M. Kastner, Phys. Rev. Lett. **28**, 355 (1972).
43. S.R. Ovshinsky and K. Sapru, in *Proc. of the Fifth International Amorphous and Liquid Semiconductors*, Garmisch-Partenkirchen, Germany (1974), p. 447.
44. S.R. Ovshinsky, Phys. Rev. Lett. **36** (24), 1469-1472 (1976).
45. S.R. Ovshinsky and H. Fritzsche, *IEEE Trans. Elect. Dev.*, **ED-20** (2) 91-105 (1973).
46. M. Kastner, D. Adler and H. Fritzsche, Phys. Rev. Lett. **37**, 1504 (1976).
47. For history and early references, see S. R. Ovshinsky, "Historique du Changement de Phase" in Memoires, Optiques et Systemes, No. 127, Sept. 1994, p. 65; in the *Proc. of the Fifth Annual National Conference on Industrial Research*, Chicago, Illinois, 18-19 September 1969; Journal de Physique **42**, supplement au no. 10 October 1981.
48. S. R. Ovshinsky, presented at the 1997 International Semiconductor Conference, Sinaia, Romania , 1997.
49. S. R. Ovshinsky, presented at High Density Phase Change Optical Memories in Multi-Media Era, 9th Conference for Phase Change Media, Shizuoka, Japan, 1997.
50. S.R. Ovshinsky and I.M. Ovshinsky, Mats. Res. Bull. **5**, 681 (1970).

New Science Publications

Amorphous Semiconductors, Science Journal 5A (1969) 73.

Amorphous Semiconductors, Electronic Material (Japan) 8 (1969) 30.

Simple Band Model for Amorphous Semiconducting Alloys (with M.H. Cohen and H. Fritzsche), Phys. Rev. Lett. 22 (1969) 1065.

Hopping Conduction in an Amorphous Chalcogenide Alloy Film (with E.A. Fagen and H. Fritzsche), Bull. Am. Phys. Soc. II 14 (1969) 311.

An Introduction to Ovonic Research, J. Non-Cryst. Solids 2 (1970) 99.

Amorphous Semiconductors, Detroit Engineers 34 (1970) 13.

Radial Distribution Studies of Amorphous GexTe1-x Alloys (with F. Betts and A. Bienenstock) J. Non-Cryst. Solids 4 (1970) 554.

Calorimetric and Dilatometric Studies on Chalcogenide Alloy Glasses (with H. Fritzsche), J. Non-Cryst. Solids 2 (1970) 148.

Electrical Conductivity of Amorphous Chalcogenide Alloy Films (with E.A. Fagen and H. Fritzsche), ibid., p. 170.

New Thin-Film Tunnel Triode Using Amorphous Semiconductors (with R.F. Shaw, H. Fritzsche, M. Silver, P. Smejtek and S. Holmberg), Appl. Phys. Lett 20 (1972) 241.

Ovonics Revisited, Industrial Research 14 (1972) 48.

Three Dimensional Model of Structure and Electronic Properties of Chalcogenide Glasses (with K. Sapru), Proc. 5th Intl. Conf. Amorphous and Liquid Semiconductors, eds. J. Stuke & W. Brenig (Taylor and Francis, London, 1974) p. 447.

Photostructural Transformations in Amorphous As2Se3 and As2S3 Films (with J.P.deNeufville and S.C. Moss), J. Non-Cryst. Solids 13 (1974) 191.

Amorphous Materials as Interactive Systems, Proc. 6th International Conference on Amorphous and Liquid Semiconductors, Leningrad (1975) p. 426.

Lone-Pair Relationships and the Origin of Excited States in Amorphous Chalcogenides, Proc. Intl. Topical Conference on Structure and Excitation of Amorphous Solids, Williamsburg, VA, APS Conf. Proc. 3 (1976) 31.

Localized States in the Gap of Amorphous Semiconductors, Phys. Rev. Lett. 36 (1976) 1471.

Chemical Modification of Amorphous Chalcogenides, Proc. of the 7th Intl. Conf. on Amorphous and Liquid Semiconductors, Edinburgh, Scotland (1977) p. 519.

Optical and Electronic Properties of Modified Amorphous Materials (with R.A. Flasck, M. Izu, K. Sapru, T. Anderson and H. Fritzsche), ibid. p. 524.

Modification of SiOx (with K. Sapru and K. Dec), Proc. Intl. Topical Conf. on the Physics of Si02 and its Interfaces, Yorktown Heights, NY (1978) p. 304.

Local Structure, Bonding and Electronic Properties of Covalent Amorphous Semiconductors (with D. Adler), Contemp. Phys. 19 (1978) 109.

Electrical and Optical Properties of Amorphous Si:F:H Alloys (with A. Madan and E. Benn), Phil. Mag. B.40 (1979) 259.

The Shape of Disorder, J. Non-Cryst. Solids 32 (1979) 17.

Some Electrical and Optical Properties of a-Si:F:H Alloys (with A. Madan, W. Czubatyj and M. Shur), J. Elect. Mat. 9 (1980) 385.

Properties of Amorphous Si:F:H Alloys (with A. Madan), J. Non-Cryst. Solids 35/36 (1980) 171.

Book Review on The Physics of Selenium and Tellurium, ed. E. Gerlach and P. Grosse, American Scientist 68 (1980) 316.

Effect of an Interfacial Oxide in Amorphous Si:F:H Alloy Based MIS Devices (with A. Madan, J. McGill, W. Czubatyj, J. Yang and M. Shur), SPIE Proc. 248 (1980) 26.

Electronic and Vibrational Properties of Glow-Discharge Amorphous Si:F:H (with R. Tsu, M. Izu and V. Cannella), J. Phys. Soc. Japan 49 (1980) Suppl. A, p.1249.

Electroreflectance and Raman Scattering Investigation of Glow-Discharge Amorphous Si:F:H (with R. Tsu, M. Izu and F.H. Pollak), Solid State Comm. 36, (1980) 817.

The Chemical Basis of Amorphicity: Structure and Function, Revue Roumaine de Physique 26 (1981) 893.

Present Status of the Science and Technology of Amorphous Solids (with D. Adler), Nikkei Science (Japanese Scientific American) (1983) p. 60.

Laser-Induced Fluorescence Detection of Reactive Intermediates in Diffusion Flames and in Glow-Discharge Deposition Reactors (with H.U. Lee and J. deNeufville, J. Non-Cryst. Solids 59/60 (1983) 671.

The Role of Free Radicals in the Formation of Amorphous Thin Films, Proc. Intl. Ion Engineering Congress, ISIAT '83 & IPAT '83, Kyoto, Japan (1983) p. 817.

Order Parameters in a-Si Systems (with R. Tsu, J. Gonzales-Hernandez and J. Doehler), Solid State Comm. 46 (1983) 79.

Properties of Amorphous Semiconducting Multilayer Films (with J. Kakalios, H.Fritzsche and N. Ibaraki), J. Non-Cryst. Solids 66 (1984) 339.

Amorphous Materials – Past, Present and Future; J. Non-Cryst. Solids 73 (1985) 395.

Basic Anticrystalline Chemical Bonding Configurations and Their Structural and Physical Implications, J. Non-Cryst. Solids 75 (1985) 161.

Chemical Bond Approach to Glass Structure (with J. Bicerano), ibid., p. 169.

Chemistry and Structure in Amorphous Materials: The Shape of Things to Come, in "Physics of Disordered Materials," edited by D. Adler, H. Fritzsche and S. R. Ovshinsky, (Plenum Press, New York, 1985) p. 37.

Critical Materials Parameters for the Development of Amorphous Silicon Alloys (with D. Adler), Mat. Res. Soc. Symp. Proc. 49 (1985) 251.

Fundamentals of Amorphous Materials, in "Physical Properties of Amorphous Materials," Eds. D. Alder, B.B. Schwartz and M.S. Steele, (Plenum Press, 1985) p. 105.

A New Role for Vacuum Technology (with D. Adler), Proc. 28th Annual Technical Conf. of the Society of Vacuum Coaters, Washington, D.C. (1985) p.1.

Intuition and Quantum Chemistry, Proceedings of the Nobel Laureate Symposium on Applied Quantum Chemistry (in honor of G. Herzberg, R.S. Mulliken, K. Fukui, W. Lipscomb and R. Hoffman), Applied Quantum Chemistry, edited by V. H. Smith, Jr. et al. (D. Reidel Publishing, 1986) p.27

Chemical Bonding and the Nature of Glass Structure (with J. Bicerano), ibid., p.325.

Progress in the Science and Application of Amorphous Materials (with D. Adler), J. Non-Cryst. Solids 90 (1987) 229.

Passivation of Dangling Bonds in Amorphous Si and Ge by Gas Absorption (with R. Tsu, D. Martin and J. Gonzalez-Hernandez), Phys. Rev. B 35 (1987) 2385.

The Quantum Nature of Amorphous Solids in "Disordered Semiconductors," Eds. M. A. Kastner, G. A. Thomas and S. R. Ovshinsky (Plenum Press, New York, 1987) p. 195.

A Personal Adventure in Stereochemistry, Local Order and Defects: Models for Room Temperature Superconductivity, in "Disorder and Order in the Solid State: Concepts and Devices," edited by R. W. Pryor, B. B. Schwartz and S. R. Ovshinsky (Plenum Press, New York, 1988) p.143.

Structural Changes Induced by Thermal Annealing in W/C Multilayers (with B.S. Chao, J. Gonzalez-Hernandez, D. Pawlik, J. Scholhamer, J. Wood and K. Parker), SPIE Proc.1547 (1991) 196.

Ion and Neutral Argon Temperatures in Electron Cyclotron Resonance Plasmas by Doppler Broadened Emission Spectroscopy (with David V. Tsu, R.T. Young, C.C. Klepper and L.A. Barry), J. Vac. Sci. Technol. A 13 (1995) 935.

The Structure of W/C ($0.15< \gamma < 0.8$) Multilayers Annealed in Argon or Air (with J. Gonzalez-Hernandez, B.S. Chao and D.D. Allred), Journal of X-Ray Science and Technology 6 (1996) 1.

Lifting the Tyranny of the Lattice: A Revolution in Progress (with I.M. Ovshinsky), Proc. of the Norbert Kreidl Symp. on Present State and Future Prospects of Glass Science and Technology Vol. 70C (1997).

Mott's Room, in "Reminiscences and Appreciations", ed. E.A. Davis (Taylor & Francis Ltd, London, 1998) p. 282.

Amorphous and Disordered Materials – The Basis of New Industries, Mat. Res. Soc. Symp. Proc. 554 (1999) 399.

Chapter III: Phase Change Memory

Memory, the preservation and transmittance of information, has been the life blood of human civilization. The beginning of writing, the printing of books and the invention of photography have been monumental milestones of the past. With the invention of the phonograph and the widespread use of plastic vinyl records the world of music and speech could be preserved for the first time. More recently, however, we have entered a rapidly expanding and unimagined new information age. The magnetic tape, offered by its inventor to General Electric Company in 1928 and ironically rejected by GE as impractical, quickly became the basic memory modem of the nascent computer industry. Soon it began to replace the LP records. The big wheels of magnetic tape memories were in turn replaced around 1956 by hard disk drive memories as computers became smaller. On hard disks the data is recorded in binary form by magnetizing tiny spots of magnetic material along the spiral track on a rapidly rotating disk. Hard disks are the hearts of present computers and of digital video and audio players.

Although semiconductor technology spawned the explosive use of personal computers, it first was difficult to design semiconductor memories which could retain their information when the electric power was removed, a property called nonvolatile. The magnetic memories are nonvolatile as well as re-writable, which means the information can be erased and new data can be written over the old. The most widely used primary memory for operating a computer is in the form of volatile random access memory (RAM) whose data are lost when the computer is shut down. The recall of this information explains the delay in getting a computer started or rebooted.

Electronic semiconductor memories which are erasable, programmable and nonvolatile are now found in memory cards of digital cameras, cell phones and many other mobile electronic devices which need the kinetic shock resistance of such semiconductor solid state circuits. The memory function of these semiconductors is the storage of a small amount of electrical charge on a floating gate in an insulator. The presence or absence of the charge affects the binary code operation of an adjacent transistor. For erasing and recording data, the charge is removed from or inserted into the floating gate by a voltage pulse.

There has been an exponential growth of the market for ever faster and higher bit-density memories during the past decades fulfilling the demands for faster and more compact computers for our increasingly sophisticated entertainment needs, for navigation, industrial automation and archival information storage and retrieval. There is hardly one aspect of modern life which is not facilitated or made possible by information storage and computers. The competition is fierce for more and cheaper data storage devices. Some of us remember backing up the relatively primitive hard drive on a floppy disk. Nowadays one would need several thousand floppy disks for that purpose. The contents of about 1000 books can be stored on a 5GB memory disk or a UBS flash memory, an incredible amount. Portable 150GB hard drives are now available that can store the contents of a library of 30,000 books. We sketch this story as background for an entirely different kind of nonvolatile memory invented by Ovshinsky in the 1960s.

Inspired by his earlier work on neurophysiology and the mystery of the human brain and memory, Ovshinsky realized that electronic functions in nature do not occur in crystalline materials and that the multifunctional cognitive processes in the human brain do not resemble the binary serial processes of modern day computers. This led him to synthesize and explore the electronic properties of thin amorphous films of mixtures of elements which had not been put together before. He found that certain mixtures containing for example tellurium, germanium, and antimony could change easily from the amorphous phase to the crystalline phase and back to the amorphous phase in response to electrical or optical energy pulses. Since the two phases of the material have very different electrical conductivities and different optical reflectance and transmission, the phase change produced in small spots can be used to store binary data which can be read as a change in electrical resistance or a change in optical reflectance. Ovshinsky's first optical phase change memories were operated with light pulses from a large water cooled gas laser of impractical size. But two decades later very small lasers became available and the first commercial phase change optical memory disks were produced by Matsushita/Panasonic in 1990. Now, all computers have a port for optical disk memories to transport and transfer a large store of information. The phase change compact disks, CDs, and the phase change digital video disks, DVDs, are the only ones which can be rewritten and erased at will. These rewrite and erase properties can be accomplished not only a few times, which would already beat the competition of read-only CD's and DVD's, but for as many as million times, which is a great advantage. These memories are of course nonvolatile because a small pulse of energy is required to change the phase in any given spot.

One advantage of phase change memories is that the spots in which the phase change is induced can be very small and closer together than for example the transistors with their floating gates of the electrical semiconductor memories thus providing a high information density. As the spots are made tinier, a smaller energy pulse is needed to write the information in the spots. This advantage of smaller bit size of phase change memories compared to that of floating gate storage memories led to the present effort by leading electronic companies including Intel, Samsung, and STMicroelectronics to further develop electrical phase change memories. Ovshinsky is recognized and celebrated as the father of phase change memories. International workshops and congresses are devoted to this important technology with names like "...on Phase Change and Ovonic Science".*

Ovshinsky is already far ahead of these developments. He and his team have discovered that his phase change materials have a rich potential for devices more interesting than binary memories even though these beat their competitors in size and speed. Instead of having just two states common for binary data recording, the phase change materials can be programmed between the extremes of amorphous and crystalline phases to many distinct resistive or optical reflectivity states having different volume fractions of crystallites mixed with amorphous material. Reliable programming to say eight different phase states increases the data recording density many-fold. Moreover, the energy pulse needed for programming a tiny spot of the material from the amorphous to the crystalline phase can be delivered not only in a single pulse but instead in small portions, for example 5, 6, or 8 smaller energy pulses. Only when the total energy is accumulated, is a crystalline percolation path established and the memory spot drops its resistance to the crystalline state. The similarity to a biological neuron is obvious whose synaptic cell fires only after receiving a sum of signals which add up to the synaptic threshold value. In

* Ovonic = Ovshinsky + Electronic

the neuron the synaptic threshold can in turn be modified by inhibiting or enhancing inputs by the multiply connecting axons. A similar control for modifying the threshold of electrical phase change memory devices was established by providing a third control electrode. Ovshinsky seems to have all the devices now at hand to realize a cognitive computer, his very early vision of an entirely new non-binary computer paradigm.

Ovshinsky's publications in this chapter describe these later developments of the optical and electrical phase change memories. The earlier discoveries and the use of phase change characteristics for non-silver photography, imaging and microfiche cards as well as the first nonvolatile electrical phase change memoriesbuilt under license by Burroughs are published in a volume of Ovshinsky's earlier publications: "Disordered Materials: Science and Technology", eds. D. Adler, B. B. Schwartz and M. Silver, Plenum Press, New York and London, (1991).

Journal of Non-Crystalline Solids 141 (1992) 200–203
North-Holland

JOURNAL OF
NON-CRYSTALLINE SOLIDS

Optically induced phase changes in amorphous materials

S.R. Ovshinsky
Energy Conversion Devices, Troy, MI, USA

The ability to modify the structure of thin films of chalcogenide alloys using light allows the construction of several types of imaging systems, most notably phase change optical memories. The benefits of phase change optical memories include performance and, as a result of the inherent simplicity of the read and write processes, low cost. The key to achieving high speed is the design of an alloy where the crystallization process involves diffusionless crystal growth in a system that does not phase segregate. These materials can be crystallized with laser pulses of 30 ns duration or less, and such high crystallization speeds allow the use of the recording media in the direct-overwrite mode. GeSbTe alloys show excellent characteristics for such application, including fast transition times and excellent stability. Stoichiometric GeTe also shows good performance, but the addition of Sb extends the performance to a wider compositional range, improving manufacturability.

1. Introduction

Iris and I first met Jan Tauc in 1967 in Prague and we had a very interesting and stimulating discussion with him on my work in threshold and memory switching in amorphous materials. It is a pleasure to be able to participate in this Festschrift in the acknowledgement of Tauc's many contributions to the amorphous field.

A major motivation for my work in optical interactions with amorphous materials was to buttress my position that Ovonic threshold and memory switching events were involved with electronic excitation and that specially designed chalcogenide materials could be reversibly switched between amorphous and crystalline by optical means, thereby encoding information. This led not only to optical memories but to new forms of non-silver photography [1].

The first optical memories used gas tube lasers, and by 1972 IBM had taken out a license for this technology, and some fine work resulted from their technical group [2,3]. In the latter period, under the leadership of Hayakawa and Suemitsu, the group at Matsushita has made important contributions [4,5].

The Ovonic structural change memory has become known ubiquitously as phase change memory and is a leading contender to fill the great need for optical storage. Advances in laser technology, particularly semiconductor diode lasers, made small, efficient and inexpensive disk drives a reality. One gigabyte of data can be stored on a single phase change erasable optical disk which is used in a drive that fits in a single drive bay of today's personal computers. The phase change recording materials are the lynch pin of the entire memory system, and recent advances in this technology have provided the operating characteristics which place these products at the forefront of the optical disk drive market.

2. Phase change technology

Phase change recording materials are designed so as to have at least two structural forms, amorphous and crystalline, which can coexist at room temperature. Alloys based upon elements of the chalcogen family (sulfur, selenium and especially tellurium) are most commonly used. They are so chosen because they can be readily converted between two structural states while simultaneously offering excellent thermal stability over a temperature range that extends well beyond the limits of anticipated storage or usage conditions for such products. Phase change materials can be optically switched between the amorphous and crystalline states by the energy contained in a laser beam [6].

Beyond the required ability to record and store data in a stable manner, phase change materials have the advantage of a simple data reading process. The crystalline structure results in a narrower band gap in these semiconductor materials compared with the more random structure of the amorphous state. Both the absorption coefficient and the index of refraction are higher when the material is crystalline. Optimum performance is achieved when these two optical changes are used in concert to form a tuned optical structure having appropriate layer thicknesses in a thin film device. This simple read process has important implications in the manufacturing cost of both the data disk and the optical disk drive that uses it.

Ovonic phase change optical memories are being chosen worldwide by leading manufacturers for new generations of commercial devices which utilize the inherent speeds possible in these materials. Our licensees, including Matsushita, Asahi Chemical, Hitachi, Sony and others, are now exploiting this technology. Phase change erasable products, which have the highest data transfer rate during recording, the highest carrier-to-noise ratio and an overwrite cycle life of up to more than a million recordings, are now generating much excitement in the data storage field.

3. High speed materials

The operational parameter which separates phase change erasable systems from competing erasable optical data storage technologies is the record speed. Whereas the currently popular magneto-optical recording products use two revolutions of the disk to record information, phase change systems use only one, effectively doubling the data transfer rate during record. Although all optical disk products share the attribute of lowest cost-per-byte of any type of random access data storage technologies, they are relatively slow compared with magnetic hard drives. Products based on phase change technology have overcome this limitation, and now one can have the capacity and stability features of optical data storage in a product having data transfer rates approaching those of the more common magnetic hard drives.

The recording process by which phase change materials have been able to achieve such high data transfer rates is referred to as direct-overwrite. Direct-overwrite, as the name implies, is the direct replacement of existing data with new data without the need for any intermediate erasure step. Phase change materials can be switched to both the crystalline and amorphous states by a laser exposure of 50 ns. This ability, which requires the use of a high atomic mobility state for the overwrite process, represents a major step forward in phase change recording technology.

Phase change materials can be of two different types based on the crystallization process: those that phase segregate and those that do not. Phase segregation involves diffusion which causes crystal growth rates to be correspondingly slow. Although nucleation control and chemical modification techniques have been used to improve the crystallization process, the crystallization time of phase segregating materials can be as long as 1 μs. By adding heterogeneous nucleation sites, the crystallization process can be speeded by reducing the distance that crystals have to grow. (Increasing the number of nucleation sites in a given volume necessarily positions them closer together.) This technique has the further advantage of creating smaller crystallites, which reduces the background noise level of the media.

Direct-overwrite materials, on the other hand, involve diffusionless crystal growth. This improvement is obtained by designing materials which have the same chemical composition in both the amorphous and crystalline states. Materials such as GeTe can exhibit both the high thermal stability of an extensively crosslinked amorphous network and the fast crystallization times needed for direct-overwrite. The only stoichiometric compound in the GeTe binary system is $Ge_{50}Te_{50}$. This composition forms a rhombohedral crystal structure with no homo-elemental bonds. This bonding configuration is also well suited for the amorphous phase, which shares similar nearest-neighbor configurations with the crystalline structure. The four-fold coordination of germanium causes the amorphous structure to be quite rigid, preventing spontaneous transformation to the crystalline phase. The inter-atomic rearrangement necessary to complete the structural transition involves minor changes in bond angles and lengths. Only when enough energy is added to the system to achieve a state of high mobility can these geometric relaxations occur and consequently allow the material to convert to the more thermodynamically stable crystalline structure.

It has been shown that the crystallization time for GeTe is strongly dependent on the composition of the material. A thin film of stoichiometric GeTe can be crystallized in as little as 30 ns. If the atomic ratio is changed from 50 : 50 to 60 : 40 or 40 : 60, the crystallization time rapidly increases to over one 1 μs. This is a result of changing the crystallization mechanism from a diffusionless transformation to one where the excess of either Ge or Te must be phase separated during the crystallization process.

The addition of antimony to the GeTe binary system increases the practicality of commercial-scale production by easing the requirement for precision in composition of the deposited film without deleterious effects in the speed or stability. Antimony can be substituted for germanium, relaxing the interatomic stresses between tellurium and germanium and allowing formation of the more symmetrical face centered cubic structure. This is true as long as the antimony level is less than the germanium level. Addition of antimony also allows the freedom to vary the tellurium level from about 45 to 55% without degradation of the crystallization speed, by contrast with the binary GeTe system.

The amorphous-to-crystalline phase transformation in the GeSbTe system is a semiconductor-to-semimetal conversion, and the changes in optical constants are correspondingly large. The real portion of the index of refraction changes from about 5 to about 7, and the absorption coefficient (measured at 830 nm) doubles from about 1.3 cm^{-1} to 2.6 cm^{-1}, depending on the exact composition. These substantial changes can be used to develop a large change in the surface reflectivity of a single layer of the material. If, however, a multiple layer structure is used, the reflection mode contrast can be even greater.

Further consideration of these devices shows that other benefits are obtained by using a multi-layer device structure. Materials that have very short crystallization times become increasingly difficult to re-amorphize. If a heat sinking layer is added to the device structure, the quench rate can be increased, and materials that show a tendency to crystallize when cooled from the melt can be reliably quenched to the amorphous structure.

When the optical coupling, quench rate and layer thickness deformations are taken into consideration, the optimum device becomes a four layer structure. First is a dielectric layer, followed by a recording layer, then a second dielectric layer and finally a heat sinking reflective metal layer. The manufacture of this structure compares favorably to that of the competing magneto-optical devices, where there are usually five layers, each having tight tolerances on thickness in order to maximize the inherently low read contrast.

The phase change erasable disk drive also benefits from the high contrast of the media. Compared with the simple read process of the phase change systems, the magneto-optical systems need polarization analysis to detect the read signal, and more elaborate electronic circuitry to separate the high frequency, low level signal from the background noise. Further, there is the additional need for an electromagnet to provide the magnetic fields that need to be applied during the record and erase processes.

As other enabling technologies associated with optical disk drives mature, phase change technology is best positioned to take advantage of them to increase system performance. For example, much progress is being made on the frequency doubling of the output of semiconductor lasers. This will translate into a four-fold increase in the storage capacity of optical disks, and phase change media have already demonstrated both the sensitivity to these short wavelengths and also the capability to record at resolutions far exceeding even the half-micron spots that these new lasers will be able to record.

There are other imaging systems that utilize optically initiated structural changes that were successfully developed at ECD. Examples include an organo-tellurium dry process film which has excellent gray scale and amplification; an instant, updateable microfilm system which uses a photodispersive imaging mechanism; a reversible recording film where scattering centers are formed that are comprised of microscopic bubbles which can be generated and re-absorbed; and a high contrast lithographic film system. All of these systems were based on non-silver imaging systems [1]. These diverse phenomena as well as the uses of amorphous semiconductors in thin-film photovoltaics and as sensors indicate the richness of the field of optical interaction with amorphous materials, a field which is now widely studied scientifically and whose technological importance worldwide is without question.

We look forward to Tauc's many more years of contribution. The best way to honor him is by continued efforts in the field by his colleagues. Iris joins me in sending Jan our best.

We would like to acknowledge our co-workers who have made so many valuable contributions: John deNeufville, Peter Klose, Sato Iwasa, Hellmut Fritzsche, Julius Feinlieb, David Adler, Rosa Young and so many other talented colleagues.

We especially would like to acknowledge the contributions of David Strand whose work in the latter period of development of the optical memory did so much to advance the commercialization process.

References

[1] David Adler, Brian B. Schwartz and Marvin Silver, eds., Disordered Materials: Science and Technology – Selected papers by S.R. Ovshinsky, 2nd Ed., Institute for Amorphous Studies Series (Plenum, New York, 1991).

[2] R.J. von Gutfield and P. Chaudhari, J. Appl. Phys. 43 (1972) 4688.

[3] A.W. Smith, Appl. Opt. 13 (1974) 795.

[4] N. Yamada, M. Takao and M. Takenaga, Proc. SPIE 695 (1986) 79.

[5] T. Ohta, M. Uchida, K. Yoshioka, K. Inoue, T. Akiyama, S. Furukawa, K. Kotera and S. Nakamura, Proc. SPIE, 1078 (1989) 27.

[6] David Strand and David Adler, Proc. SPIE 420 (1983) 200.

THE RELATIONSHIP BETWEEN CRYSTAL STRUCTURE AND PERFORMANCE AS OPTICAL RECORDING MEDIA IN Te-Ge-Sb THIN FILMS

D. STRAND, J. GONZALEZ-HERNANDEZ, B.S. CHAO, S.R. OVSHINSKY, P. GASIOROWSKI AND D.A. PAWLIK

Energy Conversion Devices, Inc., 1675 West Maple Road, Troy, MI 48084

ABSTRACT

The crystallization properties of Te-Ge and Te-Ge-Sb alloys prepared by thermal evaporation were analyzed using various characterization techniques. Similar to previous results, our data for Te-Ge shows that alloys that deviate slightly from $Te_{50}Ge_{50}$ stoichiometry show drastically slower crystallization kinetics. Raman spectroscopy and x-ray diffraction show that alloys having non-stoichiometric atomic ratios phase separate during crystallization into a $Te_{50}Ge_{50}$ phase plus pure crystalline tellurium or germanium. It is this relatively slow process of phase segregation which limits the crystallization rate. Phase segregation during crystallization of non-stoichiometric Te-Ge can be eliminated by adding antimony to samples having a tellurium concentration of from 45 to 55 atomic percent over a wide range of Ge:Sb ratios. These alloys can have laser induced crystallization times of less than 50 nsec. The thermal crystallization temperature is reduced only slightly when antimony is substituted for germanium.

INTRODUCTION

The idea of phase change technology, which was originated by S.R. Ovshinsky more than twenty years ago [1,2], is now being successfully applied in commercial optical disk based data storage systems. Although phase change technology has long been recognized for the simple record and read processes it uses, its commercial success was predicated on its direct overwrite capability. Direct overwrite is the replacement of pre-existing recorded data with new data in a single action. Compared to currently available magneto-optical data storage systems, which require a first revolution of the disk to erase existing recordings, followed by a second revolution for recording new data, information can be recorded up to twice as fast using direct overwrite phase change media.

In order to qualify for use as a direct overwrite media, phase change materials had to be developed which could be crystallized in the same duration of laser exposure as was available for making them amorphous. This means that the crystallization process has to be very rapid. Stoichiometric compounds have been shown to exhibit the necessary crystallization speeds, but often show the disadvantage of rapid loss of crystallization speed when the composition deviates even slightly from stoichiometry [3,4]. In this work, we show that adding Sb to the Te-Ge binary system greatly increases the compositional range over which rapid crystallization is achieved. Further, we correlate the crystal structure to the crystallization kinetics to explain these and more subtle effects in the relationship between composition and crystallization temperature.

EXPERIMENTAL

Te-Ge-Sb alloy films were prepared by thermal evaporation on unheated (100) crystalline silicon, glass and polymethylmethacrylate substrates. Multiple element alloy sources and single element sources were used singly and in combination for the evaporation of the films used in this study. The compositions of the films were determined using energy dispersive x-ray spectrometry (EDS). Auger compositional depth profiles were measured to determine compositional uniformity and light element impurities. The sudden decrease in the optical transmission (measured at 630 nm) which accompanies crystallization during heating at a constant rate of 60 °C/min was used to indicate the crystallization temperature (T_x) of the as-deposited amorphous films. X-ray diffraction (XRD) was used to determine the crystalline structure in annealed films. The films were annealed by heating under argon atmosphere to a pre-determined temperature at a heating rate similar to that used in the T_x measurements. For the Raman measurements the 488 nm line from an argon ion laser was focused to a power density of about 3 W/cm^2 onto the sample using a cylindrical lens. The phase transformation kinetics data used for determining crystallization rate was measured using a static tester we constructed which focuses substrate-incident 830 nm semiconductor laser radiation to a spot approximately 1 µm diameter on an as-deposited 80 nm thick amorphous film on a polymethylmethacrylate substrate.

RESULTS

Figure 1 shows the Raman spectra obtained from two Te_xGe_{100-x} films with x = 43 (curves (a) and (b)) and x = 53 (curve (c)). Spectra (a), (b) and (c) correspond to the as-prepared, 400 °C and 300 °C annealed samples respectively. In the as-prepared amorphous sample, the broad Raman lines centered at about 125, 163, 225 and 275 cm^{-1} are characteristic of the scattering from Te-Te (125 and 163 cm^{-1}), Te-Ge (225 cm^{-1}) and Ge-Ge (275 cm^{-1}) bonds in a covalently bonded random network material. The stronger intensity of lines related to the Te-Te bonds results from the high Raman cross-section of tellurium. After annealing, the broad amorphous Raman features in the spectrum of both compositions vanish and sharper peaks associated with identifiable crystalline phases appear. For the x = 43 sample the peak at 297 cm^{-1} indicates the presence of crystalline germanium. The two peaks at 124 and 142 cm^{-1} in the film with x = 53 indicate the presence of crystalline tellurium. Although during the crystallization process the major volume fraction of each of these films crystallized into a TeGe rhombohedral phase, as indicated by XRD, no obvious Raman signal from that phase is observed. This indicates that this semi-metallic compound has a low Raman cross-section.

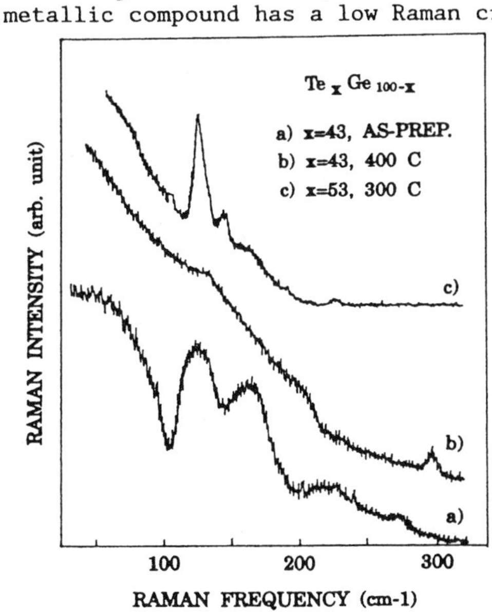

Figure 1. Raman intensity vs. Raman frequency for two Te_xGe_{100-x} films with x = 43, as-prepared (a) and annealed at 400 °C (b), and x = 53, annealed at 300 °C (c). The broad peaks in curve (a) are related with Te-Te, Te-Ge, and Ge-Ge bonds in the amorphous structure. The peak at about 300 and 100-150 cm^{-1} in curves (b) and (c) corresponds to c-Ge and c-Te, respectively.

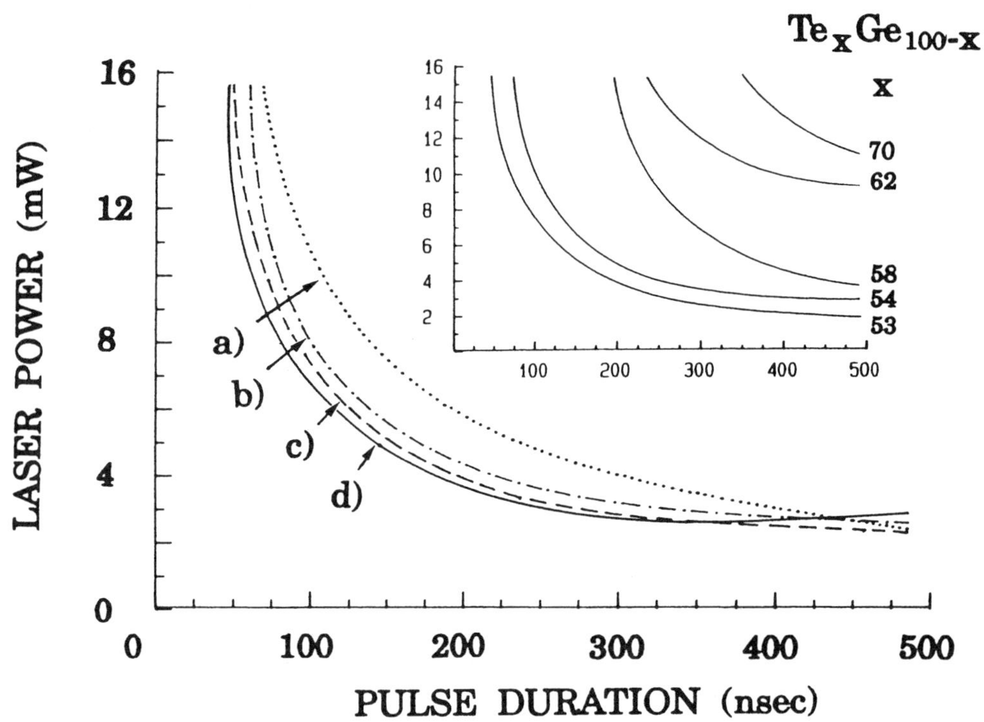

Figure 2. Phase transformation kinetics curves for the Te rich Te_xGe_{100-x} films (insert) and the Te-Ge-Sb alloy films: (a) $Te_{46}Ge_{50.5}Sb_{3.5}$, (b) $Te_{48}Ge_{50.5}Sb_{1.5}$, (c) $Te_{51}Ge_{32}Sb_{12}$, and (d) $Te_{55}Ge_{24}Sb_{21}$ and $Te_{54}Ge_{41}Sb_5$.

The curves in figure 2 indicate where the onset of crystallization occurs in a test matrix of incident laser pulses having various pulse amplitudes and durations (phase transformation kinetics, or PTK plots) [5] for Te-Ge-Sb alloy films having a wide range of compositions. In the figure insert, results of similar measurements are shown for Te rich Te_xGe_{100-x} alloys. Similar to previous results [3], we find that the crystallization speed of Te_xGe_{100-x} alloys strongly depends on the value of x. Films having a composition close to stoichoimetric TeGe can be crystallized using short laser pulses, however, much longer laser pulses are needed to crystallize films whose compositions are slightly away from stoichiometry. A different behavior is observed in the Te-Ge-Sb alloy films. The phase transformation kinetics of these films are similar to those of stoichiometric TeGe films. Further, the phase transformation kinetics of various compositions in the Te-Ge-Sb alloy system are similar regardless of the Ge:Sb ratios as long as the Te level is 45 to 55 atomic percent. Crystallization times of less than 50 nsec were observed in compositions having Sb levels from 1.3 up to 35 atomic percent.

The crystallization temperature (T_x) of the Te-Ge-Sb alloy films is summarized in figure 3(a). Our measurements show that the T_x of

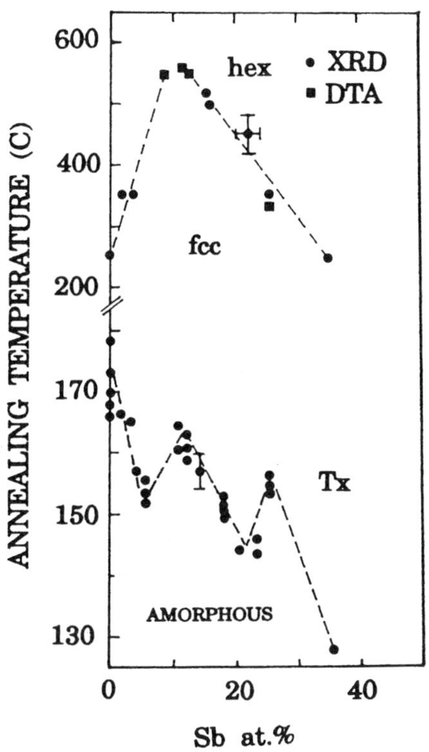

Figure 3. Dependence of the crystallization temperature, T_x, (a) and the crystalline-to-crystalline transition (fcc to hexagonal) temperature (b) as a function of the Sb content in the Te-Ge-Sb alloy films.

Te_xGe_{100-x} alloy thin films, in the vicinity of a stoichiometric TeGe, is in the range of 165-178 °C. Similar to what we observed in the crystallization speed, we found that T_x is also very sensitive to the ratio of Te:Ge [3]. Any slight deviation from stoichiometric TeGe results in a higher T_x. The T_x decreases slightly when antimony is added to Te-Ge alloy films. However, the T_x does not follow a monotonically decreasing trend with an increase of the amount of Sb in the alloys, as seen in figure 3(a). As antimony is added to the alloy, the T_x first drops from 165 °C at zero atomic percent antimony to 153 °C at 5 atomic percent antimony. The T_x then increases to about 162 °C at 11 atomic percent antimony. Adding more antimony to the composition gradually decreases the T_x again, followed by a second relative maximum at 25 atomic percent antimony.

All of the Te-Ge-Sb alloy films having up to 35 atomic percent antimony were amorphous as-prepared. Films having a composition of $Te_{50}Sb_{50}$ are crystalline (hexagonal phase) as-prepared. When heated to a temperature slightly above their T_x, the amorphous Te-Ge-Sb films crystallize into a fcc structure (see figure 3(a)). The crystalline fcc phase transforms to a more stable crystalline structure having rhombohedral symmetry (hexagonal structure) upon annealing at higher temperatures. This crystalline-to-crystalline phase transition temperature depends on the film composition, as shown in figure 3(b). The data in figure 3(b) was collected using two separate experimental techniques: differential thermal analysis (DTA) and thermal annealing as performed in conjunction with XRD measurements. A stoichiometric TeGe film has a transition temperature at about 250 °C. The transition temperature increases when antimony is added, to 560 °C at 11 atomic percent, and then gradually decreases with further increases in the Sb concentration.

Figure 4 shows XRD data measured on a $Te_{52}Ge_{23}Sb_{25}$ film in the as-prepared state (a) and after annealing at temperatures of 300 °C (b) and 400 °C (c). The broad features shown in curve 4(a) centered at an angle of about 28° and 47° in the 2Θ scale are typical of the amorphous structure of the as-deposited alloy film. In film annealed at or below 300 °C (curve (b)), all of the diffraction peaks are related to the crystalline fcc structure. Additional diffraction peaks, marked by arrows in curve 4(c), clearly indicate that the crystalline-to-crystalline phase transition occurred in the film at annealing temperatures between 300 and 400 °C.

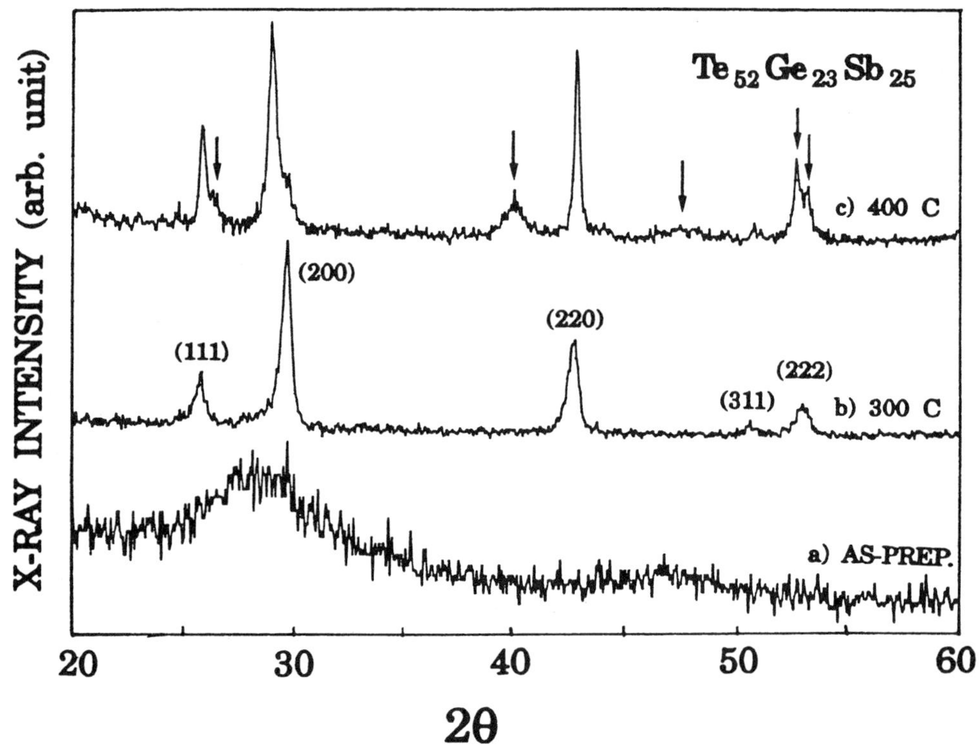

Figure 4. X-ray diffraction patterns for $Te_{52}Ge_{23}Sb_{25}$ film as-deposited (a), annealed at 300 °C and 400 °C. The arrows indicate the lines that are only associated with the hexagonal structure.

DISCUSSION AND CONCLUSION

It has been previously reported that the crystallization kinetics of Te_xGe_{100-x} strongly depend on the value of x (figure 2, insert) [3]. Deviations from stoichiometry result in a much slower crystallization rate compared to the $Te_{50}Ge_{50}$ alloy. Phase segregation with slow atomic diffusion during crystallization has been suggested as the limiting mechanism for crystallization in non-stoichiometric compounds. In the present study, we have used several characterization techniques to analyze the structure of crystalline Te-Ge and Te-Ge-Sb alloys.

Using Raman sepctroscopy we have been able to detect phase segregated crystalline tellurium in fully crystallized Te-Ge films with tellurium concentrations of 53 atomic percent or higher. Similarly, our Raman and XRD measurements clearly show the presence of crystalline germanium in alloys which have Ge concentrations of 57 atomic percent or higher and which have been annealed at 400 °C or higher (figure 1). Higher sensitivity for the detection of crystalline tellurium by Raman scattering results from a higher Raman cross-section at the energy used for excitation. Further, due to resonance effects, crystalline tellurium incorporated in an absorbing matrix can be detected when present at less than one percent of the volume. Based upon the Raman data shown in figure 1(a), we expect to be able to detect the presence of crystalline germanium at a volume fraction of about three percent in an absorbing matrix.

A similar Raman analysis has been carried out in the Te-Ge-Sb alloy system with Te concentrations ranging from 45 to 56 atomic percent and Sb concentrations from 1.3 to 50 atomic percent. In that compositional range no phase segregation has been observed, which means that if there is any of crystalline Te or Ge present, the volume fraction must be below our detection limits. The EDS data indicates that there is enough Te in some of our alloys to readily be detected by Raman, if present, and enough Ge to be detectable by Raman, if present, in other alloys. The PTK curves in figure 2, showing only slight differences in performance for various compositions, support our premise that antimony averts phase segregation.

As illustrated in figure 3(a), the Te-Ge-Sb alloys have lower T_x than stoichiometric TeGe. However, it is clear that the T_x reached local maximum values in films with Sb concentrations of about 11 and 25 atomic percent. It is expected that ordering in the amorphous alloys varies smoothly with antimony concentration. If this is so, the slight increase in T_x (~ 2%) for these two compositions is probably related to slightly different crystallization kinetics. At a composition of about 11 atomic percent antimony, crystallite size obtained from the XRD data is largest [6] and the fcc crystalline structure is most stable. This suggests that at this composition, the contribution to the growth process by growth, relative to nucleation, becomes more important.

REFERENCE

[1]. J. Feinleib and S.R. Ovshinsky, J. Non-Cryst. Sol. 4, 564 (1970).
[2]. J. Feinleib, J. deNeufville, S.C. Moss, and S.R. Ovshinsky, Appl. Phys. Lett. 18, 254 (1971).
[3]. M. Chen, K.A. Rubin, and R.W. Barton, Appl. Phys. Lett. 49, 502 (1986).
[4]. N. Yamada, E. Ohno, K. Nishiuchi, N. Akahira and M. Takao, J. Appl. Phys. 69, 2489 (1991).
[5]. K.A. Rubin, R.W. Barton, M. Chen, V.B. Jipson, and D. Rugar, Appl. Phys. Lett. 50, 1488 (1987).
[6]. J. Gonzalez-Hernandez, B.S. Chao, D. Strand, S.R. Ovshinsky, P. Gasiorowsky and D.A. Pawlik, unpublished.

OVONIC PHASE CHANGE MEMORY MAKING POSSIBLE NEW OPTICAL AND ELECTRICAL DEVICES

Stanford R. Ovshinsky

Energy Conversion Devices, Inc.
1675 W. Maple Road, Troy, MI 48084, USA

Invited Paper – Presented at Numazu City, Shizuoka Pref., Japan – 9th Symposium on Phase Change Recording November 27, 1997 – "Ovonic Phase Change Memory Making Possible New Optical and Electrical Devices"

Thank you for the honor of inviting me to present a paper at this meeting on phase change memories, a subject which is so dear to my heart. It offers me the opportunity to thank you in the audience whom I look upon as colleagues and collaborators for the important and creative role that you have played in making my invention (1-3) into such an important information area. This has led to the building of the whole new field of optical phase change memory that is the basis for the much heralded rewriteable DVD industry. All of this would not have taken place without the vision, insight and entrepreneurial skills of the Japanese companies who depend upon your talented contributions.

I particularly would like to thank the engineers and scientists at Matsushita for the pioneering role that they have played. I wish that my great friend Dr. Hayakawa as well as Dr. Suemitsu were in the audience today. I am very happy that Dr. Ohta is continuing in this tradition. I thank all of you, our other licensees in this field, for your many contributions.

I have not come here, as we say, to teach birds how to sing. You have shown great skill in tuning the layer structures to optimize the thermal environment; have done very important work in increasing the quality of the shape of the recorded spot which has improved the reproducibility of the recording process and therefore the spot size and jitter; have continued the improvements in the servocontrols, the optics, and, of course, will be utilizing shorter wave length lasers as they become available, all of which increases the storage capacity.

In this talk I wish to show where the opportunities for major advances lie. I will therefore first discuss the materials science that will permit great increases of information density, speed and lifetime, for it is in the materials that the answers are to be found.

Phase change memories are based on thin-film alloys typically incorporating one or more elements from Column VI of the Periodic Table. These chalcogenide materials have been designed to exist in two or more distinct atomic structural states. An energy barrier separates the structural states thereby providing the temporal stability required by any memory device.

The energy necessary to allow the memory materials to change their atomic configuration can be supplied in various ways, including exposure to a laser beam or application of a current pulse. Laser exposure is used for recording, erasing, and rewriting in the case of the optical memory. If the laser energy applied exceeds a threshold value, the memory material will be excited to a state of high atomic mobility. It becomes possible for the chemical bonding to rapidly rearrange by slight movement of the individual atoms, especially in phase-congruent materials.

In lone-pair materials (alloys containing chalcogen atoms such as selenium, tellurium, or sulfur) the Group VI atoms use two of their p electrons to be divalently bonded, and the atomic rearrangement that takes place during the phase change process may occur by simply shifting the other two non-bonding or weakly bonding "lone-pair" p electrons to make new connections. The flexibility allowed by divalency and the utilization of excited lone pairs to affect both conformational and configurational changes in materials, as we have explained, are of critical importance in the phase change mechanism. This permits structural changes such as altering the shape of molecules due to the strong repulsive forces that lone pairs exert. Lone-pair electrons provide a spectrum of interactions which include non-bonding and various bonding configurations including the dative bond (4).

Tellurium-based materials can be made with fewer and weaker crosslinks and therefore be susceptible to rapid crystallization/amorphization.[1] Selenium-based materials with their stronger bonds, for example, to arsenic, can be made to resist crystallization/amorphization and reversible elastomeric and rheological changes can be designed in. What is important to note is that we are utilizing atomic engineering to provide a very large number of lone pairs that can be coupled to light or an electric field. The creation of a high density of electron-hole pairs either by electric field or photon absorption is the means by which one can provide an electronic mechanism rather than a thermal one. There is no doubt that the amount of laser light absorbed by the medium causes significant heating. However, one can minimize the thermal effects and emphasize the electronic through the materials science that we have described.

We have already shown through our work in Ovonic electrical memories that the performance of the chalcogenide alloy memory materials goes far beyond what has been accomplished in optical memories. For example, optical memories typically have a cycle life of about 10^6 and they use a record pulse width on the order of 50 or 60 nanoseconds. As we will show here, we have demonstrated that the similar phase change memory materials used in electrical memory devices have a lifetime of over 10^{13} cycles (where we discontinued testing)[2], and can be recorded with pulse widths of less than three nanoseconds. These performance parameters are at the limit of our present test equipment and do not reach the limits of the materials. We think these numbers prove that the cycle lifetime of these materials is essentially unlimited and that the theoretical expectations of the time for structural change to take place can be realized in the proper device configuration. This means that the optical memory lifetime is limited by the substrate material and not by the chalcogenide media. For example, while I am not suggesting it, I am sure that changing to a glass substrate would greatly increase the lifetime of the optical memory. Shifting the emphasis from the more energetic thermal mode to the faster, electronically-driven mode will be beneficial in every way.

Selection of an appropriate composition for the memory alloys and the creation of a high mobility state during laser exposure are the underlying principles in direct-overwrite phase change erasable optical recording media. The coming generations of optical memory disks will have a huge increase in information density by storing many different states of information in one spot, no matter how small the spot.

We have been able to make unique electrical memory devices that we call Ovonic Universal Memories (OUM). By utilizing similar materials and by optical means such as a laser,

---

[1] When phase change is not desired in these tellurium-based materials, one adds more and strongly bonded crosslinks, e.g., instead of germanium only as in the memory materials, one uses silicon, and instead of antimony, one uses arsenic. In this manner, electronic excitation does not lead to phase change but to an Ovonic threshold switch.

[2] We have every reason to believe that the cycle life is as good as any conventional semiconductor device.

we can make a new family of optical phase change memories that have either two states or are multilevel in their ability to be recorded, read, erased and rewritten.

Let us look at some of the electrical memory features first to get a better understanding of its operation and its possibilities. In Fig. 1 we show an OUM test device in which we have electrically programmed 16 different states. Fig. 2 shows a device with a cycle life of over 10^{13} and Fig. 3 shows programming down to 3 nanoseconds.

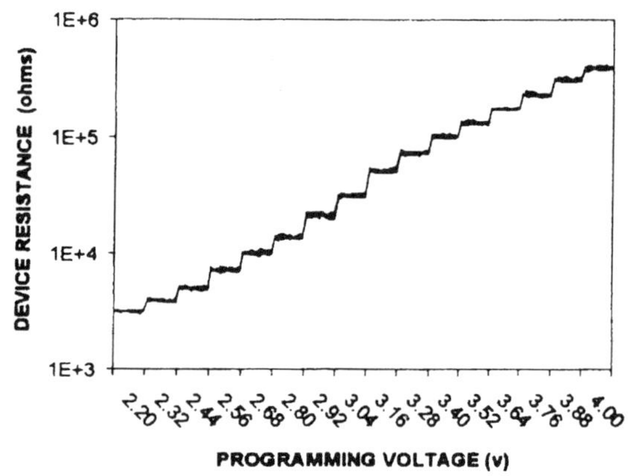

Fig. 1.Multiple-Bit Storage in One Memory Cell (Ten steps per voltage increment repeated ten times).

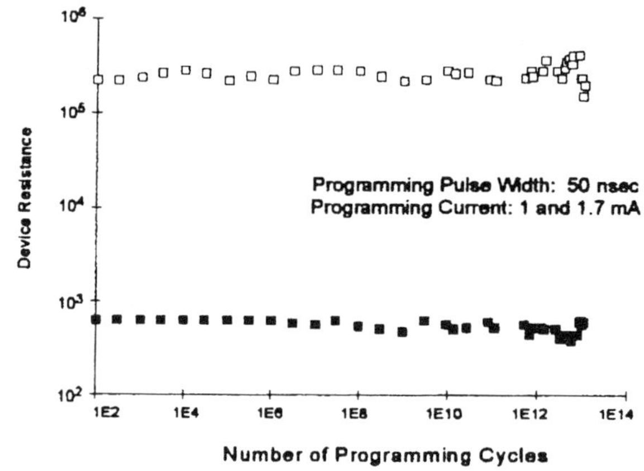

Fig. 2. Device resistance of a test cell subjected to alternating SET and RESET pulses as a function of number of pulses.

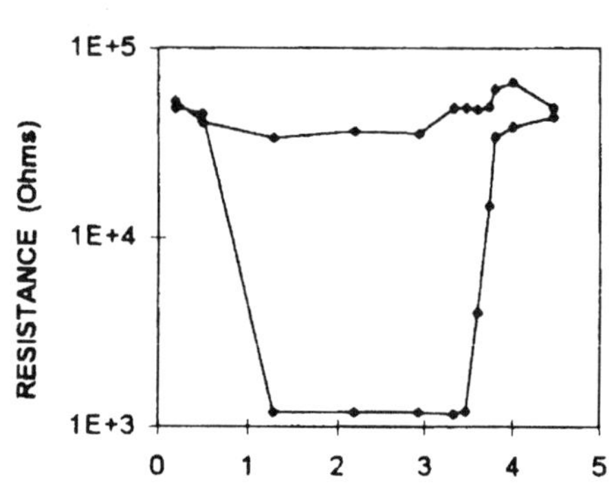

Fig. 3. Programmed resistance vs. current for very short programming pulses.

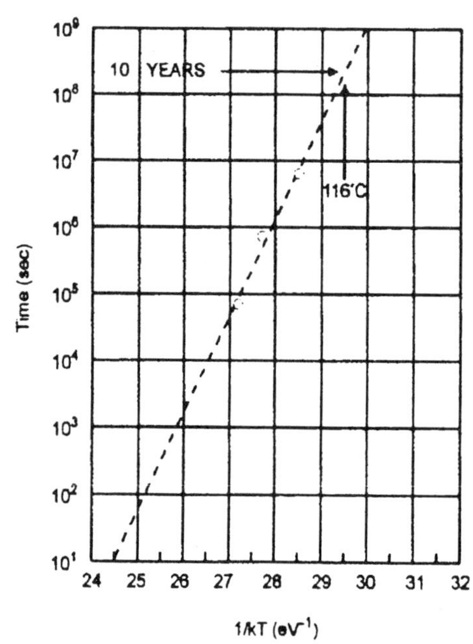

Fig. 4. Arrhenius Plot of the Crystallization Time of a Memory Alloy Film.

■**Current Advantages**

➢non-volatile information storage (unique and proprietary)
➢high speed --- nanosecond programming
➢long cycle life > 10^{13}
➢scaleable with improved lithography
➢low voltage operation < 2 volts
➢areal density > silicon devices
➢simple structure --- two terminal device
➢multi-state --- capable of storing multiple bits per cell
➢compatible with conventional silicon processing

■**Next Generations**

➢Adaptive memory/interconnect
➢3 - Dimensional -- multilayer
➢Integrated logic (OTS and OUM) and memory (OMS)
➢all thin-film computers
➢Optical interface circuits and computation
➢Artificial Intelligence

Fig. 5.

To investigate long term data retention in these memory devices, the thermal stability of programmed information has been investigated by measuring the crystallization time of the memory alloy materials as a function of temperature. These data are shown in Fig. 4 as an Arrhenius plot. Extrapolations indicate that a data storage time greater than ten years can be expected at a temperature of 116° C.

It is apparent from the data that advances in rapidly crystallizing phase-change memory materials and new thin-film device designs have permitted the development of a new type of high-performance, non-volatile semiconductor memory. The combination of rapid programming speed, extremely long cycle-life, simple fabrication process, small device footprint, and the ability to store multiple bits of data per memory cell demonstrated by these devices is unique among non-volatile memory technologies. (See Fig. 5). We call these devices universal because one single plane of our memory can replace DRAM, SRAM and FLASH memory. At the same time it can be used flexibly, for example, as a programmable and rewriteable imbedded memory. Fig. 6 compares our proprietary OUM multistate memory with Intel's multilevel FLASH memory which they announced as the "Holy Grail" which will have a "revolutionary" impact on the FLASH market.

Property	Intel Strataflash	Ovonic Multistate Memory
Density (bits)	2 bit (4 levels)	>4 bit (16 levels)
Write Cycles	10,000	$>10^{13}$ (no limitation)
Write Voltage	5V (with large 12V Charge pump)	<4V
Write Time	10 microseconds times n iterations	0.6 microseconds times n iterations
Erase Time	1 second per block	No erase required
Minimum Erasure Block	must erase 128 Kbytes Before writing	direct overwrite byte-by-byte
Read Access Time	~20 microseconds	~1.0 microseconds

Fig. 6. Comparison between Intel and Ovonic multistate memory

Utilizing the principles for optical memories that we have described for electrical memories will allow us to have multilevel (multistate) memories that can be read by measurements of the intermediate reflectivity levels. It has been shown that complicated analog signals can be successfully decoded into the digital data stream. The number of levels that can be stored will depend on the accuracy of attaining the intermediate reflectivity levels in the material and the accuracy of the drive system in reading the levels back. Proper electronic control mechanisms can assure this.

Electronic transformation of the materials will allow very accurate and predictable intermediate structural changes. Intermediate reflectivities will most likely be provided by continuous changes in the short and mid range atomic order, but other changes, like mixed partial volume fractions of amorphous and crystalline phases on a very small size scale, or crystal orientations, or even layering of the phases within the material, can be designed into the system.

Edge sharpness capabilities of phase change optical memory material are on an atomic scale. The limitation of achieving this resolution in the present devices and systems is governed by heating and heat transfer within the media. Of course, the transition from Pulse Position Modulation to Pulse Width Modulation is being made, but the capabilities of the materials are not being challenged in the current work. Rather, the limitation is the thermal structure. The transition to an electronically governed system from the current thermally governed system will open new regimes of recorded spot accuracy, and the concomitant increases in capacity and transfer rate.

It is interesting to note that there have been only two types of commercial electronic data recording processes prior to phase change. The first was the physical deformation of the surface started by Edison to make his wax drum recordings. These principles are still used in CD-ROM media. The second was magnetic recordings which, from their humble beginnings, have progressed to magneto-optic and magnetic hard drives. The third generation is phase change—the great adventure that we all have embarked on and, as I have tried to show, whose potential for advancement in materials, configuration and, in fact, low cost production technology is exciting.

To put all this in perspective, in this talk I have described ways that phase change optical memory storage can not only go down to smaller spot sizes by virtue of a photonic source which can include electron beams so that one can realistically speak of 100 angstrom spots if one desires, but of greater importance, one can have multi-sites so that there can be many states of information in one spot which can lead to many new kinds of devices including my favorite, the adaptive memory that can be used for much more real intelligence than present artificial intelligence attempts.

In any case, there is a world to win for all of us in the information field. Good luck.

Acknowledgements

It is with great appreciation that I thank David Strand and his Optical Memory Group who have been so important in this project; Wally Czubatyj, head of the Microelectronics Group in collaboration with Pat Klersy, Sergey Kostylev, and Boil Pashmakov for their important work on the electrical memory; Ben Chao and the ECD analytical laboratory; Steve Hudgens for his collaborative input; Iris Ovshinsky, the other half of our lone pair.

References

1. S. R. Ovshinsky, Storing and Retrieving Information, U.S. Patent # 3,530,441.

2. S. R. Ovshinsky, The Ovshinsky Switch, Proc. Fifth Ann. Natl. Conf. on Industrial Research, Chicago, Illinois, Sept. 1969.

3. S. R. Ovshinsky, Historique du Changement de Phase, Memoires Optiques & Systemes, No. 127, Sept. 1994.

4. Numerous papers in D. Adler, B. Schwartz, and M. Silver, eds, Disordered Materials, Science and Technology, Selected Papers by Stanford R. Ovshinsky, (Plenum Press, New York, 1991).

18 Phase-Change Optical Storage Media

Takeo Ohta and Stanford R. Ovshinsky

18.1 Introduction

In modern society, dramatic changes ar taking place in the fields of energy and information. There is a deep connection between them – information is encoded energy and it is energy in the form of light and electric fields that couples to the atomic orbital to bring about the changes in materials that represent information.

Soon after starting in the mid-1950s of an unexpected field of amorphous and disordered materials wherein the constraints of the crystalline lattice were lifted so that new degrees of freedom for the atomic design of synthetic materials were made possible. Ovshinsky announced amorphous semiconductor switching and memory effect "Ovonic" in 1968 [1].

This opened a new materials research area [2, 3] and since 1971 both OMS (Ovonic memory switching) and laser recording phase-change memory have been rapidly developing. Significant progress was made increasing the data rate, reliability and in other related areas [4, 5]. A landmark was reached in 1989 when Ohta announced over million-cycle stable phase-change optical disk technology [6]. Matsushita/Panasonic were the first to succeed in the development of new rewritable phase-change optical disk products with first shipments in 1990 [7]. Now phase-change rewritable optical disk technology has become the main stream of the optical disk world as demonstrated by the new rewritable compact discs (CDs) and digital versatile discs (DVDs), which are based upon it [8].

In the present chapter we describe how amorphous order–disorder phase-change memory became the enabling technology for DVD media, high-density technology developments and the next generation phase-change media.

18.2 Phase-Change Overwrite Optical Disk

18.2.1 Phase-Change Optical Memory Phenomena

Phase-change chalcogenide materials was first reported by Feinleib et al. in 1971 [3]. Fritzsche made a groundbreaking presentation on amorphous semiconductors at the American Physical Society Meeting held in Philadelphia, Pennsylvania (USA) and also in Japan at the conference of Japan Applied Physics (Spring) in 1971 [9]. Ovshinsky's pioneering patents in 1968 [10] and 1972 [11] established the phase-change data storage field.

Ohta proposed sub-oxide thin film phase-change media (TeO_x) which led to write-once optical disk products for video image file storage in 1982 [12, 13]. There are many thin film

Photo-Induced Metastability in Amorphous Semiconductors. Edited by Alexander V. Kolobov sub-oxide materials such as GeO_x, SbO_x, and MoO_x with additives of Sn or In that undergo a structural transformation. All of these materials, when in the as-deposited state, are amorphous. Upon heating, they exhibit a darkening effect [12]. Table 18.1 shows change in the optical constants of the sub-oxide films following heat treatment. For rewritable phase-change optical disc, chalcogenide systems including GeSbTe and SbTe eutectics with various modifiers are applied to achieve reversible direct overwrite cycle performance.

Table 18.1: Optical constants of suboxide films at $\lambda = 830$ nm [12]

Suboxide	As deposited		After heat treatment		Conditions
	$\alpha_1(\text{cm}^{-1})$	n_1	$\alpha_2(\text{cm}^{-1})$	n_2	
SbO_x	2.5×10^4	1.8	6.1×10^4	1.9	250 °C, 5 min
TeO_x	0.8×10^5	3.1	1.0×10^5	3.6	250 °C, 5 min
MoO_x	5.6×10^3	1.8	1.1×10^4	2.1	250 °C, 5 min
GeO_x(Te)	4.9×10^4	2.5	1.8×10^5	2.8	250 °C, 5 min

18.2.2 The Phase-Change Memory Mechanism

Figure 18.1 shows the model of phase-change data storage. The enthalpy of the material when in the amorphous phase is higher than when the material is in the crystalline phase. When the structure changes from amorphous to crystalline, the optical absorption edge of the material shifts to a longer wavelength. This is accompanied by a change in the complex refractive index $N = n + ik$ (n is the refractive index and k is extinction coefficient). This causes the reflectivity of the film to change, which is the basis of the storage and retrieval means. The information can be detected by changes in the reflectivity just like CD disk playback. When crystalline film at "a" is exposed to a high-power laser spot, the temperature of the irradiated region increases to over the melting temperature T_m and the state changes "b" to "c" and "d", as show in Figure 18.1. After the laser irradiation, the temperature decreases rapidly through a super cooled state between "c" and "e" to room temperature, "f". When the cooling rate is above the critical cooling rate (3.4 K/ns) [14], the portion of the film at "f" remains in the amorphous phase. When a lower power level laser spot irradiates the amorphous region "f", the temperature increases above "e", the glass transition temperature (T_g), and then the amorphous structure is transformed to the crystalline structure and the mark is erased. Above T_g, when softening – not melting – takes place, a higher mobility state initially induced by lone pair electron absorption provides the ability for atoms to move slightly at very high speed to establish the new configurations leading to either crystallization or returning to the amorphous state, depending on conditions.

18.2.3 Phase-Change Overwriting Method

The requirement for overwriting a phase-change material is a high enough crystallization so that in the time the laser spot moves the distance of its diameter, the irradiated region can be

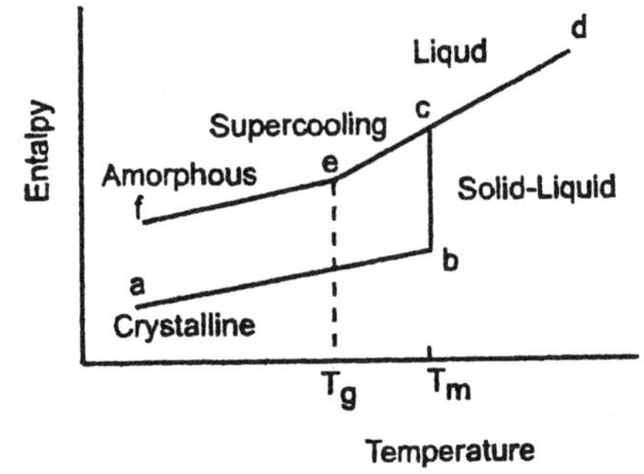

Figure 18.1: Model of the phase-change memory (T_g, glass transition temperature; T_m, melting temperature).

changed to a fully crystalline state and previously recorded information is erased [15]. Terao et al. announced the overwrite data on In–Se phase-change material in 1986 [16]. The critical cooling rate depends on the constituent atomic elements and their concentrations. Our experiments [14] have allowed us to estimate that the critical cooling rate of the GeTe–Sb_2Te_3–Sb system is 3.4 K/ns, which is sufficient for application as direct overwrite media at linear speed above 8 m/s.

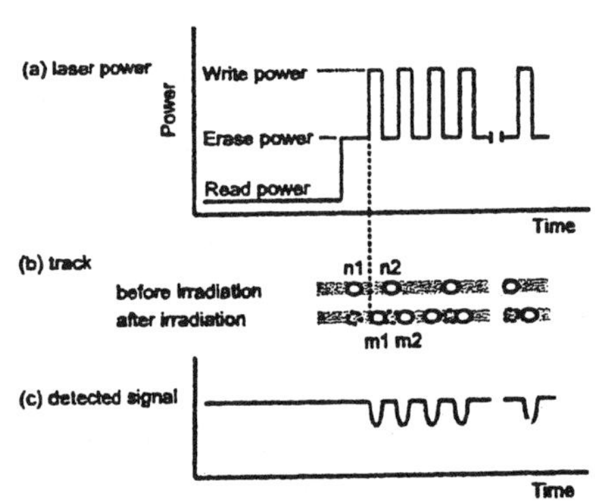

Figure 18.2: Direct overwriting method by optical mean only on phase-change optical disk [15, 16].

Figure 18.2 illustrates the operating principle of phase-change overwriting: Figure 18.2(a) shows a laser power modulation waveform, Fig. 18.2(b) shows written marks on a track before and after overwriting, and Fig. 18.2(c) shows a read-out signal after overwriting.

18.3 Phase-Change Materials

18.3.1 Bonding Features of Chalcogenide Phase-Change Materials

Chalcogenide compounds are notable for their easy transition from the crystalline state to the amorphous states. Those compounds include the elements Te, Se and/or S, which are in the group VI of the Periodic Table and show the chain structure in the amorphous state [17]. Additive of As in Te works to form the ring network structure as $As_{25}Te_{75}$ [17, 18]. The chain-like bonding in Te and Se is important in facilitating the structural changes.

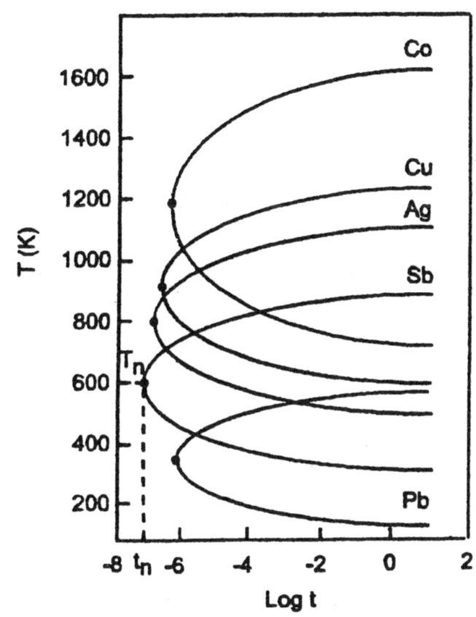

Figure 18.3: T–T–T curve (time–temperature–transformation curve) for Co, Cu, Ag, Sb and Pb elements corresponding to a volume fraction crystallization of 1×10^6 [20].

Okuda et al. reported [19] the mechanism of amorphous-crystalline phase change in terms of Avrami equation and investigated its dynamic transition using T–T–T (time-temperature-transition) curves [20]. The T–T–T curves of Co, Cu, Ag, Sb, Pb are shown in Figure 18.3. The nose temperature (T_n) and the time (t_n) of the temperature coming to T_n, the melting temperature (T_m), and the critical crystallization time (volume fraction of crystallization of 10^{-6}) are given in Table 18.2. The critical cooling rates of crystallization of these elements are given by the equation $R_c = (T_m - T_n)/t_n$. The log R_c values of these elements decrease in the order Sb > Ag > Cu > Co > Pb. These elements will control the crystallization rate of the phase-change materials.

18.3.2 Phase-Change Optical Disk Materials for Optical Disk Memory

Functional phase-change rewritable media can be made based on the $Te_{85}Ge_{15}$ eutectic composition [3, 4, 21]. The melting temperature shows a minimum (375 °C) at the eutectic com-

Table 18.2: Estimated values of T_g, T_g/T_m and calculated values of T_n, t_n, log R_c of elements [19]

Element	T_m(K)	T_g(K)	T_g/T_m	T_n(K)	t_n(s)	Log R_c (K/s)
Sb	895	182	0.20	600	7.41×10^{-8}	9.61
Ag	1235	250	0.20	800	2.98×10^{-7}	9.16
Cu	1357	298	0.22	900	3.38×10^{-7}	9.13
Co	1768	445	0.25	1200	6.54×10^{-7}	8.94
Pb	601	152	0.25	375	7.07×10^{-7}	8.50
Te	722	285	0.39	500	5.75×10^{-6}	7.54
Ge	1212	750	0.62	945	8.96×10^{-4}	5.47

position, and it is expected that the viscosity will increase at this composition. Addition of small amounts of Sb and S increases both ease of formation and stability of the amorphous phase as e.g. in the material of composition $Ge_{15}Te_{81}Sb_2S_2$ [3].

Phase-change materials for overwriting by one laser spot need to have high-speed crystallizing characteristics. Three different families of phase-change material systems that are suitable for this application are In–Sb–Te [22], Ge–Te–Sb [6,23-25], Ag–In–Sb–Te [26]. These families have different crystallization processes. The two major kinds of phase change optical disk materials are nucleation dominant materials typified by Ge–Sb–Te, and edge growth dominant materials typified by the $Sb_{69}Te_{31}$ eutectic material system.

In–Sb–Te

Crystalline growth dominates over nucleation in In–Sb–Te. When the erase laser power irradiates written marks to a point where the film just melts, after irradiation recrystallization will proceed from the mark edge and the mark is erased. In the In–Sb–Te alloy, careful choice of overwrite laser power levels allows one to achieve very high erase ratios.

GeTe–Sb_2Te_3–Sb (GST)

In GST media, the crystallization process is governed by nucleation rather than crystal growth. Recrystallized amorphous marks in GeTe–Sb_2–Te_3–Sb films have a large number of small crystallites in their grain structure. This suggests a two-step erase process. In the first step, a large number of crystalline nuclei are formed and then crystal grain growth occurs leading to erasure of the amorphous mark. The erase process proceeds in the solid state and the power tolerance becomes comfortably wide. However, the erase ratio is lower than in In–Sb–Te films.

Ohta and Yamada's groups developed the GeTe–Sb_2Te_3–Sb system and achieved a crystallization speed of 30–100 ns with small crystalline grain size [6, 23]. Figure 18.4 shows the crystallization temperature of this system. The latent heat of transformation from crystalline to liquid (16.3 kcal/kg) and amorphous to liquid (8.5 kcal/kg) was obtained from these measurements [27].

Ag–In–Sb–Te (AIST)

The AIST family is generally characterized as having growth dominated crystallization and high erase ratio [26]. The origin of the large erase ratio is ascribed to the low thermal conductivity components present in the thin film structure, which is composed of a mixture of $AgSbTe_2$ phase-change component in an amorphous In–Sb matrix. The overwrite cycle performance of Ag–In–Sb–Te system is around 10 000 or more, and the limitation of this overwrite cycle is still being investigated.

Sb-rich, Sb–Te (FGM)

At the Optical Data Storage Conference in 2000, eutectic composition systems such as $Sb_{69}Te_{31}$ were proposed [28, 29] as fast growth materials (FGM). These crystal growth dominant compositions might be superior for high-density, high data rate recording, phase-change optical disks. As the density of the phase-change optical disk increases, the recorded marks become smaller (<200 nm), and growth dominant material can lead to shorter crystallization times. The eutectic composition of $Sb_{69}Te_{31}$ is indicated on the line from Sb_2Te_3 to Sb in Fig. 18.4. The composition at that point is 0.18 (Sb_2Te_3) and 0.82 Sb. The reason why this eutectic composition shows rapid crystallization characteristics is being investigated.

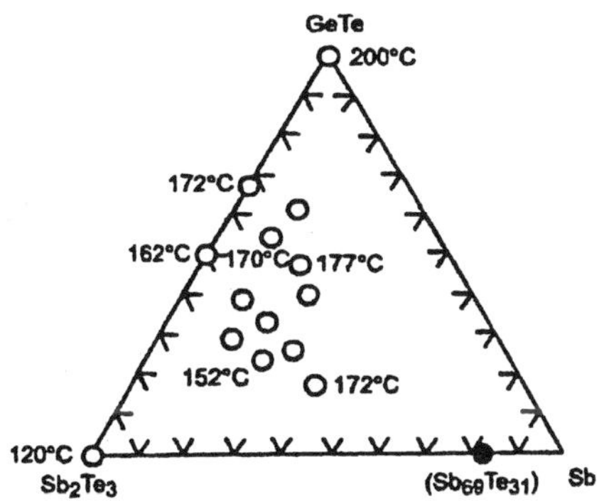

Figure 18.4: Crystallization temperature of quasi ternary GeTe–Sb_2Te_3–Sb alloy system [6]. Heating: 100 °C /min

GaSb system

At the PCOS2002 conference, Okuda suggested that explosive material elements such as Ga, Ge, and Sb will assist the crystallization process and Tashiro announced high-speed crystallization material in the GaSb system [30, 31].

18.4 Breakthrough Technologies of the Phase-Change Optical Disk Media

18.4.1 Basic Layer Structure

In the early phases of its development, the most important subject of phase-change optical disks was performance degradation over write–erase cycles. A basic phase-change optical disk is a four-layer structure. The four layers are a bottom dielectric layer (155 nm), an active layer (24 nm), an upper dielectric layer (45 nm), and a reflection layer (100 nm).

The total layer structure design is based on optical and thermal considerations. The first priority is the amplitude of the optical signal. This corresponds to the reflection difference between that of the amorphous state (R_a) and of the crystalline state (R_c). These reflectivities can be calculated using multi-layer optical methods. Next, the disk sensitivity is optimized by simultaneously considering optical absorption for the amorphous state and for the crystalline state, and the thermal structure to introduce a rapid cooling structure.

In early phase-change media, the overwrite cycle life characteristics on a spinning disk were rather small compared with the static cycle test. This was resolved by selecting materials that are thermally stable and whose thermal expansion coefficients are rather small. The optical, thermal, and mechanical characteristics of the layers used in phase-change optical disk media are summarized in Table 18.3. One difficulty in modeling the thermal characteristics of recordable optical media is use of accurate values for the thermal constants. The thermal coefficients of most materials are different when in bulk and thin-film forms. Peng and Mansuripur addressed this issue by developing a new method for measurement of the thermal coefficients of the layers in phase-change optical disk media. This method can determine the coefficients of the various layers by reflectivity measurements of heated and laser-exposed samples [32]. The principle of this elegant method is use of the basic thermal diffusion equation along with known transition temperatures for melting and the crystallization of phase-change thin films [32, 33].

The newly developed materials of GeTe–Sb_2Te_3–Sb used in the active layer and ZnS–SiO_2 used in the protective layer have resolved the issue of variations of signal amplitude and noise level with cycle life. The grain size of ZnS–SiO_2 is very small at around 2 nm [34]. The ZnS–SiO_2 dielectric layer is thermally stable and does not show grain growth even after annealing at 700 °C for 5 min.

18.4.2 Million Overwrite Cycle Phase Change Optical Disk [6]

Another degradation mechanism which has to be counteracted to extend rewrite cycle life is space deformation of the disk layers on a sub-nanometer level, which works as a motive force for sub-nanometer displacement of the active layer components. The deformation is driven by thermal expansion of the layers during the recording process. The deformation is generally asymmetric along the laser scanning direction, greater toward the forward edge and less toward the backward edge. Inoue et al. calculated the thermal deformation of the phase-change optical disk layers [35] during the recording process. This phenomenon can be reduced by adding a layer that has small thermal expansion coefficient between the phase-change layer and the upper dielectric layer. The thermal expansion coefficient of SiO_2 (5.5×10^{-7}) and of

Table 18.3: Optical, mechanical and thermal properties of materials

Material	Reflective index $\lambda = 650$ nm	Density (kg/m^3)	Young's modulus (N/m^2)	Poisson's ratio	Specific heat (J/(kg K))	Thermal conductivity W/(m K)	Coefficient of linear expansion
GeTe–Sb_2Te_3–Sb (amorphous)	4.21 + 1.89i	6150	5.49×10^{10}	0.33	0.209×10^3	0.581	1.1×10^{-5}
GeTe–Sb_2Te_3–Sb (crystal) 2:1:0.5 (mol ratio)	4.56 + 4.23i	3650	5.49×10^{10}	0.33	0.209×10^3	0.581	1.1×10^{-5}
ZnS–SiO_2 4:1 (mol ratio)	2.0	2202	7.81×10^{10}	0.2	0.563×10^3	0.657	7.4×10^{-6}
SiO_2	1.46	2750	7.81×10^{10}	0.2	0.753×10^3	1.313	5.5×10^{-7}
Al alloy	2.2 + 7.5i	1200	7.03×10^{10}	0.345	0.892×10^3	0.215×10^3	2.2×10^{-5}
Polycarbonate	1.58		2.26×10^9	0.3	0.126×10^2	0.223	7.0×10^{-5}

$ZnS–SiO_2$ (6.1×10^{-6}) serves this purpose very well. Use of an added SiO_2 layer increases the dynamic overwrite cycle characteristics to over 2×10^6 cycles [6,35]. The new five-layer structure, which has the additional SiO_2 layer, shows more than 1 000 000 overwrite cycles. Figure 18.5 shows more than 2 million cycle characteristics of the phase-change optical disk with the additional SiO_2 layer [6,35].

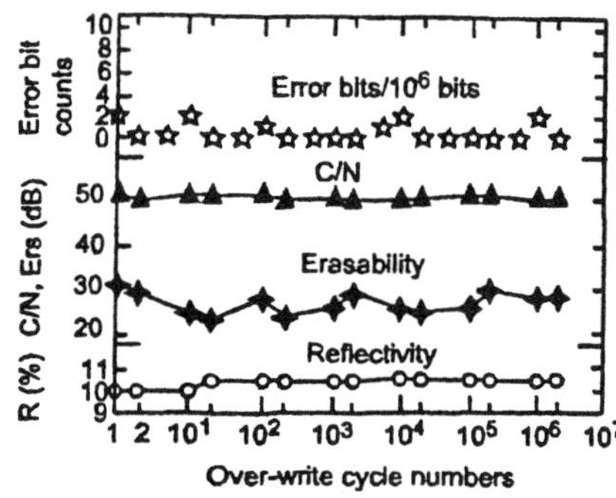

Figure 18.5: Two million overwrite cycle test results of phase-change optical disk with additional SiO_2 protection layer [6].

The sensitivity of the phase-change optical disk was rather low in the first commercial disk products, requiring 21 mW of laser power on the disk. We found that the sensitivity and the cycle characteristics are basically not a trade-off relation. It has been shown that the overwrite cycle characteristics and record sensitivity can be simultaneously optimized, and the high sensitivity PD phase-change optical disk was developed that has more than 500 000 overwrite capability and 10 mW recording sensitivity optimized through layer thickness [36].

18.5 Thin Substrate Technology of Phase Change Optical Disk Promotes DVD

Increased storage density can be achieved in optical disks by using more powerful, larger numerical aperture (NA) lens to form a smaller laser spot. However, the sensitivity to tilt angle in the disk increases dramatically with increasing numerical aperture. Satoh and Ohta were the first to demonstrate that a thin disk substrate, which is effective for resolving the disk tilt sensitivity at high numerical aperture, could be successfully made and used in an optical disk during high-density recording [37]. Their system combined emerging technologies such as a red laser diode, a large numerical aperture (0.6) lens, and thin disk substrate (0.6 mm) [37,38]. They proposed a high-density 90 mm diameter phase-change optical disk for ISO standardization in 1995 [39]. Figure 18.6 shows the crosstalk characteristics of optical disks made using 1.2 mm and 0.6 mm substrates with a high numerical aperture lens (NA = 0.6). As can be seen in the figure, the thinner substrate shows lower crosstalk than the thick substrate disk when the tilt angle increases. These technologies were adopted into DVD in 1995.

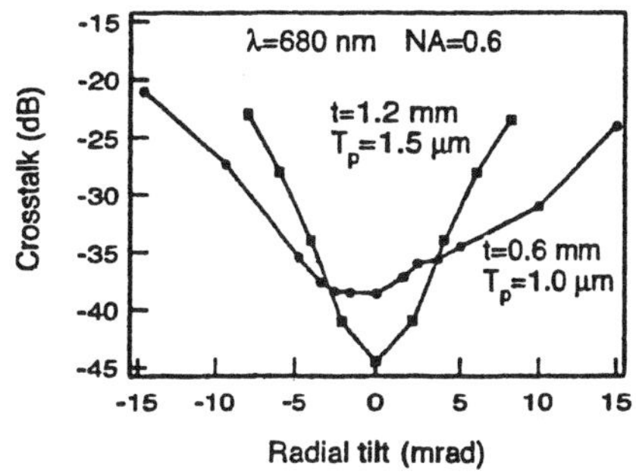

Figure 18.6: Comparison of the tilt angle dependence of crosstalk for substrate thickness of $t = 1.2$ mm and thin substrate of $t = 0.6$ mm [37]. Lens numerical aperture: NA = 0.6. Laser wavelength: 680 nm.

A simple overwrite disk function and compatibility of the phase-change with ROM disks is also featured in the DVD specification of rewritable DVD-RAMs, DVD-RWs and +RWs.

18.6 High-Density Recording Technologies for Phase Change Optical Disks

18.6.1 Short-Wavelength Blue Laser and High Numerical Aperture Lens Recording

Phase-change optical storage disks have high signal output and record sensitivity over a wide spectrum of light from infrared of 830 nm to blue–violet of 405 nm. When blue laser light of wavelength 425 nm and lens numerical aperture of 0.6 [40] are applied, the recording density is estimated to be 9 Gbit/in^2. The next increase in recording density will result from a combination of an even larger numerical aperture lens and a short wavelength (405 nm) blue laser. The newly proposed blu-ray disc technology incorporates these components with a 0.85 numerical aperture lens and a new disk structure incorporating a thin (0.1 mm) cover layer, with recording and reading of the data done through the thin, cover-layer side. This new system has a recording capacity of 22.4 GB/side on a 120 mm diameter disk [41].

18.6.2 Dual-Layer Recording

Another approach to increase the storage capacity is dual-layer recording, creating volumetric rather than two-dimensional surface recording. Dual-layer DVD optical disks that read out from one side have been commercialized for 8.5 GB disks for longer cinema titles. Nagata et al. announced a phase-change rewritable dual layer optical disk also having a capacity of 8.5 GB, and a density of 6.4 Gbit/in^2, nearly doubling the previous density [42]. The first layer in a dual-layer, phase-change optical disk should have transmission of about 50% in the crystalline state for recording in the second layer through the first layer. Using a blue laser with a 0.6 NA lens gives a capacity of 27 GB in a dual-layer disk [43].

Blue lasers and a high (0.85) NA lens can be used together in dual-layer, phase-change optical disks and this combination increases the density even more. The recording capacity of the dual-layer structure is between 40 GB and 50 GB/side, and the carrier-to-noise ratios (CNRs) of the layers are 51 dB and 49 dB, respectively [44,45].

18.6.3 Multi-Level Recording

A third approach to increased storage density is multi-level (ML) recording on phase-change optical disks. Ovshinsky demonstrated ML recording of phase-change electric switching memory in 1997, showing 16 switching levels [46]. ML recording, for example, 4-ary recording gives $\log_2 4 = 2$ bits per mark, doubling the recording density of conventional binary recording.

ML recording requires a high C/N ratio for accurate detection of the M-ary signal. At ODS2000, O'Neil and Wong demonstrated eight-level phase-change recording technology and announced a CD-RW system having a capacity of 2 GB [47]. Horie et al. proposed a new recording strategy for multi-level recording on a phase-change optical disk, the phase-change film material being FGM [48]. Flynn et al. proposed In–(SbTe) as a new material for ML recording media and found that slow melt crystallization characteristics are required for high-quality signal recording [49].

Ohta et al. proposed to divide this large reflectivity signal into multi-level signals on a phase-change optical disk using the Mark Radial Width Modulation (MRWM) method [50].

Although most multi-level recording strategies use a fixed mark length, Honguh and Murakami proposed a run-length-limited (RLL) code for multi-level recording adding spacing to the minimum and maximum run-length constraints. This code shows a magnification factor of 1.76 for a (1,7) code applied to multi-level recording [51].

Miyagawa and Mansuripur showed a different multi-level recording method using a phase-change optical medium. They define 256 distinct patterns that are each recorded in a data block length of 1.6 μm. Each block corresponds to 8 bits of information [52]. Tsu and Strand demonstrated that the short pulse write strategy improves the mark shape and mark edge quality [53].

18.6.4 Near-Field Recording and Super-RENS Recording

The storage capacity of optical media can be increased by using more powerful high NA lenses. Terris et al. proposed to use a solid immersion lens (SIL) [54] which has a numerical aperture larger than 1.0. Kishima et al. developed a near-field optical head using a super-hemispherical solid immersion lens (Super-SIL) to increase recording density of phase-change optical disks [55]. They used a spin on glass (SOG) disk medium designed to optimize the coupling of the evanescent field present in near-field systems. The laser beam is incident on the coated side of the disk, which is coated with a multi-layer stack comprised of SiN (25 nm) / SiO_2 (SOG) (82 nm groove and 57 nm land)/SiN (20 nm)/ZnS–SiO_2 (85 nm)/$Ge_2Sb_2Te_5$ (12 nm)/ZnS–SiO_2 (20 nm)/Al alloy (100 nm)/glass substrate. In order to couple efficiently the evanescent field to the medium, the air gap between the disk surface and the optical head is only 50 nm. A recording density of 50.4 Gbit/in^2 is achieved using an extremely small 160 nm track pitch and an 80 nm minimum bit length.

In 1999, Tominaga announced a new high-density recording method called Super-RENS (Super-REsolution Near-field Structure) that can record and read marks beyond the diffraction limit [56]. Super-RENS is unique for optical near-field recording whose near-field aperture is generated within the optical disk itself with an added Sb mask layer. During the recording process, the laser spot heats up the Sb layer, which shows non-linear optical characteristics at higher temperatures. When the laser beam heats the Sb layer in the center of the Gaussian spot, a small aperture opens and functions to create an evanescent field so the laser can record marks smaller than the diffraction limit. Recently Super-RENS media using AgOx as the mask layer were reported showing an improved CRN of 32 dB using reflection readout mode at a mark size of 200 nm [57]. For comparison, in a conventional 4.7 GB DVD, the minimum mark size is 400 nm. Super-RENS near-field recording shows potential for four times higher density recording.

18.6.5 High Data Rate, High-Density Recording on Phase-Change Disk

The maximum data rate of a phase-change optical disk is related to the crystallization speed of the active layer. When overwriting data on the track, the previously recorded data are erased by the laser spot within the time the spot traverses over each location on the disk. Rewritable DVD materials such as GST and AIST have a crystallization time of 30–50 ns and, therefore, have the characteristic of a more than 10 Mbps recording data rate. A second limit comes from the cooling speed of the disk structure. If the cooling speed is low, the laser spot cannot form amorphous marks because the heat does not diffuse away fast enough to quench the amorphous structure. At ISOM2001, Kato et al. announced a 140 Mbit/s recording rate [58] in the disk structure with a rapid cooling layer of Al_2O_3, and using an AIST recording material modified with germanium.

18.6.6 Combination Technology

Short-wavelength blue lasers increase the density by the ratio of the wavelength $(650/405)^2$ to produce an increase of about 2.6-fold.

Introducing a magnification factor of multi-level recording of $M = 4(\times 1.76)$, the recording density will further increase. Using a combination of multi-level recording with a dual layer disk structure and blue lasers, the density reaches 30G bit/in^2 with a 0.6 lens NA.

The second strategy to increase storage density is to use a lens of 0.85 NA and a 0.1 mm thin overcoat layer. The recording density can be double $[(0.85/0.6)^2]$ that of a conventional DVD by using the larger numerical aperture lens. The recording density is predicted to be 60 Gbit/in^2 and the capacity will rise to 83 GB on a single side of a 120 mm disk. Figure 18.7 shows the area recording density growth of phase-change optical disks.

The Super-RENS effect can be combined with the above technologies, resulting in the potential for increasing the density by a factor of four to achieve 240 Gbit/in^2 in the future.

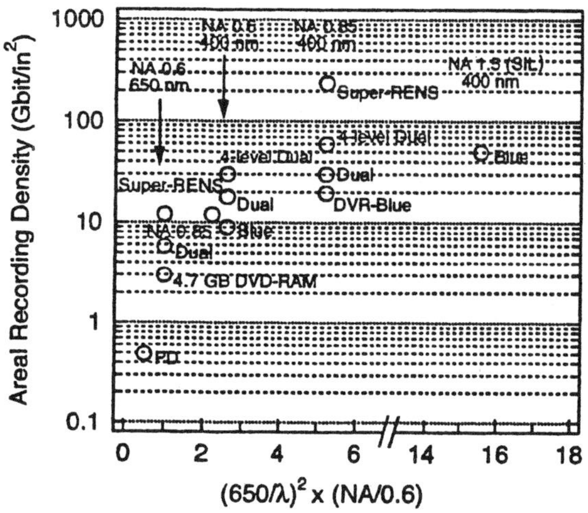

Figure 18.7: Area recording density growth of phase-change optical disk.
Version 1 (λ = 830 nm, NA = 0.5). PD (λ = 780 nm, NA = 0.5). 4.7 GB DVD-RAM (λ = 650 nm, NA = 0.6). 50 GB dual-layer disk, blue wavelength, NA = 0.85. multi-level recording. Super-RENS (640 nm, NA = 0.6).

18.7 Future Directions of the Phase-Change Storage Media

18.7.1 Ultra Short Pulse (Femtosecond) Laser Recording

Recent advances in lasers having pulse widths in the pico- and femtosecond regime have allowed new studies of materials. High-speed fiber communication, high-resolution laser processing and ultra short time resolution measurements are important applications. Femtosecond laser pulses are used to observe ultra high-speed chemical reactions and bio-molecular dynamics. Photo-induced refractive index changes in silicate glass induced by multi-photon absorption process [59] can be observed using 120 fs laser pulses at a laser wavelength of 800 nm. Laser processing on metals and ceramics can be done almost without laser ablation using femtosecond laser irradiation.

Although phase changes are initiated by direct electronic excitation through optical absorption, the thermal energy created when the electrons drop to their ground state must be controlled in order to provide for optimal recorded mark formation. The heating time of the phase-change material as a result of conventional laser spot recording on a disk is rather long and therefore there is substantial heat diffusion during the recording process. Heat diffusion accompanying conventional laser recording can limit the performance of future high-density and high data rate optical disks. Space that is heated by thermal diffusion must be left between the recorded marks. The variation of the position of the mark and its edges during writing are the major limiters jitter characteristics and hence conventional recording density.

Ohta et al. were the first to examine the response of phase-change thin film media to femtosecond laser pulses to investigate capabilities for ultra short pulse laser recording [60].

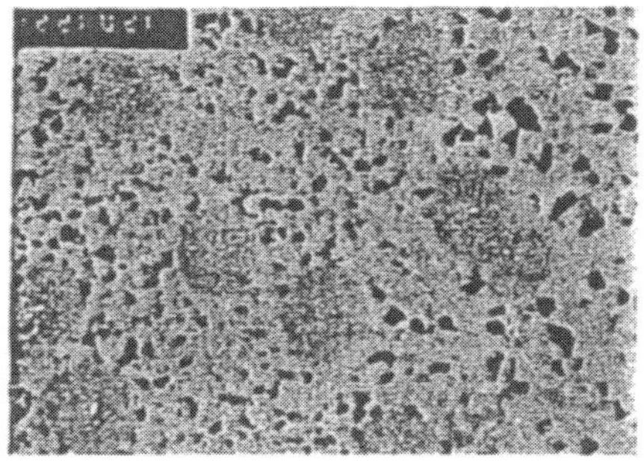

Figure 18.8: TEM observation of amorphous marks formed by femtosecond laser pulse on the phase-change optical disk media [60].

The sample disk had a structure comprised of a polycarbonate substrate/155 nm $ZnS–SiO_2$ layer/24 nm GeSbTe layer/ 45 nm $ZnS–SiO_2$ layer. The disk did not have the conventional reflection layer. The femtosecond laser beam was incident on the coated side. Figure 18.8 shows the transmission electron microscope (TEM) observation of the mark formed by the 120 femtosecond laser exposure. Shaw et al. predicted femtosecond switching in Ovonic memory devices [61], and this experiment shows that order-to-disorder phase-change can be completed using a single 120 femtosecond laser pulse. The femtosecond response experiment of the phase-change materials and the observations of the process provide mean to more fully expand the fundamental understanding of the phase-change phenomena. This first experiment shows the amorphous mark recording data rate of phase-change optical disk is expected to be more than Tbit/s.

18.8 Conclusion

An innovative proposal of Ovshinsky on amorphous phase-change memory in 1968 was attractive for the material science and research and technology development field. Due to engineering and analytical efforts the advantage of this technology became obvious. A million-overwrite cycle performance of the phase-change optical disk was a major breakthrough in the technology which realized the first product. The next major step in high-density rewritable phase-change optical disk development was the use of a thin disk substrate which led to a appearance of DVDs and rewritable DVDs in 1996. These consumer-use rewritable phase-change optical disk became the mainstream of the optical disk world. Phase-change technology can be expected to advance significantly in the coming decades. Optical disks with dramatically increased density (240 Gbit/in^2) and data-processing rate (Tbit/s) are expected in the near future.

Acknowledgments

The authors are grateful to all members of ECD (Energy Conversion Devices Inc.), in particular to Dave Strand, David Tsu and Kelly D. Flynn for their innovative work on amorphous materials. Also important contributors to optical disk development are Noboru Yamada and Isao Satoh at Matsushita. The authors appreciate important contributions to phase-change optical disk storage made by Koichiro Kishima (Sony), Michikazu Horie (Mitsubishi Chemical), Tatsuya Kato (TDK), Junji Tominaga (AIST), Naoyasu Miyagawa , Kenji Narumi (Matsushita), whose excellent results were discussed in this chapter.

Our thanks are extended to Masud Mansuripur of the Optical Science Center of the University of Arizona for profitable discussions and collaborative research on optical storage. The authors would like to express their deep gratitude to Iris Ovshinsky for her immense help and kind support of all our work on phase-change amorphous materials.

References

[1] S.R. Ovshinsky, Phys. Rev. Lett. **21**, 1450 (1968).

[2] E.J. Evans, J.H. Helbers, and S.R. Ovshinsky. J. Non-Cryst. Solids **2**, 334 (1970).

[3] J. Feinleib, J. de Neufville, S.C. Moss, and S.R. Ovshinsky, Appl. Phys. Lett. **18**, 254 (1971).

[4] S.R. Ovshinsky and H. Fritzche, Metall. Trans. **2**, 641 (1971).

[5] S.R. Ovshinsky and P.H. Klose, J. Non-Cryst. Solids, **8-10**, 892 (1972).

[6] T. Ohta, M. Uchida, K. Yoshioka, K. Inoue, T. Akiyama, S. Furukawa, K. Kotera, S. Nakamura, Proc. SPIE **1078**, 27 (1989).

[7] Panasonic Catalog, Mutlifunction rewitable optical disk, LF-7110 (1990).

[8] Panasonic Catalog, Rewritable optical disk DVD-RAM 4.7GB, LF-D100J, DY-HB47(2000).

[9] M.H. Cohen, H. Fritzsche, and S.R. Ovshinsky, Phys. Rev. Lett. **22**, 1065 (1969).

[10] S.R. Ovshinsky, Symmetrical current controlling device, USP 3 271 591.

[11] S.R. Ovshinsky, Method and apparatus storing and retrieving information, USP 3 530 441.

[12] T. Ohta, M. Takenaga, N. Akahira, and T. Yamashita, J. Appl. Phys. **53**, 8497 (1982).

[13] T. Yoshida, T. Ohta, and S. Ohara, Proc. SPIE **329**, 40 (1982).

[14] T. Ohta, K. Inoue, S. Furukawa, T. Akiyama, M. Uchida, and S. Nakamura, Electro. Commun. Tech. Res. Meet. Rep. CPM89-84 (1989) p. 41.

[15] T. Ohta, T. Nakamura, N. Akahira, and T. Yamasita, Method of overwrite optical disk media, Japanese patent No. 1668522.

[16] M. Terao, N. Nishida, Y. Miyauchi, S. Horigome, T. Kaku, N. Ohta, Proc. SPIE **695**, 105 (1986).

[17] R. Grigorovici, Amorphous and liquid semiconductors, J. Tauc (Ed.), Plenum Press, London, 1974, chapter 2.

[18] J. Cornet and D. Rossier, J. Non-Cryst. Solids **12**, 85 (1973).

[19] M. Okuda, H. Naito, T. Matsushita, JJAP Series 6, Proc. Int. Symp. on Optical Memory, 1991, p. 73.

[20] H.A. Davies, Phys. Chem. Glasses **17**, 159 (1976).

[21] W. Klemm and G. Frischmuth, Z. Anorg. Chem. **218**, 249 (1934).

[22] Y. Maeda, H. Andoh, I. Ikuta, M. Nagai, Y. Katoh, H. Minemura, N. Tsuboi, and Y. Satoh, Appl. Phys. Lett. 54, 893 (1989).

[23] N. Yamada, E. Ohno, N. Akahira, K. Nishiuchi, K. Nagata and M. Takao, Proc. Int. Symp. on Optical Memory, 1987, p. 61.

[24] M. Suzuki, I. Doi, K. Nishimura, I. Morimoto, and K. Mori, Proc. Optical Memory Symposium'88, 1988, p. 41.

[25] N. Kh. Abrikosov and G.T. Danilova-Dobroyakova, Inorg. Mater. **1**, 187 (1965).

[26] H. Iwasaki, Proc. SPIE **3109**, 12 (1997).

[27] N. Yamada, E. Ohno, K. Nishiuchi, and N. Akahira, J. Appl. Phys. **69**, 2849 (1991).

[28] G.F. Zhou, H.J. Borg, J.C.N. Rijpers, M.H.R. Lankhorst, and J.J.L. Horikx, Proc. SPIE **4090**, 108 (2000).

[29] M. Horie, T. Ohno, N . Nobukuni, K. Kiyono, T. Hashizume, and M. Mizuno, Proc. SPIE **4342**, 77 (2001).

[30] M. Okuda, H. Inaba, and S. Usuda, Proc. PCOS2002, 2002, p. 1.

[31] H. Tashiro, M. Harigaya, K. Ito, M. Shinkai, K. Tani, N. Yiwata, A. Watada, K. Makita, and K. Kato, Proc. PCOS2002, 2002, p. 11.

[32] C. Peng and M. Mansuripur, Appl. Opt. **39**, 2347 (2000).

[33] C. Peng and M. Mansuripur, Measurement of the thermal coefficients of rewritable phase change optical recording media, ODSC (Optical data storage center of The State Univ. of Arizona), Semi-annual Report, July 31, 2001.

[34] T. Ohta, K. Inoue, S. Furukawa, K. Yoshioka, M. Uchida, and S. Nakamura, Rapid cooling phase-change optical disk with ZnS–SiO2 dielectric layer, Electro. Commun. Tech. Res. Meet. Rep. CPM90-35, 1990, p. 43.

[35] K. Inoue, S. Furukawa, K. Yoshioka, K. Kawahara, and T. Ohta, Proc. ASME **2**, 593 (1992).

[36] T. Ohta, K. Yoshioka, H. Isomura, T. Akiyama, and R. Imanaka, Proc. SPIE **2514**, 302 (1995).

[37] T. Ohta, K. Inoue, T. Ishida, Y. Gotoh and I. Satoh, Jpn. J. Appl. Phys. **32**, 5214 (1993).

[38] T. Sugaya, T. Taguchi, K. Shimura, K. Taiara, Y. Honguh, and H. Satoh, Jpn. J. Appl. Phys. **32**, 5402 (1993).

[39] 1.3 GB 90 mm Phase-change optical disk, ISO/IEC JTC: Project 1.23.14760, 1995.

[40] M. Kato, Y. Kitaoka, K. Yamamoto, and K. Mizuuchi, MORIS/ISOM 97, Yamagata, Tech. Digest of Joint MORIS/ISOM 97, 1997, We-F-01, p. 46.

[41] M.J. Dekker, N. Pfeffer, M. Kuijper, I.P.D. Ubbens, W.M.J. Coene, E.R. Meinders, and H.J. Borg, Proc. SPIE **4090**, 28 (2000).

[42] K. Nagata, K. Nishiuchi, S. Furukawa, N. Yamada, and N. Akahira, Jpn. J. Appl. Phys. **38**, 1679 (1999).

[43] T. Akiyama, M. Uno, H. Kitaura, K. Narumi, K. Nishiuchi, and N. Yamada, Jpn. J. Appl. Phys. **40**, 1598 (2001).

[44] K. Narumi, S. Furukawa, T. Nishihara, H. Kitaura, R. Kojima, K. Nishiuchi, and N. Yamada, Tech. Digest ISOM2001, 2001, Fr-K-02, p. 202.

[45] K. Hayashi, K. Hisada, and E. Ohono, Tech. Digest ISOM2001, 2001, Pd-33, p. 312.

[46] S.R. Ovshinsky, Proc. 9th Symp. on Phase Change Recording, PCOS1997, 1997, p. 44.
[47] M.P. O'Neill and T.L. Wong, Tech. Digest ODS2000, 2000, WB2, p. 170.
[48] K. Kiyono, M. Horie, T. Ohno, T. Uematsu, T. Hashizume, M.P. O'Nell, K. Balasubramanian, R. Narayan, D. Warland, and T. Zhou, Tech. Digest ISOM2000, 2000, MC1, p.37.
[49] K.D. Flynn, D. Strand, and T. Ohta, Proc 14th Symp, PCOS2002, 2002, p. 43.
[50] T. Ohta, K. Nishiuchi, K. Narumi, Y. Kitaoka, H. Ishibashi, N. Yamada, and T. Kozaki, Jpn. J. Appl. Phys. **9**, 770 (2000).
[51] Y. Honguh and T. Murakami, Electron. Commun. Jpn. Part 3, **77**, 85 (1994).
[52] N. Miyagawa and M. Mansuripur, Tech. Digest ISOM2001, 2001, Fr-N-04, p. 240.
[53] D. Tsu and D. Strand, Proc. SPIE **4342**, 124, 2001, p. 124.
[54] B.D. Terris, H.J. Mamin, D. Rugar, W.R. Studenmund, and G.S. Kino, Appl. Phys. Lett. **65**, 388 (1994).
[55] K. Kishima, I. Ichimura, K. Saito, K. Yamamoto, Y. Kuroda, A. Iida, S. Masuhara, and K. Osato, Jpn. J. Appl. Phys. **41**, 1894 (2002).
[56] J. Tominaga, Tech. Digest ODS99, 1999.
[57] H. Fuji, J. Tominaga, L. Men, T. Nakano, H. Katayama, and N. Atoda, Jpn. J. Appl. Phys. Part (1) **39**, 980 (2000).
[58] T. Kato, H. Hirata, H. Inoue, H. Shingai, and H. Utsunomiya, Tech. Digest ISOM2001, 2001, Fr-K-01, p. 200.
[59] K. Miura, J. Qiu, H. Inoue, T. Mitsuyu, and K. Hirao, Appl. Phys. Lett. **71**, 3329, 1997.
[60] T. Ohta, N. Yamada, H. Yamamoto, T. Mitsuyu, T. Kozaki, J. Qiu, and K. Hirao, MRS 2001 Spring, Proc. Vol. 674 (2001) V1.1.1.
[61] M.P. Shaw, S.H. Hilmberg, and S.A. Kostylev, Phys. Rev. Lett. **23**, 521 (1973)

Japanese Journal of Applied Physics
Vol. 43, No. 7B, 2004, pp. 4695–4699

Optical Cognitive Information Processing — A New Field

Stanford R. OVSHINSKY

Energy Conversion Devices, Inc., 2956 Waterview Drive, Rochester Hills, MI 48309, U.S.A.

(Received November 28, 2003; revised March 16, 2004; accepted April 28, 2004; published July 29, 2004)

I will discuss unique electronic and structural mechanisms of Ovonic optical phase-change devices making possible orders of magnitude increase of density of memory and introducing multiple information functions in a single nanostructure spot. [DOI: 10.1143/JJAP.43.4695]

KEYWORDS: optical computing, phase change material, chalcogenide, neural computing

1. Introduction

I have always considered a keynote talk – especially on our Ovonic optical phase change memory – not to be a summary of work completed, but to point to the future by introducing new concepts which can stimulate new work so that the field that we love can grow to its full potential.

In the past, I have also presented data in a very specific manner to show that my vision is not a utopian one, but a road map with basic principles and foundations in place so that there is little question that the end result desired can be achieved. A detailed presentation of our previous work can be found in earlier publications[1–4] and two books of my collective papers.[5,6]

The commercial work in phase change memory has been devoted to binary activity, for the binary paradigm is how one stores optical memory, no matter what its mechanism is, magnetic or phase change. The strategy is always to make the spots/marks smaller and the density much higher. Hence, the blue laser.

Of course, there is always ongoing work on how to use the gradation of amorphous to crystalline with its various changes of local and intermediate range order as a means to achieve readable multistates. This has been accomplished both electrically and optically.

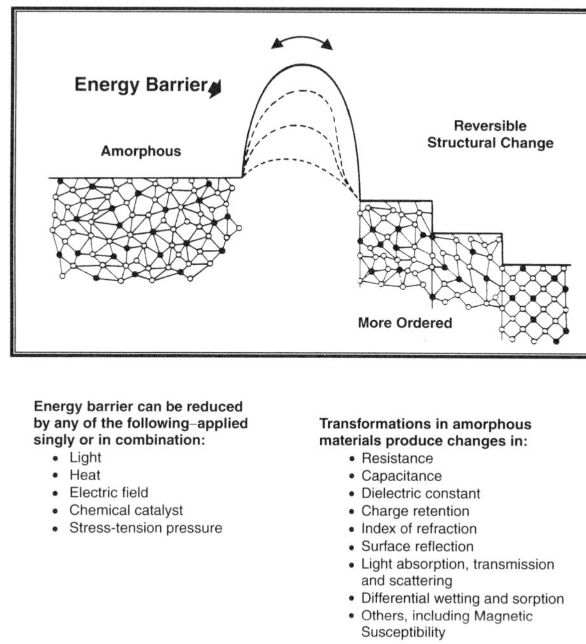

Fig. 1. Mechanism for information storage/retrieval and display by structural transformation.

2. Electrical and Optical Memory

Figure 1 shows my original concept of multistate for Ovonic phase change memory. Figure 2 shows the performance we have attained in Ovonic electrical phase change memory devices optimized for multistate storage.

The work to utilize this concept by David Strand successfully showed that such graded/multi-states could be achieved optically[7] (Fig. 3). The multi-state optical function is analogous to the electrical multi-state function in that each of these functions are achieved by delivering energy to the device in a direct overwrite fashion that first amorphizes the memory location and then by design of the pulse profile leaves the location in the desired final state of partial crystallinity. The barrier to its commercialization has not been that it doesn't work, but that today's lasers change power output as their temperature changes, and these fluctuations limit the accuracy that any given reflectivity level can be reproduced. It is the standardized optical storage products that have been the commercial successes, and industry giants will have to lead introduction of multi-level products with their increase storage capability.

Figure 4 shows the "U", which is the resistance versus current ($R(I)$) curve of an Ovonic binary memory device. The devices can be made for both binary and multistate (see Fig. 2) storage. The ability to attain intermediate states comes from the fact that the materials can exist in configurations that range from completely amorphous to completely crystalline, including a continuum of structures having partial amorphous and partial crystalline nature. The device resistance when in the intermediate states is determined by both the volume fraction in each structure and also the configuration of the regions in the two structures within the volume of the entire device. It is by control of these that we can optimize a device for binary or multistate performance.

In summary, on the left side of the "U" is the synaptic activity in the amorphous phase which is not accessible to be read until the percolation path is reached, whereas on the right side the changes can be continuously measured and available for interrogation. The right side is multi-state. The left side can have many states and they carry coherent information without being accessible. Accessibility is only available through the crystalline. The left side is cumulative while the right side is direct overwrite. All the work that I have been discussing for detectible multiple phases has been accomplished on the right hand side of this "U". But

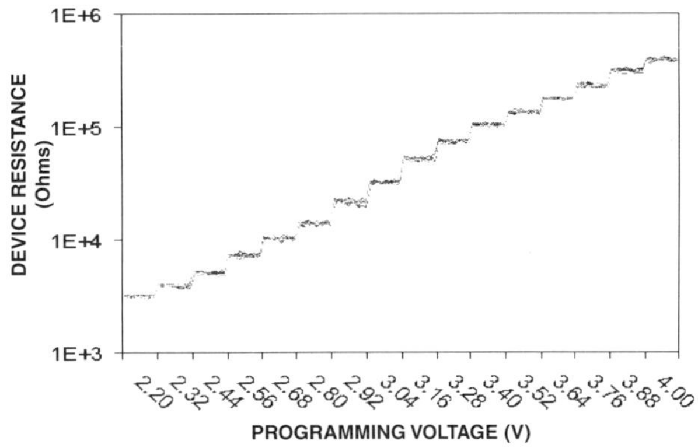

Fig. 2. Multistate electrical phase change memory. The stepwise appearance of the data in the graph arises from the testing sequence. We programmed the device a total of 1600 times, or 100 times at each applied voltage. We first applied ten programming pulses at one voltage, measuring the resulting resistance each time. We plotted these resistance values over a short distance on the horizontal axis for clarity in the presentation. We then increased the voltage to next level and programmed ten more times. After we completed programming at all 16 levels, we repeated this entire process nine more times. In this way one can see the reproducibility of the device resistance following these two programming sequences. Showing the data as steps allows visual confirmation of the clear separation between the states. The device is capable of truly continuous intermediate resistance levels. The number of states shown is arbitrary to demonstrate the principle. The number of states chosen in a particular device is determined by the specifics of its application.

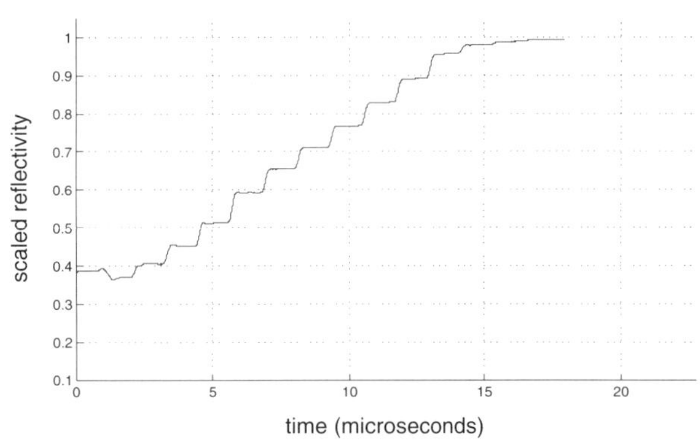

Fig. 3. Optical multistate phase change data storage.

remember that I have always said that a single spot of our "simple" tellurium-germanium-antimony material and its siblings, even in the nanostructure range, has new, rich and deep physics. The left side of Fig. 5 emphasizes the unique cognitive function.

3. Percolation Behavior

The U of an Ovonic memory device consists of two distinct regions: On the left-hand side, starting from the amorphous (reset) state, is the so-called 'pre-threshold' or 'energy-accumulation' regime. This is the region of crystal growth. If we send constant amplitude pulses in this region, we will cause crystallites to grow one step at a time with each pulse – that's why we call it energy-accumulating regime.

After sending a number of pulses, which number is determined by their amplitude and width, we reach the percolation threshold, which marks the point at which a

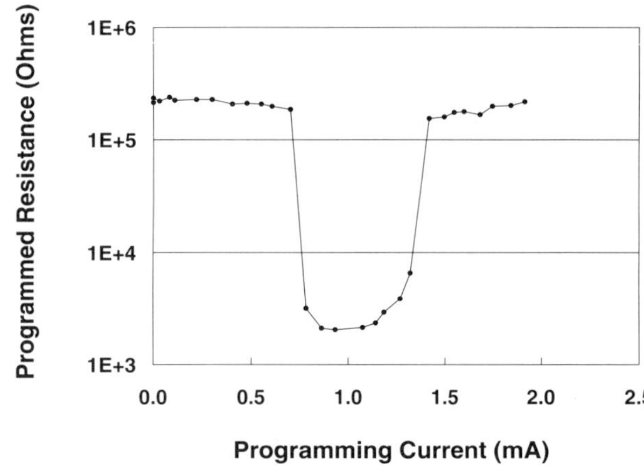

Fig. 4. Resistance as a function of current for an Ovonic phase change memory device (also referred to as the U in the text).

conducting crystalline path is being formed for the first time in the amorphous material.

At the percolation threshold, the conduction of the device drops in a fast non-linear fashion, typically by about two orders of magnitude. Because of the cumulative incorporation of the pulse energy, existence of a threshold and non-linear transition from one resistance state to the other, the device in this region operates as a solid-state analogue of the biological neuron.

This region of the U can also be used to perform arithmetic operations, store multiple bits in one device, encode information that is not forensically detectable, factor large numbers, etc. We encompass all this multi-functionality with the term 'cognitive regime', emphasizing the possibility to use such devices for building truly intelligent computers.

4. Direct Overwrite Regime

The right-hand side of the $R(I)$ curve includes the so-called 'multi-state' or 'direct overwrite, reversible' regime. Direct overwrite is the ability to go from any arbitrary recorded state to any other arbitrary recorded state without having to go through any sort of refresh state or action. This eliminates the need for an erase step, for example, and thereby increases the data transfer speed dramatically. Flash memories can be recorded in microseconds, but need on the order of a second for erasure and pumps to provide higher voltages. Flash memories also have a lifetime of no longer than 10^6 cycles, and most flash devices are not even capable of that many cycles. Optical disks can be both erased and written in 50 nanoseconds, although latency adds milliseconds to the process.

In the direct overwrite regime of the U, sending an electrical pulse to the device always puts it in a distinct resistance state uniquely determined by the amplitude of that pulse. One can go forwards and backwards in terms of pulse amplitude and always return to the same point on the curve. This regime can be utilized either as a binary, non-volatile memory (which Intel and ST are developing now as flash-memory replacement) or to store multiple bits in a single device.

5. Cognitive Functionality

At the Fall 2003 Materials Research Society (MRS) meeting in Boston,[8,9] I gave a complete description of how

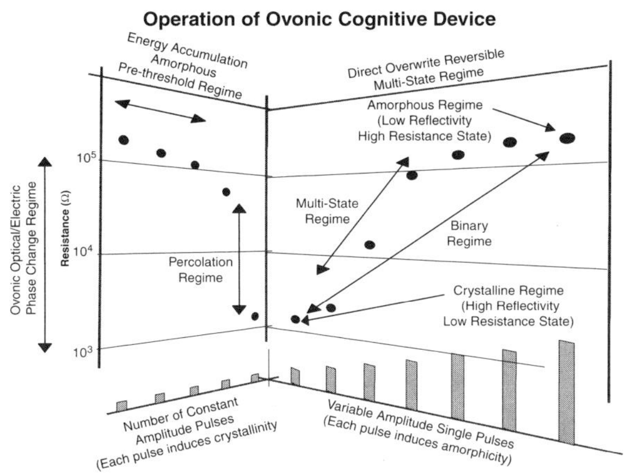

Fig. 5. Operation of an Ovonic Cognitive Device.

electronically we can use the mechanism on the left side of the U to provide a remarkable number of cognitive information states rather than binary ones in a single spot (or you would call it "mark"); a unique new behavior that I call the cognitive regime. As can be seen from Fig. 5, this regime can perform mathematical and cognitive functions. Our conventional binary regime is an amorphous to crystalline transformation and requires no pre-synaptic activity. The pre-threshold states can change the binary paradigm into an entirely new non-Von Neumann activity that is reflected in our Figs. 4 and 5.

The left side of the U is the basis of a fundamental change in computing that while it can easily do conventional Von Neumann binary operations, it is also possible in a single spot to do higher mathematical and cognitive computing. That is, the objective of building intelligence, such as learning capability and adaptability, into the computer is possible. In fact, the activity represented by the left side of the U very much resembles the neurosynaptic activity of the brain. Keep in mind that the left and right side describe memory in a single spot and cooperatively work together to perform tasks heretofore not possible in semiconductors. An amazing situation for a single device.

Sufficient here for my optical talk is to start by saying that optical computers will come; but hybrid computers, that is, optical and electronic together, will be the most likely next step. The mechanisms that I discuss here, which can be accessed optically or electronically, can be the basis for launching such an approach.

Of course, one can start by using the binary amorphous to crystalline transformation (or vice versa) and the most exciting aspect of doing so is that it can be accomplished without interfering with the non-binary mechanisms that we discuss here. Both binary and non-binary activity are permitted within a single spot.

The defining difference between binary and non-binary activity is that non-binary activity takes place in the amorphous phase through pre-threshold activity; think of the activity as being like a nerve cell surrounded by dendrites making synaptic connections. The synaptic activity carries the information. The energy of the summation of inputs then reaches the threshold and fires the neuron. This is the way the brain works and this is analogous to the non-binary activity associated with the left side of the U seen in Fig. 5, except that our synapses are pulses of energy carrying information. In this way we can emulate the behavior of the brain and build an analogous device. Our design starts with the neuron itself and builds into networks. We are currently building devices having three and more terminals. Since our devices are built in thin film form we can add layers of our devices not possible with silicon devices which require lattice matching not yet achieved. In one example implementation we can create a pattern of parallel lines, followed by a layer of Ovonic material, followed by another layer of lines orthogonal to the lower layer. The ability to program the Ovonic material at the cross points of these lines allows for the weighting required in successful neural implementations. Since we are building an analog and not a biological replacement we can achieve our design goals without having to attain the level of three-dimensionality present in the brain.

The brain operates on several biological levels of fundamental information processing. The evolution of the nervous system starts with reflex actions which are considered in neurophysiology to be hard-wired, fixed-action behavioral patterns. These always occur in a fixed manner when they are triggered and carry with them responsive actions to incoming information; that is, animals are genetically programmed to carry on certain activities such as an automatic response to stimuli to avoid danger. Such nerve cells (neurons) are specifically activated in a specific manner. Their functions do not change.

In higher vertebrates there is also adaptive behavior, which is the ability to change behaviors as a result of experience. This adaptive behavior is the result of plasticity of various nerve cells and their interacting networks that results in learning. The learning procedures require repetition so that changes occur in the nerve cells to repeated inputs which modify nerve cells through synaptic activity in response to incoming information from other nerve cells. In this way a task can be learned by strengthening the signal and encoding it into memory by, for example, having the synaptic connections change the nerve cell in such a manner that the execution of a task can be put into a memory. This repetition is called weighing in neurological terms and results in the nerve cells and the networks they are involved in to incorporate memory. This is why memory is a general property of neural nets and a basis for adaptive behavior. Learned activity is stored in memory cells. Our device incorporates the plasticity needed for learning in the same device that encodes as memory. The extent of intelligence is due to the number of cells and number of interconnections. In summary, nerve cells that have plasticity change their structure with each learning incoming "pulse" which at a certain threshold level uses these very minute changes of structure to encode it as memory.[10,11]

We make devices that duplicate this action by having incoming information make minute changes in our neural device. These minute changes are almost exactly analogous to the nerve cell. When enough pulses are accumulated the information is completed, it is encoded in our device, and, like a nerve cell, our device fires and by a dramatic change in resistance connects itself to other devices undergoing the same activity. Being non-biological gives us advantages because each pulse is not based on timing, but is coherently involved with the next pulse in space rather than time. A mathematical operation can be started in our devices and, for example, 40 years can pass by and then the same formula can be completed. In humans the formula is likely forgotten years before! Arithmetic operations in brains are similarly done in single neurons.[12,13]

In other words, the coherence of the information is a result of a series of pulses summing up to a threshold. These thresholds can be changed and we have degrees of freedom not found in the brain because that memory is reversible. This is important in that we do not require excessive numbers of synapses and the devices can be reused for other tasks. The easiest way to understand what this paper is about is that with a much smaller number of neurons and synapses we can achieve cognitive function and amazingly achieve it even in a single cell. Adaptiveness and memory become synonymous.

Inputs can be optical, electrical, or others, and outputs can be directed to various types of human interfaces. Early applications include pattern recognition and will progress to more sophisticated functions. Our device can be used to factor large numbers and therefore be applied in coding/decoding. The synaptic pulses can be optical or electrical. They operate in the amorphous state where the information is very secure and cannot be forensically investigated. Therefore, they are perfect for encryption.

6. Duality of Energy and Information

Recall also that the reason that Iris and I called our company Energy Conversion Devices is that we have used as our fundamental concept that energy and information are opposite sides of the same coin, that information is encoded energy and that pulses of energy correspond to the storage of energy/information.

I have always said that the Ovonic phase change memory and the ovonic threshold switch have new rich and deep physics. When one sees the amorphous-crystalline transition or intermediate stages that can be detected optically or electronically, one sees that the transitions that are achieved can have many different kinds of inputs and many different kinds of outputs. Paraphrasing Marshall McLuhan, the media is in fact the message.

However, if one examines the mechanism shown in Fig. 5 from another viewpoint, the richness of physics that it contains shows that there can be new phenomena that can open up the field of phase change memory, transforming it into something entirely new with exciting potential.

This makes possible the ability to expand beyond the usual memory applications and at the same time to consider new unique operations such as cognitive computing and, in fact, aspects of quantum computing; the latter a desired but far distant goal.

Figure 6 shows that it is possible to have a "silent space"; that is, the storage of energy/information by optical means very much like the storage of energy shown in Fig. 4 by electrical pulses. It also shows the possibility that optical pulses, just like electrical pulses, can encode and entangle information in the amorphous phase. This is what we mean when we say that one can have a quantum computer analog

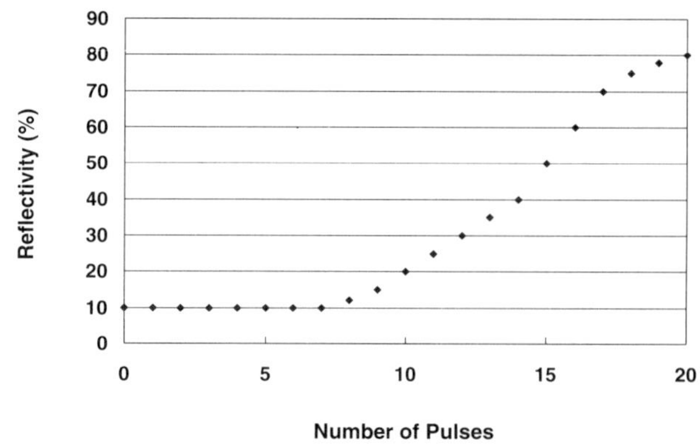

Fig. 6. Optical cognitive behavior of an Ovonic phase change material.

that is realistic and operates at room temperature and above (the activity in the amorphous cognitive phase acts very much like having non-accessible, but real portions of a wavefunction, in which at a certain point the wavefunction collapses and the data emerges). The information stored in the amorphous phase can then be read out or combined with the multistate activities of the crystalline phase; that is, combined with either the binary crystalline phase or the multistate activities of that phase.

7. Multifunctional Operation

The device that I am describing can perform in one spot a unique multiplicity of tasks. Yet that spot can be as small or smaller than a hundred angstroms. In fact, it so scales that as one makes it thinner and smaller, all parameters show increased performance. Obviously, we cannot go smaller than the overall limits imposed by the wavelengths of light that are available to us optically. Therefore, the possibility of getting more information into one mark optically or electrically would be of great value. The unique mechanism that permits us to use the amorphous state as a means of storing information is a great advance; however, we are still in the early stages in the development of the optical mode of operation, while we have been demonstrating the electronic version for some time. We are pleased to say that we have received many complimentary validations of our cognitive computer work in the electronic area. It is the electronic area that will be the first to be commercialized. The reason for thinking in new ways optically is that the optical field of data storage needs new innovative approaches beyond making the mark size smaller and smaller. We are all familiar with the above noted wavelength limitations.

The mechanisms in Figs. 5 and 6 are a means of outwitting such limitations by having one spot/mark with the capability of storing not only a one or a zero, but having multi-capabilities that can also mathematically represent a greater number of events in that one mark. Both of these mechanisms enable the cognitive function of the device, and therefore we can design devices that work with electrical pulses, optical pulses, and combinations of these energy deliveries. For example, humans have their sensory inputs from their eyes that activate nerve cells and their synapses and the cerebral cortex that responds to direct electrical stimulation, which means that one can perform their functions from the use of optical or electrical input.

The capacity and versatility of an optical phase change memory is thus increased, opening up new important fields of applications.

It is my wish to stimulate the field by introducing new concepts into the optical phase change arena, for new scientific insights and mechanisms make this field very exciting.

What I am suggesting is not only an increase in storage, but the use of the same devices (that is, the same material in its thin film form), which are so widely used and that you are so accustomed to, to offer opportunities for not only nonvolatile reversible storage of information, but, in addition, for using these ideas that I have expressed and the data that I show to explore the expansion of the optical memory field so that optical activity can be used for logic, learning, etc. with a natural link to our ongoing work in the Ovonic cognitive computer.

In other words, marrying optical and electronic together to create new industries that offer unique, low cost multi-functionality in a mark that you already use for optical phase change storage.

In summary, I am proposing that for optical applications and for the marriage of optics and electronics we understand that we have a commonality. Our optical or electronic pulses are inputs of shaped energy that contain information. We have made devices that have transparent cover layers that enable us to address the memory cells by both electrical and optical pulses.

In our binary mode, we only need to use a single pulse, but the use of multi-sub-threshold pulses in the cognitive regime offers new possibilities for energy storage and information formation and therefore gives us greater freedom to investigate new applications for the field of phase change memory.

For example, the summation of pre-threshold pulses of energy, either electrical or laser, can result in the ability to factor numbers/information and perform other mathematical functions as well as being able to have neurosynaptic, that is, adaptive, learning capabilities. After all, human optical information is closely related to the pulse firing of cells and their response to various frequencies of light.

I have not mentioned that an advantage that we have in using our nonvolatile memories that the brain does not have is that the brain depends greatly on the timing of its pre-threshold firing for its information.[14,15] While that can be a degree of freedom for us to utilize in our excitation activities, we do not require it for the mechanism described. It is just another degree of design freedom.

8. Other Members of the Ovonic Device Family

I am pleased to report that the Ovonic Unified Memory that makes it possible to have one device that has the potential to perform Flash memory, DRAM and SRAM functions is coming along very well under the Ovonyx joint venture which includes Tyler Lowrey, formerly technical head of Micron and Intel. Intel and STM have reported important progress and contributions. It is very encouraging to see that Intel and STM as well as additional powerhouse organizations such as Samsung, the Data Storage Institute and CEA-LETI are showing great interest in our field.

I also announced important new electrical devices such as an Ovonic multi-terminal device at the MRS meeting in Boston in December, 2003. I think that you can see that the amorphous and disordered field, with its unique switching and phase change activities, is living up to its potential and has many new possibilities.

Our colleague, Dr. Takeo Ohta, who has made through the years so many important contributions to the Ovonic optical phase change memory, reported on some of the OUM work at PCOS 2003. I would like to thank him deeply for presenting my paper and at the same time tell you how much I missed being able to be with you at that meeting.

There is much more to be said and much more to be done. This is a beginning and not an end. You are all my colleagues and collaborators and I feel privileged to unveil this new possibility to you. It has been an honor to have been invited here. Thank you.

My sincere and warmest best wishes and good luck to all of you.

Acknowledgements

I wish to express my deep appreciation and thanks to Boil Pashmakov, who is the co-inventor with me of the Ovonic Cognitive Computer. I would also like to thank the information team of Wally Czubatyj, David Strand, Takeo Ohta, Genie Mytilineou, Kevin Bray and the technical staff who have been involved in the preparation of our materials. David Strand has been the leader of our optical work and his important contributions are distributed throughout all of our semiconductor activities. I wish to thank Genie Mytilineou for her contributions and help as well as Kevin Bray who has been so valuable in our activities. As always, Iris is my inspiration and loving collaborator.

1) M. P. Southworth: Control Engineering **11** (1964) 69.
2) S. R. Ovshinsky: Phys. Rev. Lett. **21** (1968) 1450.
3) S. R. Ovshinsky: *Proc. 5th Annual National Conf. Ind. Res.*, (Chicago, 1969) p. 86.
4) S. R. Ovshinsky and I. M. Ovshinsky: Mater. Res. Bull. **5** (1970) 681.
5) S. R. Ovshinsky: *Disordered Materials - Science and Technology: Selected Papers*, eds. D. Adler, B. B. Schwartz and M. Silver (Plenum Press, New York, 1991).
6) S. R. Ovshinsky: Mat. Res. Soc. Symp. Proc. **554** (1999) 399.
7) K. D. Flynn and D. Strand: Joint Int. Symp. Optica Memory and Optical Data Storage (Waikoloa, 2002).
8) S. R. Ovshinsky: Fall Meet. Materials Research Society (Boston, 2003).
9) S. R. Ovshinsky: Fall Meet. Materials Research Society (Boston, 2003).
10) I. B. Levitan and L. K. Kaczmarek: The Neuron: *Cell and Molecular Biology* (Oxford University Press, NewYork, 1991) p. 395.
11) F. O. Schmitt, *et al.*: *The Neurosciences: Second Study Program* (The Rockefeller University Press, New York, 1970) p. 193.
12) M. See and M. Count: Science **297** (2002) 1607.
13) S. Dehaene: Science **297** (2002) 1652.
14) F. Morin, G. LaMarche and S. R. Ovshinsky: Anat. Rec. **127** (1957) 436.
15) F. Morin, G. LaMarche and S. R. Ovshinsky: Laval Medical **26** (1958) 633.

Mat. Res. Soc. Symp. Proc. Vol. 803 © 2004 Materials Research Society HH1.1

Innovation Providing New Multiple Functions in Phase-Change Materials To Achieve Cognitive Computing

Stanford R. Ovshinsky and Boil Pashmakov
Energy Conversion Devices, 2956 Waterview Drive, Rochester Hills, MI 48309

ABSTRACT

This paper describes a basic new scientific and technological approach for information and computing use. It is based on Ovonic cognitive devices that utilize an atomically engineered Ovonic chalcogenide material as the active medium. We demonstrate how such a device possesses many unique functions including an intrinsic neurosynaptic functionality that permits the processing of information in a manner analogous to that of biological neurons and synapses. Our Ovonic cognitive devices can not only accomplish conventional binary computing, but are capable of non-binary generation of information, storage, encryption, higher mathematics, modular arithmetic and factoring. Uniquely, almost all of these functions can be accomplished in a single nanosized device. These devices and systems are robust at room temperature (and above). They are non-volatile and also can include other volatile devices such as the Ovonic Threshold Switch and Ovonic multi-terminal threshold and memory devices that can replace transistors.

INTRODUCTION

The global computer industry is based upon silicon in a binary mode where information is processed sequentially. The transistor is fabricated from crystalline silicon where periodicity is fundamental and where doping in the ppm and above range of donor and acceptor atoms such as P and B is required.

Computers are characterized by two fundamental attributes. First, operation is based on binary logic. The storage and manipulation of data occurs through conversions to binary strings and transformations of binary strings. Second, today's computers operate sequentially in a manner first described by John Von Neumann. Completion of a computational function is inherently a step by step process. Computer programs are simply line by line instructions that outline a sequence of steps to be implemented. They are executed in a one by one fashion in which the results of preceding steps are typically forwarded to later steps.

Despite their tremendous successes, certain computations, functions and tasks remain largely unamenable to solution or implementation by conventional silicon computers. Such computers become increasingly inefficient as the complexity of computation increases. Computational problems whose time of computation scales exponentially with the input size (number of bits) become intractable with conventional computers. Examples of such problems include the factoring of large numbers and searching or sorting large databases.

Quantum computing has recently been proposed as a solution for overcoming these limitations of conventional computers. Proposed quantum computers seek to exploit the quantum mechanical principle of wavefunction superposition to achieve more than binary state computing through the massive parallelism inherent in entangled, yet coherent states. These states are not accessible for detection and utilization except upon wavefunction collapse. Quantum algorithms such as Shor's factoring algorithm [1] and Grover's searching algorithm [2] demonstrate the benefits potentially available from such computation. However, the systems involved are operable mostly at very low temperatures with states stable only for very brief periods of time.

The statistical nature of the quantum activity requires very complex error correction. Decoherence is a fundamental problem, and device states are volatile with short lifetimes. Therefore quantum computing is a topic of scientific interest, but it will be far in the future before it can become a viable industry [3]. Laboratory quantum computer experiments at exceedingly low temperatures received worldwide attention by showing that it is possible to factor the number 15. With our approach, we can factor 15 trivially and stably at room temperature with none of the above problems. Our approach is demonstrable now, not only for factoring, but also for the many functions described below.

We have taken the position since the 1950's that information is physical; it is encoded energy. Our computer principles are therefore based upon generating and storing units of energy so that they can be added, subtracted, multiplied and divided to provide simple arithmetic and to be utilizable for higher mathematics, while having the inherent plasticity needed for neurosynaptic operation, all in a stable, non-volatile manner.

Our devices are able to emulate the biological functions of memory, switching, learning, adaptability, higher mathematics etc. occurring in the brain, see e.g. refs. 4, 5. The materials of choice through which we achieve our unique mechanisms replicating these functions, are inorganic and polymeric. The devices can operate as neurons, synapses and dendrites, all in a single nanostructure. The devices are able to generate, store, and transform information within a single multifunctional entity in the nanostructure range, assuring high density and high speed. These devices and their systems, while not being quantum computers in any sense, are still able to emulate in a much improved and practical manner some of the quantum computing operations that have been proposed.

The devices must be cost effective, manufacturable and near term. To accomplish this, we utilize thin film structures in the nanoscale at room temperature and above, with both non-volatile as well as volatile operation. We construct multi-terminal devices which achieve the equivalent function of the transistor with far faster speed, increased current capacity and smaller size. One example is a device of nanometer size made of Ovonic threshold material [6] showing a normal threshold voltage of less than 2V with a third electrode that modulates the threshold voltage, while at the same time reducing the holding current to essentially zero, keeping the conducting state intact. When the third electrode is turned off, the original threshold voltage appears [7]. This behavior clearly demonstrates the electronic nature of Ovonic switching by establishing that the conducting filament is a plasma as orginally proposed [8]. The lifetimes of the threshold and memory devices have been proven to be the same as that of other semiconductor devices.

The basis of our work in chalcogenide based materials (e.g. $Ge_{22}Sb_{22}Te_{56}$) is well known [8-11]. Ge, Sb and Te have been the archetypical elements for the Ovonic memory material from its beginnings [8,12-15] to which other elements can be added. Ovonic optical phase-change materials are utilized throughout the world in devices such as rewritable DVD's. Our Ovonic electrical phase-change memories are the basis of our joint venture Ovonyx with Tyler Lowrey, Intel and others [16]. Ovonyx has several licensees, including STMicroelectronics, and the work is progressing very well [17]. Both the Ovonic optical and electrical memory are currently binary in nature, and the latter are intended as replacements for flash, DRAM and SRAM.

Current implementations of artificial intelligence utilize conventional transistor technology. We have many more degrees of freedom in material and device design that, while non-biological, permit achievement of higher level functionality of intelligence. We describe here the principles permitting this higher level functionality. We show that we are able to use our proven materials, production techniques and devices to achieve functions that cannot be achieved in any other known manner.

From a socioeconomic viewpoint, our economics are far more favorable than those of conventional computing. The capital costs of equipment can be lowered by adapting our eighth generation, continuous-web, multijunction nanostructure layer machines. These fabrication facilities can produce complex multilayer materials such as our Ovonic triple junctions by miles and tons with very high yield. The materials can have as many as eleven layers in an overall thickness of less than 0.5 μm with individual layer thicknesses of 80 - 100Å [18,19]. The cost of our machines is in the millions instead of the billions of dollars required for silicon wafers, including the cost of complex photolithography for the latter, however.

The first step along the path to full realization of the potential of cognitive computing may well be the implementation of a hybrid technology. Arrays of our devices can be compatibly fabricated upon a silicon chip engineered to contain all necessary drivers and other auxiliary circuits if desired.

A NEW COMPUTING PARADIGM

The familiar consumer computer and semiconductor industries are now cyclical, approaching important fundamental limits of the science, technology and costs. A new tranformative approach is needed. We offer one that operates in the nano-range with new physical mechanisms on various thin film substrates, assuring mechanical flexibility. We have already described the transformative potential of our technology for a new information age [20].

Figure 1 presents a comparative summary of the features of the current silicon paradigm and those of our approach. We emphasize that the functions described in Figure 1 have been and are being demonstrated on the benchtop. Since the field of application is so large, building integrated systems is required now as the first task on our critical path to commercialization. It is important to recognize that the device can be used to perform in the binary or in higher modes as required for particular tasks, emulating various functions within the brain. More explicitly, our proprietary devices can also perform as non-binary processors capable of manipulating and storing data in high level arithmetic bases (e.g. decimal, hexadecimal, base 8) which provide for additional operational capabilities via multi-valued logic. The Ovonic Cognitive Computer also has remarkable encryption possibilities and has the plasticity to show adaptive learning and cognitive functions, hence the name.

We can make full use of the unique functionality of individual devices to increase dramatically the functionality at the array level. The device principles make possible the variable interconnection strengths among cognitive devices needed to emulate the biological plasticity and complexity needed for adaptive and learning capabilities. Each cognitive device can further be connected to a very large number of devices in an array, with three-dimensional interconnectivity made possible with large fan in and fan out of the connections. **As a result, a**

Conventional Silicon Computers	**Ovonic Cognitive Computer Multifunctionality in a Single Element**
Each Element: • Computes based on single bit (binary) manipulation • Manipulates data sequentially, bit by bit	Each Element: • Manipulates, processes and stores information in a non-volatile fashion • Hardware and software are unified • Low voltage and low current operation • Performs arithmetic operations (+,-,x,÷) on multi-bit numbers (0,1,2,3...n) • Performs modular arithmetic • Executes multi-valued logic • Stores the result in a non-volatile manner • Simple, powerful encryption • Acts as a neurosynaptic cell; i.e. possesses intelligence capability • Scales down to nanoscale dimensions; huge density • Device speed is in the picosecond range • Capable of massive parallelism • Combines logic and memory in a single device • Has attributes of proposed quantum computers without their limitations, such as analogs of quantum entanglement and coherence at practical conditions and environments
Arrays of computation and storage elements are combined in a conventional computer which: • Requires separate storage and processor units or regions • Has limited parallel processing capability • Is limited to Von Neumann operations	**An Array of Ovonic Cognitive Elements working as a System:** • Easily factors large numbers • Performs high level mathematical functions (e.g. vector and array processing) • Has high 3-dimensional interconnectivity, huge density, giving rise to high speed, hyper-parallel processing (i.e. millions of interconnected processors) • Has adaptive learning capability • Interconnectivity is simply and inherently reconfigurable • Can generate dynamic activity
	The Ovonic Cognitive Devices are: • Mass produced in exceptionally dense, all thin film, uniquely interconnected arrays • Mass manufactured as a thin film, flexible device using proven technologies • Ovonic "transistor" unique high speed low cost 3-terminal device. Nanostructure capable of carrying large amounts of current both in nonvolatile and volatile modes

Figure 1. A comparison of the features and operational characteristic of conventional silicon elements and arrays with those of the Ovonic Cognitive Element and Computer.

highly dense, interactive, and massively parallel architecture is achievable. This makes possible the use of dynamical states of activity for computation, as in human brains, instead of the sequential switching from static state to state in conventional computers.

HOW DOES THE OVONIC COGNITIVE DEVICE WORK?

There are two kinds of Ovonic materials, the Ovonic threshold switch material (OTS) and the Ovonic memory switch (OMS) or phase-change material. The OTS material has a strongly crosslinked polymeric structure and strong interbond interactions which ensure its structural stability during the electronic transitions associated with switching. The OMS, on the other hand, has a different polymeric structure which is designed to have fewer, weaker crosslinks and strong interactions between lone pairs, all of which facilitates reversible structural transitions between the amorphous and crystalline states.

The active material in our cognitive device is our Ovonic, solid-state, chalcogenide phase-change material, the same material that is used in commercial applications. Those applications, however, are binary in nature. In them, the devices utilize only the reversible phase-change from amorphous (high resistance, low reflectivity state) to crystalline (low resistance, high reflectivity state) in current commercial applications. In contrast, we show here the deep and rich new physics that be utilized in single or multiple elements, especially in the amorphous state.

The basic operation of the active material is illustrated by the data presented in Figure 2, showing the resistance characteristics of a representative Ovonic chalcogenide material, $Ge_{22}Sb_{22}Te_{56}$. This is the material used in the binary mode by our optical phase-change licensees and in the electrical Ovonic memory device now called the Ovonic Unified Memory (OUM) currently so successfully pursued by Tyler Lowrey, a towering figure in the memory field, and his talented group at Ovonyx.

We show in Figure 2 the response of our Ovonic Cognitive device as a function of electrical energy (lower axis) applied to the cognitive device in the form of current pulses. The amorphous regime, which in the past has been considered silent regarding information, is where the pre-threshold pulses act. The pre-threshold states are the equivalent of the coherent and entangled states of the quantum computer. In contrast to quantum computers, they are non-volatile; new pulses needed to complete a computation or encryption can be added much later (e.g. over forty years later). The devices are also radiation hard.

The response of the material to the current pulses can be described via the two general response regimes depicted in the folded presentation format shown in Figure 2. The fold coincides with a minimum in the resistance and demarcates a low constant amplitude pulse regime to the left from a higher current variable amplitude pulse regime to the right. The higher current range shows the multistate activity of our Ovonic electrical memory [10,11]. Operation in the variable amplitude regime (VAR) requires a minimum current pulse amplitude and this minimum amplitude pulse produces the lowest resistance (highest crystallinity) state in the VAR regime. The amorphous-crystalline transition utilizes a reversible phase-change mechanism.

Our new Ovonic cognitive devices make use of new mechanisms in the deceptively simple single, amorphous, nano-dimensional spot in the low current operational regime shown to the left of the fold in Figure 2. As current pulses are applied in the cognitive regime, minute nanocrystalline regions form, the volume fraction of such crystalline phases increasing with each current pulse. Crystallization can occur through nucleation/growth upon the application of a current pulse. The microcrystallites generated by a sequence of pulses form a temporally

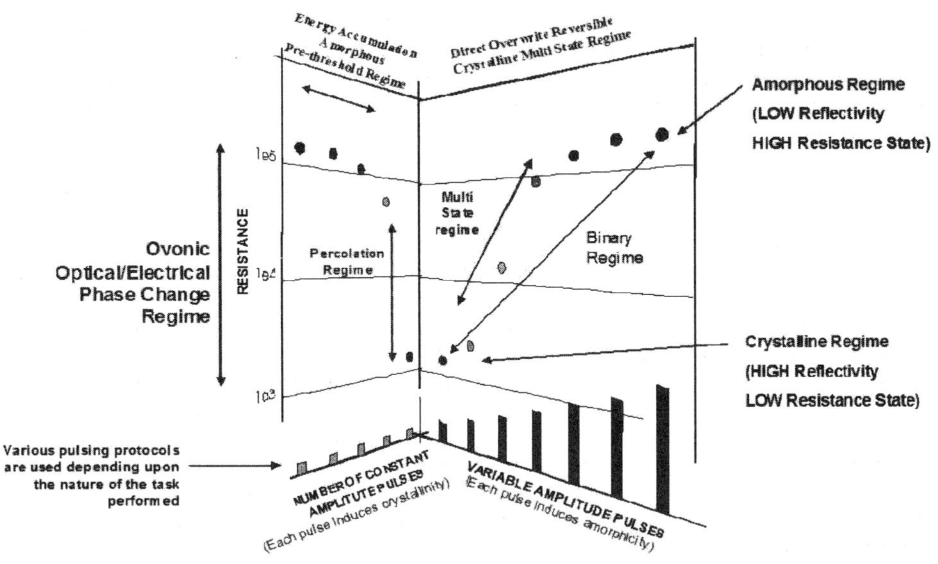

Figure 2. Resistance characteristics of an Ovonic Cognitive Device. The cognitive amorphous pre-threshold synaptic regime (left side) culminates in a percolative transition to crystalline material, the equivalent of neurosynaptic switching. The resistance change accompanying the transition to the crystalline regime can provide readout and transferring of a completed signal to other devices. The leftmost and rightmost data points of Figure 2 (the high resistance endpoints) both correspond to material that is substantially amorphous and the material becomes increasingly crystalline toward the center of the figure, with the lowest resistance states having the greatest crystallinity. The right side is the multistate crystalline cognitive regime (CCR). One should look upon the left side as being either standalone when the crystalline sums up the synaptic information or united with the activities of the right side.

coherent sequence of states.

The nanocrystallites are distributed randomly throughout the chalcogenide material. As they grow, a percolation path results, a continuous, high-conductivity pathway across the material between the contacts. Once percolation has occurred, the material exits the amorphous cognitive regime and enters the right side, the CCR regime, if desired. Otherwise, the material can be reset to an amorphous state [12].

Accumulated energy, rather than current pulse amplitude, is a more fundamental representation of the modification of the Ovonic chalcogenide material in the cognitive regime. The increment of crystallization that occurs upon application of a current pulse is dictated by the energy deposited by the pulse into the material. This is an essential feature of the cognitive functionality of our new device because the structural state (defined by its crystalline volume fraction) of the material at any point in the cognitive regime is a manifestation of the total accumulated energy applied to the material. The crystallites represent stored energy that has encoded meaning. This stored energy represents what we mean when we say information is encoded energy.

The accumulative nature of the cognitive regime also provides a close analogy to the neurosynaptic functionality essential to cognitive behavior in biological organisms [21-23]. In the cognitive regime, each application of energy to our adaptable polymeric material induces a partial crystallization of the material to an extent characteristic of the applied energy. Upon removal of the energy source, the material remains in the partially crystallized state until exposed to energy once again. Since the pulse energies in the cognitive regime are sufficiently low to prevent reversion of crystallized regions back to the amorphous phase, the crystallization process is stable until one desires to erase it or make it reversible as in the Ovonic memory.

The structural state is thus a record of the energy accumulated by the material. Continued application of energy to the structural state induces additional crystallization and further accumulation of energy until sufficient energy has been applied to reach the percolation transition. The energy required to induce percolation is a threshold for a transition from a high resistance state to a low resistance state. In optical applications, we replace resistance with reflectivity. We have proposed that an Ovonic hybrid optical-electrical memory will precede the emergence of the all-optical memory [20].

The ability of our cognitive device to undergo an abrupt change in a readily detectable manner after accumulating its threshold energy provides for neurosynaptic functionality [21-23]. A biological neuron receives energetic inputs at its dendritic synaptic terminals and accumulates them until it reaches a threshold and fires. Before firing, a neuron "acts" as if uncognizant of the signals it has accumulated, and yet it fires when the net signal reaches the threshold value. Our cognitive device exhibits analogous accumulation and threshold activated firing capabilities. The accumulation response is a series of pre-percolation structural states with altered local order having similar resistances and crystalline volume fractions, increasing in proportion to the accumulated energy. Since the resistances of the pre-percolation states are similar, these states are functionally equivalent and analogous to the pre-threshold states of a biological neuron. The abrupt reduction in resistance that occurs at the percolation transition is analogous to the firing event of a biological neuron. This apparently silent zone is the basis of our encryption and other functions.

The firing pulse, which represents crystallization (or in quantum analogy terms, collapse of the wavefunction) gives meaning to the pre-threshold events which could not be interrogated individually. These events are correlated in such a way as to provide functionality analogous to that derivable from quantum entanglement, while representing a significant number, symbol, or information value, etc. The firing pulse in effect reveals the meaning of information stored in forensically inaccessible pre-percolation states. Upon firing, that which was inaccessible becomes tangible and can be read out and interacted with other devices and functions. What was once inaccessible in the "silent" processes of information gathering and storage in a pre-percolation state becomes detectable and manipulable by other devices.

Keep in mind that we are thus far speaking of a *single* cognitive device in the *nanosize range* that can perform a wide variety of mathematical operations, neurosynaptic functions etc. Such activity is unique and exemplifies the deep and rich physics of our nanostructured amorphous material.

APPLICATIONS OF THE OVONIC COGNITIVE DEVICE

We offer a fundamental new approach to computing that enables Ovonic cognitive devices and networks to provide new strategies for doing not only conventional computing but, even more importantly, a whole new approach to informational and computational applications. It also opens up a new phase in the use of non-silicon material for semiconductors. Several illustrative examples of the dramatic increase in multifunctionality are discussed below.

Non-Binary Storage

Non-binary data storage is a unique aspect of the Ovonic cognitive device. In non-binary data storage, a single Ovonic cognitive device can be programmed to store any one of three or more numerical values. Each distinct numerical value corresponds to a distinct structural state in the cognitive regime. Programming or storage of a particular numerical value occurs by providing energy to the Ovonic cognitive device in an amount sufficient to transform the device to the structural state corresponding to the information or value (e.g. letter, number, symbol). In a typical application, the programming energy is provided to the Ovonic cognitive device in its reset state (the initial state (amorphous endpoint) in the cognitive regime) and becomes characteristic of the numerical value being stored. Distinct numerical values are assigned to each of a series of selected structural states in the cognitive regime. Since each structural state has a unique programming energy, a numerical value is encoded through the programming energy and retained by the material through its structural state in a non-volatile manner.

The assignment of numerical values to specific selected structural states can occur in many ways. From an operational point of view, it is most convenient to assign increasing (or decreasing) consecutive integer values to the structural states in order of increasing accumulated energy relative to the initial, reset state of the cognitive regime. In its simplest operation, it is desirable to separate consecutive integer values by equal intervals of accumulated energy so that repeated application of a particular pre-threshold energy pulse increases the stored value by one. This pulsing is done in the amorphous state. The energy relative to the reset state required to store an integer is proportional to the integer. This is advantageous because it renders the cognitive device inherently additive. The reproducibility of the values is assured because the materials always respond in the same way, making for a very stable computer.

Transformations of the Ovonic cognitive device from a structural state assigned to one integer to a structural state assigned to a different integer is a basic operation of the Ovonic cognitive device in mathematical computations. These transformations correspond to incrementing the device from one state to another through the application of energy, typically in the form of one or more electrical current (or optical) pulses of the same energy. Pulse energy can be varied through the pulse amplitude, pulse duration and even the shape of the pulse. In practical operation, non-binary storage and incrementing are most conveniently accomplished with pulses having a common amplitude and variable duration so that energy is proportional to pulse duration and different structural states separated equally in energy are separated by equal pulse durations. New degrees of freedom of electronic and material design can also be utilized.

An inherent feature of our cognitive device is the ability to operate it according to many different non-binary storage protocols. Figure 2 presents an example of a five state protocol in which the threshold energy separating the reset and set states is divided into five intervals so that five incrementing pulses are required to transform the material from its reset state to its set state.

The energy threshold can be divided into a desired number of intervals to provide arbitrary multistate storage in which an arbitrary number of pulses is used to transform the material from its reset state to its set state. Storage protocols based on three states, four states, etc. can be realized by dividing the threshold energy into three intervals, four intervals etc. Devices that operate using a large number of states are readily realized and operate reproducibly over a large number of reset-set-reset cycles. Our cognitive device can easily be reconfigured from one non-binary storage protocol to another. A device utilizing a three state protocol in one computation, for example, can be reconfigured to operate in a seven state protocol in another computation.

Encryption

The reconfigurable storage capability of our cognitive device provides a unique and remarkably effective mechanism for encrypting information. The encryption capabilities of our cognitive device originate from the non-uniqueness of the relation between the structural state of the active material and the information stored in the device. Reconfigurability precludes a unique one-to-one correspondence between the structural state and the stored information. The information content of a particular structural state in the cognitive regime depends on the number of states included in the non-binary storage protocol of the device. Different information can be encoded in the same structural state. In the absence of knowledge about the number of storage states and the energy increments separating energy states, knowledge of the structural state of the cognitive device provides no insight about the information value (alphanumeric, symbolic or otherwise) assigned to the structural state.

Another level of security provided by our cognitive device involves the difficulties in inferring the structural state of the device. Except for the set state, the structural states in the cognitive regime of our device are pre-percolation states that consist of a random, non-contiguous distribution of nanocrystallites within an amorphous matrix. The pre-percolation crystallites can be nanoscale particles that are below the size resolution of common analytical techniques. Furthermore, efforts to identify the structural state necessarily require exposing the device to energy in the form of electron beams, photons etc. Any manipulations or probes of the device that alter its structural state have the effect of deleting the stored information because a change in structural state corresponds to changing the information content of the device. Even if one were able to deduce the structural state, one would still be faced with the impossible task of decoding the information content of the state since a particular structural state is determined by the accumulated energy and this accumulated energy can be provided in a variety of different ways through variations in the pulse amplitude, pulse duration, and number and shape of pulses. Each of the different ways of transforming the material to a particular structural state corresponds to a different way of encoding information. The information content of our cognitive device cannot be determined merely through knowledge of the structural state.

Non-Binary Arithmetic

The multistate, non-binary storage capability of our cognitive device provides a natural basis for calculations in non-binary arithmetic systems. Whereas conventional computers are limited to binary computations, our cognitive devices can operate in a non-binary fashion and permit computations in base 3, base 4, etc., where the arithmetic base of operation corresponds to the number of states included in the multistate, non-binary protocol. Decimal (base 10) operation, for example, is a particularly intuitive mode of operation and may be accomplished using a ten state protocol in which ten current pulses are used to traverse the energy threshold of a cognitive device. The important factor is that we can go to any base up to the resolution of our ability to distinguish distinct states in the cognitive regime. We can even use base 60, the sexagesimal base of the ancient Sumerians, which persists to this day in angular and temporal measurements.

Addition

Because of its intrinsic accumulative functionality, our device is naturally suited to addition. Since each pulse applied to the device signifies the operation of incrementing by one, the structural state of our device provides a record of the cumulative number of increments applied to the device since its last reset. Addition of two numbers is accomplished by storing one of the addends in the device and subsequently applying pulses to the device in a number equal to the other addend.

Division

Division exploits the accumulative nature and reconfigurability of our cognitive device. In division, the divisor is used to define the arithmetic base of computation for a cognitive device and a number of pulses equal to the dividend is applied to the device with a requirement that the device be reset each time it sets until all of the pulses have been applied. The quotient of the division is equal to the number of times the device sets while applying the pulses corresponding to the dividend and the remainder corresponds to the final state of the device.

Modular Arithmetic

Implementation of modular arithmetic with our cognitive device is similar to the method of division described above. Determination of the modulo X equivalent of the number Z is accomplished by applying Z pulses to a cognitive device whose cognitive operational range is partitioned into X intervals, resetting the device each time it sets and reading the final state of the device to obtain the result. The modulo 7 equivalent of 17, for example, can be obtained by applying 17 incrementing pulses to a cognitive device requiring 7 incrementing pulses to progress from its reset state to its set state. Application of the 17 incrementing pulses causes the device to set twice and leaves the device in a state removed from the reset state by 3 incrementing pulses. Hence, $17(\mathrm{mod}\ 7) = 3$.

Factoring in Parallel

Our cognitive device offers a new approach to factoring that is efficient and amenable to parallel operation. During factoring, a number of pulses equal to the input number is applied to each of several devices configured to divide by a different prime number and each device is reset every time it sets. After all pulses have been applied, each device is in a state that corresponds to the modular equivalent of the input number in the modulus of a different prime number. Since any factor of the input number necessarily has a modular equivalent of zero, the prime numbers that are factors of the input number are those associated with cognitive devices that are in their set state after applying pulses in a number equal to the input number. This method can also factor information that is not numerical; for example, intelligent database searching and associative memory.

SUMMARY

We have described and demonstrated a unique new computational/information device that possesses the neurosynaptic functionality necessary to achieve cognitive computing. The cognitive device shows threshold activated firing, possesses a threshold energy that is variable, records experiential history, combines memory and processing in a single device, and responds to stimuli of many types. The cognitive devices can be connected into densely interconnected, highly parallel networks that exhibit plasticity and learning capabilities. The neurosynaptic properties of individual devices and the connection strengths between devices in a network are adjustable and permit reconfiguration and adaptation of a network as it confronts new situations. A single device can do both logic and memory.

In addition to providing a new concept in computing, our cognitive devices make possible the redefining of the manufacturing of computers. The atomically engineered chalcogenide materials used in our cognitive devices and networks can be deposited uniformly as thin films on a variety of substrate materials, including silicon, using methods such as sputtering, physical vapor deposition, and chemical vapor deposition. These processes are inexpensive and adaptable to large scale manufacturing. Post-deposition processing and patterning can be achieved using existing techniques that are well-known in silicon technology and can be incorporated into our continuous-web technology.

The era of truly cognitive computing in which machines utilize higher order reasoning capabilities to process, interpret and respond to information is now upon us. Our continuing efforts will focus on interconnecting devices, scale up of cognitive networks from the few to the many, optimizing learning protocols and answering emerging needs by developing task-specific devices that display adaptability within a bounded range of input conditions with first implementation via a hybrid technology. Space prohibits the description of our multiterminal junction devices that have the potential to replace the transistor, providing great performance advantages [6,7].

ACKNOWLEDGEMENTS

We thank the information team of Wally Czubatyj, my longtime collaborator in information; David Strand, for his leadership of our optical phase-change memory activities; Genie Mytilineou, for her contributions and continuous help; Kevin Bray, for his invaluable work on the preparation of the paper; and the technical staff of our Information Group. We are grateful to Morrel Cohen, a long-time collaborator, for his helpful comments. As always, SRO pays homage to his inspiration and loving collaborator, Iris.

REFERENCES

1. P.W. Shor, Proceedings of the 35th Annual Symposium on the Foundations of Computer Science (IEEE Computer Society Press, Los Alamitos, CA, 1994), p. 124.
2. L.K. Grover; Phys. Rev. Lett. **79**, 325 (1997).
3. M.A. Nielsen and I.L. Chang; Quantum Computation and Quantum Information; Cambridge University Press, Cambridge, 2000.
4. A. Nieder, D.J. Freedman, E.K. Miller; Science **297**, 1708 (2002).
5. S. Dehaene; Science **297**, 1652 (2002).
6. S.R. Ovshinsky, B. Pashmakov, E. Mytilineou; to be published.
7. A new volume updating our work since 1991 is in preparation.
8. S.R. Ovshinsky; Phys. Rev. Lett. **21**, 1450 (1968).
9. Disordered Materials: Science and Technology, Selected Papers by Stanford R. Ovshinsky; D. Adler, B.B. Schwartz, M. Silver, eds.; Plenum Press, New York, 1991.
10. Disordered Materials: Science and Technology: Selected Papers by S.R. Ovshinsky; D. Adler, ed.; Amorphous Institute Press, Bloomfield Hills, Michigan, 1982.
11. S.R. Ovshinsky; MRS Symp. Proc. **554**, 399 (1999).
12. S.R. Ovshinsky; Revue Roumaine de Physique, **26**, 893 (1981).
13. S.R. Ovshinsky; "The Quantum Nature of Amorphous Solids"; in Disordered Semiconductors; M.A. Kastner, G.A. Thomas, S.R. Ovshinsky, eds.; Plenum Press, New York (1987); p. 195.
14. S.R. Ovshinsky, H. Fritzsche; IEEE Trans. Elect. Dev. **ED-20**, 91 (1973).
15. S.R. Ovshinsky; Memoires, Optiques et Systemes, No. 127, Sept. 1994, p. 65.
16. Ovonyx is a joint venture between ECD, Tyler Lowrey and Intel Capital, among others.
17. T. Lowrey, C. Dennison, S. Hudgens, W. Czubatyj; "Characteristics of OUM Phase Change Materials and Devices for High Density Nonvolatile Commodity and Embedded Memory Applications"; 2003 MRS Fall Meeting, Dec. 1-5, 2003, Boston, MA; Symp. HH, paper 2.1.
18. S.R. Ovshinsky; Proc. of the Intl. Ion Engr. Cong., ISIAT '83 & Ipat '83, Kyoto 12-16 September 1983, p. 817.
19. D. Adler, S.R. Ovshinsky; Chemtech **15**, 538 (1985).
20. For a more recent discussion of our work in phase change optical memory, see: S.R. Ovshinsky, "Optical Cognitive Information Processing – A New Field"; presented at the International Symposium on Optical Memory '03; Nara, Japan; Nov. 3 – 7, 2003 (to be published in the Japanese Journal of Applied Physics).
21. F. Morin, G. LaMarche, and S.R. Ovshinsky; Anat. Rec. **127**, 436 (1957).
22. F. Morin, G. LaMarche, and S.R. Ovshinsky; Laval Medical **26**, 633 (1958).
23. S.R. Ovshinsky, I.M. Ovshinsky; Mat. Res. Bull. **5**, 681 (1970).

Reprinted from Materials Research Society
Symposium Proceedings Volume 803,
Advanced Data Storage Materials and Characterization Techniques,
SYMPOSIUM GG, Joachim W. Ahner,
Jeremy Levy, SYMPOSIUM HH,
Lambertus Hesselink, Andrei Mijiritskii, Editors

Ovonic chalcogenide non-binary electrical and optical devices

Stanford R. Ovshinsky*
Energy Conversion Devices, Inc.
2956 Waterview Drive - Rochester Hills, MI 48309, USA

ABSTRACT

Paul Davies, a highly respected theoretical physicist recently stated that "the essence of life is information." [1] I will describe how the essence of information is plasticity, that information is encoded energy requiring what neurophysiologists call plasticity. Plasticity is the ability of neurons through their synapses to have memory, learn, adapt and evolve in response to their environment. I will show that Ovonic memories, both optical and electrical, have rich and deep new physics that make them cognitive devices and therefore open up a huge new field of chalcogenide-based intelligent computers, intelligence that works in a similar manner to the brain. We have shown that the plasticity necessary for an intelligent opto-electronic computer is a function of energy pulses, whether electrical or optical.

Keywords: Chalcogenide materials, neurosynaptic computing, optical computing, phase change memories, threshold switching

1. DISCUSSION

The Ovonic optical and electrical phase change memories are well known to this audience [2]. A typical device is composed of tellurium, germanium and antimony and when exposed to optical or electrical energy, usually in the form of pulses, it changes state from crystalline to amorphous or amorphous to crystalline. It is fast, reversible, nonvolatile, and does not require large amounts of input energy. The amorphous state is a high resistivity state; the crystalline state is conducting. The Ovonic optical phase change device also has changes in terms of being able to absorb light in the amorphous state and have the crystalline state reflecting. In a rewritable optical disk, the energy absorbed by the chalcogenide layer is the same both when it is amorphous and when it is crystalline. The optical stack provides this to optimize direct overwrite. When the reflectivity of the stack is lower, the excess energy is absorbed in the aluminum layer below.

This simple description is not sufficient to explain why it also can demonstrate plasticity, the key factor of neuronal and synaptic activity of the brain. Plasticity is the ability of a biological material or its nonbiological analog to be able to adapt or change in response to incoming energy signals. The resulting changes are structural in nature very much like electrical or optical signals can make for conformal and configurational alterations in the amorphous phase of the Ovonic Phase Change Memory.

Information is encoded energy reflecting adaptation and learning as well as switching and memory. This paper will describe how the simple phase change binary memory can also show such plasticity and emulate brain function. Indeed, a single device can have many multifunctions, opening up an entirely new field of nonbiological, cognitive computing.

Figure 1 illustrates neurons (nerve cells) receiving inputs in the form of coherent energy pulses through dendrites (shown in a simplified manner) which, when the input energy is summed up to reach a threshold, cause the neurons to fire and transmit the information through nerve fibers (axons) to other nerve cells.

* ovshinsky@ovonic.com; phone 248 293 0440; fax 248 844 1922

Seventh International Symposium on Optical Storage, edited by Fuxi Gan, Lisong Hou, Proceedings of SPIE Vol. 5966
(SPIE, Bellingham, WA, 2005) · 0277-786X/05/$15 · doi: 10.1117/12.649584

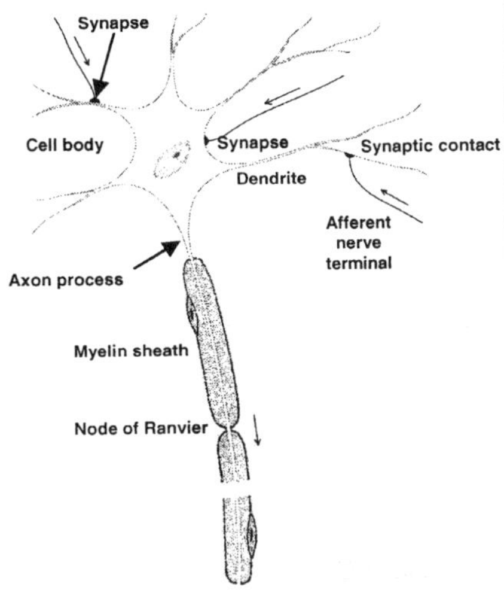

Figure 1a. Diagram of a biological nerve cell showing synapses and synaptic contacts (3)

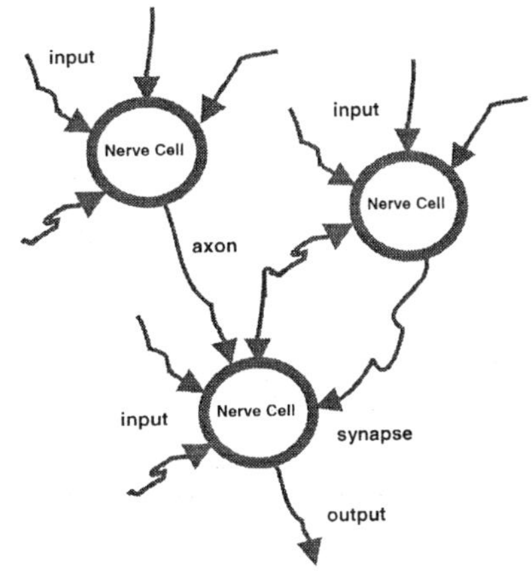

Figure 1b. A simplified schematic of the communication of information between neurons

Figure 1

Ovonic Cognitive Devices have analogous functions to the neurons and their synapses, However, the neurosynaptic system of the brain is, of course, made of organic material. How can an inorganic analog have such plasticity and intelligent, that is learning, activity?

I will describe the functions which make the Ovonic Phase Change Memory capable of being a cognitive computer and will then discuss how the Ovonic phase change memory material can perform such neurosynaptic functions.

The Ovonic Cognitive Computer is a technology that makes it possible to fulfill the long-awaited goal of achieving intelligent computing, a new paradigm. While a single Ovonic Cognitive device (or in some cases, two devices) of subnanometer size is able to have many multiple functions such as the demonstration of addition, subtraction, multiplication and division, along with the standard binary activity of any computer, it also can do nonbinary processing, modular arithmetic and encryption as well as factoring. It has the plasticity of a biological neurosynaptic cell and is based on a densely interconnected network of proprietary Ovonic Cognitive Devices where even a single device has such computing qualities.[4-6] We have also been able to emulate optically the Ovonic cognitive functions showing that electrical and optical ovonic cognitive computing are possible.[7]

A single device realistically simulates the neurosynaptic behavior of biological neurons. Like biological neurons, the device is capable of synaptic function such as receiving and weighting multiple inputs that result in threshold activation, an operational mode in which it accumulates input energy signals without responding until the total accumulated energy reaches a threshold level. Once the threshold is reached, the device undergoes an abrupt transformation from a high resistance state to a low resistance state in a process that mimics the firing of a biological neuron.

Such an individual Ovonic device can be readily interconnected to many other such devices in highly dense two-dimensional arrays or in three-dimensional, vertically integrated networks. The threshold level of individual Ovonic devices can be controlled by various means. A remarkable multi-terminal thin-film device - the Ovonic Quantum Control Device (Figure 2) - which can replace transistors as well as adding new functionalities, offers new degrees of freedom to the design of computer architecture. The plasticity of the Ovonic neurosynaptic arrays opens up possibilities of unifying software and hardware.

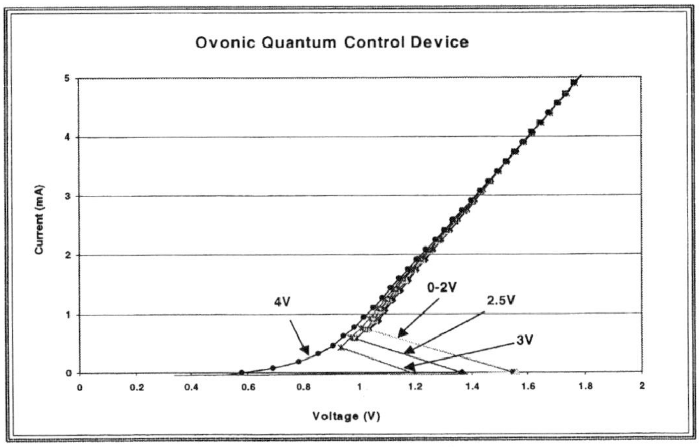

Figure 2: IV Curves of Multi-Terminal OQCD

From its beginning the conventional computer was based on the principles of VonNeumann. The Ovonic Cognitive Computer, however, need not be binary or sequential as required by VonNeumann and, therefore, opens up many new important possibilities.

The combination of small device size, speed, intrinsic neurosynaptic device functionality and dense device parallelism and interconnectivity in three dimensions offered by the Ovonic devices provides the Ovonic Cognitive Computer with a functionality and highly parallel mode of operation that follows the neurophysiological activity of the biological brain.

Inherently, individual Ovonic devices within a network are adaptive and can also be configured to function as weighting devices that can be used to control the interconnection strength between Ovonic devices configured to function neurosynaptically. Since the interconnection strength is adjustable, networks formed from Ovonic devices display learning and adaptive properties analogous to those of biological neurosynaptic networks.

The Ovonic device, singly (or in a network), is able to both process and store information in a reconfigurational nonvolatile manner and, as a result, such unique multifunctionality obviates the customary need to separate memory and logic functions in computers. Of great interest is that these devices also can operate in a manner analogous to the much-talked about quantum computer. They have several important advantages in that they, of course, operate at room temperature and higher, are robust, and they are demonstrable now. In other words, they are real world devices that can be used for various functions, for example, encryption.

To summarize, we can uniquely demonstrate addition, subtraction, multiplication, division, factoring, non-binary processing, modular arithmetic and encryption with Ovonic devices as well as neurosynaptic activity [4, 7] which, unlike present artificial intelligence, meets the criteria of true cognitive activity. The active chalcogenide material of the Ovonic devices and the Ovonic Cognitive Computer can be deposited in a low-cost, thin film fashion in a continuous manufacturing process. They can also be integrated and imbedded, that is hybridized with conventional silicon circuitry. Very importantly, they are scalable. A single device can operate at extremely small dimensions, for example under 100 angstroms. At the same time, its characteristics improve the smaller the dimension. Therefore, as photolithography goes to smaller sizes it is advantageous to our device operation.

Of great interest to this group, we have been able to duplicate these same basic functions of the Ovonic Phase Change Memory not only electrically but optically through the use of lasers. (See Figures 3 and 4.) We have shown that the sine qua non of biological intelligence is plasticity.

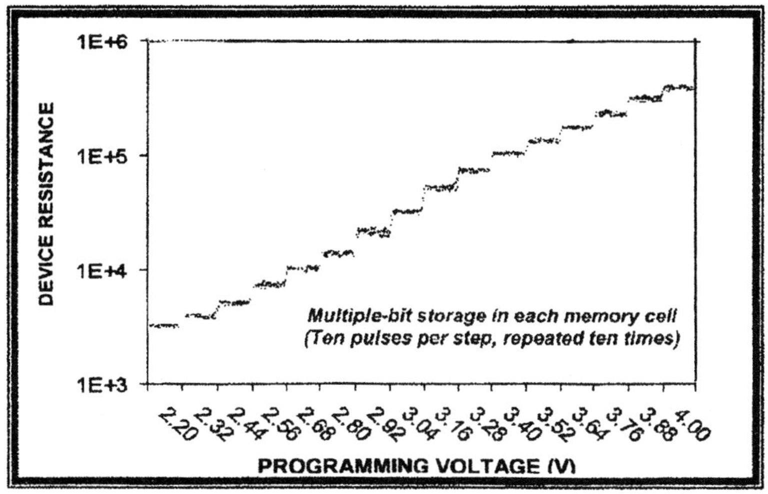

Figure 3: Ovonic Electrical Multi-State Data Storage

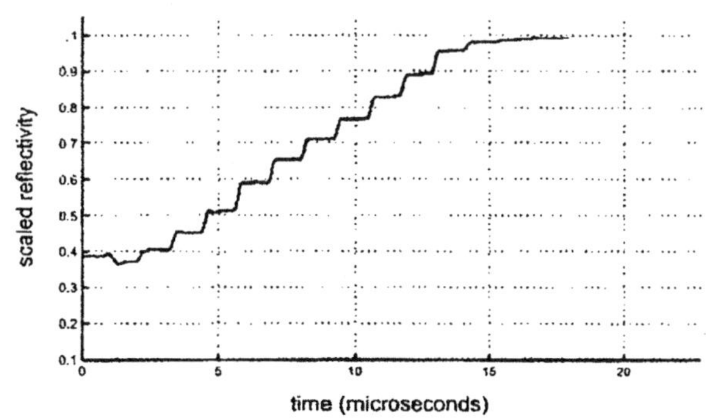

Figure 4: Ovonic Optical Multi-State Data Storage

Most importantly, we can duplicate optically the neurosynaptic Ovonic Cognitive Computer Device. Figure 5 shows that with pre-threshold pulses in the amorphous state, we can induce structural changes too small to be seen using the ultimate read technique. These pulses accumulate information which is revealed and transferred at the threshold. This is analogous to the synaptic activity taking place on the left side of the U as will be described in Figure 7.

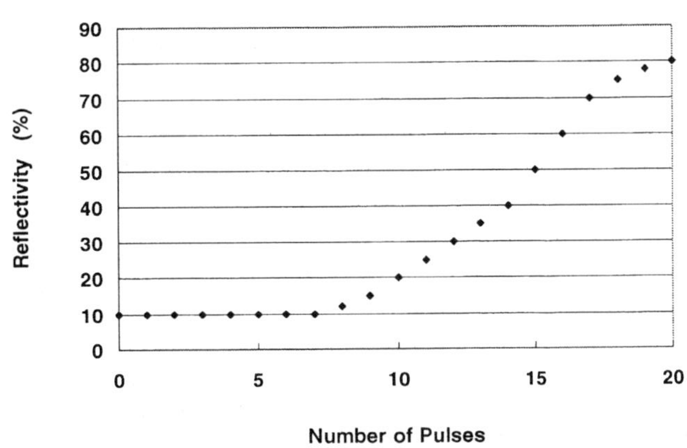

Figure 5: The Optical Ovonic Cognitive Effect

We will now describe the principles of atomic engineering which lead to the choice of elements which permit these unusual electronic and structural changes.

The divalency of tellurium, the length of its chains and particularly its free lone pairs, which are available for electronic or optical excitation, provide the flexibility of the material which is the requirement for plasticity. Germanium and antimony crosslink the tellurium chains and the material plasticity is controlled by the number and strength of crosslinks and their relatively weak bonding. This in brief is the mechanism providing reversible phase change without disturbing the plasticity. As one uses stronger and more crosslinks to tellurium such as germanium, silicon, and arsenic, the material becomes stiffer and much less flexible and the tellurium chains are much shorter.

The strongly crosslinked materials are the basis of the Ovonic Threshold Switch in which an electric field can couple to the lone pairs of tellurium and the many and strong crosslinks prevent crystallization so that the material always remains amorphous. The switching is volatile, its speed is so fast that it has never been measured and its conducting state is a highly dense solid state plasma which can carry orders of magnitude more current than a crystalline transistor and, of course, is faster and smaller. It is the basis of our 3-terminal device shown in Figure 2.

It is the ability of excited lone pairs in the Ovonic Phase Change Memory to cause conformal and configurational structural changes that is the basis for the change of phase that occurs in response to the coupling of the electrical or optical field to those lone pairs. In other words, the structure cannot withstand excitation of the large number of lone pairs and a phase change occurs. Recall that multielemental amorphous and disordered materials provide new degrees of freedom for atomic design.

Figure 6 shows the R-I curve that makes for the reversible binary state of the standard Ovonic Phase Change Memory.

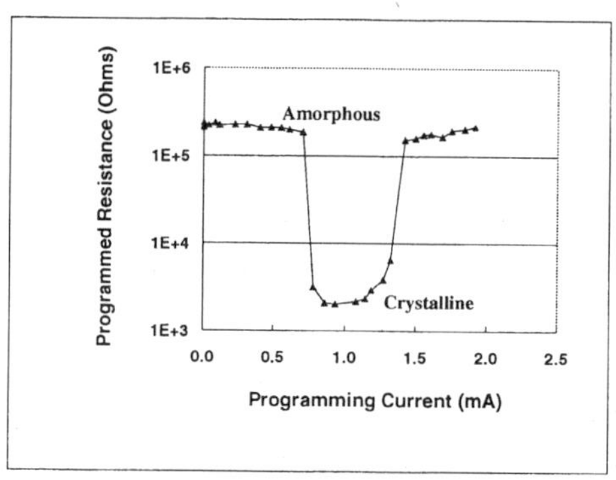

Figure 6: Resistance vs. Current for an Ovonic Phase Change Binary Memory Device

Let us now compare the electrical and optical characteristics of the Ovonic Phase Change Memory with the newly described Ovonic Cognitive Computer device.

As can be seen in Figure 7, the amorphous plasticity phase is where the synaptic activity occurs by accumulating multiple energy inputs to provide a percolation path of submicrocrystalline regions which when reaching threshold, fires the switch to change from the amorphous to the crystalline phase.

The difference between Figures 6 and 7 is that Figure 6 is binary and can have multistates on the right side of the U while in Figure 7, the left side of the U is where the synaptic and neuronal action take place in the amorphous phase. The pulses induce growth of nanocrystals to form the percolation path.

It is remarkable that a single nanostructure can have such multifunctionality and yet retain the reversible amorphous to crystalline transition including the multistates of the right side of the U. The structural changes that provide the synaptic intelligence are too subtle to be detected by forensic means, making it an exceptional encryption device.

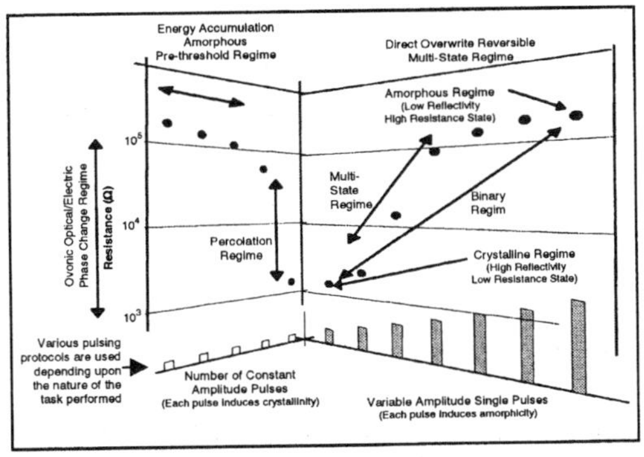

Figure 7: Operation of Ovonic Cognitive Device

2. CONCLUSION

The cognitive devices that I have described here are the beginning of a new paradigm of intelligent devices and computers. The information field is one of the most important pillars of our global economy. Conventional computers are powerful but dumb. Let us consider making all-thin-film computers powerful and intelligent. I hope that this symposium can be stimulated by what I have described here since as we have shown, optical and electrical switching, memory and computing will become unified in the future and we can be the enablers and leaders of such basic and important new industries.[8, 9]

ACKNOWLEDGEMENTS

I appreciate very much the contributions of Boil Pashmakov on the Ovonic Cognitive Computer, and of my colleagues, David Strand for his valuable collaboration through the years, especially his work on the Ovonic Optical Phase Change Memory and the Ovonic Optical Cognitive effect and Takeo Ohta for his very important contributions to the Ovonic Optical Phase Change Memory and Wally Czubatyj for his many years of contributions to the electrical Ovonic Phase Change Memory and Ovonic Threshold Switch. My special thanks to the exceptionally talented group in the ECD-Ovonics semiconducting laboratory, especially Pat Klersy. Alastair Livesey's contributions were most welcome and we thank Eugenia Mytilineou for her help. We also thank and value Tyler Lowrey, head of our joint venture Ovonyx and the great people who make up that organization. As always, Iris is the other half of my lone pair.

REFERENCES

1. P. Davies: Lecture "Did Life Come From Mars?" University of Michigan: Center for Theoretical Physics, April 16, 2004 (2004).
2. M.P. Southworth, Control Engineering, vol. 11, p. 69 (1964)
3. B. Katz: Nerve, Muscle and Synapses. London: McGraw-Hill, Inc. p. 4 (1966).
4. S.R. Ovshinsky and B. Pashmakov: Mat. Res. Soc. Symp., Proc. 803, HH1.1. p. 49 (2004).
5. B. Katz: Nerve, Muscle and Synapses. London: McGraw-Hill, Inc. p. 3 (1966). "Each nerve cell, in a way, is a nervous system in miniature."
6. S. Dehaene: Single-Neuron Arithmetic. Science 297: 1652-3 (2002). "The new findings in numerical neuroscience compel us to accept that our mathematics, sometimes heralded as the pinnacle of human activity, is really made possible by conceptual foundations laid down long ago by evolution and rooted in our primate brain. We are clearly not the only species with a knack for numbers."
7. S.R. Ovshinsky: Keynote - Int. Symp. on Optical Memory 2003 Nov 4; Nara, Japan: Jpn. J. Appl. Phys. Vol. 43, #7B, pp. 4695-4699 (2004).
8. See: S.R. Ovshinsky: In: Adler, D., Schwartz, S.S., Silver, M. editors. Disordered Materials. New York: Plenum Publishing (1991) for prior work on the Ovonic Phase Change Memory, both optical and electrical.
9. S.R. Ovshinsky: Mat. Res. Soc. Symp., Proc. 554, p. 399 (1999).

Phase Change Memory Publications

Reversible Electrical Switching Phenomena in Disordered Structures, Phys. Rev.Lett. 21 (1968) 1450.

Radiation Hardness of Ovonic Devices (with E. Evans, D. Nelson and H. Fritzsche), IEEE Trans. Nuclear Sci. NS-15 (1968) 311.

The Ovshinsky Switch, Proc. 5th Ann. National Conf. on Industrial Research, Chicago, IL (1969) p. 86.

Electronic Conduction in Amorphous Semiconductors and the Physics of the Switching and Memory Phenomena (with H. Fritzsche), J. Non-Cryst. Solids 2 (1970) 393.

An Introduction to Ovonic Research, ibid., p. 99.

Reversible Conductivity Transformations in Chalcogenide Alloy Films (with E.J.Evans and J.H. Helbers), ibid., p. 334.

Structural Studies of Amorphous Semiconductors (with A. Bienenstock and F. Betts), ibid., p. 347.

Conduction and Switching Phenomena in Covalent Alloy Semiconductors (with H. Fritzsche), J. Non-Cryst. Solids 4 (1970) 464.

A Qualitative Theory of Electrical Switching Processes in Monostable Amorphous Structures (with H.K. Henisch and E.A. Fagen), ibid., p. 538.

Radial Distribution Studies of Amorphous GexTe1-x Alloys (with F. Betts and A. Bienenstock), ibid., p. 554.

Reflectivity Studies of the Te (Ge, As)-Based Amorphous Semiconductor in the Conducting and Insulating States (with J. Feinleib), ibid., p. 564.

Time Delay for Reversible Electric Switching in Semiconducting Glasses (with K.W. Boer and G. Doehler), ibid., p. 573.

Analog Models for Information Storage and Transmission in Physiological Systems (with Iris M. Ovshinsky), Mat. Res. Bull. 5 (1970) 681.

Radial Distribution Studies of Amorphous GexTe1-x Alloys (with F. Betts and A. Bienenstock), ibid., p. 554.

Reflectivity Studies of the Te (Ge, As)-Based Amorphous Semiconductor in the Conducting and Insulating States (with J. Feinleib), ibid., p. 564.

Time Delay for Reversible Electric Switching in Semiconducting Glasses (with K.W. Boer and G. Doehler), ibid., p. 573.

Analog Models for Information Storage and Transmission in Physiological Systems (with Iris M. Ovshinsky), Mat. Res. Bulletin 5 (1970) 681.

Electrothermal Initiation of an Electronic Switching Mechanism in Semiconducting Glasses (with K.W. Boer), Appl. Phys. 41 (1970) 2675.

Reversible High-Speed High-Resolution Imaging in Amorphous Semiconductors (with P.H. Klose), Society for Information Display International
Symposium, Philadelphia, PA ; Digest of Technical Papers (1971) p. 58.

Glass Switch, McGraw-Hill Encyclopedia of Science and Technology 13 (1971) 360.

Rapid Reversible Light-Induced Crystallization of Amorphous Semiconductors (with J. Feinleib, J. deNeufville and S.C. Moss), Appl. Phys. Lett. 18 (1971) 254.

Reversible Structural Transformations in Amorphous Semiconductors for Memory and Logic (with H. Fritzsche) J. Non-Cryst. Solids 8-10 (1972) 892.

Reversible Optical Effects in Amorphous Semiconductors (with J. Feinleib, S. Iwasa, S.C. Moss and J.P. deNeufville), ibid., 909.

Amorphous Semiconductors for Switching, Memory, and Imaging Applications (with H. Fritzsche), IEEE Trans. on Electron Devices, ED-20 (1973) 91.

Mechanism of Reversible Optical Storage in Evaporated Amorphous AsSe and Ge10As40Se50 (with J.P. deNeufville, R. Seguin and S.C. Moss), Proc.
5th Intl. Amorphous and Liquid Semiconductors Conference, eds. J. Stuke & W. Brenig (Taylor and Francis, London, 1974) p.737.

Amorphous Materials as Interactive Systems, Proc. 6th Intl. Conf. on Amorphous and Liquid Semiconductors, Leningrad (1975) p.426.

An Experimental Study of Threshold Switching in Some Binary Chalcogenide-Based Glass Films (with R.A. Flasck, M.P. Shaw and K. Dec), ibid. p. 490.

Amorphous Materials as Optical Information Media, J. Appl. Photographic Engineering
3 (1977) 35.

New Amorphous Materials for Computer Use, 18th IEEE
Computer Society International Conference, San Francisco, CA (1979) p. 158.

New Experiments on Threshold Switching in Chalcogenide and Non-Chalcogenide Alloys (with K. Homma and H.K. Henisch), J. Non-Cryst. Solids 35/36 (1980) 1105.

Threshold Switching in Chalcogenide Glass Thin Films (with D. Adler, M. Shur and M. Silver), J. Appl. Phys. 51 (1980) 3289.

This Week's Citation Classic [S.R. Ovshinsky, Reversible Electrical Switching Phenomena in Disordered Structures, Phys. Rev. Lett. 21 (1968)1450], Current Contents 22 (1982) 18.

Switch, Glass (with D. Adler) McGraw-Hill Encyclopedia of Science and Technology (McGraw-Hill Book Company, 5th through 8th Editions, 1982-1994).

Reply to "Comment on 'Threshold Switching in Chalcogenide Glass Thin Films'," (with D. Adler, M.S. Shur and M. Silver), J. Appl. Physics 56 (1984) 579.

Amorphous Semiconductors for Microelectronics, SPIE Proc. 617(1986) 2.

Effects of Transition-Metal Elements on Tellurium Alloys for Reversible Optical-Data Storage (with R. Young, D. Strand and J. Gonzales-Hernandez), J. Appl. Physics 60 (1986) 4319.

Optically Induced Phase Changes in Amorphous Materials, J. Non-Cryst. Solids 141 (1992) 200.

Crystallization Studies of Ge:Sb:Te Optical Memory Materials (with J. Gonzalez-Hernandez, B. Chao, D. Strand, D. Pawlik and P. Gasiorowski), Appl. Phys. Comm. 11(1992) 557.

Ovonic Phase Chang Memory Making Possible New Optical and Electrical Devices, Japan J. Appl. Phys. (1997) 44.

The Relationship Between Crystal Structure and Performance as Optical Recording Media in Te-Ge-Sb Thin Films (with D. Strand, J. Gonzalez-Hernandez, B. Chao and P. Gasiorowski and D. Pawlik), Mat. Res. Soc. Symp. Proc. 230 (1992) 251.

Historique du Changement de Phase, Memoires Optiques & Systems, 127(1994) 65.

The Story of Phase Change for Optical Storage, Balzers Materials 9 (1999) 6.

New Developments in Optical Phase Change Memory (with W. Czubatyj),
5th International Symposium on Optical Storage (ISOS 2000),
SPIE Proc. Vol. 4085 (2001) 15.

Phase Change Optical Storage, Keynote speech "given by the great father of phase-change memory, Stanford R. Ovshinsky," E*PCOS01 European Symposium on Phase Change Optical Storage, Santis, Switzerland (September 3-4(2001).

Phase Change Optical Storage Media (with T. Ohta), Ch.18 in Photo-induced Metastability in Amorphous Semiconductors, ed. A. V. Kolobov (John Wiley&Sons, Canada Ltd. 2003) pp 310-326.

Innovation Providing New Multiple Functions in Phase Change materials to Achieve Cognitive Computing, (with B. Pashmakov), Mat. Res. Soc. Proc. 803 (2004) 49.

A New Information Paradigm-The Ovonic Cognitive Computer, Optoelectronic Materials and Devices 1 (2004) 1.

Phase Change Electronic Memories: Towards Cognitive Computing, The Encyclopedia of Materials: Science and Technology (Elsevier Science, Ltd. 2005) p.106.

Optical Cognitive Information Processing-A New Field, Japan J. Appl. Phys. 43 (2004) 4695.

Ovonic Chalcogenide Non-Binary Electrical and Optical Devices, 7th Intl. Symp. On Optical Storage Devices Proc. Vol. 5966 (2005) 1.

Electro-optical Investigations of Ovonic Chalcogenide Memory Devices (with E. Mytilineou, B. Pashmakov, D. Strand and D. Jablonski), ICANS 21 (2005).

US patents - phase change memory

Resistance switches and the like
3271584 09/06/1966

Symmetrical current controlling device
3271591 09/06/1966

Analog-to digital converter employing semiconductor threshold device &differentiator circuit
3327302 06/20/1967

Power switching circuit
3336484 08/15/1967

Transient suppressor
3343034 09/19/1967

Overvoltage protection of a.c. measuring devices
3343085 09/19/1967

Method and apparatus for storing and retrieving information
3530441 09/22/1970

Current controlling device
3571673 03/23/1971

Current controlling device including vo2
3588638 06/28/1971

Display screen using variable resistance memory semiconductor
3644741 02/22/1972

Apparatus for electrostatic printing
3659936 05/02/1972

Printing and copying employing materials with surface variations
3678852 07/25/1972

High speed printer of multiple copies for output information
3698006 10/10/1972

Switchable current controlling device with inactive material dispersed in the active semiconductor material
3715634 02/06/1973

Multi-terminal amorphous electronic control device
3748501 06/24/1973

Light emitting display array with non-volatile memory
3763468 10/02/1973

Method of imaging and imaging material
3819377 06/25/1974

Bi-directional arrangement of amorphous electronic control devices
3823331 07/09/1974

Method and apparatus for storing and reading data in a memory having catalytic material to initiate amorphous to crystalline change in memory structure
3868651 02/25/1975

Matrix of amorphous electronic control device
3876985 04/08/1975

Structure and method for producing an image
3961314 06/01/1976

Dry process production of archival microform records from hard copy
3966317 06/29/1976

Method and apparatus for recording information
3983542 09/28/1976

Recording and retrieving information in an amorphous memory material using a catalytic material
3988720 10/26/1976

Thermal imaging involving imagewise melting to form spaced apart globules
4000334 12/28/1976

Imaging and recording of information utilizing tellurium tetrahalide
4066460 01/03/1978

Continuous tone imaging film method for storage of retrievable information dispersion imaging material and method
4103044 07/25/1978

Imaging and recording of information utilizing a tellurium tetrahalide complex of an aromatic amine
4106939 08/15/1978

Dry process production and annotation of archival microform records from hard copy
4123157 10/31/1978

Method of continuous tone imaging using dispersion imaging material
4137078 01/30/1979

Organo tellurium imaging materials
4142896 03/06/1979

High temperature amorphous semiconductor member and method of making the same
4177474 12/04/1979

Data storage and retrieval system
4205387 05/27/1980

Amorphous semiconductors equivalent to crystalline semiconductors produced by a glow discharge process
4226898 10/07/1980

Method for full format imaging
4267261 05/12/1981

Tellurium imaging composition
4281058 O7/28/1981

Tellurium imaging composition
4340662 07/20/1982

Data storage and retrieval system
4346449 08/24/1982

Method for making, parallel preprogramming or field programming, of electronic matrix arrays
4545111 10/08/1985

Method of depositing semiconductor films by free radical generation energy
4615905 10/07/1986

Programmable semiconductor structures and methods for using the same
4646266 02/24/1987

Integrated circuit compatible thin filmfield effect transistor and method of making same
4668968 05/26/1987

Thin film field effect transistor
4670763 06/02/1987

Integrated circuit compatible thin film field effect transistor and method of making same
4673957 06/16/1987

Current control device
4689645 08/25/1987

Photodetection and current control devices
4703336 10/27/1987

Data storage medium incorporating a transition metal for increased switching speed
4710899 12/01/1987

Thin film electro-optical devices
4766471 08/23/1988

Thin film field effect transistor and method of making same
4769338 09/06/1988

Solid-state electronic camera including thin film matrix of photosensors
4788594 11/29/1988

Thin film overvoltage protection devices
4809044 02/28/1989

Method for making electronic matrix arrays
4818717 04/04/1989

Thin film field effect transistor and method of making same
4843443 06/27/1989

Electronic camera including electronic signal storage cartridge
4853785 08/01/1989

Semiconductor with ordered clusters
5103284 04/07/1992

Congruent state changeable optical memory material and device
5128099 06/07/1992

Vertically interconnected parallel distributed processor
5159661 10/27/1992

Electrically erasable phase change memory
5166758 11/24/1992

Thin-film structure for chalcogenide electrical switching devices and process therefor
5177567 01/05/993

Electrically erasable, directly overwritable, multibit single cell memory elements and arrays fabricated therefrom
5296716 03/22/1994

Switch with improved threshold voltage
5330630 07/19/1994

Homogeneous compositions of microcrystalline semiconductor material, semiconductor devices and directly overwritable memory elements fabricated therefrom, and
5335219 08/02/1994

Electrically erasable memory elements having reduced switching current requirements and increased write/erase cycle life
5341328 08/23/1994

Electrically erasable memory elements characterized by reduced current and improved thermal stability
5359205 10/25/1994

Electrically erasable, directly overwritable, multibit single cell memory elements and arrays fabricated therefrom
5406509 04/11/1995

Electrically erasable memory elements having improved set resistance stability and memory arrays fabricated therefrom
5414271 05/09/1995

Electrically erasable, directly overwritable, multibit single cell memory elements and arrays fabricated therefrom
5534711 07/09/1996

Electrically erasable memory elements characterized by reduced current and improved thermal stability
5534712 07/09/1996

Electrically erasable, directly overwritable, multibit single cell memory element and arrays fabricated therefrom
5536947 07/16/1996

A novel logic family employing two-terminal chalcogenide switches as the logic elements therein
5543737 08/06/1996

Optical recording medium having a plurality of discrete phase change data recording points
5591501 01/07/1997

Homogeneous compositions of microcrystalline semiconductor material, semiconductor devices and directly overwritable memory elements fabricated therefrom, and
5596522 01/21/1997

Multibit single cell memory element having tapered contact
5687112 11/11/1997

Integrated drivers for flat panel displays employing chalcogenide logic elements
5694054 12/02/1997

Liquid crystal display matrix array employing ovonic threshold switching devices to isolate individual pixels
5694146 12/02/1997

Second-layer phase change memory array on top of a logic device
5714768 02/03/1998

Liquid crystal display matrix array employing ovonic threshold switching devices to isolate individual pixels
5757446 05/26/1998

Composite memory material comprising a mixture of phase-change memory material and dielectric material
5825046 10/20/1998

Universal memory element and method of programming same
5912839 06/15/1999

Memory element with energy control mechanism
5933365 08/03/1999

Optical recording media having increased erasability
6011757 01/04/2000

Memory element with memory material comprising phase-change material and dielectric material
6087674 07/11/2000

Universal memory element with systems employing same and apparatus and method for reading, writing and programming same
6141241 10/31/2000

Multibit single cell memory element having tapered contact
re37259 07/03/2001

Method of computing with digital multistate phase change materials
6671710 12/30/2003

Methods of factoring and modular arithmetic
6714954 03/30/2004

Semiconductor with coordinating irregular structures
6723421 04/20/2004

High storage capacity alloys having excellent kinetics and a long cycle life
6726783 04/27/2004

Phase change data storage device for multi-level recording
6899938 05/31/2005

Methods of factoring and modular arithmetic
6963893 11/08/2005

Multi-terminal chalcogenide switching devices
6967344 11/22/2005

Electrically programmable memory element with improved contacts
6969866 11/29/2005

Field effect chalcogenide devices
6969867 11/29/2005

Analog neurons and neurosynaptic networks
6999953 02/14/2006

Electrically programmable memory element with improved contacts
7023009 04/04/2006

Secured phase-change devices
7085155 08/01/2006

Increased data storage in optical data storage and retrieval systems using blue lasers and/or plasmon lenses
7113474 09/26/2006

Multiple bit chalcogenide storage device
7227170 06/05/2007

Multi-terminal devices having logic functionality
7186998 03/06/2007

Error reduction circuit for chalcogenide devices
7186999 03/06/2007

Chapter IV: Conversion of Solar Energy-Photovoltaics

All life on the planet depends on the sun. Our food and essentially all the energy we use can be traced to current and past radiation from the sun. At present, coal, natural gas, and oil make up 80 percent of the world energy supply. These fuel sources are called fossil fuels because they are remnants of plant and animal life nurtured by the sun, and geologically changed by heat and pressure some 100 million years ago. Even water power and wind power can be related to cycles and effects caused by the sun. Nuclear power, about 6.3 percent of the world's energy supply, is the only energy source which does not originate from our sun. Even the element Uranium feeding nuclear reactors was produced in supernova explosions of massive suns.

Energy from the sun keeps the earth comfortably warm, grows plants and trees by photosynthesis, and circulates by evaporation the water from the oceans to provide the life sustaining rain. Nonetheless there is so much more sun energy which could be harnessed without upsetting the earth's energy balance. The earth receives in a few hours more energy from the sun than we, the 6.3 billion people inhabiting the earth, use in one year.*

Energy from the sun is plentiful, underutilized, non-polluting and renewable. A further advantage is that solar power can be generated locally on roofs instead of by big isolated power stations needing a huge and energy costly delivery system. Its major drawback of course is that the sun does not shine at night.

Sunlight can generate electricity by means of the photovoltaic effect of semiconductors in solar cells and solar panels. In its simplest form, the light absorbed in the semiconductor of a solar cell generates charge carriers (electrons and holes) which are moved by an internal electric field toward conducting contacts that transmit the electricity. The internal electric field is generated by chemically changing the two ends of the semiconductor, making one end electron rich (negative) and the other electron poor (positive). The energy of the band gap of the semiconductor determines the band of color of sunlight that is absorbed and generating charge carriers. The efficiency of a solar cell, the conversion of sunlight into electricity, is high when the absorbed color spectrum is wide and when all photo-generated charge carriers reach the conducting contacts.

**The following estimates illustrate the abundance of solar energy. Part of the solar power is reflected and absorbed by the atmosphere but about $1kW/m^2$ reach the surface near the equator at noon. Averaged over a year, including nights, the power is close to $0.25kW/m^2$ over a wide strip between +/- 30 degree latitude around the equator. Considering only the land areas, the earth receives on the average 18000 TW (18 billion billion kW) sun power. How does this huge number compare with the world's total primary energy need which includes coal, gas, and oil? The International Energy Agency predicts 12000Mtoe for the world's energy need per year in 2010. The unit one Mtoe (million ton of oil equivalent) corresponds to 12 000 million kWh per year. This requires a power source of about 17 TW. If one uses present day commercially available technology of solar power systems having only an 8% efficiency, one would need solar cells covering an area of 800,000 km^2. This area is only a small fraction (3%) of the total area of relatively uninhabited deserts lying in the zone of strong solar radiation.*

Important factors for large scale manufacture of solar panels are low cost, high efficiency, light weight, and materials which are non-toxic, and available in great abundance.

In 1983, the first solar cells made of amorphous silicon were as big as a thumbnail. Ovshinsky realized that inexpensive, large volume production of solar cells require a revolutionary manufacturing process. He maintained that photovoltaic materials should be deposited on substrates by the mile. He envisioned a process similar to the printing of newspapers with rolls of the appropriate substrate going through the deposition process. Unlike newspapers, the deposition process of amorphous silicon solar cells requires a high frequency plasma in nearly high vacuum, protected carefully from any outside contaminants.

The fact that the manufacturing of solar cells on this scale and by the roll- to-roll technique had not been tried before was a major engineering challenge. But before discovering the new field of amorphous semiconductor electronics, Ovshinsky had spent many productive years at the bench in machine shops and factories inventing machine tools. Using his machine shop background, he invented a novel manufacturing process for large area photovoltaic deposition. It is the largest vacuum-tight high- frequency plasma deposition system in the world. In his latest photovoltaic production machines, six rolls holding 14 inch wide and 1.5 mile long stainless steel webs are placed parallel to one another in the vacuum chamber. The webs move through interconnected plasma chambers for receiving different deposition layers. At the end the solar cell is complete and the finished webs are rolled up in the last vacuum chamber. This continuous roll-to-roll process is automated and the quality of the many deposition steps monitored closely. Ovshinsky chose as the semiconductor material amorphous silicon. The material is properly alloyed with hydrogen to reduce its defects which would impede the motion of photo-generated charge carriers toward the conducting contacts. The advantages of amorphous semiconductors over their crystalline counterparts are that they can be deposited over very large areas and one can deposit different kinds of compositions of amorphous materials on top of each other. Ovshinsky decided first to put two solar cells on top of one another and later three solar cell layers. The respective band gaps of the different layers of solar cells were chosen to absorb different color bands of the solar spectrum, red, green and blue. In order not to waste any light energy that had failed to be absorbed, a back reflector is deposited on top of the stainless steel web below the triple solar cell structure, which scatters light back for a second pass through the cells. The stainless steel web is very thin, only 0.005 inch thick. The total thickness of back reflector together with the triple solar cell is less than a tenth of the thickness of a newspaper page. Hence the final photovoltaic product is light, flexible, and strong. It can be shaped into large sheets to be glued on roofs or even made directly into roof shingles.

Ovshinsky's achievement found much praise and recognition. He was introduced into the US-based *Solar Hall of Fame* in 2005 for "promoting excellence in the field of solar energy utilization through public testimonials and tribute to outstanding contributors to the field". In 1999 he received the *Karl W. Boer Solar Energy Medal of Merit*, and in 2004, he was honored with the *Hoyt Clarke Hottel Award* from the American Solar Energy Society in recognition for his "significant contributions to the advancement of solar energy technologies".

Ovshinsky's publications in this section show the development of the engineering and material improvement from the first roll-to-roll machine to the most recent ones in which an area of 55,400 square feet of solar cells is manufactured in about 3 days. At the time of this writing, United Solar Ovonic Corp., a subsidiary of Energy Conversion Devices, runs 5 of these roll-to-roll manufacturing machines with an annual solar cell output of more than 120MW.

Let us mention here some of the engineering challenges and accomplishments. Each of the three solar cells deposited on top of each other consists of three layers, the negatively doped layer, the intrinsic layer whose composition determines whether its band gap absorbs red, green or blue light, and the positively doped layer. This makes for a total of 9 cell layers. The gases active in the radio frequency plasma deposition process are different for each layer and must be kept strictly apart while the stainless steel web moves continuously from one deposition region to the next one in the 300ft long vacuum tight plasma deposition machine. This problem was solved by the invention of gas gates separating the gases but allowing the passage of the webs. Some of the deposited layers are less than a hundredth of the thickness of newspaper sheet, yet this thickness must be even and uniform over the large area of the final product. This requires excellent homogeneity of the plasma power density and the composition and distribution of the active gases, problems which had not been encountered on this scale before. The possibility of producing a multilayer architecture of a triple solar cell and the synthesis of a red, green, and blue cell is a benefit of the flexible structure of the amorphous silicon hydrogen alloy. The flexibility comes with a price, however. Amorphous silicon alloy materials lose some of their properties during extensive exposure to light. As a consequence there is a decline of the initial efficiency of the solar cell to a so-called stabilized efficiency. The decline can be minimized by a structural design of the material as shown by Ovshinsky and his group in some of the publications which follow.

Ovshinsky solved the materials and engineering problems confronting the mass production of flexible solar cells and panels which are more than ten time lighter than batch produced solar cells on glass substrates. Yet, for space and airship applications even lighter solar cells are needed. By utilizing polymeric substrates or very thin stainless steel foils, Ovshinsky's team developed space cells that are even 20 times lighter and hold a record specific power greater than 1000 Watts per kilogram. The radiation hardness and superior high-temperature performance of amorphous silicon make these cells ideal for space applications.

YIELD AND PERFORMANCE OF AMORPHOUS SILICON BASED SOLAR CELLS USING ROLL-TO-ROLL DEPOSITION

K. Hoffman, P. Nath, J. Call, G. DiDio, C. Vogeli, and S.R. Ovshinsky

Sovonics Solar Systems
Troy, Michigan 48084

ABSTRACT

A production plant is being operated by Sovonics Solar Systems which makes amorphous silicon alloy based photovoltaic modules. The solar cells used in these modules are manufactured using roll-to-roll processes. Various thin film depositions are made on 1,000-foot long rolls of 14-inch wide, thin stainless steel in the deposition facility. The rolls are then cut into appropriate sizes for module fabrication. The production facility has been in operation for three years. The purpose of this paper is to detail the performance of solar cells produced at this plant.

MANUFACTURING PROCESS

A brief description of the deposition production process follows. The manufacturing sequence consists of four roll-to-roll operations (see Figs. 1 and 2). These operations are: substrate cleaning, metal reflection coating, plasma-assisted chemical vapor deposition of six amorphous silicon alloy layers, and the deposition of a transparent conductive top layer. The result is a greater than 1,000-foot roll of stainless steel, coated with generic tandem-type, thin film, amorphous silicon alloy photovoltaic material (Fig. 3). A more detailed treatment of these steps has been reported elsewhere [1]. These rolls of amorphous silicon solar cell material can be cut into convenient sizes and fabricated into power generating products. The manufacturing process has the advantages of being highly automated, with low operating cost and high run time versus cycle time. For instance, only two technicians are required to oversee the four manufacturing operations.

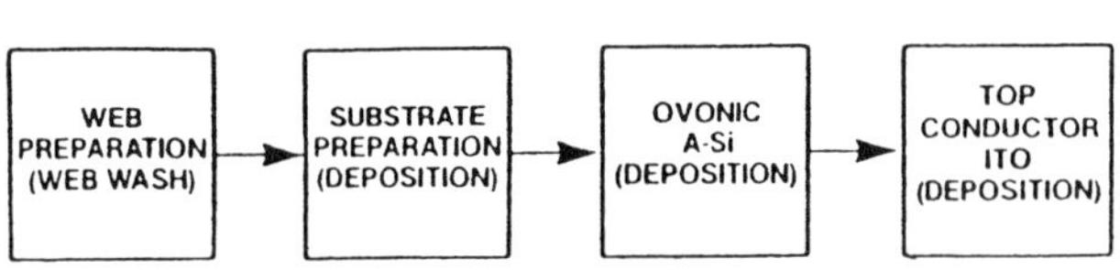

Fig. 1. Schematic of the deposition facility operations.

Fig. 2. Photograph of the deposition facility showing the roll-to-roll coating machinery.

Fig. 3. Photograph of a roll of thin stainless steel coated with the amorphous silicon alloy photovoltaic materials.

0160-8371/88/0000-0293 $1.00 © 1988 IEEE

PERFORMANCE OF THE PHOTOVOLTAIC MATERIALS AND DEVICES

Quality Analysis Procedures

For evaluation of performance, 4" x 14" samples are taken from every 40 feet of run length when the rolls are cut into slabs. These quality analysis samples are fabricated into subcells by scribing the transparent conductive top layer using a chemical etching technique. The samples are then exposed to an electrochemical, defect passivation procedure [2]. Conductive silver grids are applied by screen printing. The area of the subcells and the process used to fabricate the samples emulates the production process. The subcells in each sample may now be measured using a solar simulator and the data becomes part of a computerized database. Other tests using these samples have been devised to evaluate material quality. For example, an adhesion measurement is routinely performed which assures that when the material is flexed, no damage will result. Figure 4 is a graph of such a measurement which shows the tensile strain needed to initiate adhesion failure. Typical films may be bent to a radius of less than 1/16" without loss of adhesion.

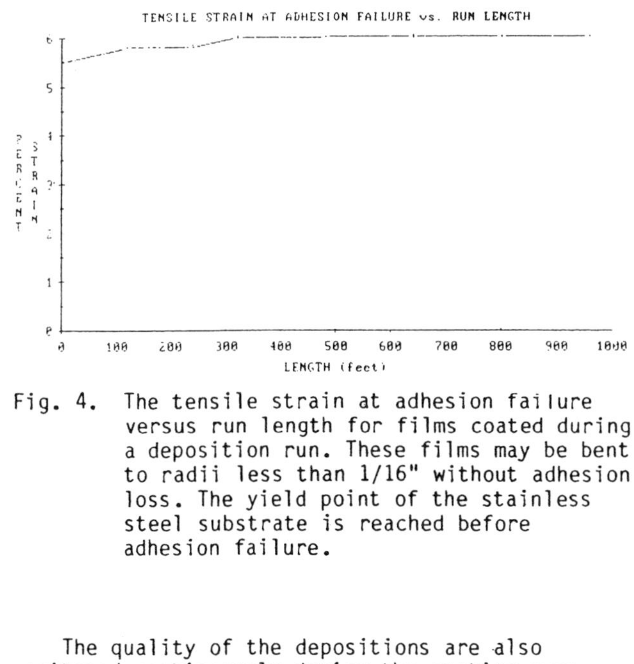

Fig. 4. The tensile strain at adhesion failure versus run length for films coated during a deposition run. These films may be bent to radii less than 1/16" without adhesion loss. The yield point of the stainless steel substrate is reached before adhesion failure.

The quality of the depositions are also monitored continuously during the coating runs. Each operation has microprocessor control. The plasma-assisted CVD processor is constantly updating a computer controlled database during run times which communicates with some 200 separate monitoring points. In addition, in situ monitors measure key parameters (such as conductive oxide resistivity or amorphous silicon layer thicknesses). Control charts, such as shown in Fig. 5 (a plot of the relative intrinsic layer thicknesses), aid quality control of the deposited materials and assure proper machine operation.

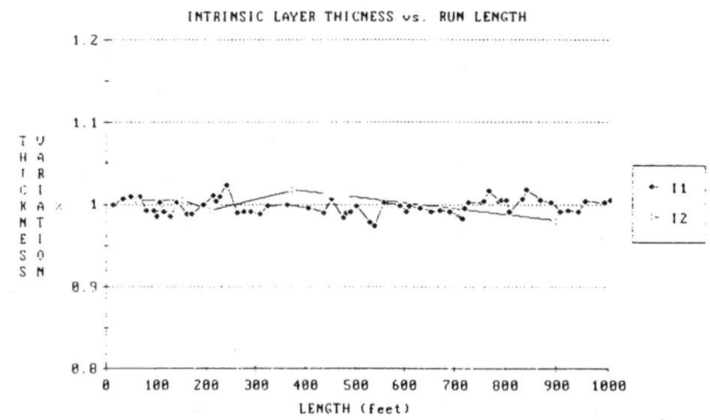

Fig. 5. Normalized thicknesses versus run length of both amorphous silicon alloy intrinsic layers (labeled I1 and I2) of the tandem device. The measurement was made in situ during an amorphous silicon alloy deposition run using a reflection spectrometer.

Photovoltaic Performance Results

A data output from a quality analysis sample is shown in Fig. 6. This result is from a production tandem amorphous silicon alloy deposition run. The sample was fabricated as outlined earlier. The figure displays several features of the photovoltaic characteristics of the film. Device parameters (e.g. Voc, Jsc, fill factor, Pmax) are given for the best cell and the average of the best 5 cells, the best 10 cells, and for cells which pass the yield criteria. The yield is based on cells achieving a fill factor (FF) greater than 0.45. The highest of each of the cell parameters is also given.

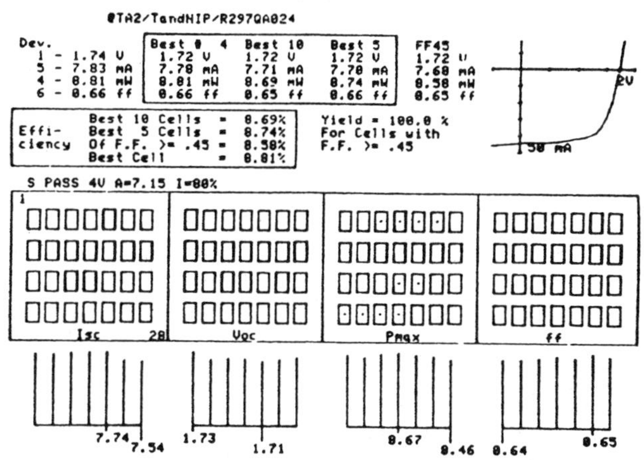

Fig. 6. Quality analysis chart of a test sample used to evaluate photovoltaic devices. Measurements were made using an Oriel solar simulator at AM1.5 Global standard test conditions. The subcell size is 7.15 sq. cm.

Profiles of various device parameters are illustrated by rectangles, whose dimensions are proportional to their respective value. Also provided is information regarding uniformity. Thus it can be seen that this sample has a yield of 100%, an average solar conversion efficiency of 8.58%, and a best subcell efficiency of 8.81%. All measurements were made using standard AM1.5 test conditions at 25 deg C.

Efficiency and subcell yield as a function of run length are illustrated in Fig. 7. This graph was obtained from the database containing sample measurements as outlined above. Such charts of measurements versus run length are available for many photovoltaic and deposition machine control parameters, as was demonstrated in Figs. 4 and 5. Future modification of the plasma-assisted CVD processor to make triple stacked cells will give this facility the capability to make state-of-the-art devices which have been demonstrated in the laboratory using depositions on stainless steel [3].

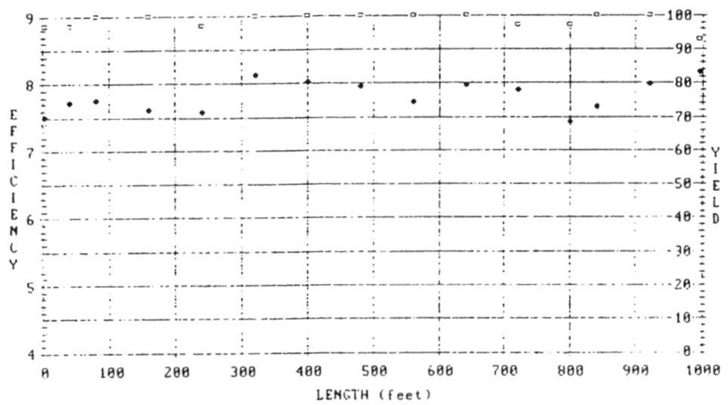

Fig. 7. Photovoltaic conversion efficiency and subcell yield versus run length for a deposition production run. Each data point represents the average control chart values (as in Fig. 6). The closed dots represent efficiency, open squares are yield.

CONCLUSIONS

A manufacturing plant for producing thin film amorphous silicon alloy based solar cells using roll-to-roll deposition is being operated by Sovonics Solar Systems in Troy, Michigan. Its capacity is greater than 1 megawatt of power producing product per year. Roll-to-roll deposition on stainless steel has been shown to be a viable method of producing high quality thin film solar cell materials. Quality analysis, in situ deposition monitoring and post-deposition measurements have enabled the detailed analysis of production run data, including such parameters as layer thickness, adhesion, conversion efficiency, yield, etc. These and other properties of the PV materials and devices have been monitored and studied in order to optimize the roll-to-roll deposition process.

ACKNOWLEDGMENTS

The authors wish to thank T. Ammar for her assistance in the preparation of this manuscript and J. Burdick for useful suggestions.

REFERENCES

1. P. Nath, K. Hoffman, J. Call, C. Vogeli, M. Izu, and S.R. Ovshinsky, proceedings of the 3rd International Photovoltaic Science and Engineering Conference, Tokyo (1987) p.395

2. P. Nath, K. Hoffman, C. Vogeli, and S.R. Ovshinsky, Appl. Phys. Lett. (in press)

3. J. Yang, R. Ross, T. Glatfelter, R. Mohr, G. Hammond, C. Bernotaitis, E. Chen, J. Burdick, M. Hopson, and S. Guha, Late News, this conference.

LIGHTWEIGHT FLEXIBLE ROOFTOP PV MODULE

M. Izu, H. C. Ovshinsky, K. Whelan, L. Fatalski, and S.R. Ovshinsky
Energy Conversion Devices., Inc.
1675 West Maple Road
Troy, MI 48084

T. Glatfelter, K. Younan, K. Hoffman, A. Banerjee, J. Yang, and S. Guha
United Solar Systems Corp.
1100 West Maple Road, Troy, MI 48084

ABSTRACT

Energy Conversion Devices, Inc. (ECD) and United Solar Systems Corp. (United Solar) are developing lightweight, flexible photovoltaic modules that can replace conventional roofing materials and be economically and aesthetically integrated into residential and commercial buildings. The modules will be fabricated from high-efficiency multi-junction a-Si alloy solar cells developed by ECD and United Solar. These cells are produced on thin, flexible, stainless steel substrates. Two types of products, 1 ft by 10 ft overlapping PV shingles and 1.3 ft by 20 ft PV roof panels are being developed by United Solar and ECD, respectively. United Solar's shingle type design uses a roof mounting procedures similar to those used with conventional asphalt shingles, while ECD's PV panel uses mounting procedures conforming to metal roof systems.Thus, they can be installed on roof sheathings, replacing ordinary shingles or metal roofing panels, on a standard wood roof construction.

INTRODUCTION

During the past fifteen years, ECD has made important progress in the development of materials, device designs and manufacturing processes required for the continued advancement of photovoltaic technology. Among these accomplishments, ECD has pioneered, and continues to develop, two key proprietary technologies with significant potential for achieving the cost goals necessary for widespread growth of the photovoltaic market: (1) a low-cost, roll-to-roll continuous substrate thin-film solar cell manufacturing process; and (2) a high-efficiency, multiple-junction, spectrum-splitting thin-film amorphous-silicon alloy device structure [1-10].

Commercial production of multiple-junction a-Si alloy modules has been underway at ECD and its American joint venture company, United Solar, for a number of years using ECD's proprietary roll-to-roll process and numerous advantages of this technology have been demonstrated. These advantages include relatively low semiconductor material cost, relatively low process cost, and the productoin of lightweight, rugged and flexible solar modules with reduced installed PV system costs.

ECD and United Solar are now developing lightweight flexible PV modules that can replace conventional shingles or other roofing materials and be economically and aesthetically integrated into residential and commercial buildings. The modules will be fabricated from high-efficiency, multiple-junction a-Si alloy solar cells developed by ECD and United Solar. These cells are produced on thin, flexible stainless steel substrates and encapsulated with polymer materials.

SCOPE AND METHOD OF APPROACH

Two types of products, 1 ft by 10 ft overlapping PV shingles and 1.3 by 20 ft PV roof panels are being developed by United Solar and ECD, respectively. United Solar's shingle type design utilizes a mounting procedure similar to that for conventional asphalt shingles, while ECD's PV panel design utilizes a mounting procedure similar to that for metal roofing systems. Both types of rooftop PV modules are lightweight and flexible. Consequently they can be installed on roof sheathings, replacing ordinary shingles or metal roofing panels, on a standard wood roof construction. The product design features and installation methods have been refined through discussions and reviews with building industry experts including National Association of Home Builders (NAHB) Research Center, Solar Design Associates, and Minoru Yamasaki Associates.

UNITED SOLAR'S SHINGLE MODULE

United Solar's roofing module is designed to emulate the conventional asphalt shingle in form, structure, function, and installation. This module's form is illustrated conceptually in Fig. 1. The module is meant to be used as the roofing material; additional shingles are not required to be underneath. We anticipate that the shingle module can be installed at the same time as a building is roofed with standard asphalt shingles, using similar techniques and achieving aesthetically appealing integration between the PV and asphalt shingles. Thus, the

First WCPEC; Dec. 5-9, 1994; Hawaii

module will become the roofing material, as well as supplying PV power. The installation procedures envisioned for this module have two distinct advantages: (1) a roofer can physically install the PV shingle using essentially identical procedures to that for conventional asphalt (roofing brads), and (2) the electrical interconnection of the PV modules may be performed as a completely independent task, thereby separating trades during installation. These two advantages combined offer the opportunity of introducing PV into the market without the need to develop a new trade.

A shingle roofing module offers considerable economic advantages when compared to other applications of photovoltaics since a mechanical support structure is not needed for the shingle module. The structure is present at no additional cost in the form of the existing roof deck. The need for module framing is also eliminated for the same reason. United Solar's thin, flexible PV material combined with module construction technology developed in this work can produce environmentally protected modules without the need for edge protection and/or framing.

Module Design Considerations

The new module's geometry is nominally 12 inches by 132 inches, the same width but longer than a conventional asphalt shingle. The module length can be somewhat arbitrary, but is dictated by factors such as weight, ease of handling, electrical connections, and integration with asphalt shingles. As is the case of conventional shingles, the PV shingles will also be overlapped in such a manner as to expose only 5 inches when installed.

A shingle module is composed of several PV sub-panels integrated by laminating the elements together. The sub-panels consist of various thin-film coatings deposited onto thin sheets of stainless steel. The materials used in the module are illustrated in Fig. 2. Environmental protection is provided by ethylene vinyl acetate (EVA) and Tefzel. A polyester-based material is also used on the back surface. The shingle also includes interconnecting wires and current bypass diodes (not shown). Electrical inter-module connections will be made below the sheaves of the roof. There will be one roof penetration for each module via the use of a sealed insert. The result will be a flexible PV module that bears a close resemblance to an asphalt shingle. The PV shingles will protect the roof in the same manner as standard shingles, that is by providing water run off. As indicated above, a clear separation of the building trades is envisioned for the installation of these modules. The roofer would tack down the shingles with customary roofing brads. Later, an electrician would subsequently complete the electrical connections in the attic spaces beneath the roof.

ECD'S METAL ROOF-TYPE MODULE

ECD has designed an aesthetically and mechanically integrated rooftop system using standard metal roofing materials and installation techniques already available in the metal roofing industry. Fig. 3 shows a module schematic. Fig. 4 shows cross-sections of construction components.

As illustrated in Fig. 3, the module size is approximately 1.3 ft by 20 ft. Depending on the application, the width may be varied from 1 to 2 ft and the length may be varied from 10 to 22 ft. Typically, the module consists of seventeen series-connected stainless steel strip cells encapsulated with EVA/Tefzel onto an approximately 1.3 ft by 20 ft standard metal roof panel. Both edges of the module will be rolled into 3/4 inch standing seams after encapsulation so that it will accept the standard metal roof mounting hardware. The module will be installed vertically on the slope of the roof. The length of the module can be adjusted so that it can fit the slope of the roof. The installation procedure is summarized in Table 1. The metal roof-type PV installation does not have horizontal seams. This maximizes watertight integrity. The vertical seams of the metal roof utilize standard metal roof watertight construction. This design does not require any holes on the roof for wiring. All the wiring is made on top of the roof in the ridge vent. These design features will result in ease of installation and reliability of the system, as well as low installed cost. The rooftop PV modules will be mechanically and aesthetically integrated with the adjacent standard metal roof as is seen in Fig. 4. The features and benefits of ECD's d1.3 ft by 20 ft metal rooftop module are summarized in Table 2.

ECD has designed and constructed a large-area 2.5 ft by 22 ft laminator to fabricate its 1.3 ft by 20 ft metal rooftop module.

Table 1. 1.3 FT X 20 FT Metal Roof Module Basic Installation

- → **Prepare Roof Using Standard Roofing Construction**
- → **Locate And Mount First 1.3' X 20' Module And Standard Ridge Vent Bracket With 3 Fasteners At Top Of Module**
- → **Attach Standard Clips Along Sides Of Standing Seam**
- → **Add Standard Metal Roofing, Ends And Flashing To Complete Roof Installation**
- → **Electrically Connect Modules, Accessible From The Top With Ridge Vent Removed Or From The Attic Space**

Table 2. 1.3 FT X 20 FT Metal Roof Module Features

- → **Mechanically And Aesthetically Integrates With Existing Metal Roofing Systems**
- → **Leakproof And Durability Featues Of The Best Metal Roofing Systems**
- → **Minimum Wiring Connections – Wiring Can Be Totally Above The Roof**
- → **Installed By Metal Roofers Using Same Techniques As Metal Roofing**
- → **Installs On Any Roof**
- → **Metal Roof Market Has Grown 270% In 5 Years**

ACKNOWLEDGMENT

The project is supported by DOE/NREL's PV Bonus Program, Building opportunities in the U.S. for

Photovoltaics under subcontract number, DE-FC02-93CH10571. Members of the project team include NAHB Research Center, Detroit Edison, Arizona Public Service Company, Bechtel Corporation, Solar Design Associates, Minoru Yamasaki Associates and others.

The authors wish to express thanks to Joe Weihagen of NAHB Research Center and Steve Strong of Solar Design Associates for their helpful discussions and collaboration.

REFERENCES

[1] M. Izu and S. R. Ovshinsky, Production of Tandem Amorphous Silicon Alloy Cells in a Continuous Roll-To-Roll Process, SPIE Proc. 407, 42 (1983).

[2] M. Izu and S. R. Ovshinsky, Roll-To-Roll Plasma Deposition Machine for the Production of Tandem Amorphous Silicon Alloy Solar Cells, Thin Solid Films 119 55 (1984).

[3] H. Morimoto and M. Izu, Amorphous Silicon Solar Cells Production in a Roll-To-Roll Plasma CVD Process, JARECT 16 (1984); Amorphous Semiconductor Technology & Devices, North Holland Publishing Company, Edited by Y. Hamakawa, 212 (1984).

[4] S. R Ovshinsky, Roll-To-Roll Mass Production Process for Amorphous Silicon Solar Cell Fabrication, Proc. International PVSEC-1, 577 (1988).

[5] High efficiency Multiple-Junction Solar Cells using Amorphous Silicon and Amorphous Silicon-Germanium Alloys, J. Yang, R. Ross, T. Glatfelter, R. Mohr, G. Hammond, C. Bernotaitis, E. Chen, J. Burdick, M. Hopson and S. Guha, Proc. 20th IEEE PV Spec. Conf. 241 (1988).

[6] J. Yang, R. Ross, T. Glatfelter, R. Mohr, and S, Guha, Amorphous Silicon-Germanium Alloy Solar Cells with using Amorphous Silicon and Amorphous Silicon-Germanium Alloys, Proc. 20th IEEE PV Spec. Conf. 241 (1988).Profiled Bandgaps, MRS Symposium Proc. Vol. 149, 435 (1989).

[7] M. Izu, X. Deng, A. Krisko, K. Whelan, R. Young, H. C. Ovshinsky, K. L. Narasimhan, and S. R. Ovshinsky, Manufacturing of Triple-Junction 4 ft^2 a-Si Alloy PV Modules, Proc. of 23rd IEEE PVSC, 919, (1993).

[8] M. Izu, S. R. Ovshinsky, X. Deng, A. Krisko, H. C. Ovshinsky, K. L. Narasimhan, and R. Young, Continuous Roll-to-Roll Amorphous Silicon Photovoltaic Manufacturing Technology, AIP Conf. Proc. 306, 12th NREL PV Program Rev., 198, (1993).

[9] S. Guha, J. Yang, A. Banerjee, T. Glatfelter, K. Hoffman, S. R. Ovshinsky, M. Izu, H. C. Ovshinsky, and X. Deng, Amorphous Silicon Alloy Photovoltaic Technology - From R&D to Production. Proc. of MRS Spring meeting (1994).

[10] X. Deng, M. Izu, K. L. Narasimhan and S. R. Ovshinsky, Stability Test of 4 ft^2 Triple-Junction a-Si Alloy PV Production Modules, Proc. of MRS Spring meeting (1994).

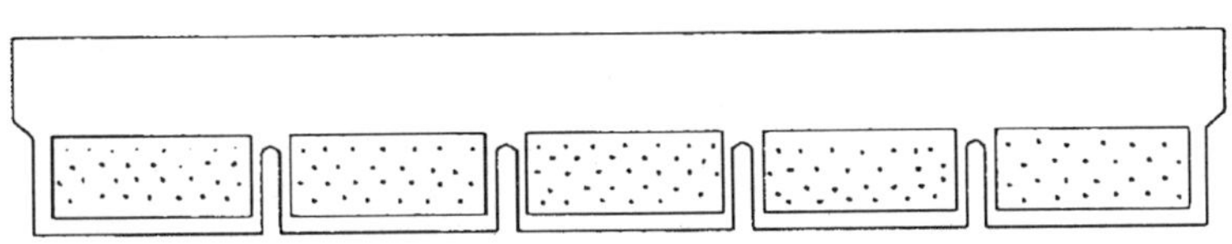

Figure 1 **Schematic of shingle module footprint. The shaded area represents the exposed PV.**

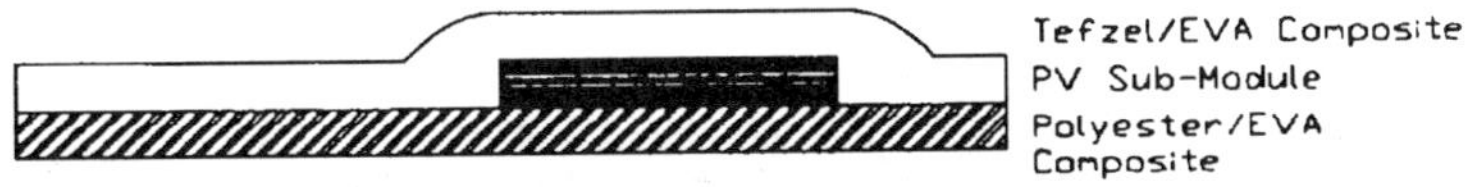

Figure 2 **Shingle module materials stack-up**

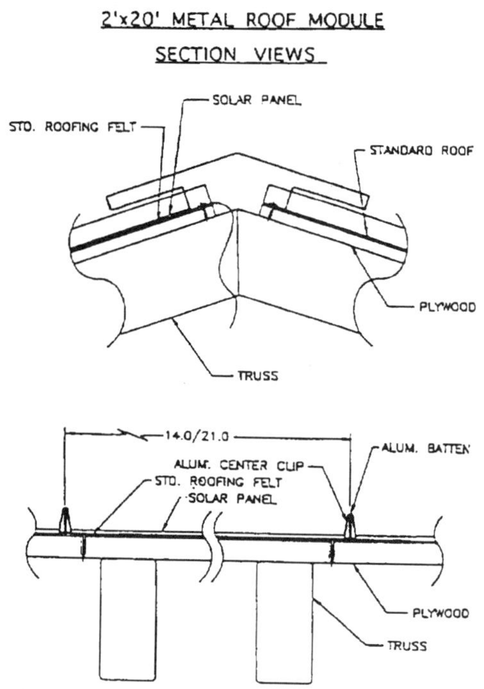

Figure 4 Metal Roof Module Section Views

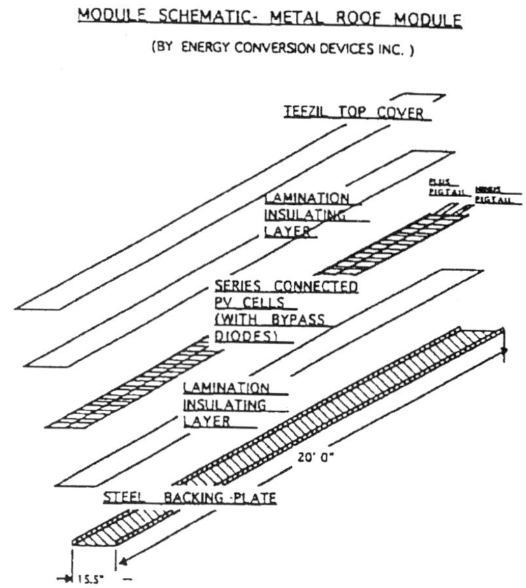

Figure 3 Module Schematic Metal Roof Module

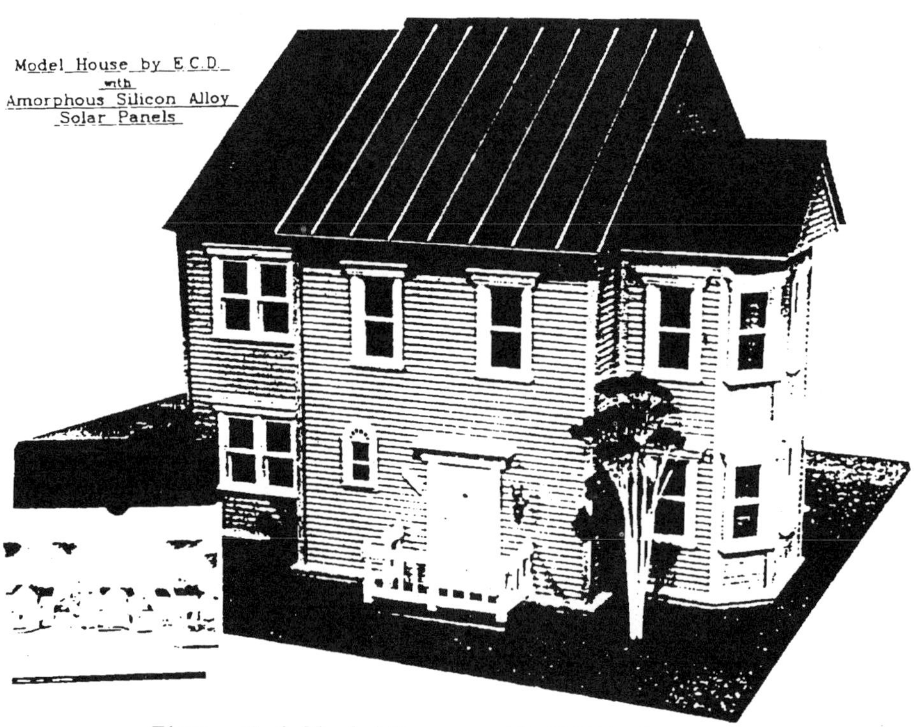

Figure 5 A Model House with ECD's Rooftop Modules

ELSEVIER Solar Energy Materials and Solar Cells 32 (1994) 443–449

Solar Energy Materials and Solar Cells

The material basis of efficiency and stability in amorphous photovoltaics

Stanford R. Ovshinsky *

Energy Conversion Devices, Troy, MI 48084, USA

Abstract

Amorphous thin film tetrahedrally based alloys have recently achieved an efficiency which permits them in volume production to become competitive to fossil fuels. In this paper, the next stage of development which is to raise stabilized efficiencies to 18% is discussed. It is shown how the scientific and technological problems can be viewed and understood and point to means of solution.

1. Introduction

Amorphous tetrahedrally based alloys have led to recent photovoltaic modules of a one square foot commercial type which reached the 10% stabilized efficiency [1] which has been the holy grail of workers in this field.

The 10% figure has been of vital importance because when such material efficiency is combined with the continuous wide web, triple junction, roll-to-roll production process of a suitable size, for example, a 50 to 100 megawatt machine, it is possible for the first time for solar energy to compete head to head with conventional fossil fuels and nuclear energy at approximately 7 cents per kilowatt hour. The figures used to validate such an assumption are based upon a US government formulation [2].

These advances are of historical importance in the struggle to minimize pollution, to reduce the energy role of depletable fossil fuels, especially oil whose control is a major cause of war, and to create new industries which will generate desperately needed jobs without which social and economic disorder is a given.

It has been shown that proven technology is already available to meet these important objectives (see Fig. 1). Amorphous silicon based alloys are the basis for

Correspondence to: Energy Conversion Devices, 1675 W. Maple Rd., Troy, MI 48084, USA).

SSDI 0927-0248(93)E0128-Z

Fig. 1. ECD's fifth generation, continuous web, roll-to-roll, triple (three band gap), photovoltaic production plant in Troy, Michigan. In its initial operation in the US before shipment to its Sovlux joint venture in Russia, it had already achieved world records for efficiency.

the only thin film solar cell technology that has been proven in production and these products have through the years met the test of use. However, rather than resting on our laurels, the author would like to discuss in this paper the strategies that can raise the efficiencies to the 18% stabilized range and beyond. As of now, the small area initial efficiency record is 13.7% which was achieved at our laboratory [3].

The attainment of this next transition step would make photovoltaics ubiquitous and do so much to achieve Bernie Seraphin's vision of freeing the developing world from the ever-deepening energy crisis that envelops it, preventing the achievement of industrial development so necessary for the dignity inherent in a civilized standard of living.

It is fitting that this paper is dedicated to Bernie Seraphin who is an exemplary figure and represents the best part of science. He has made important contributions and at the same time has recognized that science and technology are not value free but have responsibilities to serve our world societal needs. Bernie has done more than anyone to develop scientists from Third World countries so that they can be nucleating agents for change in their native lands as well as to integrate them into the international scientific community. Iris and I remember with great pleasure our visits to Trieste at Abdus Salam's International Centre for Theoretical Physics to address Bernie's meetings. We interacted with scientists from various parts of the world and felt that we influenced their thinking about the possibilities of much desired nonpolluting energy for the building of their countries and introduced them to the scientific and technological potential of amorphous and disordered materials and thin film technology.

2. Amorphous and disordered materials problems

While the advantages of amorphous and disordered materials in the field of photovoltaics are now quite obvious, the remaining problems that have acted as barriers to their further development can be clearly stated. By understanding their physical and chemical causes, one can develop strategies to minimize or indeed eliminate them. For example, the Staebler–Wronski Effect, which concerns itself with the decline of initial efficiency within a period of time, is such a problem. Amorphous materials by nature have a degree of freedom available for relaxation not found in rigid crystalline structures such as crystalline silicon/germanium [4]. Another problem is the high density of states that is generated when germanium is added to silicon to form alloys of lower band gap materials which incorporate hydrogen. The same problem holds true when carbon is added to silicon in the higher bandgap materials.

A triple-junction device of three different bandgaps (altogether no more than one micron thick) is the optimal configuration to utilize the solar spectrum for the highest efficiency. If the density of states can be lowered in the two other layers to the same degree ($5 \times 10^{15}/cm^3$) as the amorphous silicon alloy layer containing silicon, hydrogen and/or fluorine (the author has a great affection for fluorine [4]) and stabilized, then a stable 18% photovoltaic commercial product is assured.

3. Influence of local chemical configurations

The triple band gap device is already being put into production. Here we would like to discuss the basis for new plasma chemistry. Rather than discussing the transient chemical species which are the precursors that become part of the film-making process, we wish to address the basic and fundamental problem which when understood can lead to the generation of the proper chemical species necessary to achieve our goals.

Why should the addition of germanium and/or carbon, which are band controlling elements so necessary in a multijunction, high efficiency device, cause critical limiting problems and if one has an understanding of these factors, are there solutions? The author believes so and discuss here such approaches. From a materials point of view, much work has been based upon a flawed understanding

Table 1
Bond strengths

Bond type	Bond energy (kcal/mole)
Si–Si	53
Si–H	76
Si–F	135
Si–O	190
Ge–Ge	45
Ge–H	69
Ge–F	108
Ge–O	158

of the structural/chemical basis of the semiconducting properties of amorphous tetrahedral materials. The treatment of the Column IV elements, particularly silicon and germanium, had as its basic assumption that in the amorphous state these two elements were equally made tetrahedral in new configurations by the introduction of hydrogen. A glance at Table 1 shows that the silicon–silicon bond is quite weak, the silicon–hydrogen bond is far from optimal and the silicon–fluorine bond is the preferred configuration. When we look at the germanium–germanium bond, we find it to be even weaker than the silicon–silicon bond. The germanium–hydrogen bond is weaker than the silicon–hydrogen bond and again we see that the germanium–fluorine bond is the preferred one.

We can also understand that the stronger bonding elements will compete with the weaker ones in an alloy, for example, hydrogen and fluorine would prefer a silicon mate rather than a germanium one during the conventional deposition process. We can appreciate the perils of an insufficient vacuum which would allow contaminants with strong bonding energies such as oxygen to fill up sites and generate defects that should be available to truly compensating elements such as hydrogen and/or fluorine. Fluorine's small size and particularly its unique extreme electronegativity make it ideal for expanding orbitals, for example, in utilizing the "inert" orbitals (lone pairs) of germanium so as to assure tetrahedral bonding. However, due to its high reactivity, it is difficult to utilize fluorine in the plasma.

Other differences between silicon alloys and alloys containing germanium have to do with the larger size and heaviness of the germanium atom. This not only affects the mobility of the germanium atom on the growing surface, but requires higher temperatures during deposition which increase hydrogen evolution. This lack of surface mobility in films containing amorphous germanium has other negative aspects since there is a higher degree of disorder in the network and a higher porosity in the deposited film. This is a subject which requires greater discussion and a detailed paper is presently being prepared showing how this problem can be minimized [5]. The Staebler–Wronski Effect is intimately tied to weak bond formation and the relaxations available to unoptimally cross-linked amorphous materials [4]. The answer to this effect lies in stronger and more tetrahedrally bonded configurations and denser materials.

When we examine carbon containing alloys, we should assume that it is no easy matter to create tetrahedral materials out of carbon for if this were true, obviously diamond would be quite common. The problem with carbon is simply the beauty of carbon that allows life to exist that it has many different configurations. Forcing it to assume a uniform tetrahedral structure, one has to fight this tendency. I have shown many years ago that fluorine is an element which has the chemical strength to force carbon into a more tetrahedral configuration for density of states is a reflection of the lack of ability to complete tetrahedral bonding in carbon-based materials [6]. The alloying elements chosen to control the band gap play an obvious role but must be chosen with the basic premise in mind of assuring completed four-coordinated bonding sites without introducing unwanted states in the gap.

The problem of completing tetrahedral configurations is less obvious in germanium than in carbon but has basic similarities. It seems so simple. Alloy silicon with germanium, narrow the band gap, add some hydrogen to take care of dangling bonds and the result will be a low band gap tetrahedral material with a low density of states. The basic point here is that, particularly in the amorphous state, there is a tendency for some germanium atoms to assume trivalent and divalent configurations. These, of course, add to the inherent density of states. (By the way, as one goes down to tin, this tendency to some lower coordination becomes even more pronounced.) Many of these configurations are silent defects since germanium has lone pairs and when tetrahedralness is not assured, these lone pairs can be involved in new configurations. Even if their lone pairs are in their more normal, nonbonding state, they still cannot be detected with ESR since they are spin up and spin down configurations [4]. This has been a source of misunderstanding in the semiconductor community which assumes that in these materials, increase of density of states should be associated with conventional dangling bonds.

What we mean by tetrahedral bonding in amorphous materials is often misunderstood. If all silicon or germanium were truly tetrahedrally bonded, we would have crystalline materials and as I have pointed out [7], there is an important factor that must be taken into account which is that unlike, for example, chalcogenide materials, wherein their divalency allows flexible restructuring and completion of local configurations, we must treat amorphous tetrahedral materials in terms of their manufacturing criteria with the same sort of care that one would manufacture crystalline materials, for any deviation from tetrahedralness introduces states in the gap. It is important to note that amorphous tetrahedral materials are process dependent.

The role of hydrogen and fluorine is not only to assure the lack of dangling bonds by completing the tetrahedral structure, but by so doing, silicon/hydrogen/fluorine generate new configurations with differing local geometries than silicon/germanium atoms alone would assume. In fact, amorphous materials of the tetrahedral type are dependent upon deviations from local order [8,9] as periodic materials are affected by perturbations of periodicity. One may look at amorphous silicon alloys as materials in which disorder is minimized yet utilized for the freedom it permits so that local order and total interactive environment around the local order become the important parameters.

An important factor chemically is that one atom does not fit all sizes. The use of hydrogen has been automatically extended to germanium without taking into account the differences in bond strengths. It is by this time well known that unlike the original hope, it is much more difficult to dope amorphous tetrahedral material than crystalline. Doping not only acts substitutionally but can generate additional defects. As the materials are made into forms that have lower and lower density of states, doping efficiency increases. Also, the tendency for intermediate order increases. In my opinion, there is a great future for the semiconducting industry by utilizing intermediate range and nanostrucally ordered materials in devices [10].

4. Influence of speed of deposition

We have discussed local chemical configurations, both bonding and structural. There is another area of topology that is vital to be addressed if one is to achieve optimum materials. Guha, an important contributor to our field, was able to show that the speed of deposition was becoming a barrier to good electronic quality materials by virtue of porosity [11]. Since porosity in conventional plasma deposition techniques is associated with higher speeds of deposition reflected in greater number of defects leading to increased density of states, we must again turn to a structural/chemical solution. That is, we must generate a different type of plasma than is presently being utilized. This requires the control of free radicals [12] in the plasma. We have and are developing proprietary means of achieving this goal which depend upon the understanding of the gas phase environment and the controlling of the elements within that plasma, the electrons, various types of ions and the chemically active neutral species. The author believes very much that the plasma has been less of a "zoo" and more of a "jungle" and in sorting out and controlling the various desired and friendly species, we have a new approach to plasmas as a deposition means which will assure the proper growth kinetics and surface reactions which control the film growth and density as well as generating local structural/chemical configurations which reduce the density of states.

Thin film technology with its ability to do atomic engineering predates the present emphasis on nanostructures. There are many new device possibilities that have been and are being generated by the investigation of amorphicity, disorder and local order. It is important to point out that local order and the localized total interactive environment are materials design parameters which form the path to highly efficient photovoltaics as well as the forerunners to new areas of semiconductor use.

Iris and Stan wish Bernie many more happy birthdays.

References

[1] S. Guha, J. Yang, A. Banerjee, T. Glatfelter, K. Hoffman and X. Xu, 12th NREL PV Program Meeting, Denver, October 1993 (to be published). Initial efficiency of 11.4% on one square foot validated by NREL August 1993. Stabilized efficiency of 10.2% on one square foot validated by NREL December 1993.

[2] G.J. Jones, H.N. Post and M.G. Thomas, 19th IEEE Photovoltaic Specialist Conf., 1987, p. 25.
[3] S. Guha, J. Yang, A. Pawlikiewicz, T. Glatfelter, R. Ross, and S.R. Ovshinsky, Appl. Phys. Lett. 54 (1989) 2330.
[4] S.R. Ovshinsky, in: D. Adler, B.B. Schwartz and M. Silver (Eds.), Disordered Materials: Science and Technology — Selected Papers, 2nd ed. (Institute for Amorphous Studies Series, Plenum, NY, 1991).
[5] S.R. Ovshinsky, R. Young and D. Tsu (to be published).
[6] S.R. Ovshinsky and J. Flasck, US Patent 4-663-183, 5 May 1987.
[7] S.R. Ovshinsky, Proc. 6th Int. Conf. on Amorphous and Liquid Semiconductors, Leningrad, 1975, p. 426.
[8] S.R. Ovshinsky, Proc. 9th Int. Conf. on Amorphous and Liquid Semiconductors, Grenoble, 1981, p. 1095.
[9] S.R. Ovshinsky, Rev. Roum. Phys. 26 (1981) 893.
[10] S.R. Ovshinsky and R. Young, US Patent 5-103-284, 7 April 1992.
[11] S. Guha, J. Yang, S.J. Jones, Y. Chen and D. Williamson, Appl. Phys. Lett. 61 (1992) 1444.
[12] S.R. Ovshinsky, Proc. Int. Ion Engineering Congress, ISIAT '83 and IPAT '83, Kyoto, Japan, 1983, p. 817.

AMORPHOUS SILICON ALLOY PHOTOVOLTAIC TECHNOLOGY - FROM R&D TO PRODUCTION

S. GUHA,* J. YANG,* A. BANERJEE,* T. GLATFELTER,* K. HOFFMAN,*
S.R. OVSHINSKY,** M. IZU,** H. C. OVSHINSKY,** AND X. DENG**
*United Solar Systems Corp., 1100 West Maple Road, Troy, MI 48084
**Energy Conversion Devices, Inc., 1675 West Maple Road, Troy, MI 48084

ABSTRACT

The key requirements for photovoltaic modules to be accepted for large-scale terrestrial applications are (i) low material cost, (ii) high efficiency with good stability, (iii) low manufacturing cost with good yield and (iv) environmental safety. Thin films of amorphous silicon alloy are inexpensive; the products are also environmentally benign. The challenge has been to improve the stable efficiency of these modules and transfer the R&D results into production. Using a multijunction, multi-bandgap approach to capture the solar spectrum more efficiently, we have developed one-square-foot modules with initial efficiency of 11.8%. After 1000 h of one-sun light soaking, a stable efficiency of 10.2% was obtained. Both the efficiency values were confirmed by National Renewable Energy Laboratory. The technology has been transferred to production using an automated roll-to-roll process in which different layers of the cell structure are deposited in a continuous manner onto stainless steel rolls, 14" wide and half a mile long. The rolls are next processed into modules of different sizes. This inexpensive manufacturing process produces high efficiency modules with subcell yields greater than 99%. The key features of the technology transfer and future scope for improvement are discussed.

INTRODUCTION

There is a great need for a renewable, non-polluting energy source which can be used for generation of electricity for large-scale terrestrial applications. Photovoltaic (PV) modules are being used extensively for generation of electricity in remote areas; the cost, however is still much higher than that produced by conventional fuels. In order for PV to be economically viable, the modules must have low material cost, they must show high efficiency with good stability and they must be easily manufacturable with good yield. Amorphous silicon (a-Si) alloys have attracted a great deal of attention [1] because only a thin layer (less than 1 μm) is needed to absorb the solar photons. The material cost of amorphous silicon PV panels is therefore low. The challenge has been how to improve stable efficiency and demonstrate manufacturability. In this paper, we describe our work to address these issues.

EFFICIENCY CONSIDERATIONS

Cell Efficiency

It is now well recognized that a multi-bandgap, multijunction approach offers the opportunity to obtain the highest stable efficiency for a-Si alloy solar cells [2]. A schematic

Mat. Res. Soc. Symp. Proc. Vol. 336. ©1994 Materials Research Society diagram of a triple-junction structure is shown in Fig. 1. The top cell which captures the blue photons uses a-Si alloy with an optical gap of ~1.8 eV for the intrinsic (*i*) layer. The *i* layer for the middle cell is a-SiGe alloy with about 10% Ge. The optical gap is ~1.6 eV which is ideally suited for capturing the green photons. The bottom cell captures the red and the infrared photons and uses an *i* layer of a-SiGe alloy with about 40-50% Ge corresponding to an optical gap of ~1.4 eV. Light which is not absorbed in the cells gets reflected from the Ag/ZnO back reflector which is usually textured to scatter the light at an angle to facilitate multiple internal reflections.

The requirements to obtain high efficiency multijunction cells are the following: (1) high quality back reflector for efficient light trapping, (2) high efficiency component cells, (3) high quality doped layers to obtain good internal "tunnel" *p n* junctions with low electrical and optical losses and (4) optimum matching of the component cells.

The back reflector should perform two important functions. It must be highly reflecting; it should also scatter light at an angle higher than the critical angle for total internal reflection which, for a-Si alloy, is 16.6°. Ag is usually used to obtain high reflectivity. The interface between Ag and Si, however, is not highly reflecting because of intermixing of the two elements, and a buffer layer of ZnO is deposited in between to prevent intermixing. The required texture for optimum scattering is obtained by depositing Ag and ZnO at a high temperature in the range between 100 to 400 °C.

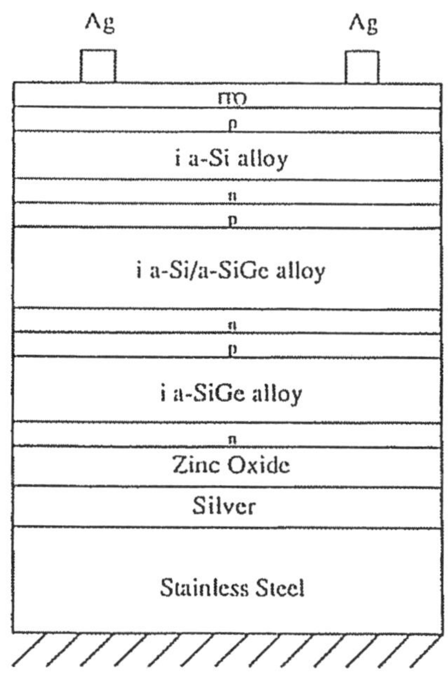

Fig. 1. Schematic diagram of a triple-junction cell structure.

The use of Ag/ZnO back reflector has resulted in significant gain [3,4] in short-circuit current density (J_{sc}) over specular stainless steel substrate. For a 4000 Å *i* layer of a-Si alloy, the gain is as much as 4 to 5 mA/cm^2. Theoretical calculations, however, show [5] that there is scope for much larger gain if there is total internal reflection without any loss at the reflecting surface. Experiments with different kinds of texture with different specular to diffusive reflection ratio show a remarkable insensitivity of J_{sc} to the degree of texture once a certain amount of texture is achieved [6]. It appears that there is a parasitic loss at the Si/ZnO and ZnO/Ag interface which limits J_{sc}. Elimination or reduction of this loss will have a large impact on cell efficiency, especially for the multijunction structure.

High efficiency component cells need high quality *i* layers. As mentioned earlier, we use a-Si alloy for the top cell and a-SiGe alloy for the middle and the bottom cells. In order to obtain high open-circuit voltage (V_{oc}) with good stability, we use hydrogen dilution of the gas mixture. Since the first [7] report of observation of improved stability with hydrogen dilution in a-Si alloy films, significant work has been carried out [8,9] to understand the role of excess hydrogen in the growth process. It is believed that hydrogen coverage of the growing surface gives the impinging species longer time to be incorporated and as a result gives a denser structure. For a-SiGe alloy we use [10] hydrogen diluted gas mixture of Si_2H_6 and GeH_4. We have shown earlier that since dissociation rates of Si_2H_6 and GeH_4 are similar, a gas mixture of Si_2H_6 and GeH_4 produces higher quality material than SiH_4 and GeH_4. Although significant advances have been made to improve the quality of a-SiGe alloy, the best quality a-SiGe alloy still has poorer transport properties than a-Si alloy. To achieve better collection from a given material, profiling of Ge-content in the cell has been used [11] to (a) provide increased built-in field and (b) generate the holes closer to the p-contact so that they do not have very far to move.

Typical initial performances for state-of-the-art component cells for the triple-junction structure are shown [12] in Table I. Also shown are the values after filtered one-sun (metal-arc lamp), 50 °C, 600 h light soaking. In this experiment, the component cells were degraded under open-circuit condition at 50 °C for 600 h and measured at 25 °C. The top cell was degraded under one sun and measured under AM1.5 illumination; the middle cell was degraded under one sun with a 530 nm cut-on filter and measured under AM1.5 illumination with the same filter; the bottom cell was degraded under one sun with a 630 nm cut-on filter and measured under AM1.5 illumination with the same filter. The top and middle cells were deposited on textured substrate without any back reflector, since in the multijunction configuration these cells do not see much reflected light. The bottom cells were deposited on our conventional Ag/ZnO textured back reflector. We notice a degradation of 10% to 20% after light soaking. We should mention that one can improve the initial performance by making the component cells thicker, but this results in larger degradation and lower light-degraded efficiency.

All the component cells in this study show true saturation in efficiency after prolonged light exposure. A typical example for the top and the bottom cell is shown in Fig. 2. The degradation is much lower than those obtained under intense light illumination, demonstrating the importance of thermal annealing of defects under normal operating conditions.

The doped layers play an important role in terms of providing high built-in potential in the bulk and also facilitating junctions between the adjacent cells without resistive loss. Since the doped layers are inactive as far as conversion of light to electricity is concerned, they must also be optically transparent. B-doping of a-Si alloy usually results in larger absorption in the material; the conductivity is also low resulting in large junction loss between the *p* and the *n* layers. We have developed [13] microcrystalline *p* layer with low optical loss. The layers are also highly conducting so as to provide high built-in potential and low tunnel junction loss [14].

Table I. Present status of typical initial and degraded cell parameters for component cells degraded and measured under conditions described in the text. The high- and the mid-bandgap cells use Cr as back reflector. Use of Ag/ZnO as back reflectors for these cells increases J_{sc} by 30% to 40%.

			J_{sc} (mA/cm^2)	V_{oc} (V)	FF	P_{max} (mW/cm^2)
a-Si	high-bandgap cell	initial	7.3	1.01	0.75	5.53
		degraded	7.2	0.98	0.71	5.01
		degradation (%)	1.4	3.0	5.3	9.4
a-SiGe	mid-bandgap cell	initial	7.02	0.77	0.65	3.51
		degraded	6.85	0.74	0.57	2.89
		degradation (%)	2.4	3.9	12.3	17.7
a-SiGe	low-bandgap cell	initial	7.8	0.67	0.64	3.34
		degraded	7.7	0.65	0.56	2.80
		degradation (%)	1.3	3.0	12.5	16.2

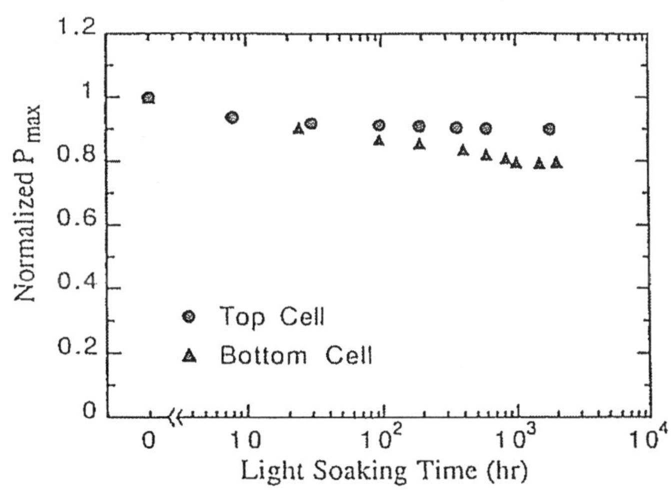

Fig. 2. Saturation in light-induced degradation of component cells.

For the fabrication of the multijunction cell, the next requirement is the matching of the component cells. The stability of the multijunction cell is primarily dictated by the performance of the component cell that limits the current. Since the top cell shows the maximum stability, it is desirable to design the structure top-cell limited, and small-area stable efficiency exceeding 11% has been achieved [15] using this approach.

Module Efficiency

In order to translate the small-area cell performance into modules, several key requirements need to be fulfilled. There has to be good uniformity over the deposited area; the optical loss due to encapsulation and current carrying grid-lines and the electrical loss due to transparent conducting oxide, grid-lines, etc. should be low. We use a monolithic approach in which the unit cell is a large-area cell of one-square-foot area. The grid design and the encapsulant are optimized to give an optical loss of ~2.1% and an electrical loss of ~2.4%. The total loss from average small-area efficiency to module is thus expected to be 4 to 5%.

A large number (about 200) of multijunction modules has been made over the last two years incorporating the optimizations outlined above. This resulted in a significant progress in the improvement of module efficiency as indicated in Fig. 3. The program on fabrication of one-square foot modules on Ag/ZnO back reflector started in September 1991, and in a period of about two years, the initial module efficiency has increased from 7.5% to 11.8%. The best results achieved to date are shown in Table II where the initial efficiencies for one-square-foot modules as measured at National Renewable Energy Laboratory (NREL) under a Spire 240A simulator are shown. Also shown are the measurements at NREL on modules from the same batch which have been light soaked under one-sun condition for 1000 h at 50 °C.

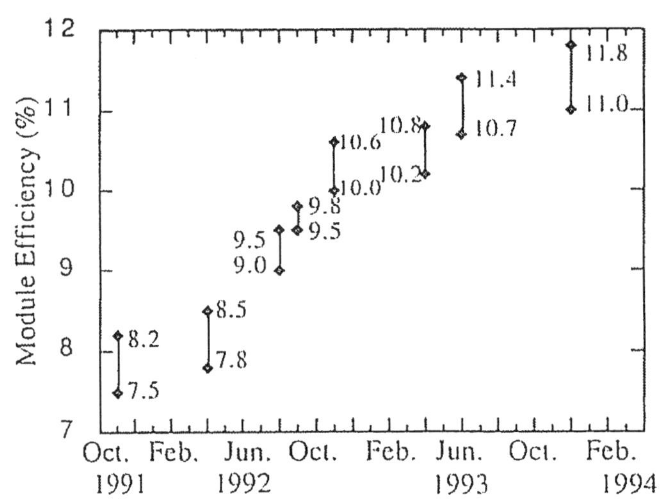

Fig. 3. Progress in initial module efficiency.

Table II. Summary of Module Results as Measured at NREL.

Sample	State	J_{sc} (mA/cm^2)	V_{oc} (V)	FF	η (%)
2452	Initial	7.25	2.400	0.675	11.75
2465	Initial	7.48	2.395	0.652	11.69
2437	Final	6.86	2.354	0.629	10.16
2445	Final	7.04	2.349	0.607	10.04
2447	Final	7.17	2.318	0.612	10.17

The stable module efficiency of 10.2% meets the important milestone of 10% stipulated for thin film PV modules to be acceptable for large-scale terrestrial applications. We should mention that as per NREL guidelines [16], the term "stable" refers to the performance level reached after 600 h of one-sun light soaking at 50 °C. As we have demonstrated earlier [17], our multijunction modules show saturation in degradation after about 600 h of one-sun light soaking at 50 °C. Because of thermal annealing effects, the degradation is lower at higher light soaking temperature and higher at lower temperature.

MANUFACTURING ISSUES

In order to translate the R&D results into production, the efficiency targets must be met with using a low-cost process with high yield. Energy Conversion Devices (ECD) pioneered a roll-to-roll method [18] of deposition of solar cells in which rolls of stainless steel, half-a-mile long, 14" wide and 5 mil thick move in a continuous manner in four machines that serve the purpose of (i) washing, (ii) depositing the back reflector, (iii) depositing the a-Si alloy layers and (iv) depositing an antireflection coating. At United Solar, a fully automated roll-to-roll manufacturing line has been operational for many years [19] for depositing same bandgap double-junction cells. The coated web with the deposited cell is next processed to make a variety of lightweight, flexible and rugged products. The processing steps involve (i) cutting of the web into 16" x 14" slabs, (ii) scribing of ITO by screen printing, (iii) short and shunt passivation, (iv) screen printing of silver grid and (v) final assembly involving cell cutting and interconnection and lamination.

Typical stable module efficiency of the products is about 6% [20]. In order to improve the efficiency further, it is necessary to introduce many of the innovations described in the previous section into the manufacturing line. A triple-junction module manufacturing facility with an annual capacity of 2 MW has recently been designed and built [21] for Sovlux, a joint venture between ECD and Kvant, Moscow. The manufacturing line incorporated a Ag/ZnO back reflector sputtering machine to facilitate improved light trapping in the module. It also uses the triple-junction cell design in which the bottom cell uses a-SiGe alloy with bandgap profiling. The middle and the top cell use a-Si alloy deposited at different temperatures to change the bandgap.

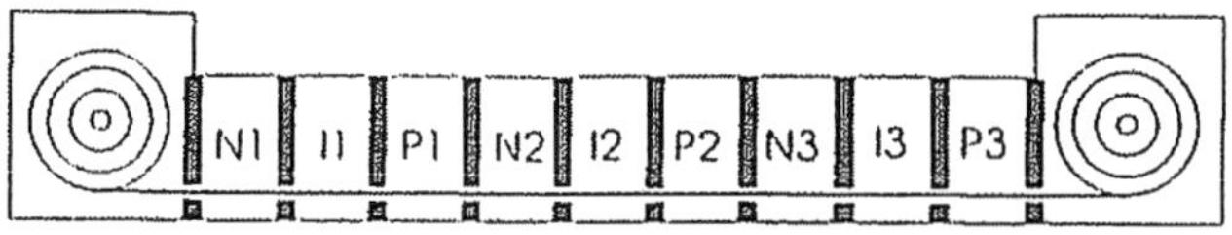

Fig. 4. Schematic diagram of roll-to-roll triple-junction a-Si alloy deposition machine.

The schematic diagram for the a-Si alloy deposition machine is shown in Fig. 4. The machine consists of a web payoff chamber, nine plasma-CVD chambers for the nine layers of the triple-junction cell and a take-up chamber. Stainless steel web, coated with Ag/ZnO, moves continuously through the chambers depositing the various layers sequentially. The process gas mixtures in each section are dynamically isolated from the adjacent chambers depositing the doped and undoped layers by proprietary "gas gates." The "gas gates" utilize laminar gas flow through constant geometrical cross-section conduits in a direction opposite to the diffusion gradient of the dopant gas concentration. SIMS analysis shows that one can eliminate migration of dopants to the undoped layer to a level below $10^{16}/cm^3$.

One of the key features in the a-Si alloy deposition machine is the provision for bandgap profiling of a-SiGe alloy in the bottom cell. This is achieved by using a proprietary gas distribution manifold and cathode configuration so that a gas mixture containing different amounts of GeH_4 is delivered into different parts of the chamber. The design was optimized to obtain any desired profile of Ge-concentration in the intrinsic layer.

For a PV production line, it is important to have appropriate evaluation of the cell performance such as cell efficiency, yield and uniformity through the entire production run. In our quality analysis (QA) process, statistical samples of 4″ x 14″ are selected from the production roll of 2500′ at an interval of 60′. Twenty-eight test solar cells of 7.35 cm^2 area (7 rows and 4 columns) are processed on each sample by the following procedures: (1) TCO scribing by screen printing of etching paste, heat curing and rinsing, (2) short and shunt passivation, and (3) screen printing of Ag paste grid.

The J-V data of 28 cells in a typical sample, sample 23 of run 109, is summarized in Table III. The efficiencies of all 28 cells are above 10%. The uniformity is excellent. The average V_{oc}, J_{sc}, FF and η are 2.37 V, 6.51 mA/cm^2, 0.673 and 10.41%, respectively, as shown in the table. With a subcell yield criterion of FF $\geq$ 0.55, the subcell yield of the sample is 100%. Figure 5 is a three-dimensional plot of subcell efficiency for every cell from the statistical QA/QC samples in a production run. Out of 1176 cells, only 3 show shunts or shorts. The results represent the excellent consistency, uniformity and yield achieved in a continuous roll-to-roll manufacturing process.

Table III. Performance Data of 28 Cells in a QA/QC Sample.

Cell No.	V_{oc} (V)	J_{sc} (mA/cm^2)	FF	Eff (%)	Cell No.	V_{oc} (V)	J_{sc} (mA/cm^2)	FF	Eff (%)
1	2.38	6.45	0.706	10.49	15	2.38	6.24	0.698	10.36
2	2.38	6.45	0.694	10.64	16	2.38	6.47	0.663	10.20
3	2.37	6.66	0.658	10.37	17	2.37	6.55	0.679	10.54
4	2.37	6.88	0.602	10.46	18	2.37	6.52	0.672	10.39
5	2.37	6.65	0.670	10.59	19	2.37	6.67	0.654	10.33
6	2.38	6.54	0.676	10.52	20	2.38	6.49	0.674	10.40
7	2.38	6.52	0.668	10.36	21	2.39	6.63	0.663	10.50
8	2.39	6.36	0.670	10.19	22	2.38	6.39	0.684	10.39
9	2.38	6.32	0.684	10.29	23	2.37	6.27	0.704	10.47
10	2.37	6.35	0.686	10.30	24	2.37	6.43	0.695	10.59
11	2.37	6.88	0.633	10.31	25	2.37	6.53	0.679	10.50
12	2.37	6.52	0.671	10.36	26	2.37	6.66	0.652	10.30
13	2.38	6.39	0.688	10.45	27	2.38	6.67	0.651	10.34
14	2.39	6.71	0.669	10.72	28	2.38	6.37	0.674	10.21
					Average	2.37	6.51	0.673	10.41

Yield for cells with FF ≥ 0.55: 100%

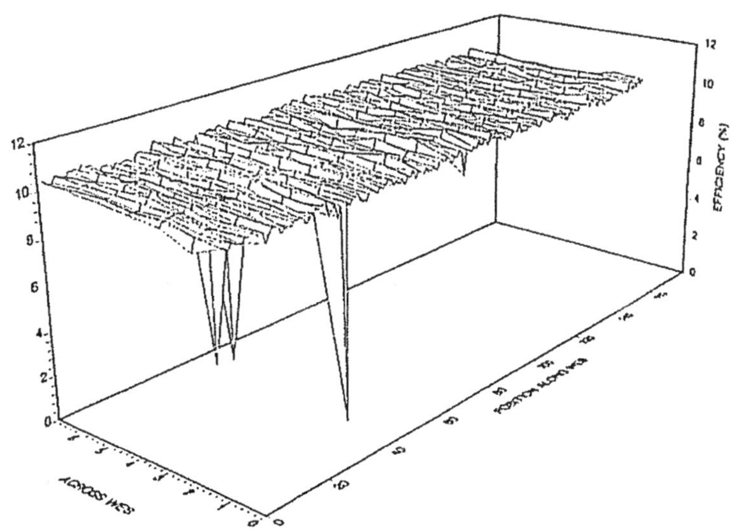

Fig. 5. Three dimensional plot of cell efficiency of test cells in a production run.

Table IV. Initial Module Performance Data of Four-square-foot Modules Produced in the Manufacturing Line.

Module	V_{oc} (V)	I_{sc} (A)	FF	η	Measurement Lab
23	21.62	2.68	0.628	9.31	ECD
	21.56	2.72	0.63	9.46	NREL
27	21.6	2.72	0.632	9.48	ECD
28	21.3	2.75	0.634	9.47	ECD
30	21.61	2.68	0.63	9.36	ECD
	21.51	2.74	0.627	9.47	NREL

Modules of four-square-foot area (1′ x 4′) were fabricated by interconnecting 9 strip cells for obtaining approximately 16 V at maximum power point. Typical results are shown in Table IV. Note that initial aperture area efficiencies in the range of 9.4 to 9.5% have been confirmed both at ECD and NREL. The modules were light soaked under one sun for 2000 h at a temperature between 50 to 60 °C. The degradation behavior is shown in Fig. 6. We note that the stabilized efficiency is 8%.

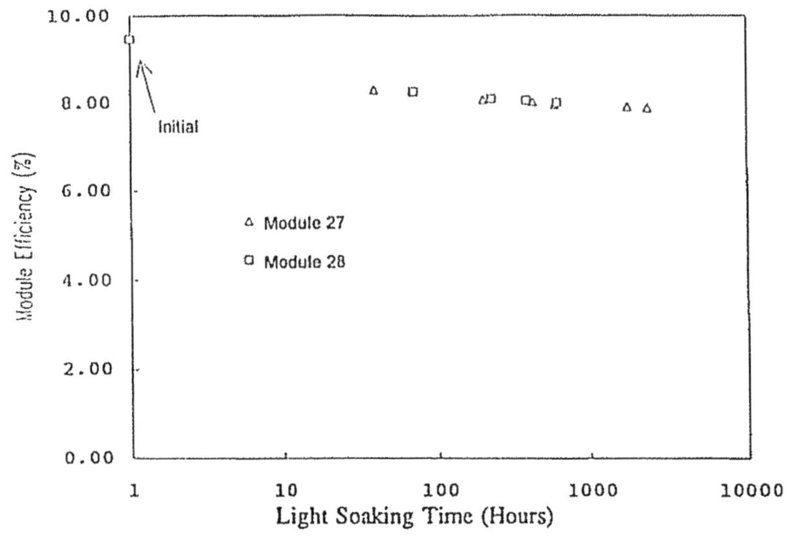

Fig. 6. Module efficiencies of two 4 ft^2 modules as a function of light soaking time under one-sun illumination.

The manufacturing line is still being optimized, and it is important to discuss how the gap between 10% stabilized efficiency in R&D and 8% from the manufacturing line can be bridged. The middle cell in the latter case does not contain any germanium. This limits the short-circuit current from the cell and lowers the efficiency. The deposition rate of the top cell is high (> 10 Å/s). Use of somewhat lower deposition rate and higher hydrogen dilution as discussed earlier will improve the module efficiency further.

FUTURE DIRECTIONS

Although the achievement of 10% stable module efficiency addresses the important issue of near-term market acceptability, the multijunction approach holds promise for much higher efficiency. The component cell requirement for 15% module efficiency has been addressed by Guha et al. [17]. Typical characteristics necessary vis-a-vis the current status at United Solar are shown in Table V. Improvement in material quality, especially for the low bandgap alloy, will be necessary to meet that goal. Use of novel plasma deposition methods [22] in which the growth kinetics can be controlled by selecting suitable species to impinge on the growing surface will play a key role in improving the stable material quality. Further investigation of the role of microstructure, hydrogen bonding and impurities on stable device performance for materials deposited under different conditions will help us to have a better understanding of the issues involved. Reduction of losses in the back reflector will also improve the efficiency. With a well-focussed sustained approach encompassing both material and device research, the progress in improvement of stable module efficiency can surely be maintained.

Table V. Component cell stabilized parameters for 15% module efficiency. (Present status at United Solar shown in parentheses.)

	Voc (V)	FF	J_{sc} (mA/cm^2)	Power (mW/cm^2)
Top Cell$^+$	1.1 (0.98)	0.75 (0.71)	8.2 (7.7)	6.8 (5.3)
Middle Cell^{*+}	0.89 (0.74)	0.70 (0.57)	8.4 (6.9)	5.2 (2.9)
Bottom Cell$^\Delta$	0.68 (0.65)	0.68 (0.56)	8.6 (7.7)	4.0 (2.8)
Devices	2.67	0.72	8.3	16.0

$^+$ Without back reflector
* Under λ>530 nm light
$^\Delta$ Under λ>630 nm light

CONCLUSIONS

General considerations for obtaining high stabilized solar cell and module efficiencies using a-Si alloy are discussed. Using a multi-bandgap, multijunction approach, a stable efficiency of 10.2% has been achieved on a one-square-foot panel. A manufacturing line has been designed and built to translate the R&D results into a low-cost production process with high yield. Further scope for improvement in efficiency is discussed.

ACKNOWLEDGEMENT

The authors would like to thank X. Xu, A. Krisko and K.L. Narasimhan for discussions and collaboration, V. Trudeau for preparation of the manuscript, and National Renewable Energy Laboratory for supporting the program under Subcontract Nos. ZM-1-19033-2 at United Solar Systems Corp. and ZM-2-1104-7 at Energy Conversion Devices, Inc.

REFERENCES

1. See, for example, Mat. Res. Soc. Symp. Proc. **297** (1993).
2. S. Guha, Mat. Res. Soc. Symp. Proc. **149**, 405 (1980).
3. A. Banerjee and S. Guha, J. Appl. Phys. **69**, 1030 (1991).
4. A. Banerjee, J. Yang, and S. Guha (to be published).
5. E. Yablonovitch and G. Cody, IEEE ED-**29**, 300 (1982).
6. S. Guha, Optoelectronics **5** (2), 201-207 (1990).
7. S. Guha, K.L. Narasimhan, and S.M. Pietruszko, J. Appl. Phys. **52**, 859 (1981).
8. K. Tanaka and A. Matsuda, Mat. Sci. Reports **2**, 139 (1987).
9. A. Gallagher, SERI Technical Report, SERI/TP-211-3747 (1990).
10. S. Guha, J.S. Payson, S.C. Agarwal, and S.R. Ovshinsky, J. Non-cryst. Solids **97-98**, 1455 (1987).
11. S. Guha, J. Yang, A. Pawlikiewicz, T. Glatfelter, R. Ross, and S.R. Ovshinsky, Appl. Phys. Lett. **54**, 2330 (1989).
12. X. Xu, J. Yang, and S. Guha, Proc. 23rd IEEE PVSC, Louisville, KY, 971 (1993).
13. S. Guha, J. Yang, P. Nath, and M. Hack, Appl. Phys. Lett. **49**, 218 (1986).
14. A. Banerjee, J. Yang, T. Glatfelter, K. Hoffman, and S. Guha, Appl. Phys. Lett. **64**, 1517 (1994).
15. J. Yang and S. Guha, Appl. Phys. Lett. **61**, 2917 (1992).
16. W. Luft, B. Stafford, and B. von Roedern, in Amorphous Silicon Materials and Solar Cells, AIP Conf. Proc. No. 234, edited by B. Stafford (American Institute of Physics, New York, 1991), p. 3.
17. S. Guha, J. Yang, A. Banerjee, T. Glatfelter, K. Hoffman, and X. Xu, PVSEC-7, 43 (1993).
18. M. Izu and S.R. Ovshinsky, SPIE Proc. **407**, 43 (1983).
19. P. Nath, K. Hoffman, J. Call, C. Vogeli, M. Izu, and S.R. Ovshinsky, PVSEC-3, 395 (1987).
20. W. Luft, B. von Roedern, B. Stafford, D. Waddington, and L. Mrig, Proc. 22nd IEEE PVSC, Las Vegas, NV, 1393 (1981).
21. M. Izu, X. Deng, A. Krisko, K.Whelan, R. Young, H.C. Ovshinsky, K.L. Narasimhan, and S.R. Ovshinsky, Proc. 23rd IEEE PVSC, Louisville, KY, 919 (1993).
22. S.R. Ovshinsky, Solar Energy Materials (in press).

This article appears in Mat. Res. Soc. Symp. Proc. Vol. 345

PV METAL ROOFING MODULE

Tim Ellison, Larry Fatalski, Rob Kopf, Herb Ovshinsky, Masat Izu, Ramona Souleyrette, Ken Whelan, Stanford R. Ovshinsky
Energy Conversion Devices, Inc. (ECD), 1675 W. Maple, Troy MI 48084

Joe Wiehagen, Larry Zarker
NAHB Research Center. (Research Center), 400 Prince Georges Blvd., Upper Marlboro MD 20772

ABSTRACT

Energy Conversion Devices, Inc. (ECD) has developed a building integrated PV (BIPV) metal roofing module under the DOE/NREL PV:BONUS program. The module has been designed to be aesthetically and mechanically integrated into residential and commercial buildings. The module is an exact replacement for a metal roofing "pan" manufactured by ATAS International, requires no electrical feedthroughs or additional fasteners, and uses standard metal roofing materials and installation techniques already in use in the metal roofing industry. Consequently this PV roofing retains the watertight features of the best 30 year roofing systems and can be installed by metal roofing contractors with no additional specialized training. The modules are fabricated from multi-junction a-Si alloy solar cells produced on a thin, flexible stainless steel substrate. The polymer encapsulated modules are large-area (typically up to 22 ft long and 1.5 ft wide), lightweight, and flexible. An array of these modules was installed on one of the NAHB Research Center 21st Century Townhouses on 06 May '96, a week prior to this conference. This paper summarizes: the PV module and demonstration array design; the installation procedure; the demonstration program; and the projected cost.

BACKGROUND

During the past fifteen years, ECD has made important progress in the development of materials, device designs and manufacturing processes required for the continued advancement and commercialization of photovoltaic technology. Among these accomplishments, ECD has pioneered and continues to develop two key proprietary technologies with significant potential for achieving the cost goals necessary for widespread growth of the photovoltaic market: (1) a low-cost, roll-to-roll continuous substrate thin-film solar cell manufacturing process; and (2) a high-efficiency, multiple-junction, spectrum-splitting thin-film amorphous-silicon alloy device structure [1-11]. Commercial production of multiple-junction a-Si alloy modules has been underway at ECD and its American joint venture company, United Solar Systems Corp. (United Solar), for a number of years using ECD's proprietary roll-to-roll process and numerous advantages of this technology have been demonstrated. These include relatively low semiconductor material and process costs and a lightweight, rugged and flexible substrate that results in lowered installed costs of PV systems.

Under the PV:BONUS Program, ECD and United Solar have developed, and are demonstrating and commercializing, two new lightweight flexible BIPV modules specifically designed as exact replacements for conventional asphalt shingles and standing seam metal roofing, respectively. These modules shall incorporate the new triple-junction multi-bandgap cells manufactured in United Solar's new 5 MW manufacturing line. This paper describes ECD's prototype metal roofing module.

PV MODULE AND ARRAY DESIGN

The specifications for the prototype cells, modules and array are summarized in Table I, and a mechanical schematic is shown in Fig. 1. The triple-junction multi-bandgap cells were manufactured on the 2 MW ECD/SOVLUX roll-to-roll manufacturing line [7-8].

Table I. NAHB metal roofing PV specifications.

PARAMETER	SYMBOL	VALUE	UNITS
Cell Properties			
Efficiency	E	7.1	%
Cell Width	w_c	0.316	m
Cell Length	l_c	0.275	m
Cell Open Circuit Voltage	$V_{c,oc}$	2.1	V
Cell Short Circuit Current	$I_{c,sc}$	4.7	A
Module Properties			
Length	L	19	ft
Width	w	15 3	in
Number of Cell/Module	N_m	19	
Module Open Circuit Voltage	$V_{m,oc}$	40	V
Module Short Circuit Current	$I_{m,sc}$	4.7	A
Nominal Module Voltage	$V_{m,p}$	29	V
Nominal Module Current	$I_{m,p}$	4	A
Nominal Module Power	P_m	0.115	kW
Array Properties			
Number of Modules/Array	N_a	18	
Array Open Circuit Voltage	$V_{a,oc}$	80	V
Array Short Circuit Current	$I_{a,sc}$	42	A
Nominal Array Voltage	$V_{a,p}$	58	V
Nominal Array Current	$I_{a,p}$	36	A
Nominal Array Power	P_a	2.1	kW

0-7803-3166-4/96/$5.00 © 1996 IEEE

25th PVSC; May 13–17, 1996; Washington, D.C.

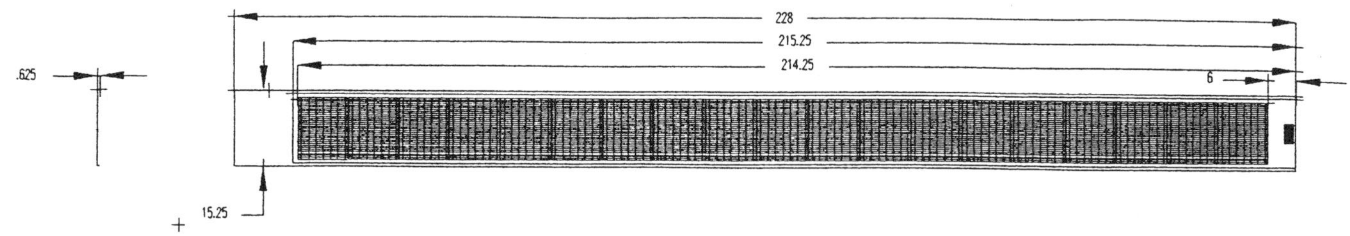

Figure 1. Mechanical schematic of ECD's prototype metal roofing module designed for the NAHB 21st Century Townhouse (all dimensions are in inches).

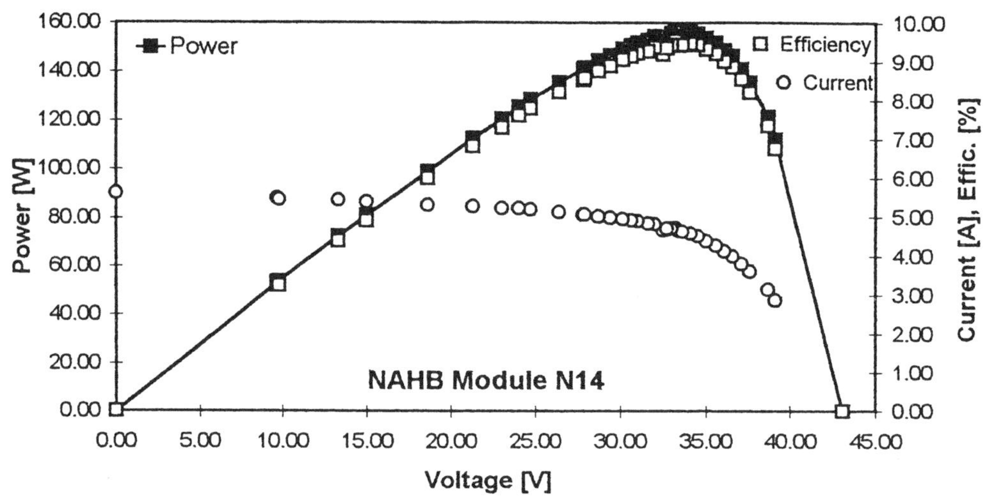

Figure 2. V-I curve for 19 ft long metal roofing module at AM 1.5 conditions.

Each cell was exposed to AM 1.5 conditions, and then tested under 10% light conditions and assembled into 5-cell submodules with flat chip bypass diodes. These submodules were then connected together and laminated to 19 ft long, 16.5 in wide standard Kynar/Hylar-coated 24 gauge galvalume (AZ-55) metal roofing material in a lamination machine capable of making modules up to 23.5 ft long and 2.8 ft wide.

After lamination, the modules were tested outdoors. Figure 2 shows the I-V curve for one of the modules. Finally, the modules were shipped to ATAS International in Allentown PA and roll-formed into standard metal roofing "pans". The roll-forming machine bends a 5/8 inch border up on each side of the module, as shown in Fig. 1, resulting in a 15 ¼ in wide module.

PROTOTYPE TEST ARRAYS

ECD fabricated 3 test arrays using its previous generation of triple-junction dual bandgap PV material to test the module design and installation methods. Two of these test arrays were sent to NREL, a third was kept at ECD and used for outdoor testing. The modules were also tested in United Solar's environmental chambers using the IEEE par 1262 test specifications.

INSTALLATION DESIGN

As mentioned previously, the ECD metal roofing module is an exact replacement for an ATAS International standing seam metal roofing pan. Consequently, the array retains all the watertight features of this 30 year roofing system, integrates perfectly with the rest of the roof, and most importantly, can be installed by metal roofing tradesman with no additional or specialized training. NAHB has contracted the same roofing company which installed the original metal roofs on the 21st Century Townhouses to install this array as well.

There are no electrical feedthroughs or conduits on the roof; rather, each module has a pair of leads and a ground wire which the roofer lets fall through the ridge vent at the top of the module during installation. The electrical hookup is performed independently by an electrician working in the attic area.

The array installation took place on Monday 06 May. The roofers were given about 5 minutes of instructions. Approximately 5 minutes was required for installing each of the 18 panels; the bulk of the roofer's time was spent with trim work (independent of the PV installation). The roofers reported that: (1) installing the PV was no more time-consuming than installing a conventional metal roof; and (2) the PV panels, although somewhat heavier than conventional metal pans, were somewhat easier to manage due to the increased stiffness from the lamination. The electrician performed the hookup to the Balance of System (BOS) on Tuesday and Wednesday 07 - 08 May.

DEMONSTRATION PROJECT

A set of four townhouse units have been constructed by the Research Center demonstrating advanced energy efficiency and alternatives to dimensional lumber. An end unit, lot #10, hosts the PV metal roofing as shown in Fig. 3 with the PV installed. Advanced building technologies at or near commercialization are demonstrated in the research homes. Each of the four townhouse units were constructed using different wall materials; major themes are summarized in Table II.

All of the units are fitted with metal roofing. ATAS International supplied the material for lot #10 and also trained the roofers. Due to different schedules for the townhouse construction and the PV metal roofing module fabrication, the modules were installed as retrofit for 18 of the existing pans.

Two sections were used for the PV metal roofing: the modules cover the entire south-facing side of the center section (about 290 ft^2) using 12 modules; the left south-facing section uses 6 modules (141 ft^2) with an additional 60 ft^2 of standard roofing. The townhouse faces 20° east of south.

Table II. NAHB 21st Century Townhouse summary

Lot	Wall Material	HVAC System
#7	Structural Insulated Panels	Natural gas engine heat pump
#8	I.C.E. block insulated concrete forms	Integrated gas furnace and hot water heater
#9	Steel framing with Icynene foam insulation	Geothermal Heat Pump with Hot water desuperheater
#10	Hebel precast autoclaved aerated concrete	Integrated gas furnace and hot water heater

The output of the PV metal roofing modules is used to charge a set of sealed lead-acid batteries or is converted to ac and fed to the house loads. The inverter is able to connect to the utility grid. Various operation regimes will include the following:

- disconnection from the utility grid of a number of house loads which will be supplied through the inverter during a portion of the day; off-peak charging of the batteries as needed each night;
- float charging the batteries at all times with the PV output feeding either the loads or the utility grid;
- active battery discharge to the grid to a pre-determined state-of-charge with the rate set to a maximum based upon a base load for the home;
- other operation scenarios as determined in conjunction with the utility.

Figure 3 NAHB Research Center Townhouse Project with ECD PV-Metal Roofing Modules on Left and Center Roof Sections

The BOS components include a series-connecting junction box with a blocking diode for each pair of modules; the output is then fed to a fuse panel serving as a string combiner for three pairs of modules. Three output circuits are extended to the Amanda power center containing over- and short-circuit protection. A 4 kW Trace inverter is used to convert the dc 48 V to 120 V_{ac}. All wiring was performed by residential electricians.

PROJECTED COST

These modules have a number of features which lead to lower installed costs: they are very large area modules (about 20 ft^2); they have no frame; and the backing plate is the metal roofing pan. Consequently, ECD projects the cost of these modules to be about 30% less than standard framed modules.

The project large volume (> 50 MW/yr) costs for a 1.9 kW_{ac} metal roofing system are shown in Table III. For comparison, a 2,000 ft^2 metal roof costs about $16,000.

Table III. Project large manufacturing volume cost

	Total Cost	Cost/W_{ac}
PV Modules (1.5 $/$W_{dc}$)	$3204	$1.68
Inverter (2 kW)	$1000	$0.53
Wire, Junction Box's, etc.	$ 250	$0.13
Installation Cost	$1000	$0.52
Total Cost	$5274	$2.78
Avoided Cost	($1364)	($0.72)
Net Cost	$4353	$2.05

SUMMARY

It is expected that the BIPV systems developed under the PV:BONUS program will become the principle PV product marketed in this country for the next several years. These systems, which compete in the retail (8 - 15¢/kW-hr) rather than the wholesale (3 - 5¢/kW-hr) power markets, appear to be the most economically favorable near- and medium- term markets as the production volume increases and PV cost decreases.

United Solar has technical and marketing programs underway to integrate these BIPV modules, developed under the PV:BONUS program, into their product line on a time-line to dovetail with the start-up of the new 5 MW Troy MI production facility. Product launch of *Uni-Solar™* roofing is scheduled for early 1997.

ACKNOWLEDGEMENTS

This project is supported by the DOE/NREL PV:BONUS (PV Building Opportunities in the US) under subcontract number DE-FC36-93CH10571. Members of the project team include: United Solar System Corp.; the NAHB Research Center; Arizona Public Service Company; Detroit Edison, Minoru Yamasaki Associates and Solar Design Associates.

Special thanks to our colleagues at ATAS International (Dick Buss, President; Jim Buss, and Bob Booth) for help in all phases of this project and to Dante Freson for help in fabricating the modules.

REFERENCES

1. M. Izu and S.R. Ovshinsky, Production of tandem amorphous silicon alloy cells in a continuous roll-to-roll process, SPIE Proc. 407, **42** (1983).
2. M. Izu and S.R. Ovshinsky, Roll-to-roll plasma deposition machine for the production of tandem a-Si alloy solar cells, Thin Sol Film 119, **55** (1984).
3. H. Morimoto and M. Izu, Amorphous silicon solar cell production in a roll-to-roll plasma CVD process, JARECT 16 (1984); Amorphous Semiconductor Technology & Devices, North Holland, Edited by Y. Hamakawa, 212 (1984).
4. S.R. Ovshinsky, Roll-to-roll mass production process for a-Si solar cell fabrication, Proc. International PVSEC-1, 577 (1988).
5. J. Yang, R. Ross, T. Glatfelter, R. Mohr, G. Hammond, C. Bemotaitis, E. Chen, J. Burdick, M. Hopson and S. Guha, High efficiency multiple-junction solar cells using amorphous silicon and amorphous silicon germanium alloys, Proc. 20th IEEE PV Spec. Conf., 241 (1988).
6. J. Yang, R. Ross, T. Glatfelter, R. Mohr and S. Guha, Amorphous silicon-germanium alloy solar cells using amorphous silicon and amorphous silicon-germanium alloys, Proc. 20th IEEE PV Spec. Conf., 241 (1988). Profiled Bandgaps, MRS Symposium Proc. Vol. 149, 435 (1989).
7. M. Izu, X. Deng, A. Krisko, K. Whelan, R. Young, H.C. Ovshinsky, K.L. Narasimhan and S.R. Ovshinsky, Manufacturing of 3-junction 4 ft^2 a-Si PV modules, Proc. of 23rd IEEE PVSC, 919 (1983).
8. M. Izu, S.R. Ovshinsky, X. Deng, A. Krisko, H.C. Ovshinsky, K.L. Narasimhan and R. Young, Continuous roll-to-roll amorphous silicon photovoltaic manufacturing technology, AIP Conf. Proc. 306, 12th NREL PV Prog. Rev.,198 (1993).
9. S. Guha, J. Yang, A. Banerjee, T. Glatfelter, K. Hoffman, S.R. Ovshinsky, M. Izu, H.C. Ovshinsky and X. Deng, Amorphous silicon alloy photovoltaic technology - from R&D to production, Proc. of MRS Spring Meeting (1994).
10. X. Deng, M. Izu, K.L. Narasimhan and S.R. Ovshinsky, Stability test of 4 ft^2 3-junction a-Si alloy PV production modules, Proc. of MRS Spring Meeting (1994).
11. M. Izu, H.C. Ovshinsky, X. Deng, A. Krisko, K.L. Narasimhan, R. Crucet, T. Laarman, A. Myatt and S.R. Ovshinsky, Continuous roll-to-roll serpentine deposition for high throughput a-Si PV manufacturing, First World Conf. on Photovoltaic Energy Conversion, Hawaii, December (1994).

Effect of hydrogen dilution on the structure of amorphous silicon alloys

D. V. Tsu, B. S. Chao, and S. R. Ovshinsky
Energy Conversion Devices, Inc., Troy, Michigan 48084

S. Guha and J. Yang
United Solar System Corp., Troy, Michigan 48084

(Received 29 April 1997; accepted for publication 3 July 1997)

We investigate why high levels of hydrogen dilution of the process gas lead to enhanced light soaking stability of amorphous silicon (*a*-Si) alloy solar cells by studying the microstructural properties of the material using high-resolution transmission electron microscopy (TEM) and Raman spectroscopy. The TEM results show that *a*-Si alloy (with or without hydrogen dilution) is a heterogeneous mixture of amorphous network and linear-like objects that show evidence of order along their length. The volume fraction of these ordered regions increases with increasing hydrogen dilution. © *1997 American Institute of Physics.* [S0003-6951(97)02736-8]

Light-induced degradation in amorphous silicon (*a*-Si) alloy has received a great deal of attention in recent years.[1] Since the first observation[2] of improvement in stability of *a*-Si alloy films prepared from a dilute mixture of silane and hydrogen, hydrogen dilution of the process gas mixtures has been extensively used to improve the stability of *a*-Si alloy materials and devices. Recent results indicate[3] that significant improvement in the stability of *a*-Si alloy cells takes place when the intrinsic layers are prepared from a gas mixture heavily diluted with hydrogen. While the hydrogen content of this material is similar to that in the alloy prepared with low hydrogen dilution, hydrogen effusion from this film takes place at a much lower temperature. Small-angle x-ray scattering[4] also showed that a more stable material was characterized by an anisotropic structure with low void density, and it was postulated[5] that an improvement in the structure of the material was contributing to the improved stability.

In this letter, we investigate films prepared with increasing hydrogen dilution using Raman spectroscopy and high-resolution transmission electron microscopy (TEM) and demonstrate that a more ordered structure is obtained as the hydrogen dilution increases.

Three *a*-Si alloy films were made from a gas mixture of disilane and hydrogen under H_2 dilution levels, from low to medium to high, while one standard *a*-Si alloy film was made from silane without any hydrogen dilution to serve as reference. The hydrogen to disilane ratios were varied from 5:1 to 150:1. All of the samples were prepared in a radio-frequency glow-discharge diode reactor on 7059 glass, and the thickness of each sample was kept near 500 Å. Details of deposition conditions are as follows[3]: Pressure—1 Torr, temperature—200–300 °C, and power density—50–150 mW/cm^2.

We begin with the Raman scattering measurements where we first look for qualitative differences, then quantify those differences by Gaussian deconvolution. The Raman spectra of the Si TO region were collected in the near-backscattering geometry using the 4880 Å line of an Ar^+ laser and a double monochrometer. Details are given elsewhere.[6] Trace (a) of Fig. 1 shows the spectrum of the standard *a*-Si alloy sample, while traces (b) through (d) are the spectra for the modified *a*-Si alloy samples using the low, medium, and high levels of H_2 dilution, respectively. All of the spectra in Fig. 1 were normalized with respect to the baseline (~ 560 cm^{-1}) and the LO band (~ 400 cm^{-1}). The sharper band in trace (d) at 516.5 cm^{-1} is due to microcrystalline inclusions as found in the TEM images (Fig. 2). This film was made with high hydrogen dilution. Besides this feature, the most obvious trend with increasing H_2 dilution is a shift in the peak position of the principal Si TO band from 474.5 cm^{-1} (no H_2) to 482.6 cm^{-1} (high H_2), as shown by the dashed line in Fig. 1. The two very sharp features at 529.64 and 560.80 cm^{-1} are laser plasma lines.

In order to qualitatively account for the trend in the Si TO peak position, we perform spectral subtractions using the Raman spectrum of the standard *a*-Si alloy film as reference to reveal any underlying spectral component. We compared the difference spectra of many different standard *a*-Si alloy samples, and find that there is little net signal above the noise. In contrast, when compared to any of the modified H_2

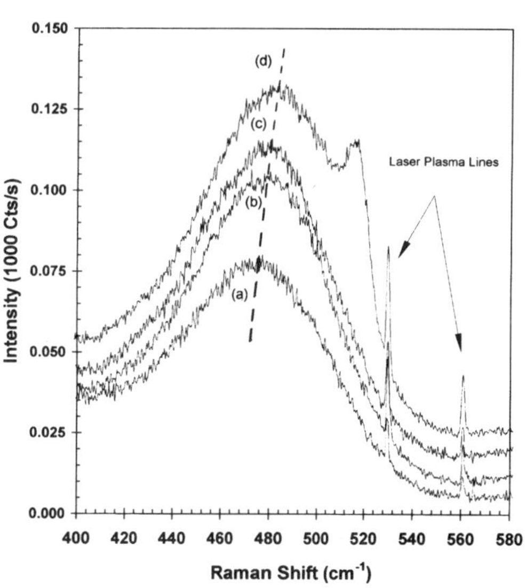

FIG. 1. Raman spectra vs hydrogen dilution of the process gas. 500 Å films on 7059 Corning glass: (a) standard *a*-Si alloy, sample RF5349 (no dilution); and modified *a*-Si alloy with (b) low dilution, US8014; (c) medium dilution, US8035; and (d) high dilution, US8013.

 0003-6951/97/71(10)/1317/3/$10.00

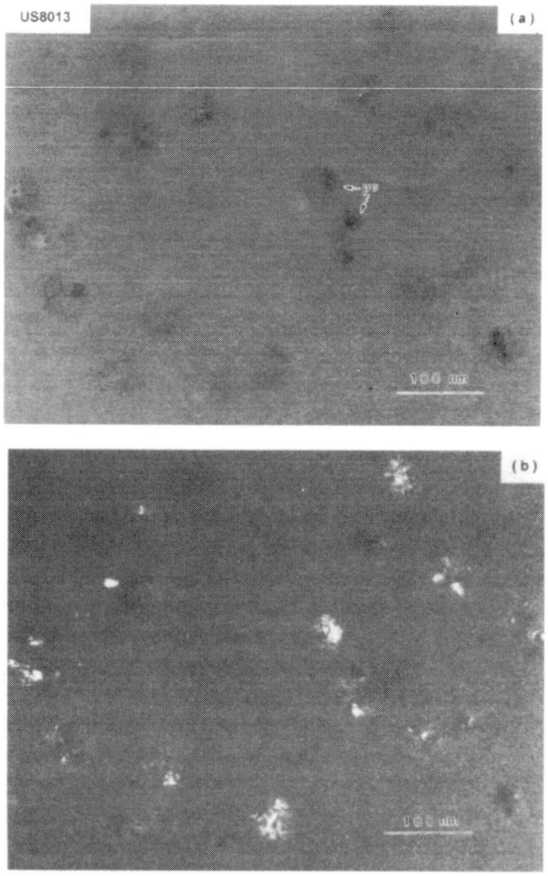

FIG. 2. (a) Low magnification bright-field TEM micrograph of modified *a*-Si alloy, sample US8013 (high hydrogen dilution). Arrows indicate the μc-Si inclusions. (b) Dark-field TEM micrograph of same region.

dilution samples, the difference spectra always show a significant signal centered at 490 cm^{-1} having a full width of 40 cm^{-1}. Although all modified samples show the residual 490 cm^{-1} band in their difference spectra, the subtraction method cannot reliably quantify its magnitude. We thus turn to Gaussian deconvolution.

In our deconvolution of the Raman data, we use a minimal number of Gaussian functions that is consistent with the entire data set. For all of the spectra, one Gaussian is used to represent the LO shoulder at ~400 cm^{-1}. For this band, its position, amplitude, and full width are all allowed to vary during the fit. Except for sample US8013, all of the other samples use two Gaussians to represent the principal TO band. Based on detailed analysis of the spectra, we find that two Gaussians represent the minimal number that is consistent with the entire data set. They should represent, then, two distinct environments within the material. For US8013, two additional Gaussians are required (at 510.2 and 517.4 cm^{-1}) to accurately fit the μc-Si TO band.[7] In all these cases, two more Gaussians are used for the two laser plasma lines to accurately position the peaks of the Raman bands during the fit.

Experimentally, we measure the peak position and width of the TO band of the standard sample to be 475 and 70.5 cm^{-1}, respectively. The position and width of the primary TO Gaussian are then fixed to these values for all of the samples. The difference spectra demonstrate that the excess Raman activity for all of the H_2 diluted samples is centered at 490 cm^{-1}. Therefore, this position is chosen for the secondary TO Gaussian for all of the samples, and its width is 40 cm^{-1}. The band areas of each component are summarized in Table I for each sample.

TABLE I. Gaussian deconvolution of the Raman spectra of Fig. 1.

		Raman band area (%)		
Sample	H_2 dilution	Primary TO (475 cm^{-1})	Secondary TO (490 cm^{-1})	μc-Si (510/517 cm^{-1})
RF5349	none	95.5	4.5	...
US8014	low	90.0	10.0	...
US8035	medium	90.4	9.6	...
US8013	high	76.1	13.8	10.1
US8013[a]	high	84.6	15.4	...

[a]Normalized without the microcrystalline band.

It is the growth of the 490 cm^{-1} band relative to the 475 cm^{-1} band that causes the peak of the overall TO band to shift to higher frequencies as observed in Fig. 1. However, the results of this analysis indicate that even the so-called standard *a*-Si alloy, i.e., made with no H_2 dilution, contains some excess scattering in the 490 cm^{-1} region. One could argue that this 4.5% residual scattering is just an indication of some non-Gaussian behavior that can be simulated by the addition of a weak second Gaussian. However, based upon our TEM analysis, it is apparent that even standard *a*-Si alloy is not completely homogeneous. We, therefore, conclude that this residual 4.5% scattering is caused by a secondary phase.

The thin *a*-Si alloy films were lifted off their glass substrates by etching with a weak HF solution, and the films were supported by a copper TEM grid. The microstructure of the films was analyzed by a Jeol 2010 STEM microscope. Figure 2(a) shows the bright-field micrograph of the modified high H_2 dilution film in relatively low magnification. Here we observe that the gross microstructure of the sample consists of distinct microcrystalline inclusions imbedded in the "featureless amorphous" matrix. The typical dimension of each inclusion is ~50 nm consisting of spherical microcrystallites ranging from 2.5 to 10 nm in diameter, as shown in the corresponding dark-field micrograph in Fig. 2(b). The presence and size of these smaller crystallites result in the Raman band (Fig. 1) that is shifted down to ~517 from the 521 cm^{-1} position of bulk crystalline Si.[7] The diffraction pattern recorded from the same region shows two distinct halos, which are normally assigned to the so-called "amorphous" silicon component. In addition, we observe[6] small, bright diffraction spots which were identified earlier[7] to be due to the microcrystalline inclusions which possess long-range order. On the other hand, no such diffraction spots are found in the standard material, nor in modified (low or medium) H_2 dilution material.

On a much finer scale in Fig. 3 of the high H_2 dilution sample, we clearly observe the long parallel lines separated by 0.31 nm that correspond to the *d* spacing of Si(111) planes of the microcrystallites. Aside from this obvious

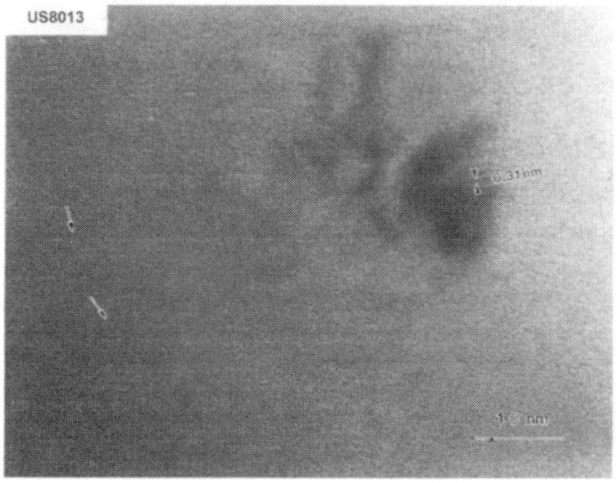

FIG. 3. High magnification bright-field TEM micrograph of ''amorphous'' matrix, sample US8013. Arrows show ordered linear-like objects.

structure, we also observe that the ''amorphous'' matrix is not entirely featureless. In particular, as indicated by the arrows, we find linearlike objects (i.e., one dimensional, having widths 2–3 nm, and lengths up to a few 10's of nm) that appear to have some degree of order along their length. These ''linear'' objects appear to meander throughout the matrix in no particular direction. Similar linear objects are also found in the rest of the samples, even in the standard material! The ''amorphous'' matrix then consists of a heterogeneous mixture of a truly amorphous component, as described by the short-range order of a continuous random network (CRN), and objects that are intermediate in order between the CRN and crystalline phases, i.e., the somewhat ordered linear-like objects. It is this newly identified structure that we believe produces the 490 cm^{-1} band, and if so, the results of the Raman deconvolution indicate that the volume fraction occupied by them within the ''amorphous'' matrix increases with increasing H_2 dilution.

The presence of the ordered linear-like regions and its impact on stability merits some discussion. As the volume fraction of these ordered regions grows, the quality of the material improves as evidenced by the increased stability. The best stability is obtained for the high hydrogen dilution case where the volume fraction of these ordered regions is the highest. It is interesting to point out that this material actually includes a small amount of microcrystalline inclusion, and effusion of hydrogen occurs more readily from this material. Further increase in hydrogen dilution increases the microcrystalline volume fraction further, and the solar cell shows lower open-circuit voltage and fill factor characteristic of microcrystalline solar cells.

What is the role of hydrogen? We believe that the presence of excess hydrogen in the gas mixture passivates the dangling bonds on the growing surface, and the impinging species have more time to find favorable sites. This allows for formation of more ordered structure. Excess hydrogen also etches the growing surface, and in the extreme case of heavy hydrogen dilution, strong etching eliminates part of the disordered configurations.[8] Since the crystalline phase is the lowest energy configuration, it is often the surviving structure. High dilution, therefore, favors the growth of microcrystalline material. The presence of microcrystalline regions in large quantities affects the performance of the material because of problems associated with grain boundaries. The best material is obtained under the conditions just below the threshold of microcrystalline growth. The research challenge will be to obtain conditions that will inhibit microcrystalline growth but will still promote the growth of the ordered regions. Other precursor gases could be useful. In fact, we have been able to grow materials with similar ordered regions using deuterated gases.

In addition to creating conditions for the growth of a more ordered material, hydrogen can also etch away weak or strained bonds which will enhance stability. Although we see a threefold increase in the volume fraction of the ordered region as we go from the ''standard'' to the best quality material, the fraction of the ordered region is still small. It is possible that hydrogen dilution is causing other changes in the structure or the bonding that are not detected by the structural tools we have used.

In conclusion, we have shown that the volume fraction of an ordered microstructure in *a*-Si alloy films increases with increasing hydrogen dilution in the process gas mixture. We suggest that the enhanced stability of solar cells prepared with hydrogen dilution is caused by the improved microstructure.

The authors thank X. Xu and S. Sugiyama for discussions, Dr. John Bradley at MVA, Inc. for TEM examinations, and V. Trudeau for manuscript preparation.

[1] For a recent review, see H. Fritzsche, *Amorphous and Microcrystalline Silicon Technology–1997*, Materials Research Society Symposium Proceedings, 1997 (to be published).
[2] S. Guha, K. L. Narasimhan, and S. M. Pietruszko, J. Appl. Phys. **52**, 859 (1981).
[3] X. Xu, J. Yang, and S. Guha, J. Non-Cryst. Solids **198–200**, 60 (1996).
[4] D. L. Williamson, in *Amorphous Silicon Technology–1995*, Materials Research Society Symposium Proceedings Vol. 377, edited by M. Hack, E. A. Schiff, A. Madan, M. Powell, and A. Matsuda (Materials Research Society, Pittsburgh, 1995), pp. 251–262.
[5] S. Guha, Proc. 25th IEEE PVSC, 1017 (1996).
[6] D. V. Tsu, B. Chao, S. R. Ovshinsky, S. Guha, and J. Yang (unpublished).
[7] R. Tsu, S. S. Chao, M. Izu, S. R. Ovshinsky, G. J. Jan, and F. H. Pollack, J. Phys. (France) **C4**, 269 (1981).
[8] C. C. Tsai, in *Amorphous Silicon and Related Materials*, edited by H. Fritzsche (World Scientific, Singapore, 1988), Vol. 1, p. 123.

PHYSICAL REVIEW B, VOLUME 63, 125338

Heterogeneity in hydrogenated silicon: Evidence for intermediately ordered chainlike objects

David V. Tsu, Benjamin S. Chao, Stanford R. Ovshinsky, and Scott J. Jones
Energy Conversion Devices, Troy, Michigan 48084

Jeffrey Yang and Subhendu Guha
United Solar Systems Corporation, Troy, Michigan 48084

Raphael Tsu
Electrical Engineering Department, University of North Carolina at Charlotte, Charlotte, North Carolina 28223
(Received 23 December 1999; revised manuscript received 6 November 2000; published 13 March 2001)

Hydrogen (H_2) dilution of the source gas is known to be a key factor in producing hydrogenated amorphous silicon films that demonstrate a high degree of optoelectronic stability. In this work, we investigate, using Raman spectroscopy and high-resolution transmission electron microscopy (TEM), whether microstructural differences exist between such films and those made with no H_2 dilution (i.e., that have greater instabilities). The key variable is the H_2 dilution, which ranges from none to very high levels, producing amorphous and microcrystalline silicon films. The TEM results show that embedded within the amorphous matrix are chainlike objects (CLO's) having ~3 nm widths, ~30 nm lengths, and showing a high degree of order along their length. Such order implies vanishing levels of bond-angle distortion (BAD). These CLO's are present in all samples investigated, but their density increases with the level of H_2 dilution. The Raman spectra show a TO band centered at 490 cm^{-1} (37 ± 3 cm^{-1} full width). Quantitative analysis shows this band to exist in *all* samples investigated, but increases in magnitude with increasing H_2 dilution. In the highest dilutions when microcrystallites are observed, the band is distinctly evident. Its position and width are also consistent with very low (crystallinelike) levels of BAD ~0°. It is thus likely the 490 cm^{-1} Raman band is a signature of the intermediate ordered CLO's.

DOI: 10.1103/PhysRevB.63.125338 PACS number(s): 61.43.−j, 61.43.Dq, 61.44.Br, 61.46.+w

I. INTRODUCTION

The question[1] whether a structural state exists in hydrogenated amorphous silicon (*a*-Si:H) between the amorphous state of short-range order and the microcrystalline state of long-range order has stimulated renewed interest. This state of intermediate order might be disordered enough to spoil the crystalline selection rules for optical transitions, hence providing for the strong optical absorption desired in photovoltaic devices (e.g., solar cells). It may yet be ordered enough to inhibit any structural changes and photoinduced defects that decrease the photocarrier lifetimes. Such a decrease ultimately leads to reduced efficiencies of these photovoltaic devices during long light exposures. The decrease or elimination of this degradation, which is called the Staebler-Wronski degradation (SWD) after its discoverers,[2] is one of the major goals in the research of *a*-Si:H and related materials.

The understanding of device stability is increasingly being focused on the nature of heterogeneity within the amorphous silicon material.[3,4] It is now commonly accepted that an early *homogeneous* model[5] of the amorphous state, based upon the continuous random network (CRN), is inadequate. The difficulty lies in identifying in more precise terms the nature of heterogeneity,[3] i.e., how a particular sort of heterogeneity might be favorable to stability. Thus, the issue of an intermediate ordered phase of silicon is centrally connected with gaining a more precise description of heterogeneity.

Experimentally, deposition conditions are being found that result in materials that have substantially lower SWD's. Recently,[6] a significant reduction of the SWD of *a*-Si:H solar cells was accomplished by modifying the plasma-enhanced chemical vapor deposition (PECVD) conditions of the *a*-Si:H layers. This modification primarily involved the use of fairly high levels of hydrogen dilution of the disilane (Si_2H_6) process gas. This led to the development of solar cells[7] with ~15% initial efficiency and with a stabilized efficiency of ~13%, representing a degradation of only ~10%. The observed improvement of the *a*-Si:H material thus motivated us to investigate whether this achievement could be attributed to a more ordered microstructure, i.e., to explore the nature of the heterostructure within this improved material. In particular, we revisit the possible existence of an intermediate ordered phase of silicon[8–10] being one of the heterogeneous elements.

Early work on the existence of an intermediately ordered state of silicon was based on electroreflectance and Raman spectroscopic analyses[8–10] of fluorinated amorphous hydrogenated silicon (*a*-Si:F:H). The electroreflectance data[8,9] indicated the presence of a structural component that could not be identified with either crystalline or amorphous states, and thus they interpreted that other component to have intermediate range order (IRO). Although a Raman peak at ~508 cm^{-1} was first thought to be a signature of IRO,[8] transmission electron microscopy (TEM) and electron diffraction analyses[9] later showed this peak to be associated with very small (2–6 nm) microcrystallites, where small-particle effects cause the measurement of the zone-center phonons to be shifted down from the 522 cm^{-1} value of crystalline silicon.

Raman analysis of the transverse optical (TO) band[9,10] showed that its peak position and width could be used to classify the relative order of the amorphous material. The position of the TO Raman band in sputtered amorphous silicon (a-Si) shifts from about 465 cm^{-1} to roughly 475 cm^{-1} upon annealing. This peak shift was interpreted to indicate an increase in the local order of the material,[10] as a consequence of a reduction of the average bond-angle deviation (BAD), from the 109.5° of tetrahedral coordination, within the random network. Hydrogenated material (a-Si:H) could be deposited directly in the state that results in the 475 cm^{-1} Raman peak, but annealing of a-Si:H was shown to be ineffectual in raising the frequency higher than 475 cm^{-1}. Peak positions below 475 cm^{-1} were thus deemed to indicate short-range order. Only by the introduction of fluorine in a-Si:F:H alloys could TO peak positions above 475 cm^{-1} be measured. In fact, positions up to 483 cm^{-1} were measured. TO frequencies this high was thought to indicate the existence of IRO.[9,10]

Since those early studies, Raman spectroscopy has been extensively used to study the quality of microcrystalline silicon (μc-Si).[11] It is noteworthy that in addition to the main microcrystalline Raman band, whose peak position can range from 508 to 522 cm^{-1} (depending on the microcrystallite dimension),[9] a distinct bump on the low-frequency side at ~490 cm^{-1} is generally observed. High-resolution TEM analysis of primarily microcrystalline material was performed by Wang.[12] He makes two conclusions about the 490 cm^{-1} band: (1) it does *not* derive from any amorphous content, and (2) it may be related to the "tissue" between microcrystallites. We believe, however, that this 490 cm^{-1} Raman band is associated with an intermediately ordered phase of silicon.

An intermediately ordered phase of silicon has been implicated in both amorphous and microcrystalline material. To clarify the nature of intermediate order, this paper extends our earlier work,[13] where we examined only a-Si:H. Here, we extend the hydrogen dilution into a regime that also results in μc-Si. Following those earlier studies,[8–12] we also use the TO Raman peak as a measure of the microstructure as well as TEM and electron diffraction. We will show that the intermediate order, which is clearly evident in the μc-Si material, has it roots in the amorphous material.

II. EXPERIMENTAL DETAILS

The a-Si:H films were deposited by PECVD using 13.56-MHz rf power onto Corning 7059 glass substrates at a temperature (T_s) of either 150 or 300 °C and at a deposition rate of about 1 Å/s. The film thickness was close to 50 nm to allow both Raman and TEM measurements on the same sample. In this work, two classes of samples were produced: (1) reference samples prepared using pure silane gas (SiH_4), i.e., without the use of any H_2 dilution, and (2) modified samples prepared with disilane (Si_2H_6) diluted with various amounts of H_2 gas. These samples are listed in Table I. In this table, the undiluted reference samples (i.e., made without H_2 dilution) are indicated as the Ux series, while the H_2 diluted samples made at the higher and lower substrate temperatures are labeled as DHx and DLx, respectively. Also, while the Ux series was made with silane, and the DHx and DLx samples made with disilane, this difference is not expected to be significant.

TABLE I. Summary of samples investigated. All samples are ~50 nm thick, except sample U1, which is ~500 nm thick. The H_2 dilution factor is defined as $(H_2+Si_2H_6)/Si_2H_6$.

Sample name	Gas	H_2 dilution	T_s (°C)
U1	SiH_4	1	300
U2	SiH_4	1	300
U3	SiH_4	1	300
U4	SiH_4	1	300
DH1	$Si_2H_6+H_2$	Low	300
DH2	$Si_2H_6+H_2$	Medium	300
DH3	$Si_2H_6+H_2$	High	300
DL4	$Si_2H_6+H_2$	250	150
DL5	$Si_2H_6+H_2$	285	150
DL6	$Si_2H_6+H_2$	330	150
DL7	$Si_2H_6+H_2$	400	150
DL8	$Si_2H_6+H_2$	500	150

For TEM and the selected-area transmission electron diffraction (TED) studies, the a-Si:H films were lifted from their glass substrates by first scoring, then etching in a weak HF solution. The films were then floated onto a thin copper TEM grid. Planar images were taken with a Jeol 2010 scanning TEM, operating at 200 keV. In this way, the films were subjected to a minimum amount of preparation.

Raman spectra were collected in the near-backscattering geometry using the 488 nm line of an Ar-ion laser at 300 mW. A cylindrical lens focused the laser light into a line that matches the entrance slit of the double monochromator (Spex model 1403, 0.85 m). The laser plasma lines were unfiltered so that their positions could be used to accurately calibrate the absolute Raman shifts. Detection is by a thermoelectrically cooled photomultiplier having about 4 dark counts/s. Scans were made between 400 and 580 cm^{-1} at slow scans rates (0.01 cm^{-1}/s) and long integration times (10 s) to obtain high signal-to-noise ratio spectra.

III. EXPERIMENTAL RESULTS

A. High-resolution

Figures 1(a) and 1(b) are the TEM bright-field and dark-field photographs of sample DH3 in relatively low magnification. One observes a gross microstructure consisting of microcrystalline inclusions embedded in an amorphous matrix. The microcrystalline inclusions typically have a diameter of ~30–50 nm and are conglomerations of smaller microcrystallites that range from 2.5 to 10 nm in diameter, as shown in the corresponding dark-field micrograph of Fig. 1(b). The heterogeneous nature of this sample is further revealed by the TED pattern shown in Fig. 1(c). Here, we note two distinct halos that result from the amorphous regions, as well as small bright diffraction spots. The diffraction spots are found in three distinct locations: (1) within the boundaries of the first inner halo, and (2) on the inner and (3) outer

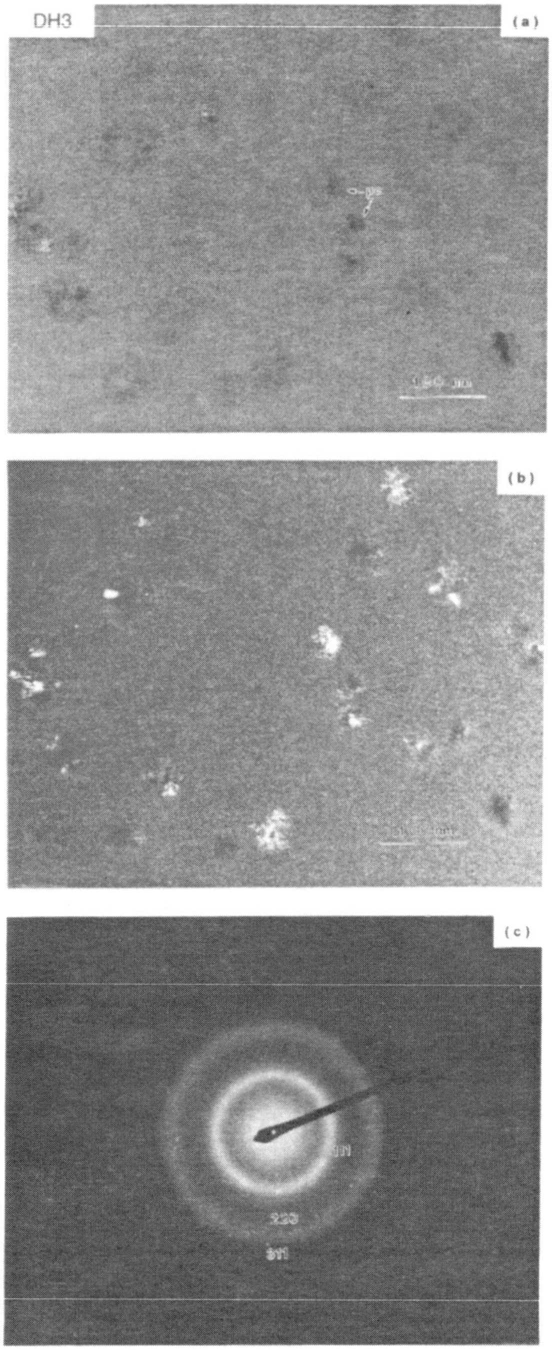

FIG. 1. Low magnification planar TEM of sample DH3 showing the (a) bright-field and (b) dark-field micrographs and (c) the corresponding TED pattern.

edges of the second halo. These correspond to the (111), (220), and (311) diffraction planes of crystalline Si, respectively. The well-defined splitting on the (220) and (311) diffraction rings indicates that the microcrystallites possess long-range order. No such diffraction spots are found in the undiluted reference samples (U1–U4) nor in the samples modified with lower levels of H_2 dilution (DH1 and DH2).

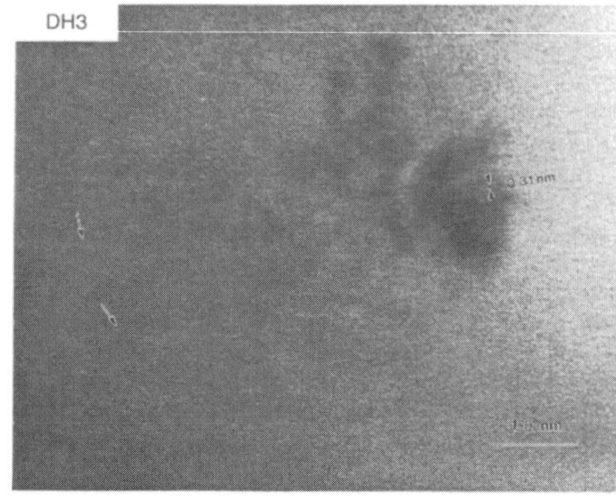

FIG. 2. High-magnification bright-field TEM of one of the microcrystallites seen in Fig. 1.

One of the microcrystallites imbedded in the amorphous matrix is shown in the bright-field micrograph of Fig. 2. The long parallel lines separated by 0.31 nm correspond to the spacing of Si(111) planes. Aside from the microcrystallites, one observes that the amorphous matrix is not entirely featureless. As indicated by the arrows, one finds meandering structures that we call "chainlike objects" (CLO's) that are quasi-one-dimensional, having widths of 2–3 nm (~10 Si atoms) and lengths up to 30 nm, which have a rather high degree of order along their length as evidenced by the short (~2 nm) repeating segments perpendicular to the length of the CLO's, e.g., like III . . . III. Although the precise nature of this order cannot be discerned from these images, it is clear that they must consist of bonding units having very small deviations from the ideal angle of 109.5° of tetrahedral coordination. For even minor deviations from this ideal angle must destroy this observed regularity. It is known that an average BAD of ~10° exists in well annealed *random* networks and that BAD accounts for virtually all of the elevated strain energy found in amorphous silicon networks. These objects, then, must represent *local* regions having substantially reduced internal strain in comparison to the surrounding random network. Thus, an important property that the CLO's conveys onto the heterogeneous material is to introduce local regions having very low strain energy. Moreover, this sort of heterogeneity, i.e., between the CRN amorphous matrix and the meandering CLO's, should have beneficial interfacial qualities in comparison to the interface between rigid microcrystallites and the CRN matrix.

Figure 3 is a closeup view of one of these chainlike objects. They appear to meander throughout the matrix in no particular direction. Some CLO's are also found in the reference samples (U1–U4) prepared without any H_2 dilution. However, their concentration increases with an increase in the level of H_2 dilution in the process gas. Such structures have also recently been identified in a-Si:H by Kamei, Stradins, and Matsuda.[14] The concentration of the microcrystallites also increases with H_2 dilution, but becomes noticeable only in the very-high-dilution regime.

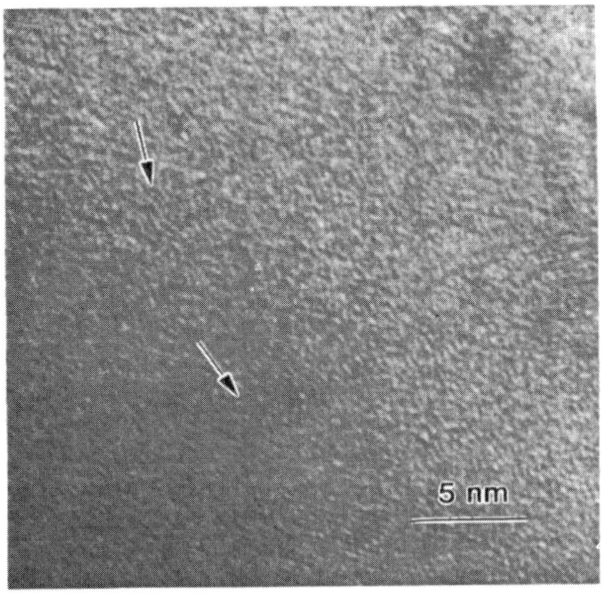

FIG. 3. High magnification of the "featureless amorphous region" (as seen in the low-magnification image of Fig. 1) away from the microcrystallites. The chainlike objects are meandering across this planar image.

B. Raman spectra

Figure 4 shows the Raman spectra (a) of an undiluted reference sample (U3) and traces (b)–(d) are of the H_2 diluted samples DH1, DH2, and DH3, respectively. All of the spectra were normalized with respect to the baseline near $\nu = 560\,\text{cm}^{-1}$ and the peak of the longitudinal optical (LO) band near $\nu = 400\,\text{cm}^{-1}$. With increasing H_2 dilution, the peak position of the TO band increases from 475 to 482.6 cm^{-1} as shown by the dashed line in the figure. The narrow band in trace (d) at 516.5 cm^{-1} is due to microcrystalline inclusions as confirmed by TEM and shown in Fig. 1(a). The absolute Raman shifts were calibrated using the plasma lines, i.e., the two very sharp features at 529.64 and 560.80 cm^{-1}.

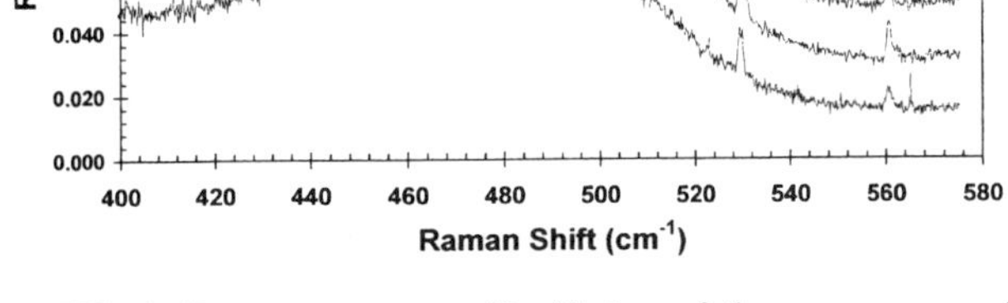

FIG. 4. Raman spectra vs H_2 dilution of the process gas of ~50-nm films on 7059 Corning glass: (a) U3 $s(\times 1.0)$, no dilution; (b) DH1 ($\times$ 0.68), low dilution; (c) DH2 ($\times$ 1.05), medium dilution; and (d) DH3 ($\times$ 0.55), high dilution.

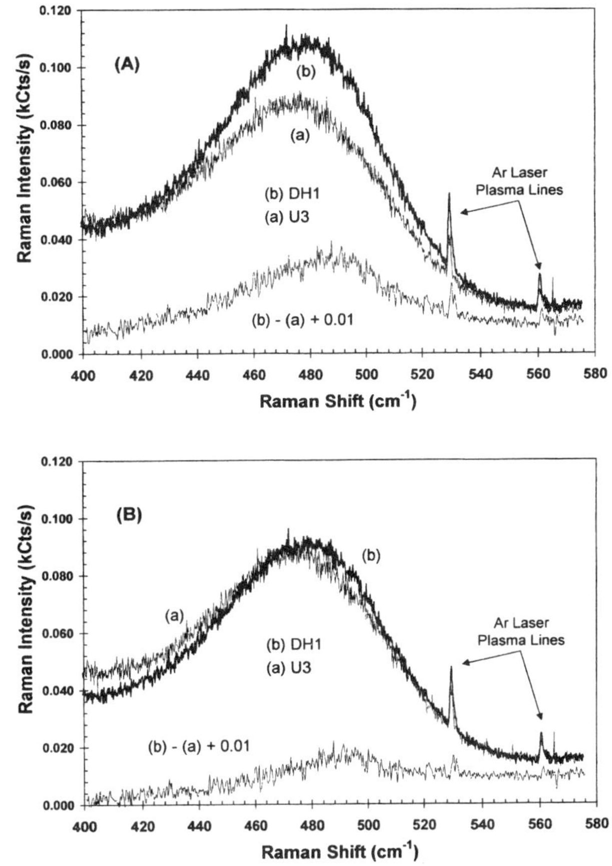

FIG. 5. Difference Raman spectra between reference (U3) sample and modified (DH1) sample with different normalization conditions, either at (A) LO or at (B) TO positions. Excess Raman scattering centered at 490 cm^{-1} is observed in either case.

Considering the fact that the TEM micrographs revealed the amorphous region to be a heterogeneous mixture of the chainlike objects as well as the amorphous matrix, we subtracted the normalized Raman spectrum of an undiluted sample (U3) from those of the samples prepared with H_2 dilution. We hoped thereby to observe the Raman signature of the two structural components separately. Figures 5(A) and 5(B) show the spectral subtraction for sample DH1 using in panel (A) a normalization at $\nu = 400\,\text{cm}^{-1}$ and in panel (B) a normalization of the TO band peak. In both cases, one finds in the difference spectrum a Raman band centered at about $\nu = 490\,\text{cm}^{-1}$ and whose width is 30–40 cm^{-1}. The *same* extra Raman band, albeit of larger amplitude, appears in the difference spectra of samples DH2 and DH3. In contrast, difference spectra performed among the four undiluted reference samples (U1–U4) showed no feature above the noise level. The H_2 diluted samples DL4–DL8, prepared at the lower substrate temperature ($T_s = 150\,°\text{C}$), showed again the extra $\nu = 490\,\text{cm}^{-1}$ Raman feature in their difference spectra. This is shown for sample DL4 in Fig. 6.

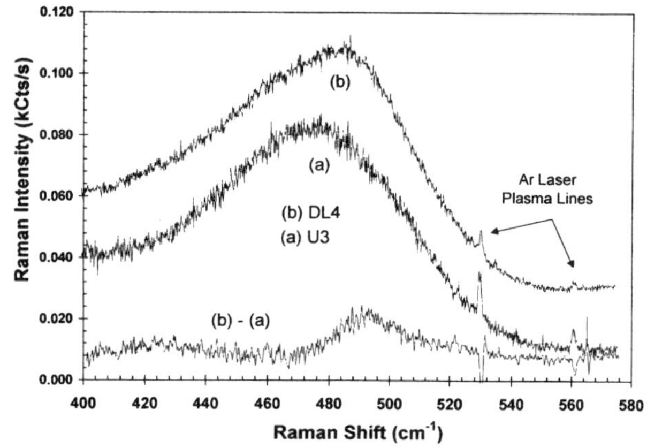

FIG. 6. Difference spectra between samples (a) U3 and (b) DL4. Traces (a) and (b) are offset for clarity.

Although the *presence* of the extra $\nu = 490\,\mathrm{cm}^{-1}$ Raman band does not depend on the particular normalization procedure, the *magnitude* does, making the spectral subtraction method somewhat unreliable for the purpose of quantification. We therefore discuss two quantification methods in the Appendix.

C. Raman spectra of microcrystalline material

Figure 7 shows the Raman spectra of samples DL4–DL8 that were prepared in successively increasing H_2 dilutions at the lower substrate temperature ($T_s = 150\,°\mathrm{C}$). The spectra show the development of the μc-Si Raman band at ~515 cm^{-1}, as well as a distinctly evident 490 cm^{-1} band. These Raman data were also subjected to the Gaussian deconvolution analysis discussed in the Appendix. Figure 8 summarizes the relative areas of the 490 cm^{-1} band and the total microcrystalline band. The 490 cm^{-1} band already contributes an appreciable 16% of the total Raman scattering *before* the onset of the microcrystalline formation and continues to rise with H_2 dilution.

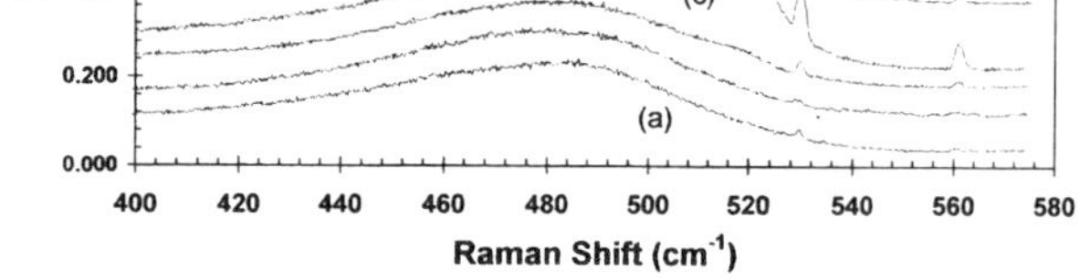

FIG. 7. Raman spectra of the very high H_2 dilution material made at a $T_s = 150\,°\mathrm{C}$, where the H_2 dilution factor increases from samples DL4 to DL8, traces (a)–(e), respectively (see Table I).

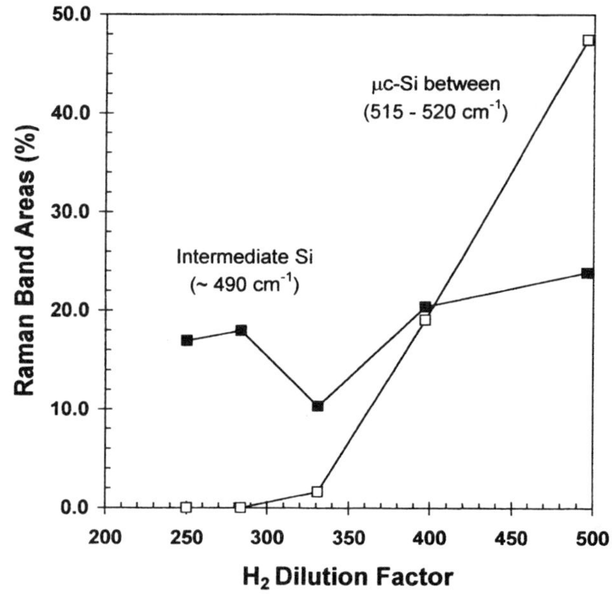

FIG. 8. Summary of the deconvolution of the Raman spectra of the DL series (very high H_2 dilution) material. The μc-Si band area consists of the two Gaussian components required to accurately fit the feature.

IV. DISCUSSION

The TEM data reveals that heterogeneity of a-Si:H involves a coexistence of chainlike structural features along with the random network of the amorphous matrix. These data differ from other forms of heterogeneity identified by others. In particular, the 2–3 nm width and up to 30 nm length of the CLO's rule out the possibility that they are related to the voids detected by low-angle x-ray scattering[15] or inferred by infrared measurements.[16] Their dimension also rules out any relation to polysilane $(SiH_2)_n$ structures whose backbone and width consist of *single* Si atoms. Moreover, infrared measurements of the a-Si:H films studied here show a negligible concentration of the polysilanes. Even films that show no infrared absorption of polysilane units have about 4 at. % of their hydrogen clustered and bonded to 6–8 Si atoms according to nuclear magnetic resonance studies.[3] These clusters again are too small to be identified with the CLO's observed here. Moreover, the hydrogen clusters are present in essentially all a-Si:H films, while the density of the CLO's is small in ordinary films, but increases with hydrogen dilution of the processing gas or with fluorination.

Our analysis of the Raman data shows that a shift of the TO peak from $\nu = 475\,\mathrm{cm}^{-1}$ of the reference amorphous material to higher wave numbers with hydrogen dilution (or fluorination) (Refs. 8–10) should be interpreted as resulting from the growth in magnitude of the $\nu = 490\,\mathrm{cm}^{-1}$ Raman band. Figure 8 shows that this additional Raman band is well established before the onset of microcrystallinity and is even present to a small degree in high-quality a-Si:H films before H_2 dilution. Thus the 490 cm^{-1} band is *not unique* to microcrystalline material nor to the conditions in which that material is formed. The structure that causes the 490 cm^{-1} band can be found in all hydrogenated silicon material. The difference is only in their production effectiveness, the higher the H_2 dilution level during film growth, the greater chance those structures are created.

It is noteworthy that the 490 cm^{-1} Raman TO band is associated with high H_2 dilution levels. One of the important roles that hydrogen plays in the deposition process is an etchant role, where depositing Si atoms having high bond-strain energy are preferentially removed (etched) from the growth surface. By this etchant mechanism, a more relaxed material is produced. Thus dilution, using high levels of H_2 gas during deposition plays the same role in achieving structural relaxation as previously found[9] by the use of fluorine. With this association, the 490 cm^{-1} band must be related to a structural component that has reduced bond-strain energy. The improvement in the optoelectronic performance of *a*-Si:H with H_2 dilution is further evidence that the 490 cm^{-1} TO band is connected to a relaxed structure. Reduction of the SWD has been attributed to a more ordered microstructure in *a*-Si:H films that allows the recombination of photoexcited electron-hole pairs without structural changes that result in Si dangling-bond defects.[17]

The Raman scattering having a TO position of 490 cm^{-1} is clearly associated with a higher level of structural relaxation. We propose that the ordered CLO's (observed in the TEM data) are in fact those relaxed structural elements, i.e., that the Raman scattering of the CLO's have a TO band with a 490 cm^{-1} peak position. The case for making this association can be fortified by considering the reason the TO band shifts position in silicon materials. By far, the dominant mechanism for shifting the TO peak position is by variations in strain energy.

We have assumed that the 490 cm^{-1} Raman band does in fact represent a TO phonon vibration. This assumption is based on the following factors. Early modeling, especially the Steinhard-Alben-Weaire model,[18] quite successfully modeled the radial distribution function of amorphous Si and Ge by considering the amorphous materials to be a CRN and included three features that attempt to minimize strain energy, including (1) restricting the first-neighbor bond length to vary by less than 0.1%, (2) allowing the second-neighbor distance and bond angle to vary by up to 10%, and (3) allowing odd-member ring statistics. In addition, the phonon structure of *a*-Si was theoretically obtained using the CRN framework by the use of *only two* "Keating" force constants,[19] one for the bond stretching and the other for the bond-bending forces. Even with such limited use of parameters, the Keating constants can describe a wide range of structures from the crystalline state to the CRN of the amorphous state. These widely differing structures have basically the *same* phonon character because that character is rather insensitive to the environment beyond the second nearest neighbor, i.e., because their near-neighbor environments are nearly indistinguishable. However, differences in strain between these states cause the positions of the various vibrational bands, e.g., TO, LO, TA, and LA bands, to shift, the difference being in the position of the bands (the most sensitive of which is the TO band) and not the character of the bands. Thus, a structure that is neither CRN nor crystalline, but has some intermediate (i.e., other) composure, can also be expected to have the same basic phonon character for it is not possible to alter the near-neighbor environment of Si atoms in any meaningful way. Thus Raman spectroscopy is somewhat limited in its ability to deduce any further characteristics that a particular atomic arrangement might have, e.g., as in the dihedral angle dependence. It is therefore evident that the 490 cm^{-1} band does in fact represent a TO vibration.

It is evident that deviations of the bond angle from the ideal 109.5° of tetrahedrally coordinated systems is an import factor governing the strain and thus the peak position of the TO Raman band. Having found by TEM that the CLO's are highly ordered, one is forced to conclude that they must be composed of bonding units having very little if any BAD, for even moderate levels of BAD destroys order.

The relationship between BAD and Raman TO peak position and peak width has been studied in detail.[20–22] We find that not only is our measured TO peak position of 490 cm^{-1} consistent with a vanishing level of BAD, so is our determination of its width (from 35 to 40 cm^{-1}). Of particular interest is that model calculations[21] showed that the linear dependence of peak position and width did not continue below a BAD$\sim$4°, but tends to flatten out to a width of $\sim$33 cm^{-1} for BAD approaching zero. In addition, it was shown[23,24] that as the BAD$\rightarrow$0°, the Raman TO width$\rightarrow$33 cm^{-1}, and the TO peak position extrapolates to $\sim$487.5 cm^{-1}. Thus, our experimentally determined values of the TO peak position and width, of 490 and 37 cm^{-1}, respectively, are entirely consistent with those early theoretical predictions of *strain-free* phonon behavior.

The close agreement in position and width of the 490 cm^{-1} TO band with those of the model calculations support our view that this TO band is associated with a structure that is more ordered and relaxed than what the amorphous matrix could ever be. Since the amorphous matrix is primarily a CRN, and since CRN's have a minimum BAD of $\sim$10°, the TO band of CRN's can never rise above $\sim$475 cm^{-1}. [We have shown that any shift in the measured (composite) peak position above this is a result of a growth in scattering strength of the 490 cm^{-1} TO band.]

We are left with two quite independent pieces of experimental data showing the existence of regions having extremely low levels of bond-angle distortion. On the one hand, the 490 cm^{-1} Raman TO band reveals a BAD approaching 0°. On the other hand, the TEM data reveal that the CLO's contain a high degree of order along their length. The achievement of this order is quite inconsistent with the finite levels of BAD found in the CRN structure. Thus, this too reveals the presence of regions containing very low levels of BAD. It is therefore natural to assign these two together, i.e., to conclude that the CLO's have a TO phonon of 490 cm^{-1}. So the three (and only three) structures reveled by TEM, i.e., the CRN amorphous network, the CLO's and the microcrystallites, have three distinct Raman signatures in the 475 cm^{-1}, 490 cm^{-1}, and 508–522 cm^{-1} scattering, respectively. The intermediacy of the CLO order is not just one of range, since their length can be as large as the largest microcrystallite cluster. Nor is it just one of intermediate Raman

TO frequency. It also includes one of dimensionality, the CLO's being quasi-one-dimensional.

It is evident that the high-quality *a*-Si:H material used in the production of solar cells having superior photoelectronic performance, especially in regard to the SWD effect, has elevated levels of CLO's and high levels of 490 cm^{-1} Raman scattering. We propose that this enhanced performance is a result of the intermediate order of the CLO's, which are essentially strain-free. With this sort of heterogeneity, an *a*-Si:H film can reach an overall more relaxed state than it could if it were composed of a homogeneous amorphous matrix. Moreover, this sort of heterogeneity, comprised of the CRN and the quasilinear CLO components, should offer few interfacial defects, compared to (say) the CRN and the rather rigid three-dimensional microcrystalline structure.

V. CONCLUSIONS

Amorphous hydrogenated silicon of improved quality for photovoltaic devices can be prepared by hydrogen dilution of the processing gas. This increase in quality, as measured, for example, by a decrease in the Staebler-Wronski degradation, is accompanied by an increase in the concentration of the chainlike objects in TEM micrographs. These CLO's are quasi-one-dimensional, having 2–3-nm widths and lengths of ~30 nm or more. Significantly, they show a high degree of order *along* their length, implying very low levels of bond-angle distortion (~0°). We associate these with a TO Raman band at 490 cm^{-1} with a full width of 37 ± 3 cm^{-1}; they also independently imply vanishing levels of bond-angle distortion. Intermediate order is revealed not only by its TO peak position being intermediate in frequency between those of the CRN amorphous matrix (475 cm^{-1}) and the smallest microcrystallites (~508 cm^{-1}), but also by the fact that the CLO's are ordered along their length compared to the zero- and three-dimensional orders of the CRN and crystalline conditions, respectively.

APPENDIX

In order to quantify the Raman spectra, we examine two different methods, both of which rely on the use of Gaussian functions: (A) Gaussian construction and (B) Gaussian deconvolution. In subsection A, we simulate the observed shifts in peak position and narrowing of the TO band by a construction of two Gaussian functions. This is done to make a historical connection between those two (position and width) commonly measured Raman parameters. In subsection B, the full Raman trace is fitted by performing a Gaussian deconvolution of the actual data.

In order to create traces that more closely resemble the measured shape of the Raman spectrum of *a*-Si:H, it is common practice to broaden the calculated *c*-Si vibrational density of states (v-DOS) by convolution with a Gaussian function.[25] A simplified description of the v-DOS can be had based on four Gaussian functions for the TA, LA, LO, and TO bands. This simplification is especially appropriate, however, for the crystalline TO v-DOS (our primary interest) since it is quite narrow (~30 cm^{-1}) and intense, i.e., Gaussian-like, compared to the neighboring LO band. Since the convolution of two Gaussians is also purely Gaussian, we expect that the convolution of the sharp (Gaussian-like) crystalline TO band with the broadening Gaussian to very closely approximate *just one* Gaussian function. This one Gaussian then represents an amorphous phase. Therefore, if the material contains multiple distinct phases, where each is represented by a distinct *level of strain*, each phase may contribute one Gaussian to the measured TO band in this description.

In the following analyses, we make use of *two* Gaussians to describe the TO Raman band. The justification for this is ultimately based on two fundamental experimental observations: (1) the spectral differences for all the H_2 diluted samples have their peaks at the same location and have ~ the same full widths, i.e., 490 and 35–40 cm^{-1}, respectively, and (2) the TEM shows the existence of two structural components (not including the obvious microcrystallites), i.e., the amorphous matrix and the CLO's. We thus interpret the observed TO band to be a composite of two sub-bands: a primary band at 475 cm^{-1} and a secondary band at 490 cm^{-1}.

A. Quantification by Gaussian construction

In this section, we examine the changes in the peak position and full width of the composite (i.e., measured) TO band and relate these changes to differences in relative magnitudes of the two subbands. The primary Gaussian represents the TO Raman band of *a*-Si:H made without H_2 dilution. Its peak position and full width are taken directly from the experimental data, i.e., 475 and 70.5 cm^{-1}, respectively. As is standard practice in the field,[10,22] the full width at half maximum is taken to be twice the half-width on the high-frequency side of the band. This is to avoid interference with the LO band centered near 400 cm^{-1}. The area of the primary band is fixed at 1. The peak position of the secondary band is fixed at 490 cm^{-1}, but its full width is not known precisely. We therefore select a few different widths from 30 to 50 cm^{-1} and for each width vary the area of the secondary band. This results in unique trajectories, in a plot of the full width versus peak position, for each secondary bandwidth. A comparison between those trajectories and the measured experimental points, shown in Fig. 9, reveals that the secondary 490 cm^{-1} band should have a full width of ~35 cm^{-1}. This is in agreement with the qualitative aspects found previously. Using this width, we thus determine the relative area of the 490 cm^{-1} band that best fits each experimental data point by minimizing the combined deviation of both the peak position and width, between the experimental and calculated values. The results are given in Table II.

B. Quantification by Gaussian deconvolution

In our deconvolution[26] of the experimental Raman traces, we use a *minimal* number of Gaussian functions that is consistent with the entire data set. For all of the spectra, one Gaussian is used to represent the LO band at ~400 cm^{-1}. For this band, its position, amplitude, and full width are allowed to vary during the fit. The TO band is generally represented by two Gaussians when no features due to μc-Si

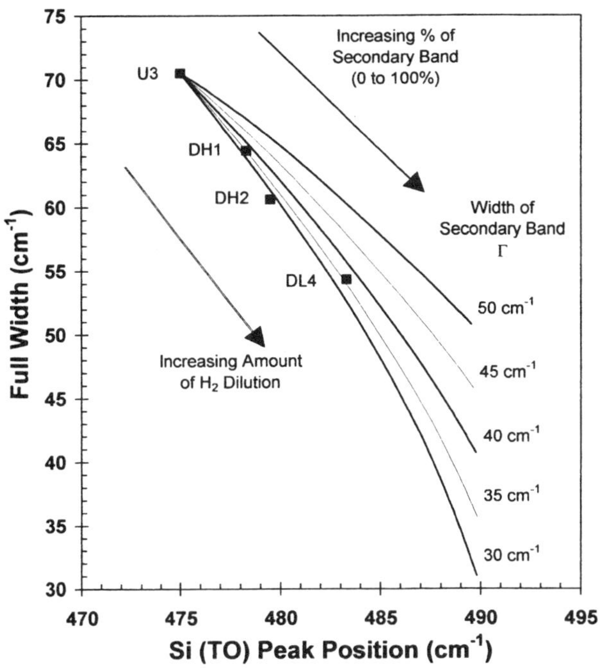

FIG. 9. Calculated relation between width and peak position of a composite curve based on the construction of two Gaussian functions, where the primary Gaussian represents the reference a-Si:H parameters taken from sample U3 and the secondary Gaussian represents the 490 cm^{-1} band vs the full width of the secondary band. Also shown are the experimental points.

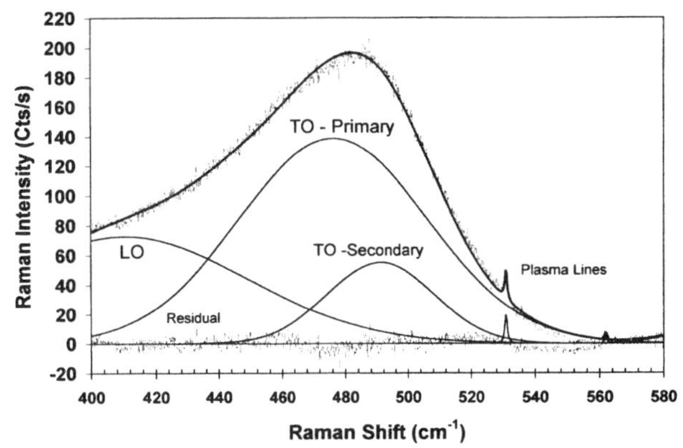

FIG. 10. Gaussian deconvolution of sample DL4 (dilution factor=250, $T_s=150\,°C$).

material are found. In such a case when the μc-Si band is indeed observed, e.g., at ~516 cm^{-1} for sample DH3, two additional Gaussians are required to fit that μc-Si band as one Gaussian results in a noticeably poor fit. In addition to these Raman features, two more Gaussians are used to locate the two plasma lines of the laser, which are important in establishing accurate Raman shifts.

Although one Gaussian can be used to adequately fit the TO band of the no H_2 dilution samples, one Gaussian is clearly insufficient in fitting the curves of the H_2 diluted samples. This is especially true the higher the H_2 dilution. At the other extreme, if we attempt to use three Gaussian functions to fit the TO band of the no H_2 dilution sample, we find that the amplitude of one of them always converges to zero during the fit. We therefore find that *two* Gaussians represent the minimal number that is consistent with the entire data set. The two Gaussians then represent two distinct environments within the material containing different levels of strain.

TABLE II. Calculated area of the secondary 490 cm^{-1} band for the samples deposited at 300 °C by (A) Gaussian construction and (B) Gaussian deconvolution. Values are given as the percentage of the total Raman scattering.

Sample name	Secondary 490 cm^{-1} band	
	(A)	(B)
U3	0.0	4.5
DH1	4.4	10.1
DH2	7.1	9.8
DH3	12.0	15.3
DH4	13.0	18.0

In performing multiple (m) Gaussian fits to a particular data set, there are $3m$ different parameters that can be varied, representing for each function its peak position, amplitude, and width. However, it is not our intent to minimize the resulting fit quality (χ^2) at all costs, but rather to restrict as much as possible the parameters that are allowed to vary using the experimental evidence as a guideline. Experimentally, we measure the peak position and width of the TO band of the no H_2 dilution sample to be 475 and 70.5 cm^{-1}, respectively. The position and width of the primary TO Gaussian are then fixed to these values for all of the samples. The difference spectra shown in Figs. 5 and 6 demonstrate that the excess Raman activity for all of the H_2 diluted samples is centered at 490 cm^{-1}. Therefore this position is chosen for the secondary Gaussian for all of the samples. In the previous section, the width of the secondary Gaussian that best fit the data by the construction analysis was ~35 cm^{-1}. However, we find that in the deconvolution analysis, a width of 40 cm^{-1} offers a slightly better fit, which is nevertheless consistent with the qualitative observations.

For sample DH3, we find that a single Gaussian for the μc-Si feature results in a very poor fit and that two are required to adequately fit that feature. Since we have no prejudices regarding the μc-Si components in sample DH3, all six of the parameters for the two Gaussians are allowed to vary. We find that those peaks are located at frequencies of 510.2 and 517.4 cm^{-1}, where the lower-frequency band represents μc-Si particles that are smaller in dimension than those in the higher-frequency band.[9] The need for two Gaussians therefore indicates the existence of a range in microcrystallite sizes, as already revealed by the TEM study. The result of the fit for sample DL4 is shown in Fig. 10. Table II gives the results for the samples prepared at 300 °C, and Fig. 8 for the samples prepared at 150 °C. Table II shows by this method that even the sample made with no H_2 dilution (U3) has a finite level of scattering in the 490 cm^{-1} band. It seems reasonable to expect that a certain amount of heterogeneity exists even in that sample. Indeed, the TEM study showed that LCO's can even be located in such samples, albeit at low concentrations. Thus, except for the ~5% offset, the two methods give quite similar results.

ACKNOWLEDGMENTS

The authors would like to thank Professor H. Fritzsche for a critical review of the manuscript and for his invaluable advice and comments. The work at United Solar Systems Corporation was supported in part by the National Renewable Energy Laboratory (NREL) under Contract No. ZAK-8-17619-09, while the work at Energy Conversion Devices was partially supported by NREL Contract No. ZAK-8-17619-18.

[1] Stanford R. Ovshinsky, J. Non-Cryst. Solids **32**, 17 (1979); Stanford R. Ovshinsky, Rosa Young, Wolodymyr Czybatyj, and Xunming Deng, U.S. Patent No. 5,103,284, April 7, 1992.
[2] D. L. Staebler and C. R. Wronski, Appl. Phys. Lett. **31**, 292 (1977).
[3] J. A. Reimer and M. A. Petrich, in *Amorphous Silicon and Related Materials*, edited by H. Fritzsche (World Scientific, New Jersey, 1989), p. 3.
[4] R. A. Street, J. Kakalios, C. C. Tsai, and T. M. Hayes, Phys. Rev. B **35**, 1316 (1987).
[5] W. H. Zachariasen, J. Am. Chem. Soc. **54**, 3841 (1932).
[6] J. C. Yang, X. Xu, and S. Guha, in *Amorphous Silicon Technology B 1994*, edited by E. A. Schiff *et al.*, MRS Symposia Proceedings No. 336 (Materials Research Society, Pittsburgh, 1994), p. 687; also Xixiang Xu, J. C. Yang, and S. Guha, J. Non-Cryst. Solids **198–200**, 60 (1996).
[7] J. Yang, A. Banerjee, and S. Guha, Appl. Phys. Lett. **70**, 2975 (1997).
[8] R. Tsu, M. Izu, S. R. Ovshinsky, and F. H. Pollak, Solid State Commun. **36**, 817 (1980).
[9] R. Tsu, S. S. Chao, M. Izu, S. R. Ovshinsky, G. J. Jan, and F. H. Pollack, J. Phys. (Paris), Colloq. **C4**, 269 (1981).
[10] R. Tsu, J. Gonzalez-Hernandez, J. Doehler, and S. R. Ovshinsky, Solid State Commun. **46**, 79 (1983).
[11] See, for example, C. C. Tsai, in *Amorphous Silicon and Related Materials*, edited by H. Fritzsche (World Scientific, Singapore, 1988), Vol. 1, p. 123.
[12] Cheng Wang, Ph.D. thesis, North Carolina State University, Raleigh, NC, 1991.
[13] D. V. Tsu, Ben Chao, S. R. Ovshinsky, S. Guha, and J. Yang, Appl. Phys. Lett. **71**, 1317 (1997).
[14] Toshihiro Kamei, Paul Stradins, and Akihisa Matsuda, Appl. Phys. Lett. **74**, 1707 (1999).
[15] D. L. Williamson, in *Amorphous Silicon Technology B 1995*, edited by M. Hack *et al.*, MRS Symposia Proceedings No. 377 (Materials Research Society, Pittsburgh, 1995), p. 251.
[16] A. H. Mahon, P. Raboisson, and R. Tsu, Appl. Phys. Lett. **50**, 335 (1987).
[17] S. Guha, J. Yang, D. L. Williamson, Y. Lubianiker, J. D. Cohen, and A. H. Mahan, Appl. Phys. Lett. **74**, 1860 (1999).
[18] P. Steinhard, R. Alben, and D. Weaire, J. Non-Cryst. Solids **15**, 199 (1974).
[19] P. N. Keating, Phys. Rev. **145**, 637 (1966).
[20] J. S. Lannin, in *Hydrogenated Amorphous Silicon*, edited by J. I. Pankove, Vol. 21 of *Semiconductors and Semimetal* (Academic, New York, 1984).
[21] R. Tsu, J. Gonzalez-Hernandez, and F. H. Pollack, J. Non-Cryst. Solids **66**, 109 (1984).
[22] D. Beeman, R. Tsu, M. F. Thorpe, Phys. Rev. B **32**, 874 (1985).
[23] R. Tsu, in *Disordered Semiconductors*, edited by M. A. Kastner, G. A. Thomas, and S. R. Ovshinsky (Plenum, New York, 1987) p. 479.
[24] R. Tsu, M. A. Paesler, and Dale Sayers, J. Non-Cryst. Solids **114**, 199 (1989).
[25] M. H. Brodsky, in *Light Scattering in Solids*, edited by M. Cardona (Springer, New York, 1975), p. 205.
[26] We used PeakSolve™ software by Galactic Industries, Salem, NH.

25/30 MW Ovonic Roll-To-Roll PV Manufacturing Machines

Stanford R. Ovshinsky and Masat Izu
Energy Conversion Devices, Inc., 2956 Waterview Drive, Rochester Hills, Michigan 48309; email: mizu@ovonic.com

Abstract: Energy Conversion Devices, Inc. (ECD Ovonics) is building a new a-Si thin film PV plant, which is a clone of the existing 25/30 MW United Solar Ovonic plant with improvements. The plant, which will be built in Auburn Hills, Michigan, for United Solar Ovonic, will be completed and commissioned for production in 2006. The proprietary production machines have been developed, engineered, and constructed by ECD Ovonics' Production Technology and Machine-Building Division (Machine Division). The machines are automated and designed to produce large, flexible rooftop PV modules that use triple-junction triple-bandgap a-Si alloy solar cells produced on stainless steel. When planned improvements are incorporated into the production line, the plant is expected to produce 30 MW PV modules per year with a module efficiency above 9%

Key Words: Amorphous Silicon, Roll-to-Roll PV Manufacturing, Building Integrated PV (BIPV), Flexible PV Modules

1 Introduction

Photovoltaic (PV) energy generation is one of the fastest growing industries in the world. The ECD Ovonics proprietary multi-junction, triple-bandgap, continuous web, thin-film technology is the ultimate solution for PV energy production [1]. ECD Ovonics, through its wholly owned subsidiary, United Solar Ovonic, has set a new standard for PV products. Compared to traditional crystalline PV, ECD Ovonics' thin-film PV can provide more overall energy since its triple-bandgap technology can capture about 30% more of the sun's energy. Our proprietary photovoltaics, manufactured by the miles, are:

- lighter,
- more durable, and
- more attractive (Figs. 4-5) than crystalline products.

In addition, our PV products can be produced in high volume from abundant and affordable raw materials. ECD Ovonics' roll-to-roll process is a most cost-effective process for high-volume production [2].

The United Solar Ovonic flexible thin-film photovoltaic material is made on proprietary high-output, football field-sized roll-to-roll machines (Fig. 2) developed by ECD Ovonics' Machine Division. The Machine Division

- has developed, designed, built, installed, started-up, and optimized the latest United Solar Ovonic roll-to-roll PV production plant;
- has over 25 years PV technology experience from R&D and pilot-scale to large-scale automated manufacturing through eight generations of PV production equipment;
- develops, designs, and fabricates production equipment and pilot lines for all ECD technologies: hydrogen, battery, etc.; and
- has been chosen by General Electric to develop roll-to-roll manufacturing for OLED lighting products.

2 Specifications of Machines

Key specifications of the new 25/30 MW United Solar Ovonic PV production plant include:

- Annual Production Capacity: 25 MW (30 MW when fully optimized), or approximately 4 million ft^2/yr.
- Substrate: Rolls of 14 inch wide, 8500 foot long, 5 mil thick stainless steel (Fig. 3).
- Device Structure: Two layers of backreflector, nine layers of a-Si alloys, and a layer of ITO (Fig. 1).
- Real-time in-line device performance monitoring.
- Lamination: Tefzel/EVA polymer encapsulation.

3 Solar Cell Manufacturing Line

The Ovonic production line [3,4,5] includes a :

Roll-to-Roll Deposition Line consisting of a

- Washing Machine,
- Backreflector (BR) Deposition Machine,
- a-Si Alloy Deposition Machine, and an
- ITO Deposition Machine.

The Automated Module Assembly Line, includes a:

- Roll-to-Sheet Cutting Line,
- Cell Line,
- Interconnect Line,
- Lamination Line, and
- Finishing/Testing Line

4 ECD Ovonics Solar Cell Structure

Figure 1 shows the structure of the ECD Ovonics solar cell.

Grid		Grid
Transparent Conductive Oxide		Sputtering
P3	Nanocrystalline Si Alloy	PECVD
I3	a-Si	PECVD
N3	a-Si Alloy	PECVD
P2	Nanocrystalline Si Alloy	PECVD
I2	a-Si/Ge Alloy	PECVD
N2	a-Si Alloy	PECVD
P1	Nanocrystalline Silicon Alloy	PECVD
I1	a-Si/Ge Alloy	PECVD
N1	a-Si Alloy	PECVD
Textured Metal/ZnO Back-reflector		Sputtering
Stainless Steel Substrate		

Fig. 1. Structure of a-Si Solar Cells.

5 Diagnostic Systems for the Roll-To-Roll Machines

The online non-contacting diagnostic systems [6] for the Ovonic roll-to-roll processors are summarized in Table I.

Table I. Diagnostic Systems for 25/30 MW Line.

Processor	Diagnostic System	No.	Measurement
BR	Scatterometer	1	Specular/diffuse reflection
	Reflectometers	2	ZnO thickness
a-Si	Reflectometers	15	Thickness of each n-, i-, and p-layer
	PVCD	4	Component and device electrical properties
ITO	Film Conductivity	1	ITO conductivity
	Reflectometers	5	Film thickness/uniformity
	PVCD	1	Device electrical properties; degree of physical shunts

Fig. 2. 25/30 MW Ovonic Roll-to-Roll PV Manufacturing Machine

Fig 3. Roll of a-Si solar cell material.

6 Acknowledgements

The authors thank Herb Ovshinsky and his group for many years of contributions in designing and constructing the ECD Ovonics production machines. Also, we would like to express our thanks to Dr. Scott Jones, Dr. Vin Cannella, Dr. Tim Ellison, Dr. Joe Doehler, and Dr. Hellmut Fritzsche along with their collaborators for contributing the design and technical specifications for the machine. We also express our gratitude to Dr. Subhendu Guha, Kevin Hoffman, Gary Didio, Jon Call, and Dr. Prem Nath for their assistance on this project.

Fig. 4. A United Solar Ovonic BIPV roofing system helps power Grand Valley State University's 22,500 ft^2 Michigan Alternative and Renewable Energy Center.

Fig. 5. United Solar Ovonic was chosen to provide a buildig-integrated photovoltaic roofing system for the Beijing New Capital Museum. *Photo courtesy of Beijing New Capital Museum.*

7 References

[1] S. R. Ovshinsky, *Proceedings of the International PVSEC-1*, 1988, p.577.

[2] M. Izu, S.R. Ovshinsky, SPIE Proc. 407 (1983) **42**.

[3] S. Guha, J. Yang, A. Banerjee, K. Hoffman, S. Sugiyama, S. Call, S.J. Jones, X. Deng, J. Doehler, M. Izu, H.C. Ovshinsky, *Proc. 26th IEEE PV Specialist Conference*, Anaheim, CA, 1997, p.607.

[4] S.R.Ovshinsky, R. Young, W. Czubatyj, X. Deng, Semiconductor with Ordered Clusters, U.S. Patent 5,103,284, April 7, 1992.

[5] S.R. Ovshinsky, S. Guha, C. Yang, X. Deng, S. Jones, Semiconductor Having Large Volume Fration of Intermediate Range Order Material, U.S. Patent 6,087,580, July 11, 2000.

[6] T. Ellison, Proc. 28th IEEE PV Specialist Conference, Anchorage, AK, 2000, p.732.

Photovoltaics Publications

Photovoltaic Solar Energy Conference, ed. A.S. Strub, book review, American Scientist 66 (1978) 616.

Solar Electricity Speeds Down to Earth, New Scientist 80 (1978) 674.

A New Amorphous Silicon-Based Alloy for Electronic Applications (with A. Madan), Nature 276 (1978) 482.

An Innovative Approach to New Sources of Energy Through Amorphous Materials, Proc. of UNITAR Conference on Long Term Energy Resources, Montreal, Canada (1979) p. 783.

Electrical and Optical Properties of Amorphous Si:F:H Alloys (with A. Madan and E. Benn), Phil. Mag. B40 (1979) 259.

Some Electrical and Optical Properties of a-Si:F:H Alloys (with A. Madan, W.Czubatyj and M. Shur), J. Elect. Mat. 9 (1980) 385.

Properties of Amorphous Si:F:H Alloys (with A. Madan); J. Non-Cryst. Solids 35/36 (1980) 171.

The Chemistry of Glassy Materials and Their Relevance to Energy Conversion, J. Non-Cryst. Solids 42 (1980) 335.

Metal-Insulator-Semiconductor Solar Cells Using Amorphous Si:F:H Alloys (with A. Madan, J. McGill, W. Czubatyj and J. Yang), Appl. Phys. Lett. 37 (1980) 826.

The Immediacy of Alternative Energy, Japanese Economic Journal and Science, Japanese Scientific American (1981).

High Efficiency, Large-Area Photovoltaic Devices Using Amorphous Si:F:H Alloy (with A. Madan, W. Czubatyj, J. Yang, J. McGill) J. de Physique 42, Suppl.10 (1981) C4-463.

The Nature of Intermediate Range Order in Si:F:H:(P) Alloy Systems (with R. Tsu, S.S. Chao, M. Izu, G.J. Jan and F.H. Pollak), ibid. p. C4-269.

Progress in Large Area Photovoltaic Devices Based on Amorphous Silicon Alloys (with J.P. deNeufville and M. Izu), Photovoltaics, The Solar Electric Magazine 3 (1982) 2217.

Commercial Development of Ovonic Thin Film Solar Cells, SPIE Proc. 407 (1983) 5.

Production of Tandem Amorphous Silicon Alloy Solar Cells in a Continuous Roll-to-Roll Process (with M. Izu), ibid. p. 42.

Roll-to-roll plasma deposition machine for the production of tandem a-Si alloy solar cells (with M. Izu), Thin Solid Films 119 (1984) 55.

Amorphous Photovoltaics (with D. Adler), Chemtech 15(1985) 538.

Low Pressure Microwave Glow Discharge Process for High Deposition Rate Amorphous Silicon Alloy (with S.J. Hudgens and A.G. Johncock), J. Non-Cryst. Solids 77/88 (1985) 809.

The Chemical and Configurational Basis of High Efficiency Amorphous Photovoltaic Cells, Proc. 18th IEEE Photovoltaic Specialists Conference (1985) p.1365.

A Figure of Merit Evaluation of Amorphous Silicon Alloy Solar Cells (with J.A. Yang), Proc. of the 1985 Intl. Conf. on Solar and Wind Energy Applications, China (Academic Publishers) p. 75.

The Breaking of the Efficiency-Stability-Production Barrier in Amorphous Photovoltaics (with J. Yang), SPIE Proc. 706 (1986) 88.

1 MW Amorphous Silicon Thin-Film PV Manufacturing Plant (with P. Nath, K. Hoffman, J. Call, C. Vogeli and M. Izu), PVSEC-3 (1987) 395.

Fluorinated Amorphous Silicon-Germanium Alloys Deposited from Disilane-Germane Mixture (with S. Guha, J.S. Payson and S.C. Agarwal), J. Non-Cryst. Solids 97&98 (1987) 1455.

Passivation of Dangling Bonds in Amorphous Si and Ge by Gas Absorption (with R. Tsu, D. Martin and J. Gonzalez-Hernandez), Phys. Rev. B 35 (1987) 2385.

A New, Inexpensive, Thin Film Photovoltaic Power Module (with P. Nath, K. Hoffman, C. Vogeli and K. Whelan), 20th IEEE Photovoltaic Specialists Conference (1988) p. 1315.

Yield and Performance of Amorphous Silicon Based Solar Cells Using Roll-to-Roll Deposition (with K. Hoffman, P. Nath, J. Call, G. DiDio and C. Vogeli), ibid. p. 293.

Conversion Process for Passivating Current Shunting Paths in Amorphous Silicon Alloy Solar Cells (with P. Nath, K. Hoffman and C. Vogeli), Appl. Phys. Lett. 53 (1988) 986.

A Novel Design for Amorphous Silicon Alloy Solar Cells (with S. Guha, J. Yang, A. Pawlikiewicz, T. Glatfelter and R. Ross), Proc.of the 20th IEEE PVSC (1988) p. 79.

Roll-to-Roll Mass Production Process for a-Si Solar Cell Fabrication, Proc. Intl. PVSEC-1 (1988) 577.

Band Gap Profiling for Improving the Efficiency of Amorphous Silicon Alloy Solar Cells (with S. Guha, J. Yang, A. Pawlikiewicz, T. Glatfelter and R. Ross), Appl. Phys. Lett. 54 (1989) 2330.

Production of 20 A Sec-1 a-Si Alloys for Use in Solar Cells (with P. Nath, K. Hoffman, J. Call and G. DiDio), Proc. of the 21st IEEE PVSC (1990) 1167.

Toward the Elimination of Light-Induced Degradation of Amorphous Si by Fluorine Incorporation (with X. Deng, E. Mytilineou and R. Young), Mat. Res. Soc. Symp. Proc. 258 (1992) 491.

Continuous Roll-to-Roll Amorphous Silicon Photovoltaic Manufacturing Technology (with M. Izu, X. Deng, A. Krisko, H.C. Ovshinsky, K.L. Narasimhan and R. Young), AIP Conf. Proc. 306 (1993) 198.

Manufacturing of Triple-Junction 4 ft2 a-Si Alloy PV Modules (with M. Izu, X. Deng, A. Krisko, K. Whelan, R. Young, H.C. Ovshinsky and K.L. Narasimhan), Proc. 23rd IEEE PVSC (1993) 919.

.

Amorphous Silicon Alloy Photovoltaic Technology – From R&D to Production (with S. Guha, J. Yang, A. Banerjee, T. Glatfelter, K. Hoffman, M. Izu, H. Ovshinsky and X Deng), Mat. Res. Soc. Symp. Proc. 336 (1994) 645.

Continuous Roll-to-Roll Serpentine Deposition for High Throughput a-Si PV Manufacturing (with M. Izu, H.C. Ovshinsky, X. Deng, A.J. Krisko, K.L. Narasimhan, R. Crucet, T. Larman and A. Myatt), IEEE First World Conf. on Photovoltaic Energy Conversion, (1994) p.820.

Dependence of a-Si Solar Cell Voc on Deposition Temperatures (with X. Deng, K.L. Narasimhan, J. Evans and M. Izu), ibid., p. 678.

Lightweight Flexible Rooftop PV Module (with M. Izu, H.C. Ovshinsky, K. Whelan, L. Fatalski, T Glatfelter, K. Younan, K. Hoffman, A. Banerjee, J. Yang and S. Guha), ibid., p. 990.

The Material Basis of Efficiency and Stability in Amorphous Photovoltaics, Solar Energy Materials and Solar Cells 32 (1994) 443.

Stability Test of 4 FT2 Triple-Junction a-Si Alloy PV Production Modules (with X. Deng, M. Izu and K.L. Narasimhan), Mat. Res.Soc.Symp. Proc.336 (1994) 699.

New Evaluation Technique for Thin-Film Solar Cell Back-Reflector Using Photothermal Deflection spectroscopy, (with X. Deng and K.L. Narasimhan), 1994 IEEE First World Conference. Vol. 1 (1994) 555.

Ion and Neutral Argon Temperatures in Electron Cyclotron Resonance Plasmas by Doppler Broadened Emission Spectroscopy (with David V. Tsu, R.T. Young, C.C. Klepper and L.A. Barry), J. Vac. Sci. Technol. A 13 (1995) 935.

PV Metal Roofing Module (with T. Ellison, L. Fatalski, R. Kopf, H. Ovshinsky, M. Izu, R. Souleyrette, K. Whelan, J. Wiehagen and L. Zarker), the 25th IEEE PVSC (1996) 1437.

Amorphous Silicon Alloys – The Optoelectronic Materials that Set the Trend for Photovoltaic Applications (with J.C. Yang), International Materials Research Congress, Cancun, Mexico, Superficies y Vacio 7 (1997) 1.

Effect of Hydrogen Dilution on the Structure of Amorphous Silicon Alloys (with R. Tsu, B.S. Chao, S. Guha and J. Yang) Appl. Phys, Lett. 71 (1997) 1317.

Roll-to-Roll Microwave PECVD Machine for High Barrier Film Coatings (with M. Izu and B. Dotter), Intl. Conf. of Vacuum Web Coating (November 1994).

.

New High Speed, Low Cost, Roll-to-Roll Antireflectivity Coating Technology, (With T. Ellison, B. Dotter, M. Izu) Proc. Society for Vacuum Coaters (1997).

Improved μc p-Layer and a-Si i-Layer Materials Using VHF Plasma Deposition, (with X. Deng, S.J. Jones, T. Liu and M. Izu), Proc. of the 26th IEEE Photovoltaic Specialists Conference (1997) p.591.

Heterogeneity in Hydrogenated Silicon: Evidence for Intermediately Ordered Chainlike Objects, (With D. Tsu, B.S. Chao, S. Jones, J. Yang, S. Guha, R. Tsu), Phys. Rev. B63 (2001) 125338-1.

25/30MW Ovonic Roll-to-Roll Manufacturing Machine (with M. Izu) PVSEC-15, Shanghai, China, 2005.

US patents - photovoltaics

Amorphous semiconductors equivalent to crystalline semiconductors produced by a glow discharge process
4226898 10/07/1980

Method for optimizing photoresponsive amorphous alloys and devices
4342044 01/27/1982

Current enhanced photovoltaic device
4379943 04/12/1983

A method of making p-doped silicon films and devices made there from
4400409 08/23/1983

Amorphous semiconductors equivalent to crystalline semiconductors
4409605 10/11/1983

Continuous amorphous solar cell production system
4410558 10/18/1983

Photo-assisted cvd
4435445 03/06/1984

Optimized doped and band gap adjusted photoresponsive amorphous alloys and devices
4492810 01/08/1985

Method of making amorphous semiconductor alloys and devices using microwave energy
4504518 03/12/1985

Continuous amorphous solar cell production system
4519339 05/28/1985

Compositionally varied materials and method for synthesizing the materials
4520039 05/28/1985

Method and apparatus for making layered amorphous semiconductor alloys using microwave energy
4521447 06/04/1985

Method for optimizing photo responsive amorphous alloys and devices
4522663 06/11/1985

Method of forming photovoltaic quality amorphous alloys by passivating defect states
4569697 02/11/1986

Tandem junction solar cell devices incorporating improved microcrystalline p-doped semiconductor alloy material
4609771 09/02/1986

Method of depositing semiconductor films by free radical generation energy
4615905 10/07/1986

Gas mixtures for the vapor deposition of semiconductor material
4637895 01/20/1987

Methods of using selective optical excitation in deposition processes and the detection of new compositions
4637938 01/20/1987

Method of making a photovoltaic panel
4677738 07/07/1987

Continuous deposition of activated process gases
4678679 07/07/1987

Gas mixtures for the vapor deposition of semiconductor material
4696758 09/29/1987

Vapor deposition of semiconductor material
4698234 10/06/1987

Method of depositing thin films using microwave energy
4701343 10/20/1987

Wide band gap semiconductor alloy material
4710786 12/01/1987

P and n-type microcrystalline semiconductor alloy material including band gap widening element devices utilizing same
4775425 10/04/1988

Thin film solar cell including a spatially modulated intrinsic layer
4816082 03/28/1989

Fluorinated precursors from which to fabricate amorphous semiconductor material
4839312 06/13/1989

Method for the high rate plasma deposition of high quality material
4883686 11/28/1989

Multiple cell photoresponsive amorphous alloys and devices
4891074 01/02/1990

A method of fabrication n-type and p-type microcrystalline semiconductor alloy material including band gap widening elements
4891330 01/02/1990

Method of creating a high flux of activated species for reaction with a remotely located substrate
4937094 06/26/1990

Multiple cell photoresponsive amorphous photovoltaic devices including graded band gaps
4954182 09/04/1990

Method of depositing directly activated species onto a remotely located substrate
5093149 03/03/1992

High quality photovoltaic semiconductor material and laser ablation method of fabricating same
5231047 07/27/1993

Chapter V: Batteries

In Paris in 1801 the Italian physicist Allessandro Volta demonstrated his battery's generation of an electric current. The emperor Napoleon was fascinated and made him Count and Senator of the kingdom of Lombardy. This rare recognition of a scientific discovery illustrates both the active interest of Napoleon in new technology and the importance of the first source of a continuous electric current. Volta placed silver and zinc electrodes into a fluid of brine and connected them in pairs to form a "voltaic pile." He further found that any two different metals placed in a conducting fluid, which we now call an electrolyte, form a battery. The unit of the electromotoric force, the Volt, was named in his honor.

The familiar lead-acid battery, the starter battery in our cars, was invented by the French physicist Gaston Planté in 1859. This battery differs fundamentally from the previous ones in that it can be recharged many times. There is a huge market for both battery types, the primary battery which can be discharged only once and the secondary battery whose electrode reactions are reversible and thus can be recharged.

For many decades the market for secondary batteries was dominated by lead-acid and nickel-cadmium (Ni-Cd) batteries. In the 1980s the demand for lightweight, high energy density batteries increased at a record pace with the massive demand for portable electronics, laptop computers, camcorders, cell phones, and power tools. Until the 1980's, the components of typical batteries contained toxic materials such as lead and the carcinogen cadmium.

Fortunately, this changed with Ovshinsky's invention of the Nickel-Metal hydride (NiMH) battery. Its use saved the world from the management of unacceptable amounts of hazardous toxic waste. In the 1990s billions of Ovshinsky's batteries were produced and sold for portable electronics world wide through his licensees, which included all the major battery manufacturers. In 1992 the Ovonic Battery Company (OBC), a subsidiary of Energy Conversion Devices (ECD), was selected by the US Advanced Battery Consortium to scale up and further develop its NiMH technology for electric vehicle applications. The major US automobile companies were seriously preparing at that time for all electric and hybrid gasoline-electric vehicles.

Through an active research program OBC continued to improve its technology, battery design and manufacturing techniques. The first Ovonic NiMH batteries were installed in EV1 (Electric Vehicle 1) of General Motors and restricted only to lease costumers. The sad and inexplicable demise of these beloved cars was documented in the movie "Who Killed the Electric Car" which prominently featured the achievements of Stan and Iris Ovshinsky. In 1994, a NiMH battery-powered car, the Solectria won the first of seven Tour de Sol road rallies with a record 214 mile range on a single charge, in 1995, it won again with a 238 mile range, and in 1996 it won the Tour de Sol road rally with an astonishing 373 mile range on a single charge using a 33kWh NiMH battery pack. Ovshinsky's batteries have made possible and now dominate the market for hybrid cars and electric vehicles and with it the technical foundation for a cleaner environment and a reduction in climate changing pollution. Stan Ovshinsky and his wife Iris were named "Heroes for the Planet" by *Time Magazine* in 1999. Ovshinsky received the 2005 Innovation Award for Energy and the Environment by the prestigious magazine *The Economist* for "his pioneering work in the development of the high-powered NiMH battery". For many this breakthrough alone would be a proud life-time achievement, for Ovshinsky, it was one of many.

The success story of the NiMH battery technology demonstrates the remarkable ingenuity of Ovshinsky and his material scientists. It is a fascinating story of how a multitude of challenges were met with a series of inventive insights utilizing multi-element metallurgy, the ability to design a desired micro-structure and porosity of the electrode materials, taking advantage of their surface chemistry, and new catalytic reactions of nano-particles. In contrast to other batteries whose metal electrode surfaces just transition back and forth from a metal to an oxide during charging and discharging, the metal hydride negative electrode and the nickel hydroxide positive electrode materials of the NiMH battery are multi-element constructs and therefore ideal for taking advantage of Ovshinsky's atomic and structural engineering ability for optimizing improvements.

A simple understanding of the science of the NiMH battery is as follows: It is a hydrogen ion battery. During charging, hydrogen leaves the bulk of the positive electrode, donates an electron and moves as an ion through the electrolyte. At the negative electrode the ion accepts an electron, and the hydrogen atom diffuses into the negative electrode where it gets bound and stored until the discharge reverses the process. There is fortunately no change in electrolyte quantity or concentration. The stored energy density obviously increases with the density of hydrogen binding sites and the total current involved in the charge and discharge increases with electrode area and the activity of the electrode electrolyte interface. The battery is made of relatively inexpensive and abundant materials which are non-toxic and totally recyclable. These processes and properties sound almost too good and simple, but before Ovshinsky started to work on this battery it was declared unsuitable for commercialization and practically abandoned after 25 years of research by several battery companies. The previous problems included very slow charge and discharge rates, a short cycle life and a lack of production scale methods for producing homogeneous alloys and powders for the negative electrode.

The story of the inventions, insights, and developments that solved these problems and led to a powerful and reliable new NiMH battery can be found in the selected papers of this section and the relevant patents listed at the end. Here we highlight some of the major steps which made NiMH the battery of choice for electric and hybrid vehicles.

The electrochemical use of a hydrogen storage electrode in an alkaline electrolyte defines the voltage range and thus limits the hydrogen bonding energy to a narrow range around 8-10 kcal/mol. The bonding energy depends not only on the transition metal element in the alloy but also on the elements surrounding them: it depends on the charge density in the spaces between the atoms, the hydrogen bonding sites. The energy density per weight or volume of the battery must of course be high. By judiciously choosing the mix of elements in the alloy and the proper microstructure, Ovshinsky optimized the density of H bonding sites and hence the battery energy density.

Hydrogen diffuses quite rapidly inside the negative electrode alloy. The difficulty arises at the electrode interface with the electrolyte. During manufacture one cannot avoid the forming of a surface oxide layer on the electrode. Unfortunately, this oxide is an effective barrier which prevents any electrochemical action. This layer has to be completely modified in the presence of the alkaline electrolyte. This is accomplished by dissolution of some elements of the alloy and dissolution and segregation of others. As a result a porous oxide layer is created with an abundance of nano-particles of mostly nickel. Because of their small size, the nickel surface atoms are active catalysts. The catalytic action inside the porous oxide layer is essential for fast charging and discharging the battery and can be understood as follows: As mentioned before, hydrogen ions arrive at the negative electrode during charging. They pick up an electron and many of these hydrogen atoms would like to react with each other and form gaseous hydrogen molecules, H_2, rather than diffuse into the electrode. This undesired hydrogen molecule formation is prevented by the catalytic activity of the nickel alloy nano-particles in the porous oxide layer. Large current densities can now move between the electrolyte and the negative electrode without hydrogen gas formation.

The negative electrode alloy consists of V-Ti-Zr-Ni-Cr and other high melting point elements. Once molten, homogenized and cooled the material ends up as a very hard block, yet, the final electrode is a compressed powder of micron-size particles. The challenge is how can one transform the block of metal into a compressed powder in an inexpensive, highly efficient production process?

This metallurgical engineering problem was solved by Michael Fetcenko of OBC. The metal alloy ingot is placed in a large vacuum reactor. Hydrogen backfilled into the reactor is absorbed by the block of metal, thereby causing a volumetric expansion of the metal lattice which breaks apart the material. Several hydride/dehydrate cycles yield the desired powder particles. The powder particles would oxidize violently if exposed to the atmosphere causing a fire hazard. Controlled oxidation prepares the particle surfaces for subsequent treatment which produces the desired micro structure and porosity. Moreover, charge-discharge cycles during the use of the battery create fresh micro-surfaces and hydrogen bonding sites.

The positive electrode is essentially nickel hydroxide changing between $Ni(OH)_2$ and Ni(OOH). The charge/discharge reaction is usually considered to be a one-electron transfer reaction. By judicious elemental modification Ovshinsky's group increased the electron transfer to 1.3 electrons per Ni atom with a corresponding increase in energy density. While the negative electrode remains highly conductive, the positive electrode becomes nearly non-conducting in its $Ni(OH)_2$ state. A non-conducting electrode is of course intolerable for the function of the battery and requires other clever metallurgical processes to provide the required electrical conductivity.

The current NiMH battery is guaranteed in hybrid cars for 100,000 miles or the life of the car. It is not only a metallurgical and electrochemical masterpiece but indispensable for a cleaner environment and a more livable world.

The selected papers following this section highlight the step by step advances in NiMH battery technology made by Ovshinsky and his team.

Presented at the 17 International Power Sources Symposium
Bournemouth, England (April, 1991)

16

ALLOY EFFECTS ON CYCLE LIFE OF Ni-MH BATTERIES

M. A. Fetcenko, S. Venkatesan, S. R. Ovshinsky, K. Kajita*, M. Hirota*, and H. Kidou*

Ovonic Battery Company, Troy, Michigan, USA
**Hitachi Maxell Ltd., Ibaraki-shi, Osaka, Japan*

ABSTRACT

Nickel-metal hydride (Ni-MH) batteries have been developed which exhibit overall performance suitable as drop-in replacements for nickel-cadmium(NI-Cd) batteries. Ni-MH batteries, with substantially higher energy density and no cadmium, offer extended device run time and address the urgent environmental concern over cadmium toxicity.

To meet widespread commerical applications, improvements in performance characteristics such as fast charge capability, high rate discharge capability, self-discharge, low temperature capability, and especially cycle life have been studied together with high volume cost and manufacturing process viability.

This study will report overall performance of Ni-MH cells using Ovonic Battery Company's proprietary V-Ti-Zr-Ni alloy system. Advances in internal resistance, self-discharge, and low temperature capability will be addressed. Results from electrical testing of these cells will be presented, as well as discussion on the approaches utilized.

Cycle life data under a variety of test conditions will be presented, and the dramatic influence of the metal hydride alloy on improved cycle life, will be discussed.

INTRODUCTION

Environmental concerns over the manufacture and disposal of all battery systems have recently become of paramount importance. In particular, cadmium toxicity and the implications of production and disposal of Ni-Cd batteries is clearly a world wide problem. Nickel-metal hydride rechargable batteries have been the focus of ongoing development programmes at many battery companies for over 25 years, primarily investigating $LaNi_5$ based systems. (Van Rijswick, 1977; Bittner and Babcock, 1983; Van Beek *et al.*, 1984). Metal hydride batteries were considered to be unsuitable for commercial application due to one or more of the following: poor cycle life, poor high rate discharge, inability to fast charge, high cost, and lack of production scale methods for producing homogeneous alloys, powders, and electrodes.

Ovonic Battery Company (OBC) has developed its patented metal hydride batteries, based upon V-Ti-Zr-Ni based alloys (Sapru *et al.*, 1985; Venkatesan *et al.*, 1988 a), for commercial application to address the environmental problem of cadmium by replacement with non-hazardous materials. OBC batteries also provide performance advantages over Ni-Cd, such as higher energy density and elimination of memory effect.

Previous publications have described performance of our 3·5 Ah C size cell in detail (Venkatesan *et al.*, 1988 b; Fetcenko *et al.*, 1990 c). This paper will present updated performance; with particular focus on cycle life improvements. Improvements in high rate discharge capability, low temperature performance and charge retention will also be addressed. Further, to illustrate the versatility of this technology, recent development activities in Ni-MH batteries for aerospace applications and large industrial batteries for electric vechicle applications will be described.

ENVIRONMENTAL ASPECTS

Development of a new battery system for the 1990's requires environmentally safer materials as a prerequisite. Not only exposure to workers in the manufacturing facilities, but disposal in public landfills and incineration must be addressed.

Recently, metal hydrides of the type used by OBC were tested for leachability according to the EPA Toxic Characteristic Leachate Procedure (TCLP), along with cadmium from Ni-Cd batteries (Knoll *et al.*, 1990). There were three conclusions drawn from the testing:

1. Cadmium will leach from Ni-Cd batteries at hazardous levels.
2. The metal alloy (V-Ti-Zr-Ni-Cr type) powders are less of an environmental risk than Ni-Cd and lead-acid batteries. In fact, the elements in the Ovonic alloys were found to be at least an order of magnitude below Hazardous Waste Limits (HWL).
3. A battery made of the Ovonic type metal hydride alloy powders is not considered a hazardous waste and could be legally disposed of in a municipal landfill.

The TCLP results illustrate that the environmental characteristics of OBC type metal hydride alloys are inherently safer than cadmium. Along with good performance characteristics and competitive costs, Ni-MH technology is expected to replace Ni-Cd in many applications.

CYCLE LIFE

For widespread use in consumer applications, a battery system must be relatively insensitive to normal usage conditions. Continuous overcharge is standard practice in many applications. Since cells are almost always used in series, where high depth of discharge is common, tolerance to deep discharge is important. In common practice, reduction in Ni-Cd battery life due to reversal is usually compensated by the fact that most cycles are less than 100% depth of discharge.

To satisfy customer requirements, OBC cells have been designed with fast charge capability. While many applications still utilize C/10 continuous overcharge, there is a growing trend towards fast charge in many applications. In a higher energy density battery system like Ni-MH, when compared to Ni-Cd, the "C rate" philosophy becomes a dilemma. The "one hour" charge and discharge rate is 2·0 A for a 2·0 Ah Ni-Cd C cell, for a 3·5 Ah Ni-MH C cell, the one hour charge and discharge rate is 3·5 A. However, most applications now seem to be directed towards longer run time. Indeed, if we use a drop-in replacement of Ni-MH for Ni-Cd, the device power requirements do not change, but charge/discharge rates closer to C/2 are perhaps more relevant. Consequently, most of our cycle life testing uses C/2 rates for both charge and discharge.

Fast Charge to Temperature

The most routine cycle test done at OBC consists of C/2 charge to a temperature cut-out, followed by a small amount of trickle charge. This is followed by a C/2 discharge to voltage cut-out. Again, the philosophy is that this test condition provides rapid feedback of results, and is a good simulation of realistic field test conditions for a wide variety of applications.

Each 3·5 Ah C cell, is individually monitored for voltage, current, and temperature and is equipped with a temperature thermistor to indicate overcharge and automatically reduce charge current. The C cells are charged at 1·8 Amps to a temperature of about 32°C, and then automatically switched to a trickle charge current of 180 mA. The total charge time is 3 h. The cells are automatically switched to discharge at 2·0 Amps. Each cell voltage is monitored by strip chart recorder on every cycle. Capacity is measured to a 1·0 volt cut-out although each cell is actually cut-out at 0·97 V. The total discharge time is 2 h allowing some period of open circuit voltage before the next charge cycle. It should be noted that every cycle is to the above test regime, and therefore distinguishes these tests from the "capacity measuring cycle" method where the cell is cycled normally to 60% depth of discharge (DOD) but every 50 cycles is subjected to 100% DOD.

Figure 1 presents cycle life data under these conditions for our "24 series" cells. The data shown here is for a group of 4 cells randomly selected from a group of several hundred. Previous results for our "15 series" cells showed a life of about 450 cycles under these conditions (Fetcenko *et al.*, 1990 a). In the results for the "24 series" cells, life has been greatly extended, attaining over 800 cycles and testing is still continuing (Fetcenko *et al.*, 1990 b).

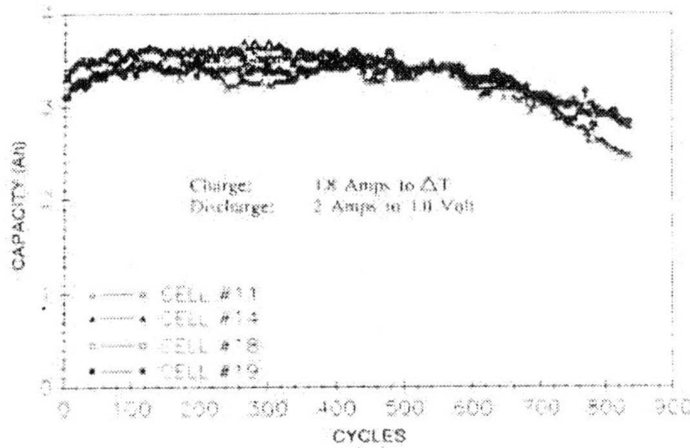

FIG. 1 "24 Series" C cells under C/2 to Temperature Charge, C/2 to 100% Discharge.

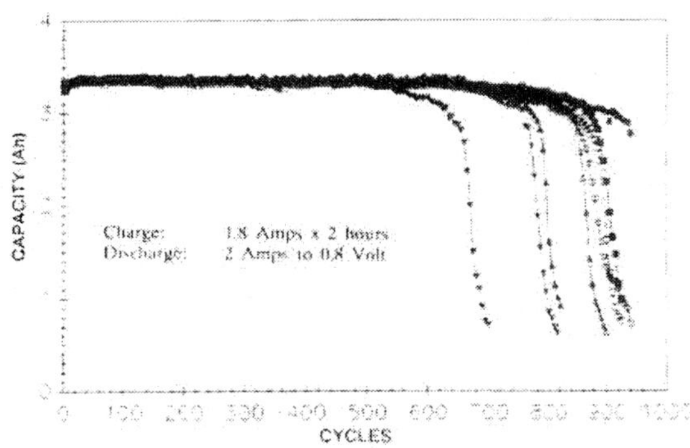

FIG. 2 "24 series" C cells under C/2 to 105% Charge, C/2 to 100% Discharge.

Fast Charge to Time

Since no single test regime can simulate all applications, the effect of overcharge was investigated with a C/2 charge to a fixed charge input. Similar to the test conditions specified above, this test method introduces approximately 5% overcharge. 3·5 Ah C cells of the construction and alloy type shown in Fig. 1 were charged at 1·8 A for 2 h. The discharge current was also 1·8 A with the capacity measured to 0·9 V and actual cut-out at 0·8 V. The capacity discharge is about 3·4 Ah, so a charge input of 3·6 Ah provides an excess charge of about 5%. The cells reach a temperature of over 40°C in overcharge.

Figure 2 presents cycle life data for the "24 series" C cells under these conditions. It can be seen that the nominal discharge capacity is slightly higher in this test due to the 5% overcharge. The capacity is also more consistent due

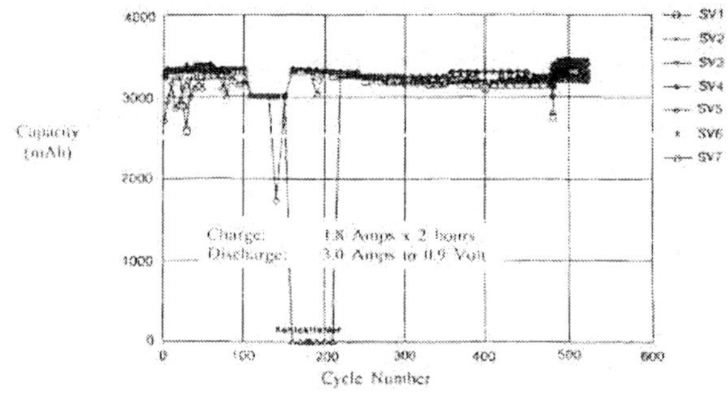

FIG. 3 "24 Series" C cells under C/2 to 105% Charge, C to 100% Discharge.

to the fixed charge input. When discharged to 80% of original capacity, cycle life typically exceeds 800 cycles compared to about 400 cycles with the "15 series" cells. It is noteworthy that the capacity decline at the end of life is much faster under this test condition than for the C/2 to temperature cut-out test method. The explanation is that when the cell capacity began to decline, no adjustment was made to the charge input. Consequently, the degree of overcharge increases, becoming quite abusive. Venting is likely to occur, further hastening the capacity loss.

Similarly, Fig. 3 presents cycle life under C/2 to 105% charge with a C rate discharge. Each cycle is 100% depth of discharge, with a 0·9 V cut-out. For the "24 series" cells, over 500 cycles have been achieved and the test is still running.

Three Hour Charge

Another factor investigated was more extensive overcharge. Three hour (C/3) charging was selected, which offers reasonable turnaround time for results and is also quite aggressive, especially for a high energy density battery. The actual conditions for the C cells were 1·2 A charge for 3·25 h, followed by 1·8 A discharge to a cut-out voltage of 0·75 V. Cell capacity is measured to 0·9 V.

Figure 4 presents cycle life data for the "24 series" cells under these conditions. Previously, "15 series" cells achieved 250-380 cycles in this test. The quality of the seal was a significant factor, as weight loss and impedance changes tended to correlate. For the "24 series" cells, over 500 cycles have been achieved with testing still continuing.

In Figure 5, cycle life data is presented where the only change from Fig. 4 was to lower the discharge cut-out voltage to 0·0 V. The capacity was still only measured to 0·9 V. Cycling behaviour is virtually the same as the foregoing, with over 500 cycles for the "24 series". Previously, the "15 series" cells had achieved about 250-300 cycles under these conditions.

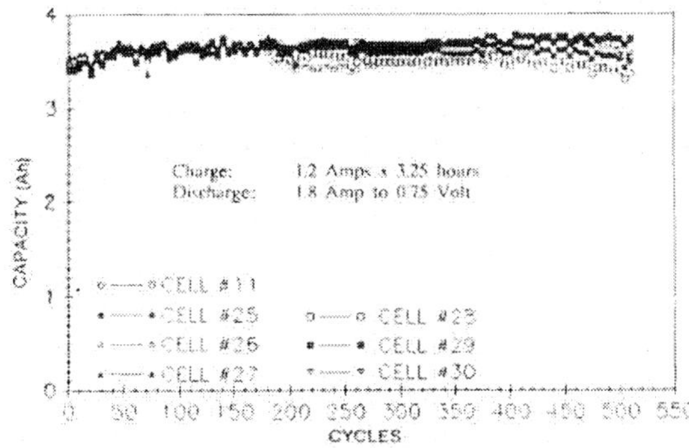

FIG. 4 "24 Series" C cells under C/3 to 115% Charge, C/2 to 100% Discharge.

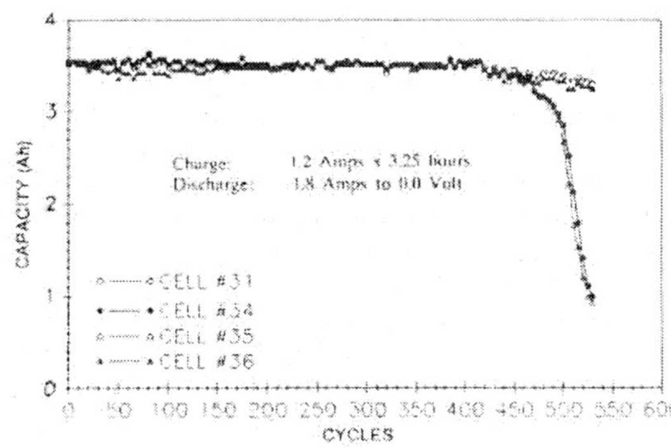

FIG. 5 "24 Series" C cells under C/3 to 115% Charge, C/2 to 0·0 Volt Deep Discharge.

Failure Mode

Detailed failure mechanisms have been discussed previously (Fetcenko *et al.*, 1988 b; Reichman *et al.*, 1987). In those studies, oxidation resistance of the metal hydride alloys and electrolyte redistribution were identified as key factors in the ultimate cell failure. Both mechanisms contributed to a gradual reduction in capacity, primarily due to voltage polarization.

The dramatic improvement in cycle life from the first generation cells to the "15 series" and later to the "24 series" cells has been related to an improvement in oxidation resistance and a reduction in electrolyte redistribution upon cycling.

Through alloy modification, corrosion resistance has been improved by a factor of 10. The improved oxidation/corrosion behaviour has been observed with a concurrent increase in catalytic behaviour. Consequently, gas recombination capability and discharge rate capability are also influenced by changes to the basic alloy.

The electrolyte distribution stability has been improved through an increase in mechanical stability of both electrodes. Additionally, improvements in sealing techniques have resulted in reduced electrolyte and gas escape. Finally, electrolyte redistribution has been reduced by the development of improved separators.

INTERNAL RESISTANCE AND RATE CAPABILITY

First generation OBC C cells had an a.c. impedance of 9 mΩ to 11 mΩ (1 kHz) and a ΔV/ΔI internal resistance of ~25 mΩ to 30 mΩ. Under high rate discharge, the resultant voltage drop was larger for these cells than for Ni-Cd cells, yielding a lower capacity in applications where the cut-out voltage is critical.

Factors which contribute to a higher internal resistance for OBC cells when compared with Ni-Cd cells include current collection, negative alloy material composition, and postive electrode polarization. To improve current collection, positive and negative electrode substrates and tab connections were investigated. It was found that the wire mesh substrate for the negative electrode and the perforated sheet used for the positive electrode had minimal impact on the impedance, but that tab connections were more important. Standard OBC C cells utilized a single tab connector for each electrode, with the positive tab integral and the negative tab physically staked. For the metal hydride electrode, it did not make any difference whether the tab was staked, integral, or welded. The most critical factor for both electrodes was the number of tabs. Along the lines of traditional edge welding, multi-tab connections to both electrodes were introduced. A.C. impedance dropped from about 9-11 milliohms to about 5-6 milliohms on cells incorporating this improved current collection. The a.c. impedance was reduced from between 9 mΩ and 11 mΩ to ~5 mΩ to 6 mΩ.

Polarization from both electrodes also contributes to a higher internal resistance for OBC Ni-MH C cells as compared to Ni-Cd cells. Generally, the internal cell design for Ni-MH cells is similar to that for Ni-Cd cells utilizing the spiral wound "jelly roll" configuration. The capacity of an OBC metal hydride electrode is higher than that of a conventional cadmium electrode by a factor of 3 and allows the replacement of a cadmium electrode with a much thinner hydride electrode. This added volume allows more positive electrode to be used, although the OBC nickel hydroxide electrode also has a higher energy density than that in conventional Ni-Cd cells. The cell design requires a thicker positive electrode, which contributes to some of the increase in polarization. On the other hand, it was found that processing variables for the

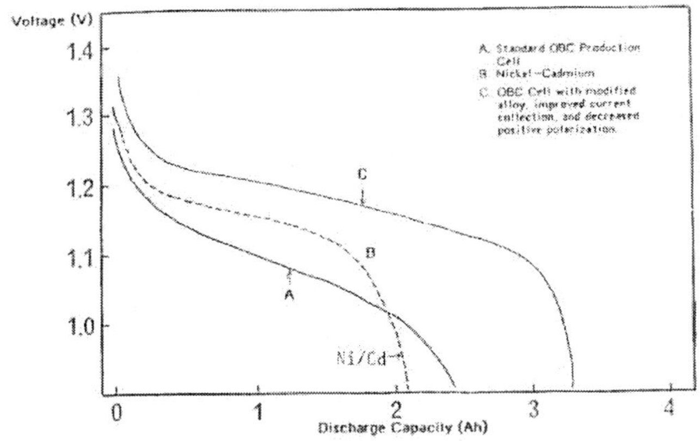

FIG. 6 Effect of Alloy Modification, Current Collection, and Positive Polarization on High Rate Discharge (C size cells at 7 Amp Discharge).

positive electrode, especially the nickel sinter structure, have an important role in polarization. By optimizing the pore size distribution of the nickel sinter structure, polarization for the positive electrode was significantly reduced.

For the metal hydride negative electrode, the most critical factor affecting polarization and discharge rate capability is the alloy composition. The type of materials used at OBC are especially well suited to custom design for given applications. Through the use of modifier elements, polarization due to the negative electrode was reduced by 50%. The improvement can be traced to more optimized surface area and pore size distribution as well as significantly greater oxidation/corrosion resistance providing enhanced bulk and surface catalysis. Figure 6 illustrates the high rate discharge behaviour for our C size cell incorporating improved current collection, decreased positive polarization, and a modified metal hydride composition under a 7 A discharge.

CHARGE RETENTION

The complete mechanisms regarding self-discharge usually inspire vigorous debate. In Ni-Cd cells, oxygen instability of the nickel hydroxide electrode and the "nitrate shuttle" are commonly used to describe self-discharge. In Ni-H_2 batteries reaction at the positive electrode has been ascribed to high pressure hydrogen. For Ni-MH cells, variations of these same mechanisms are also relevant, as well as other possible mechanisms such as separator decomposition.

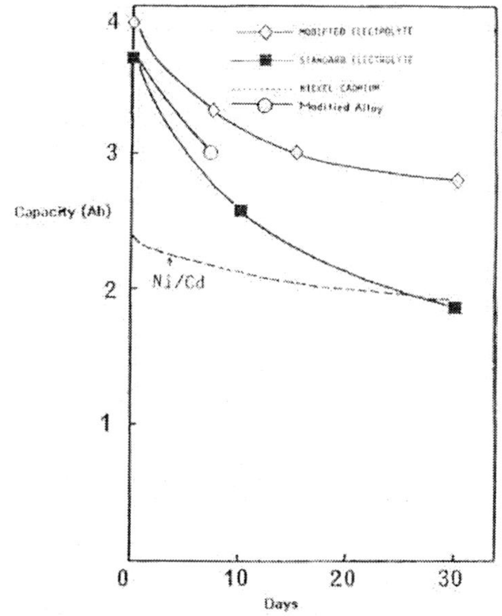

FIG. 7 Effect of Electrolyte and Alloy Modification on Self-Discharge Rate (20°C).

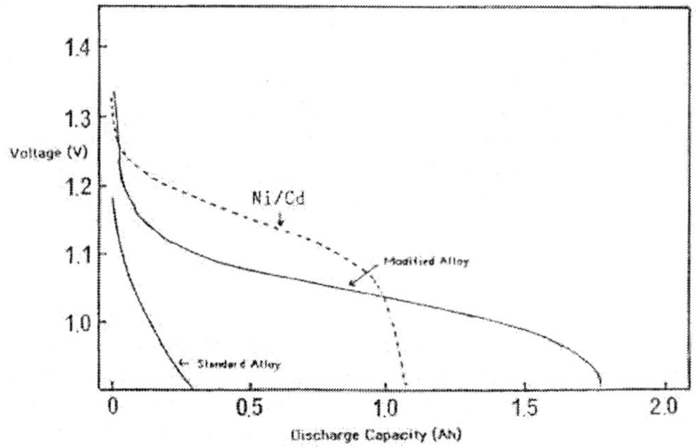

FIG. 8 Effect of Alloy Modification on Low Temperature Discharge (C cells at −20°C, 1·75 Amp).

Ovonic Ni-MH cells usually experience 50% loss in capacity over a 30 day storage period at room temperature whereas most Ni-Cd batteries only lose about 25% to 30% capacity under the same conditions. The higher self-discharge rate is attributed to several factors. The highly loaded, chemically impregnated nickel hydroxide positive electrode introduces the same ion redox species and contaminants affecting oxygen stability as in Ni-Cd cells. The separator is conventional nylon, so that some decomposition can take place. Finally, the metal hydride introduces some new concerns such as soluble vanadium which is multivalent and also non-soluble species both of which may influence the oxygen stability of the positive electrode.

Previously published results showed that chemically modified electrolyte reduced self-discharge to about 30% in 30 days (Fetcenko *et al.*, 1990c). However, the same result has also been achieved with standard electrolyte using a metal hydride alloy with a modified composition. Figure 7 compares the self-discharge rate for standard cells with that of cells improved as a result of electrolyte and alloy modifications.

LOW TEMPERATURE

Some of the same improvements achieved in high rate discharge and charge retention by the use of metal hydride alloy modification and chemical modification of the electrolyte have also improved low temperature behaviour. The basic problem in low temperature discharge with OBC Ni-MH cells is that their polarization exceeds that of Ni-Cd cells at decreased temperatures. The increased polarization is dominated by the metal hydride electrode as compared to the cadmium electrode, and is related to decreased catalytic activity at the metal/electrolyte interface, lower voltage due to thermodynamic considerations, and water generation. Low temperature discharge performance with improved pore size distribution and catalysis due solely to modification of the alloy is shown in Fig. 8.

COMMERCIAL PROGRESS

OBC has been making and providing samples from its pilot plant in Troy, Michigan since 1987. We have developed not only proprietary materials and cell design, but also patented process technology which can be easily scaled up to a high volume production, with high process yields and low cost.

In order to ensure the large scale production required to meet the world wide needs of Ni-MH cells, the Ovonic Battery Company is collaborating in various ways with other companies in the United States, the Soviet Union, Europe and the Far East in the production of these cells and batteries covering a wide range of sizes.

Manufacturing is easily adapted to large scale production. Alloys are produced in 60 kg quantities, using conventional melting techniques (Fetcenko *et al.*, 1989). The raw materials are widely available, with raw material cost in the \$6 per pound range. Size reduction is fast and efficient. Yields of virtually 100% have been achieved when comparing incoming raw materials to usable battery grade powder (Fetcenko *et al.*, 1988 a). Electrode manufacture is a continuous roll-to-roll process using no binders or additives to the active material (Wolff *et al.*, 1989). For the overall metal hydride electrode, raw material costs are by far the most dominant, with low labour and processing costs. The photographs in Fig. 9 illustrate material in various stages of processing.

Performance data presented both here and previously was established using conventional cell construction. The positive electrode is of the sintered type, using conventional chemical impregnation. One of the key advantages of Ovonic Ni-MH batteries is the elimination of cadmium from the negative electrode and indeed from the entire cell. Accordingly, no cadmium additives are used in the nickel hydroxide electrode. Since the use of cadmium as an additive in the nickel hydroxide electrode is a well established practice for reducing electrode swelling, the excellent cycling performance without cadmium anywhere in the cell is even more significant.

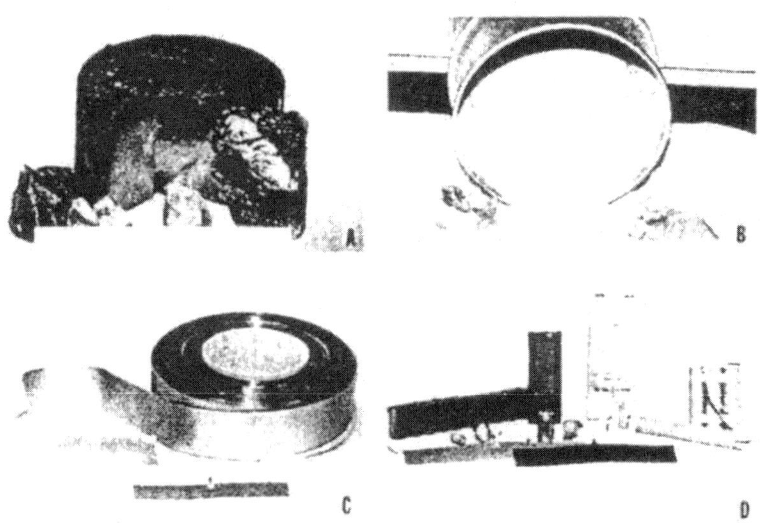

FIG. 9 Processing Stages for Metal Hydride Material and Ni-MH Battery (A) Alloy Ingot, (B) After Size Reduction, (C) Metal Hydride Negative Electrode After Sintering, (D) C Size Cell Components, Battery Packs, and other Size Cells.

The electrolyte used in our Ni-MH cells is conventional 30% KOH with a LiOH additive. The separator is standard nylon and cell construction is the typical spiral wound configuration. Other components are all of the type commonly found in Ni-Cd batteries.

The utilization of a virtually all standard cell construction with the exception of the metal hydride electrode serves to highlight the potential for further advances in Ni-MH cell performance. In the laboratory, it has been shown that modified positive electrodes using fibre or foam substrates can be easily implemented. Separators have been evaluated which offer greater electrolyte retention and chemical stability. Electrolyte modification has been shown to be useful in advancing performance attributes such as self-discharge. Edge welding and multiple tab current collection can reduce internal resistance to levels similar to the best Ni-Cd cells.

The manufacturers often make several versions of the same size Ni-Cd cell; emphasizing one or more features such as capacity, internal resistance, fast charge, temperature capability. This is an important factor in commercial markets which are becoming more specialized and, since Ni-MH batteries exhibit similar capabilities, it follows that the versatility generally associated with Ni-Cd is also applicable to Ni-MH batteries.

AEROSPACE APPLICATIONS

Ovonic metal hydride technology is also applicable in the aerospace field offering higher energy density than aerospace Ni-Cd cells and the elimination of cadmium. Ni-MH cells operate at low pressure and offer significant cost advantages over the nickel-hydrogen system.

OBC prototypes are presently undergoing accelerated life testing at Rockwell International. Under 30% depth of discharge, OBC cells have exceeded 2700 cycles with test still running. Based on end of discharge voltage modelling, over 10,000 cycles have been projected. Cycle life and characteristics test results have so far indicated that Ni-MH batteries may be good replacement candidates for aerospace systems using Ni-Cd batteries and Ni-H_2 batteries (Otzinger, 1991).

OBC is also working with Eagle-Picher Industries, Inc. to produce batteries for aerospace applications, combining OBC's metal hydride technology with Eagle-Picher's aerospace proven positive electrode, cell assembly, and quality assurance expertise. So far, prototypes ranging from 3·5 Ah to 20 Ah have been assembled and are undergoing testing with encouraging results (Brill et al., 1990).

ELECTRIC VEHICLES

Worldwide uncertainties over oil supplies and the environmental need to reduce vehicle emissions have once more thrust electric vehicles into the

FIG. 10 Various Battery Sizes under test at OBC

spotlight. After decades of development, lead-acid batteries are still the only system generally considered ready for deployment in commercial electric vehicles. Consequently, vehicle range and battery life are still unacceptable due to the low energy density and low cycle life under high depth of discharge conditions attributed to lead-acid batteries. This has been the classic problem preventing electric vehicles from becoming a commercial reality. OBC's technology addresses and solves these problems.

The performance of our Ni-MH batteries is proven in small size batteries. The cells have been tested and evaluated by several battery companies and have also been evaluated at Argonne National Laboratories. In that evaluation energy density was verified at 55 Wh kg^{-1}, specific power was verified at over 200 W kg^{-1} at full charge, and acceptability under SFUDS driving profile tests was established where the necessary power requirements (79 W kg^{-1}) were maintained over nearly the full rated capacity of the battery (Tummillo *et al.*, 1990).

Prototype large batteries of 150–200 Ah capacity have already been constructed at OBC and are being tested. Figure 10 illustrates some of the various battery sizes we have at present under test. The high energy density of the Ni-MH battery is one of its key advantages. The same energy density established in small cells has already been confirmed in the large battery prototypes. The cell design is being optimized for power and life, and all batteries are being tested under a variety of conditions.

The metal hydride electrode can be fabricated in a variety of ways, including sintered or pasted configuration. In addition, the positive electrode can also utilize fibre or foam substrates using pasted loading methods. Consequently, energy density can be raised even higher and, along with projected long cycle life for the EV application, reasonable operating costs are anticipated. Since a primary motivation for EV's is the environmental consideration, the fact that the materials in the Ovonic Ni-MH battery are safe and can be recycled makes metal hydride battery technology an even more attractive EV candidate.

CONCLUSIONS

Ovonic nickel-metal hydride rechargeable battery technology has shown extremely versatile system capability. OBC's patented V-Ti-Zr-Ni alloy system has already been demonstrated in commerically available C cells. Dramatic improvements in cycle life with the "24 series" cells have been achieved, with over 800 cycles attained under 100% DOD conditions. The improved cycle life has been achieved with no trade-offs in other performance characteristics.

Manufacturing processes have been fully scaled up for pilot production, demonstrating high yield and low processing cost. OBC has been providing sample C cells for several years, and several licensing arrangements for high volume manufacture have already been established. The performance of the batteries has already been verified by independent test organizations.

Prototype batteries are under test for aerospace and electric vehicle applications. Preliminary cycle life data under aerospace conditions is excellent and tests are continuing. In large size batteries, the energy density available in small size cells has already been attained.

From a performance perspective, our Ni-MH batteries have high energy density, high specific power, high cycle life, no memory effect, and ability to be fast charged. Further, the fact that Ovonic Ni-MH batteries are cost competitive with Ni-Cd batteries on an energy basis offers commercial acceptability. Finally, the fact that these Ni-MH batteries utilize environmentallly safe materials makes this technology truly attractive for widespread commercial application.

REFERENCES

BITTNER, H. F., and BADCOCK, C. C. (1983). *J. Electrochem. Soc.*, 130, (5), 193C.

BRILL, J., COATES, D., BEMIS, P., VENKATESAN, S., FETCENKO, M. A., and OVSHINSKY, S. R. (1990). AIAA/Utah State University Conference on Small Satellites, (August).

FETCENKO, M. A., KAATZ, T., SUMNER, S., and LAROCCA, J. (1988a). U.S. Patent 4,893,756.

FETCENKO, M. A., SUMNER, S., and LAROCCA, J. (1989). U.S. Patent 4,948,423.

FETCENKO, M. A., VENKATESAN, S., HONG, K. C., and REICHMAN, B. (1988 b). "Power Sources 12" (T. Keily and B. W. Baxter eds), International Power Sources Committee, Leatherhead, England, p. 411.

FETCENKO, M. A., VENKATESAN, S., OVSHINSKY, S. R., and HIROTA, M. (1990 a) IBA Conference, October, Seattle, Washington.

FETCENKO, M. A., VENKATESAN, S. OVSHINSKY, S. R. (1990 b). Power Sources Symposium, Cherry Hill, New Jersey.

FETCENKO, M. A., OVSHINSKY, S. R., VENKATESAN, S., KAJITA, K., KIDOU, H., and JEFFRIES, K. (1990 c). "Third International Rechargeable Battery Seminar", Deerfield Beach, Florida (March).

KNOLL, C. R., TNOMINEN, S. M., PETERSON, J. R., and MCQUEARY, T. R. (1990). "Battery Waste Management Seminar", (S. Wolsky ed.).

OTZINGER, B. (1991). "The Sixth Annual Battery Conference on Applications and Advances", Long Beach, California, (January).

REICHMAN, B., VENKATESAN, S., FETCENKO, M. A., JEFFRIES, K. STAHL, S., MAGNUSON, D., BENNETT., C., and SUMNER, S. (1987). U.S. Patent 4,716,08.

SAPRU, K., HONG, E. C., FETCENKO, M. A., and VENKATESAN, S. (1985). U.S. Patent 4,214,043.

TUMILLO, A. F., DELUCA, W. H., and MILLER, J. F. (1990). October Meeting of the Electrochemical Society, Seattle, Washington.

VANBEEK, J. R., DONKERSLOOT, H. C., and WILLEMS, J. J. G. (1985). "Power Sources 10" (L. J. Pearce, ed.) The Paul Press London, International Power Sources Symposium Committee, Leatherhead, England, p. 317.

VAN RIJSWICK, M. H. J. (1977). Proc. Int. Symp. Hydrides for Energy Storage, Geilo, Norway (A. Andresen, ed.), p. 261.

VENKATESAN, S., REICHMAN, B. and FETCENKO, M. A. (1988 a). U.S. Patent 4,728,586.

VENKATESAN, S., FETCENKO, M. A., REICHMAN, B., MAGNUSON, D., and DHAR, S. (1988 b). "The Second International Rechargeable Battery Seminar" (S. Wolsky, ed.).

WOLFF, M. T., NUSS, M., FETCENKO, M. A., LIJOI, A., SUMNER, S. P., LAROCCA, J., and KAATZ, T. (1989). U.S. Patent 4,915,898.

Presented at Electrochemical Society,
Phoenix, Arizona, October 15, 1991
Electrochemical Society Proceedings Volume 92-5

SELECTION OF METAL HYDRIDE ALLOYS FOR ELECTROCHEMICAL APPLICATIONS

M.A. Fetcenko, S. Venkatesan, S.R. Ovshinsky
Ovonic Battery Company
1826 Northwood Drive, Troy, MI 48084

ABSTRACT

Metal hydride alloys of the V-Ti-Zr-Ni based system have been characterized for electrochemical applications by a variety of methods. In particular, the metal/electrolyte interface was investigated by AES, ESCA and BET analysis. The role of the oxide was studied using STEM and SAED techniques. Results indicate this oxide is thin, stable and disordered. The conductivity, catalytic activity and microporosity characteristics of the oxide are important factors affecting high rate discharge performance.

INTRODUCTION

Nickel- metal hydride (Ni-MH) batteries are presently being commercialized and developed for a variety of applications including the replacement of Ni-Cd batteries for portable consumer use, aerospace and electric vehicles. Ni-MH batteries utilize environmentally safe materials and offer premium overall performance including high energy density and power, long life, and fast charge capability. Performance results and product announcements have been reported by several manufacturers (1,2,3).

The development of metal hydride batteries has been hampered by basic materials limitations of the hydride alloys themselves. Over the last 25 years, it seems that every battery company in the world has tried to develop Ni-MH batteries by working almost exclusively on the $LaNi_5$ type of alloy. The choice of $LaNi_5$ type materials was influenced by suitable metal-hydrogen bond strength for use as an electrode in aqueous media, reasonable hydrogen storage capacity and the possibility of chemical stability in the alkaline electrolyte. Through the use of extensive alloy modifications, microencapsulation of powder particles and special hydrophobic electrode constructions, $LaNi_5$ based products have recently become available which offer good overall performance (4,5). In some respects, these technical approaches can be characterized as efforts to compensate for a nonideal base material and in effect living with inherent material limitations.

In contrast, our approach has been directed to the development of fundamentally different alloy materials. Our alloy system is based on V-Ti-Zr-Ni materials which can be heavily alloyed for enhancement of chemical and electrochemical properties (6). These alloys are sometimes referred to as AB_2 or Laves phase materials, although the alloys are typically multiphase having multiple crystal structures.

These V-Ti-Zr-Ni based materials have significant advantages over $LaNi_5$ based materials including:

- higher hydrogen storage capacity
- superior oxidation and corrosion resistance
- higher volumetric electrode capacity
- easier processing
- plentiful, inexpensive raw materials
- ability for custom design for given applications

This paper will characterize these materials by several methods with a particular emphasis on surface properties. We will also discuss performance properties of these materials as well as selection criteria related to alloy design.

DISCUSSION

Development Background

There has always been some curiosity regarding our original motivation to develop new metal hydride materials rather than the more conventional approaches based on $LaNi_5$ materials. Our company has always been first and foremost a materials science organization in the fields of optical memories, photovoltaics, superconductivity, thermoelectricity, semiconductor devices and of relevance to the battery field, hydrogen storage (7,8,9).

In the late 1970's, we intensively studied all aspects of the "hydrogen economy". Development programs in fuel cells, electrolyzers, photoelectrochemistry, and catalysis resulted in the development of advanced materials for each application (10,11,12,13). In the field of hydrogen storage, we intensively studied and developed materials for fuel storage, heat pumps, gas separation, automotive applications and batteries (14,15).

Our pioneering work with chemical and structural modification and our understanding of long and short range atomic order led to the development of multi-element alloys with varied metal-hydrogen binding energies. Simply stated, this approach allowed the synthesis of new materials utilizing elements which would not normally be considered suitable for electrochemical applications (16).

Performance Characteristics

Development activities of OBC and its licensees have mainly focussed on small cylindrical size cells such as AA and C cells, although 200 Ah prismatic prototypes are already under test (17,18,19,20).

The nominal capacity for C size Ni-MH cells is 3.5 Ah, which corresponds to an energy density of about 56 Wh/kg. Prototype C cells of 4.5-5.0 Ah (65-70 Wh/kg) have been constructed using higher capacity metal hydride alloys, improved nickel hydroxide electrodes and an optimized cell design. We expect to achieve an energy density of 80 Wh/kg primarily through the use of lighter substrate materials in the positive electrode.

The power capability for this system is excellent. AA and C size cells provide 5C continuous discharge capability, with an internal resistance of about 17-25 milliohm (5-10 milliohm AC impedance). In larger size cells with increased electrode surface areas, impedance values are lower. A peak power analysis conducted by Argonne National Laboratories on 30 Ah cells demonstrated a peak power of 240 W/kg at full charge, 210 W/kg at 50% charge and high peak power over the complete depth of discharge exceeding all other rechargeable systems evaluated (20,21).

The cycle life for Ni-MH cells using V-Ti-Zr-Ni based alloys is exceptional. Up to 1400 cycles was achieved under 100% DOD testing using a temperature terminated charge. When using a time terminated charge, it is routine to achieve 800-1000 cycles. Extensive overcharge is not a problem. Over 500 cycles under three hour charge with 15% overcharge are attained and over 600 cycles are achieved even under repetitive 20% reversal conditions (22).

As can be observed from the cycling data, Ni-MH cells using V-Ti-Zr-Ni based alloys can achieve high cycle life under a variety of conditions and are tolerant to abuse. In fact, OBC cells are presently undergoing accelerated life testing under 30% DOD aerospace conditions. To date, over 8500 cycles have been achieved with testing still continuing (23,24).

Alloy Selection Criteria

The composition of the metal hydride alloy is the greatest single design factor for cell performance parameters such as pressure, rate, cycle life, low temperature and charge retention. An important alloy selection criteria is the amount of hydrogen absorbed per gram of alloy. Accordingly, a key advantage of the V-Ti-Zr-Ni based alloy system is higher capacity, typically ranging from about 320-350 mAh/g compared to $LaNi_5$ alloys which usually provide about 250 mAh/g (25). Advanced alloys in the V-Ti-Zr-Ni system presently under test have achieved over 400 mAh/g. However, hydrogen storage capacity is only one important factor. In the selection of an alloy system and in particular the design of a specific alloy composition, the following factors are also important:

- Thermodynamic Properties
- Oxidation Resistance
- Corrosion Resistance
- Metallurgical Properties
- Discharge Kinetics
- Gas Recombination Capability
- Manufacturability
- Cost

In the discussion of hydrogen storage capability, it is important to distinguish electrochemically active capacity from inactive capacity. There can be many reasons why a hydrogen storage site may be inactive for electrochemical application such as poor charge acceptance, insufficient discharge kinetics, instability in electrolyte, or unsuitable thermodynamic properties.

Thermodynamic properties, have a practical as well as historical significance with respect to Ni-MH batteries. The thermodynamic properties can be characterized by:

$$\Delta G = \Delta H - T\Delta S$$

where ΔH is the heat of formation and a measure of the M-H bond strength. The free energy can be further broken down to predict the pressure and voltage characteristics of a particular material according to:

$$\Delta G = RT \ln P \quad \text{and} \quad \Delta G = -nFE$$

In a thermal system, the M-H bond strength allows some latitude and can be adjusted by a simple adjustment of operating temperature. For example, many thermal applications of metal hydrides operate anywhere from 100-400°C (26). In contrast, electrochemical use in alkaline media requires the hydride to operate between discrete voltage limits. Since the strength of the metal to hydrogen bond controls the operating voltage, it can be concluded that the heat of the formation must also be of a certain range, usually considered about 8-10 kcal/mol, in order to be useful as a battery alloy. If the metal hydrogen bond is too weak, the operating voltage will be above the hydrogen evolution potential causing poor charge acceptance. If the metal hydrogen bond is too strong, the operating voltage will be below the metal oxidation potential and the metal will be electrochemically unstable.

Historically, the requirement for a battery alloy to have discrete thermodynamic properties has been the single largest determining factor in the intensive study of the $LaNi_5$ system. Well known among thermal systems as an ambient temperature hydride with well established properties, the $LaNi_5$ system was an obvious choice to consider for electrochemical use. Originally intended only to lower the hydrogen pressure in Ni-H_2 batteries, later evaluation as an actual electrode material (27,28,29) revealed many problems such as mechanical embrittlement, severe oxidation and corrosion, and difficult processing. These problems provided the focus for development activities of several organizations over the last 25 years.

The problem of oxidation resistance will be discussed in more detail later, but probably nothing is a greater obstacle to practical use of a metal hydride in a battery. The high pH alkaline electrolyte places a severe constraint on these hydrogen storage alloys, but perhaps even more severe is the requirement for sealed cell operation. To gain widespread commercial acceptance in portable batteries, totally sealed operation is an absolute requirement. Even users of large industrial batteries and EV batteries in particular consider recombinant systems superior. Knowing that hydrogen recombination on a nickel hydroxide electrode is impractical, design of a sealed Ni-MH cell must utilize the classic oxygen recombination overcharge reaction. The implication of this reaction is that oxygen recombination is occurring on the surface of the metal hydride electrode. Our studies have shown that a material with adequate oxidation resistance to KOH electrolyte for over hundreds of cycles may be totally degraded by only a few cycles of sealed operation due to oxygen recombination.

It is also important for a metal hydride alloy to have good corrosion resistance. First, there is a finite amount of electrolyte in the cell. Excessive oxidation/corrosion with a resultant consumption of electrolyte can adversely affect cycle life due to cell "dryout". Consumption of electrolyte can also affect cell performance by altering the electrolyte concentration and affecting the charge balance in sealed operation. Another implication of corrosion is the potential poisoning of the nickel hydroxide. Some oxidation products from a metal hydride alloy can decrease the oxygen overvoltage of the positive electrode, decreasing charge acceptance particularly under fast charge and affecting the self discharge of the nickel hydroxide electrode under stand conditions. Finally, we have observed that some soluble corrosion products from the metal hydride alloy are multivalent and can form redox shuttle mechanisms similar to that of nitrate ion, which can influence self-discharge (30).

A primary objective of Ni-MH batteries is the replacement of Ni-Cd for environmental reasons as well as higher energy density. One key feature of Ni-Cd batteries is the formidable power capability required for many practical applications. Accordingly, the discharge rate capability of metal hydride batteries has always been heavily scrutinized. For V-Ti-Zr-Ni based alloys, the bulk diffusion rate of hydrogen in the metal is extremely fast. Rather, the primary obstacle in achieving high rate discharge is related to surface catalytic activity. For a particular alloy, this may involve current density considerations which may be addressed by a electrode with sufficient surface area. Concentration polarization at room temperature is usually a secondary factor in an electrode having a well designed porosity and pore size distribution. It is common that discharge rate capability is finally determined by the state of the reaction surface. If one considers that the reaction of hydrogen and hydroxyl ions at the metal/electrolyte interface is taking place through an oxide, it is not surprising that properties of the oxide such as porosity, thickness, conductivity and catalytic properties greatly influence discharge rate capability.

The implications of oxygen recombination have been noted. Another requirement of the metal hydride electrode is the need for hydrogen recombination. An advantage of Ni-MH cells over Ni-Cd is a greater tolerance to reversal. Under overdischarge, hydrogen gas generated by the nickel hydroxide can be recombined on the metal hydride surface provided electrolyte film thickness and catalytic activity factors are sufficient. The other requirement for hydrogen absorption relates to overcharge. In Ni-MH cells, although oxygen recombination is very fast, some pressure of hydrogen gas is present in overcharge. The actual pressure of hydrogen is related to many factors including PCT behavior, charge rate, and temperature. Once overcharge is concluded, either on open circuit or discharge, the hydrogen gas will be reabsorbed into the metal hydride electrode. How well this occurs depends on the dissociation of molecular hydrogen to atomic hydrogen at the electrode surface. Again the controlling factors are the metal hydride surface properties.

Manufacturing and Cost

Metal hydride batteries are intended to replace nickel-cadmium in high volume commercial applications. In addition, electric vehicle material requirements are so large that should Ni-MH become the major EV battery system as expected, material and electrode manufacturability issues are indeed relevant. For example, melting of V-Ti-Zr-Ni based alloys introduces constraints of extremely high melting point, highly reactive metals of as high as 10 elements where no single element makes up more than perhaps 30% of the alloy. The resultant alloy ingot which may have a hardness of tool steel must ultimately be made into a fine powder. Finally, the final metal hydride electrode must be processed on cost effective, high volume equipment which consistently produces material having acceptable activation, porosity and pore size, surface area and state of charge.

V-Ti-Zr-Ni based alloys are currently processed in 60 kg ingots in a simple, one step melting process (31). Commercially available raw materials are used in various forms including scrap. No additional homogenization or heat treating steps are required. Size reduction is accomplished by a single hydride/dehydride cycle with a typical resultant particle size of less than 200 micron (32). Further size reduction is accomplished using a high speed impact mill at a rate of 100 kg/hour. The yield from raw materials to battery grade powder is virtually a 100% yield. Unlike $LaNi_5$ style alloys which usually require an etch treatment to the powder or even an expensive 10-20 percent microencapsulation treatment which further reduces the electrode capacity, no other treatment of the V-Ti-Zr-Ni based alloy powder is required. Electrode fabrication is fast and simple. No binders or additives which significantly reduce volumetric capacity are required. Any sort of a substrate like wire mesh, perforated metal or expanded metal can be used. Electrode belt is made by compaction of 100% active material at a speed of 60 inches per minute. This is followed by a sintering step which promotes particle-to-particle bonding and adherence to the substrate (33). Since this type of electrode is highly conductive and the hydride material is conductive in both the charged and discharged state, a single mechanically staked tab is adequate for current collection. The electrode is assembled into the cell with no further treatment (34).

Additionally, V-Ti-Zr-Ni based alloys use inexpensive, commercially available raw materials. It is important to note that even the substantial material quantities required for electric vehicles do not present any availability concerns. It is also noteworthy that metal hydride scrap, out of specification electrode, and even electrode trim scraps can be recycled in the original melting process. Even though the processes are already extremely high yield, recyclability offers further cost advantages. Since no recycling plan is 100% effective, it is important to note that V-Ti-Zr-Ni based alloys have already been subjected to TCLP tests where all elements were at least an order of magnitude below hazardous waste limits and can, therefore, be legally disposed of in municipal landfills (35).

Role of Specific Elements

The basic V-Ti-Zr-Ni-Cr alloy can be heavily modified to emphasize one or more important performance characteristics. The specific function of particular base elements as well as of modifier elements is quite different. The primary role of vanadium, titanium and zirconium is as hydrogen storage elements. However, the question becomes: "Why all three? Why not just titanium which is the least expensive?" The answer is not simple and involves many factors. Whereas titanium and zirconium both form thick, dense, passive oxides in alkaline, vanadium forms an easily soluble oxide. This characteristic is important in activation as well as in providing a degree of porosity to the oxide surface. Another factor in the selection of the hydrogen storing elements is thermodynamic, since each element has a different M-H bond energy. It is possible to custom adjust the particular alloy in terms of voltage, pressure, and rate capability by manipulation of elemental ratios. Finally, metallurgical implications enter into the design of the particular alloy. Since the alloys are multiphase and polycrystalline, there can be a degree of phase segregation associated with too much of one element compared to another based on solubility constraints during cooling. Phase segregation or the formation of particular phases may be intentional in some alloys and undesirable for other alloys depending on the final alloy requirements.

Another metallurgical implication relates to mechanical properties such as hardness, ductility, toughness and degree of embrittlement upon charge discharge cycling. Perhaps an oversimplification, a major function of zirconium is to provide a controllable degree of embrittlement to the metal. Through fine adjustment of the alloy microstructure, it is easy to form a high degree of surface area insitu to the cell, avoiding expensive processing steps and the formation of oxides during atmospheric exposure. It is even possible to adjust the rate of surface area formation during cycling. These factors are important in that surface area formation plays a large role in rate capability and that it is practically important that workup cycles be minimized. The attainment of steady state surface area is important in that unrestrained embrittlement can cause mechanical disintegration and loss of conductivity within the electrode structure.

Nickel has several functions. The first is thermodynamic in that vanadium, titanium, and zirconium by themselves have too high a heat of formation for electrochemical application. Nickel serves to destabilize the hydride or adjust the M-H bond strength within desired limits. Another function of nickel is resistance to oxidation. It is incorrect to consider these materials as mixtures of elements. Rather, these V-Ti-Zr-Ni-Cr based materials are alloys in every sense. As an alloy with nickel, the easily oxidized V-Ti-Zr elements become relatively insensitive to oxidation. The resistance of nickel to oxidation has several implications on performance. Even at the metal/electrolyte interface nickel is found in a metallic state. This provides a degree of conductivity to the oxide and also functions as a catalyst to the hydrogen/hydroxyl reaction. Further, the presence of metallic nickel at the electrode surface catalyzes the dissociation of H_2 which is important for gas reabsorption. Finally, nickel easily forms intermetallics with the other base elements and, therefore, strongly contributes to particular phases having a specific crystal structure.

The role of chromium is slightly less complicated in that the major function is to inhibit the unrestrained corrosion of vanadium (30). A certain amount of VO_x

solubility is desirable, but can be detrimental if uncontrolled. Chromium also can be used to adjust the M-H bond strength if required.

Alloy Microstructure

While V-Ti-Zr-Ni alloys are predominantly polycrystalline and multiphase by design, there is a tremendous effect of the alloy composition on the microstructure as well as on the various roles for particular phases. A representative SEM micrograph is shown in Figure 1. This material contains four distinct phases having cubic (bcc), hexagonal, and C14 Laves phase structures. It is also quite common for these materials to contain C15 type structures.

From synthesis and testing of separate, individual phases from a particular alloy, we concluded that the bcc phase absorbs large quantities of hydrogen (2.5% by weight), but lacks sufficient catalytic activity to be discharged at reasonable rates. On the other hand, surrounding phases which may store less hydrogen but contain greater catalytic activity can channel the hydrogen for discharge. This model of "storage and catalyst phases" has been reported previously for the Ti-Ni binary system when referring to the Ti_2Ni and TiNi intermetallic phases (36).

As important as the compositional effect on microstructure is the effect of processing as shown in Figure 2. The upper SEM micrograph exhibits a grain size and phase distribution characteristic of conventional solidification while the lower micrograph illustrates the microstructure of the same material prepared under rapid solidification. Through the use of rapid solidification by melt spinning or gas atomization, it is possible to prepare microcrystalline or amorphous alloys (37). This technique can be extremely useful in reducing corrosion and embrittlement, and in extending cycle life in materials where the precipitation of a particular phase may be undesirable.

PCT Behavior

The radical effect of alloy composition on performance can be observed from PCT Behavior as shown in Figure 3 (6) . It can be observed that both the plateau pressure and shape have been altered which indicates a change of M-H bond strength. The shifting of the curve on the X-axis can be interpreted as hydrogen storage capacity of which only a portion is electrochemically reversible.

It should be further noted that the role of the "plateau region" is less critical in electrochemical use than it is for thermal or gaseous applications. In thermal usage, the plateau pressure is usually an engineering design parameter. For batteries, voltage is certainly the design parameter of concern. Since voltage is proportional to the log of pressure, it is not necessary to have a distinct plateau region on a PCT curve in order to achieve a stable voltage.

Metallurgical Implications

Metallurgical properties of metal hydrides are usually considered primarily for processing reasons. However, cell performance properties like discharge rate and gas recombination can be greatly altered by the material metallurgical properties (Figure 4). In the upper photograph of a hydride electrode, a composition which is fairly soft and ductile is shown to have a smoother surface after compaction whereas, the alloy in the lower photograph is harder and has different fracture properties during size reduction. After compaction, this alloy provides a much rougher surface, thereby increasing the gas recombination surface.

This approach to fast gas recombination is quite different than methods which utilize a hydrophobic additive or binder to the electrode to reduce electrolyte film thickness. The approach of binders and additives has the disadvantages of lower volumetric capacity, reduction in discharge rate, and low temperature capability. In addition, scrap and trim are more difficult to recycle into the melt operation.

Metal hydride electrode structure changes occur as a result of cycling (38). Large particles become an agglomeration of 1 to 10 particles after cycling. The key observation is that significant size reduction occurs within 1-5 cycles, and while further size reduction continues to over 1000 cycles, where a limiting particle size of about 1 to 3 microns is reached. These very small particles still have an active role in overall cell performance, given the physical restraint placed on the negative electrode in the cell and the need to minimize oxidation.

Based on BET data, the change in surface area of the metal hydride electrode is very high. Based on SEM micrographs, we were unable to explain such a dramatic increase in surface area due to the formation of cracks alone. Comparing the uncycled and cycled electrodes with the BET surface area data led us to conclude that there is some element of microporosity to the oxide surface. As mentioned previously, these materials and electrodes are designed to embrittle insitu. The embrittlement, or cracking, can be thought of as new surface area. These new surfaces are very critical in that no exposure during processing occurs and we believe these surfaces play a great role in discharge kinetics (38).

Surface Properties/Oxidation

Surface properties of the metal affect many of the electrochemical processes in the battery. Substantial oxidation decreases charging efficiency of the metal hydride causing hydrogen gas evolution by reducing the hydrogen overvoltage. By covering metallic sites with oxide, catalytic activity and gas recombination are decreased.

The hydrogen storage capacity is also reduced by oxidation. One possible mechanism involves decreased charge efficiency, although the discharge efficiency is also reduced by polarization. If oxidation is excessive, there can be an overall decrease in conductivity or an isolation of powder particles. While the many detrimental effects of oxidation are clear, the role of the oxide layer itself on these effects is less understood.

Activation

Before addressing the change in the oxide layer during cycling, it is important to begin with a discussion of the activated surface. In this case, the activated surface is to be distinguished from the surface as a result of manufacturing.

The metal surface as a result of fabrication can be characterized as a thin, dense, passivated oxide film. The oxide results from atmospheric exposure during various processing steps and high temperature exposure to impurities. It should be noted that titanium and zirconium in particular form very stable oxides and are sometimes used as oxygen getters in high vacuum processing.

The "as fabricated" surface is not suitable for electrochemical operation without an etch treatment primarily because it has poor charge acceptance (39). There is perhaps some confusion on this point. The object of the etch treatment is not only the attainment of an electrochemically active surface, but to create a surface which allows charge acceptance. During charge, the absorbed hydrogen expands the metal lattice which in turn creates new surface area due to cracking. The degree of new surface area is large and oxide-free since these surfaces are created insitu under an electrochemically negative potential. Through gas composition analysis in overcharge, we have determined that oxygen recombination on the surface is very fast. Consequently, it can be assumed that gas recombination occurs primarily on the outer surfaces of the metal hydride electrode. Due to the fact that the insitu created surfaces on the electrode interior have also not been exposed to oxygen recombination, they are believed to have a major role in high rate discharge capability.

Using Auger Electron Spectroscopy (AES), a comparison of the "as fabricated" electrode surface oxide and that of the etched surface is shown in Figure 5. From an original oxide thickness of 40 Angstroms, the etched surface has grown to 180 Angstroms. It is important to note in that some have interpreted the etch treatment as removal of oxide. If viewed as a peeling off of an oxide layer, this interpretation is not correct. Rather, the thin dense oxide layer from fabrication actually increases in thickness due to oxidation, but becomes electrochemically active by the creation of microporosity through the dissolution of soluble oxide components like vanadium. In essence, the problem of the initial, impenetrable oxide is dealt with by the creation of a new kind of oxide by intentional corrosion or leaching.

Effect of Chemical State on Oxide Depth

A typical working oxide for V-Ti-Zr-Ni based materials is about 500 Angstroms. This steady state oxide surface is usually present after 2 or 3 cycles. The composition and state of the surface oxide can be analyzed using x-ray photoelectron spectroscopy (XPS) as shown in Figure 6. Instead of existing as a uniform consistent "layer", the oxide shows a gradual change in oxidation state.

At a depth of 50 Angstroms, all components are oxide. Titanium and zirconium are both divalent. Moving away from the surface to the bulk, nickel is quickly found to be metallic. However, at a depth of 500 Angstroms both titanium and zirconium are found in both divalent and metallic states. By 500 Angstroms, nickel can only be found in the metallic state. Vanadium, being multivalent, can be found in varied forms throughout the profile.

The AES analyses demonstrated that there was no distinct oxide layer with a sharp interface. Further, the fact that elements can be found in various oxidation states suggested that the oxide is not uniform even at any depth.

Metal/Electrolyte Interface

A greater understanding of the oxide layer was essential to the development of materials with improved performance in terms of cycle life, high rate discharge, low temperature capability and charge retention. In order to establish a baseline, at the beginning of our work we selected samples from a V-Ti-Cr-Ni sample lot where an unexplainable variance in performance was observed. The two types of samples had similar capacity and performance in every way except high discharge. "Good" samples had been capable of 3C discharge whereas "poor" samples had been only capable of C/5.

It was clear that some sort of fabrication variance had occurred but it was unclear whether it was related to oxidation, electrode mismatch, electrolyte, etc. With other variables eliminated, our attention focused on oxidation. AES profiles for the samples showed similar oxide thickness, but since AES techniques analyze the outer surface of the electrode and since the outer surfaces are largely influenced by oxygen recombination, we desired a technique to investigate interior surfaces (38).

We selected a scanning transmission electron micrograph (STEM) with the capability for selected area electron diffraction (SAED) and X-ray analysis. Experimental details have been disclosed previously (38). Samples thin enough to be electron transparent were microtomed from cross sectional samples of the electrode. An SEM was used to find appropriate sections from the microtomed cuttings that represented areas of true porosity rather than artifacts from the microtome.

Figures 7 and 8 illustrate relatively low magnification views of the "poor" and "good" rate capability specimens. For the sample with poor rate capability, we observed thick, passive oxides which did not appear on the sample with good rate capability. To ensure we were not misinterpreting the photograph, we used x-ray analysis to confirm the location of the bulk alloy which contained all elements in the proper ratios in contrast to the suspected oxide rich in titanium. To further clarify this important difference, we utilized SAED of the bulk alloy and of the oxide as shown in figures 9 and 10. The given lattice constants were consistent with the bulk alloy and titanium oxide.

Finally, we investigated the observation that samples with good high rate capability appeared oxide free. Some oxide was expected and should be beneficial in preventing corrosion and for fast gas recombination. The sample of Figure 8 was studied under extremely high magnifications as shown in Figure 11, providing our model of the true metal/electrolyte interface for a metal hydride electrode. Rather than visualizing an "oxide layer", there is instead a gradual interface which appears heterogeneous and disordered. We believe this interface is microporous and contains conductive components such as nickel which catalyze discharge. We believe an important role of the multi-element alloys with varied binding energies involves short range order and a disordered aspect to the bulk alloys as well as the surface (7,8,16).

CONCLUSIONS

Ni-MH batteries using V-Ti-Zr-Ni based alloys are being widely commercialized for small portable applications. These batteries provide higher energy density and are environmentally safer than Ni-Cd batteries while maintaining excellent power and cycle life performance. Production batteries have achieved an energy density of 56 Wh/kg. In prototype batteries, energy densities of 70 Wh/kg have been achieved with further improvements expected. EV prototype batteries of 200 Ah capacity are under extensive testing.

The V-Ti-Zr-Ni based alloy system offers significant advantages over conventional $LaNi_5$ alloys in terms of capacity, life, processing and cost. By a sufficient understanding of all performance aspects, custom alloys have been designed which utilize particular elements to emphasize hydrogen storage capacity, thermodynamic behavior, oxidation and corrosion resistance, kinetic behavior, and/or metallurgical properties. For these alloys, both the composition and microstructure are important as well as the role of particular phases.

Finally, the role of the surface properties and in particular the oxide has been identified with respect to electrochemical application. High capacity alloys of the V-Ti-Zr-Ni type can be custom designed to emphasize hydrogen storage and catalysis. The surface area is designed to be mostly created insitu to limit oxidation. The oxide surface can be designed for thickness, corrosion resistance, conductivity, and microporosity. Refinements to the interpretation of oxide layer have been proposed. It is believed that the role of interior surfaces and particular insitu created surfaces have a great impact on high rate discharge. It is further proposed that the metal/electrolyte interface is a gradual transition form oxide to metal which is heterogeneous .and disordered. The microporous surface which contains both conductive and catalytic components catalyzes the hydrogen/hydroxyl reactions.

REFERENCES

1. *M. Tadokoro, K. Moriwaki, M. Nogami, T. Ise, N. Furakowa, Abstract 106, The Electrochemical Society Extended Abstracts, p. 161, Phoenix, Arizona, October 13-17, 1991.*

2. *M.A. Fetcenko, S. Venkatesan, S.R. Ovshinsky, M. Hirota, IBA Conference, Seattle, Washington, October 1990.*

3. *M. Hirota, S. Wada, R. Magai, K. Kajita, M.A. Fetcenko, Abstract 114, The Electrochemical Society Extended Abstracts, p. 172, Phoenix, Arizona, October 13-17, 1991.*

4. *J.R. Van Beek, H.C. Donkersloot, and J.J.G. Willems. "Power Sources 10" (L.J. Pearce, ed.) The Paul Press London, International Power Sources Symposium Committee, Leatherhead, England, p. 317 (1985).*

5. *H. Ogawa, M. Ikoma, H. Kawano, I. Matsumoto, 26th International Power Sources Symposium, p. 393, Bournemouth, England, September 1988.*

6. *K. Sapru, K.C. Hong, M.A. Fetcenko, S. Venkatesan, (1985), U.S. Patent 4,551,400.*

7. *S.R. Ovshinsky, "Disordered Materials: Science and Technology", D. Adler, ed., Amorphous Institute Press, ISBN 0-910331-00-6, 1982.*

8. *S.R. Ovshinsky, "Disordered Materials: Science and Technology", 2nd Edition, D. Adler, B. Schwartz, and M. Silver, ed., Institute for Amorphous Studies Series, Plenum Publishing Corporation, New York, 1991.*

9. *S.R. Ovshinsky, K. Sapru, K. Dec, K. Hong, (1984) U.S. Patent 4,432,561.*

10. *S.R. Ovshinsky, K. Sapru, S. Venkatesan (1984). U.S. Patent 4,430,391.*

11. *S.R. Ovshinsky, K. Sapru, E. Yee (1985). U.S. Patent 4,545,883.*

12. *K. Sapru, B. Reichman, G. Liang (1985). U.S. Patent 4,511,638.*

13. *S.R. Ovshinsky, K. Sapru (1985). U.S. Patent 4,544,473.*

14. *K. Sapru, B. Reichman, A. Reger, S.R. Ovshinsky, (1986) U.S. Patent 4,623,597.*

15. *K.C. Hong, K. Sapru (1989), U.S. Patent 4,832,913.*

16. *S.R. Ovshinsky, Unpublished Lecture, May 20, 1980.*

17. *M.A. Fetcenko, S.R. Ovshinsky, S. Venkatesan, K. Kajita, H. Kidou, and K. Jeffries, Third International Rechargeable Battery Seminar, Deerfield Beach, Florida, March 1990.*

18. *S. Venkatesan, M.A. Fetcenko, B. Reichman, D. Magnuson, and S. Dhar. "The Second International Rechargeable Battery Seminar", S. Wolsky, ed (1988).*

19. *S.R. Ovshinsky, S. Venkatesan, M.A. Fetcenko, S. Dhar, Proceedings of the 24th ISATA International Symposium on Automotive Technology and Automation, Electric/Hydrid Vehicles, p. 29, Florence, Italy, May 1991.*

20. *W.H. DeLucca, A.F. Tumillo, J.E. Kulaga, C.E. Webster, K.R. Gillie, and R.L. Hofgrefe, Proc. Intersoc. Energy Conver. Eng. Conf. 25th (Vol 3), 314-319, (1990).*

21. *W.H. DeLucca, Annual Automotive Technology Development Contractors Coordination Meeting, Dearborn, Michigan, October 29, 1991.*

22. *M.A. Fetcenko, S. Venkatesan, S.R. Ovshinsky, K. Kajita, H. Hirota, and H. Kidou, 17th International Power Sources Symposium, Bournemouth, England, April 1991.*

23. *B. Otzinger, "The Sixth Annual Battery Conference on Applications and Advances", Long Beach, California, (January 1991).*

24. *J. Brill, D. Coates, P. Bemis, S. Venkatesan, M.A. Fetcenko, S.R. Ovshinsky, AIAA/Utah State University Conference on Small Satellites, (August 1990).*

25. *T. Sakai, H. Miyawura, N. Kuriyama, H. Ishikawa, Abstract 105, The Electrochemical Society Extended Abstracts, p. 159, Phoenix, Arizona. October 13-17, 1991.*

26. *J.J. Reilly, Abstract 101, The Electrochemical Society Extended Abstracts, p. 155, Phoenix, Arizona, October 13-17, 1991.*

27. *H.F. Bittner, and C.C. Badcock. J. Electrochem Soc., 130, (5), 193C (1983).*

28. *M.H.J. Van Rijswick. Proc, Int, Symp. Hydrides for Energy Storage, Geilo, Norway (A. Andersen, ed.), p. 261 (1977).*

29. *A. Percheron-Guegen, (1978). U.S. Patent 4,107,405.*

30. *S. Venkatesan, B. Reichman, M.A. Fetcenko, (1988). U.S. Patent 4,728,586.*

31. *M.A. Fetcenko, S. Sumner, and J. LaRocca, (1989). U.S. Patent 4,948,423.*

32. *M.A. Fetcenko, T. Kaatz, S. Sumner, and J. LaRocca, (1988a), U.S. Patent 4,893,756.*

33. *M.T. Wolff, M. Nuss, M.A. Fetcenko, A. Lijoi, S.P. Sumner, J. LaRocca, and T. Kaatz, (1989), U.S. Patent 4,915,898.*

34. *M.A. Fetcenko, (1991). U.S. Patent 5,002,730.*

35. *C.R. Knoll, S.M. Tuominen, J.R. Peterson, and T.R. McQueary. "Battery Waste Management Seminar", (S. Wolsky ed.) (1990).*

36. *M.A. Gutjahr, H. Buckner, K.D. Beccu, H. Saufferer, "Power Sources, Vol. IV, D.H. Collins, ed. p. 79, Oriel Press, (1973).*

37. *S.R. Ovshinsky, R.A. Flasck, (1982). U.S. Patent 4,339,255.*

38. *M.A. Fetcenko, S. Venkatesan, K.C. Hong, and B. Reichman, 16th International Power Sources Symposium, Bournemouth, England, September 1988.*

39. *B. Reichman, S. Venkatesan, M.A. Fetcenko, K. Jeffries, S. Stahl, D. Magnuson, C. Bennett. and S. Sumner, (1987). U.S. Patent 4,716,088.*

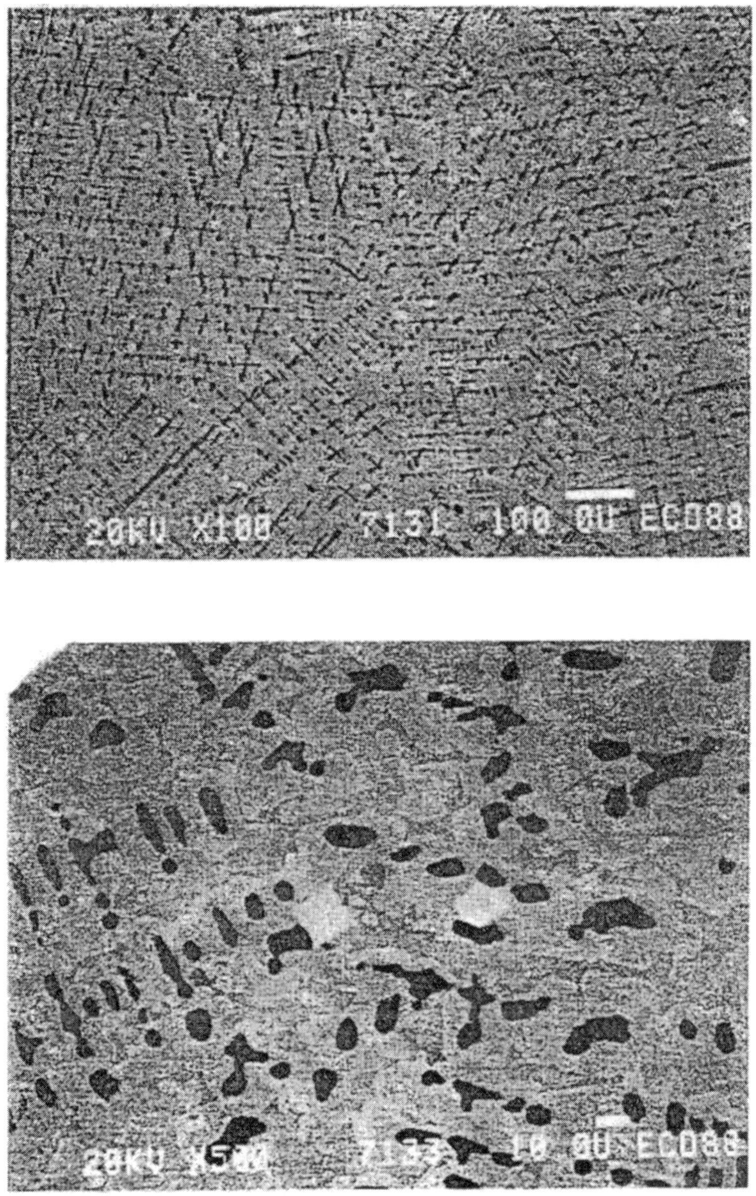

Figure 1. SEM Micrograph of $V_{22}Ti_{16}Zr_{16}Ni_{39}Cr_7$ Alloy Under Backscattered Electron Imaging. Top - 100x. Bottom - 500x.

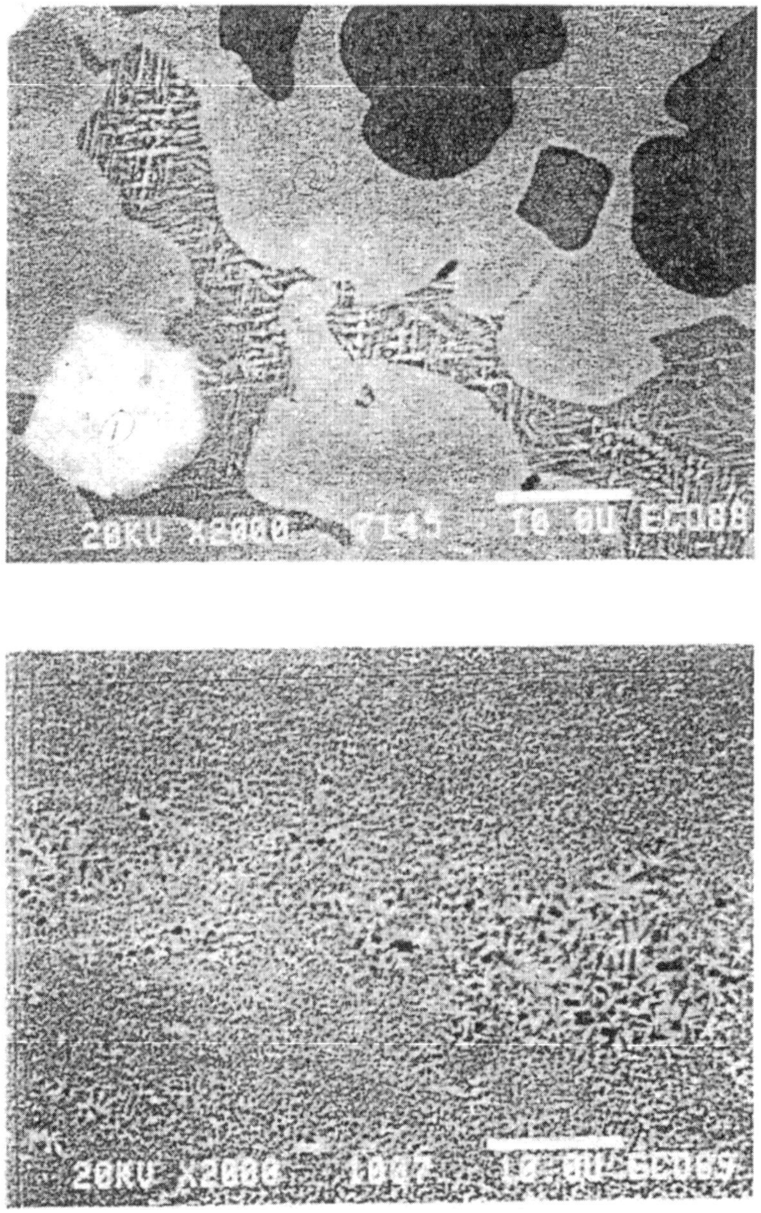

Figure 2. SEM Micrographs at 2000x of $V_{22}Ti_{17}Zr_{16}Ni_{39}Cr_7$ Alloy. Top- Conventional Solidification. Bottom - Rapid Solidification.

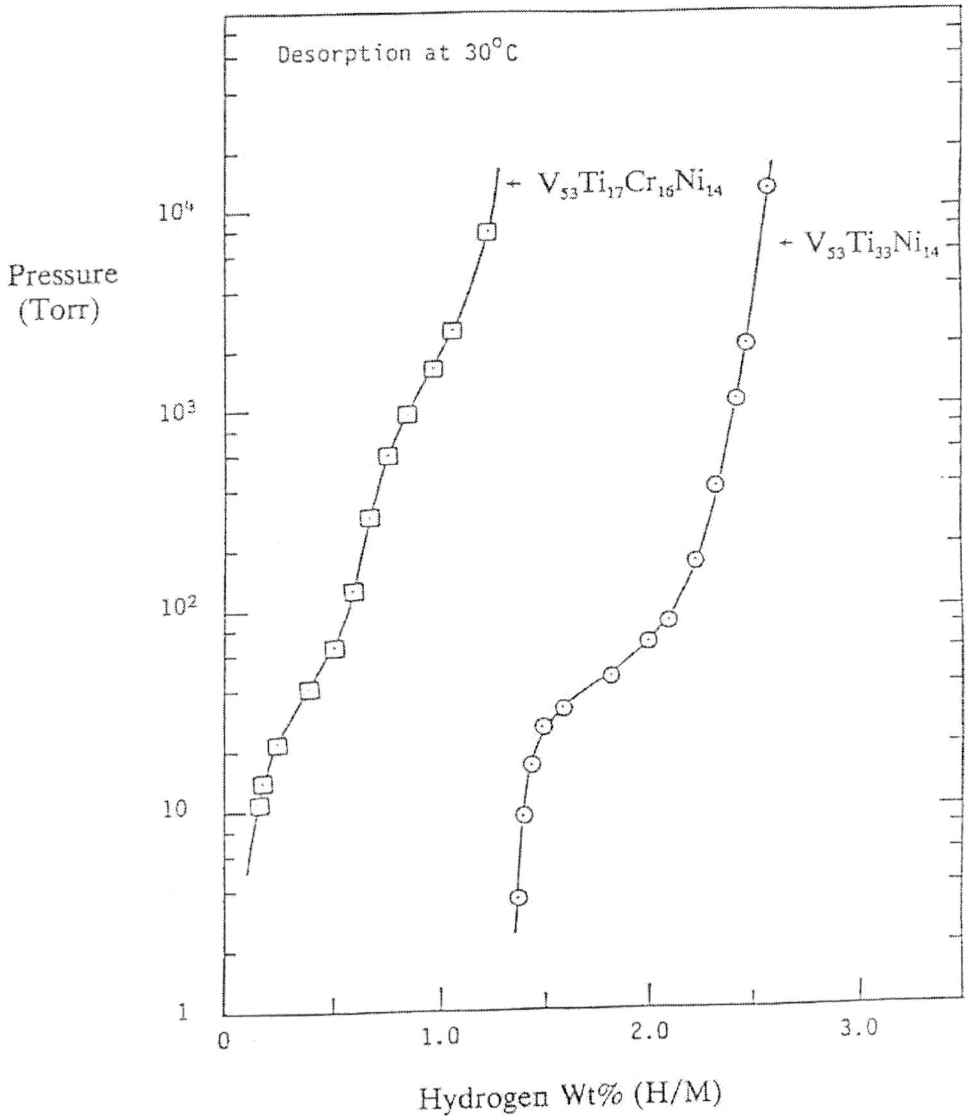

Figure 3. PCT Behavior of $V_{53}Ti_{17}Cr_{16}Ni_{14}$ and $V_{53}Ti_{33}Ni_{14}$ Alloys at 30°C Desorption.

Figure 4. SEM Micrographs at 300x with Samples 60° from Electron Beam to Illustrate Surface Topography. Top - $V_{53}Ti_{17}Cr_{16}Ni_{14}$ Bottom - $V_{25}Ti_{17}Zr_{16}Ni_{43}$.

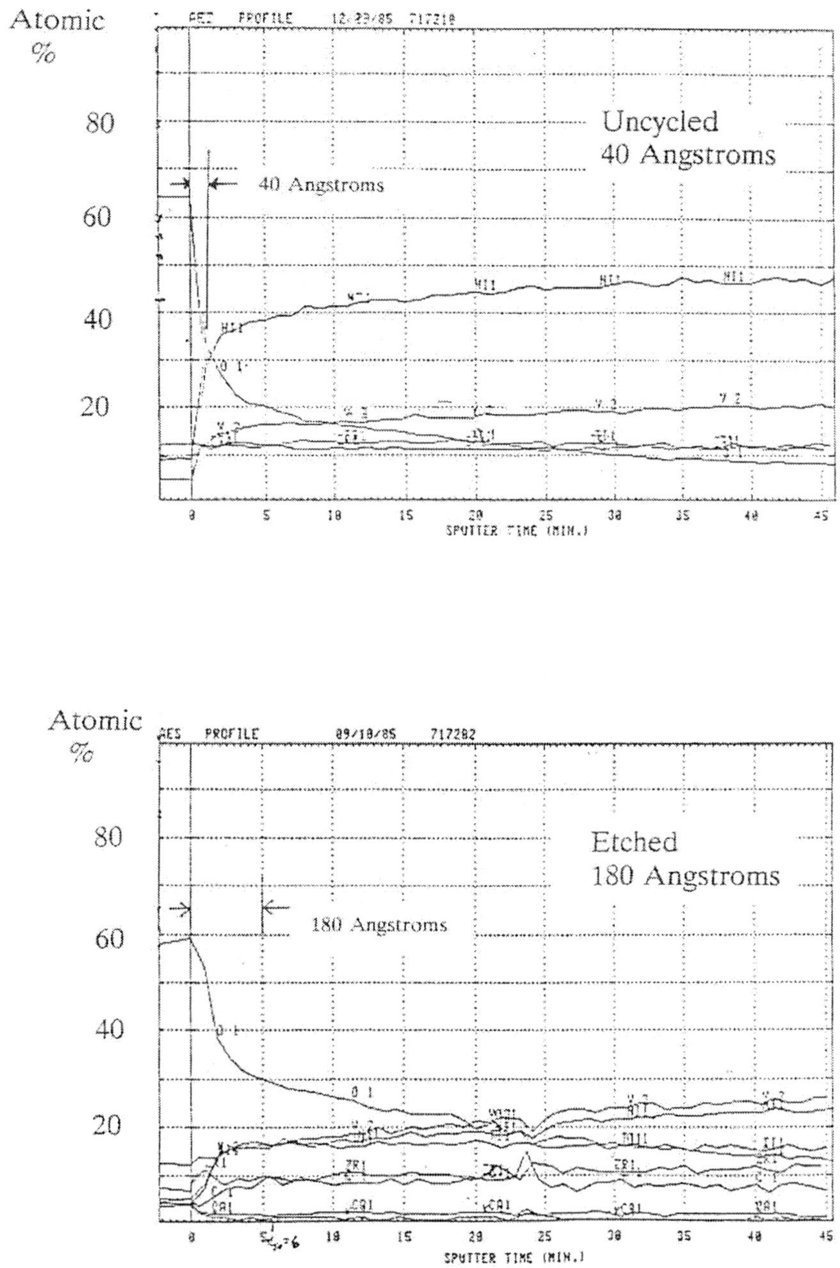

Figure 5. Auger Electron Spectroscopy of $V_{33}Ti_{17}Zr_{16}Ni_{14}$ Alloy Electrodes. Top - as fabricated. Bottom - after alkaline etch.

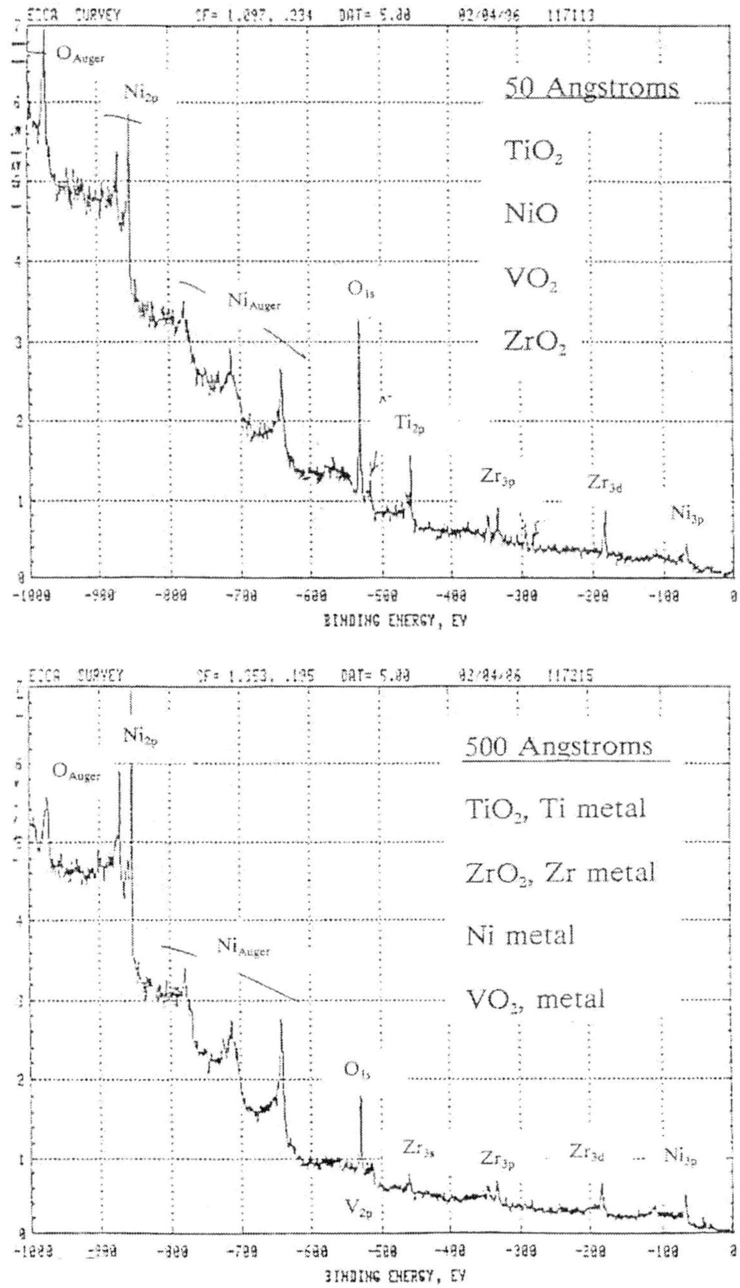

Figure 6. X-ray Photoelectron Spectroscopy of $V_{25}Ti_{17}Zr_{16}Ni_{43}$ Alloy Electrode Cycled 10 times in a Sealed Cell. Top - 50 Angstroms Depth. Bottom - 500 Angstroms.

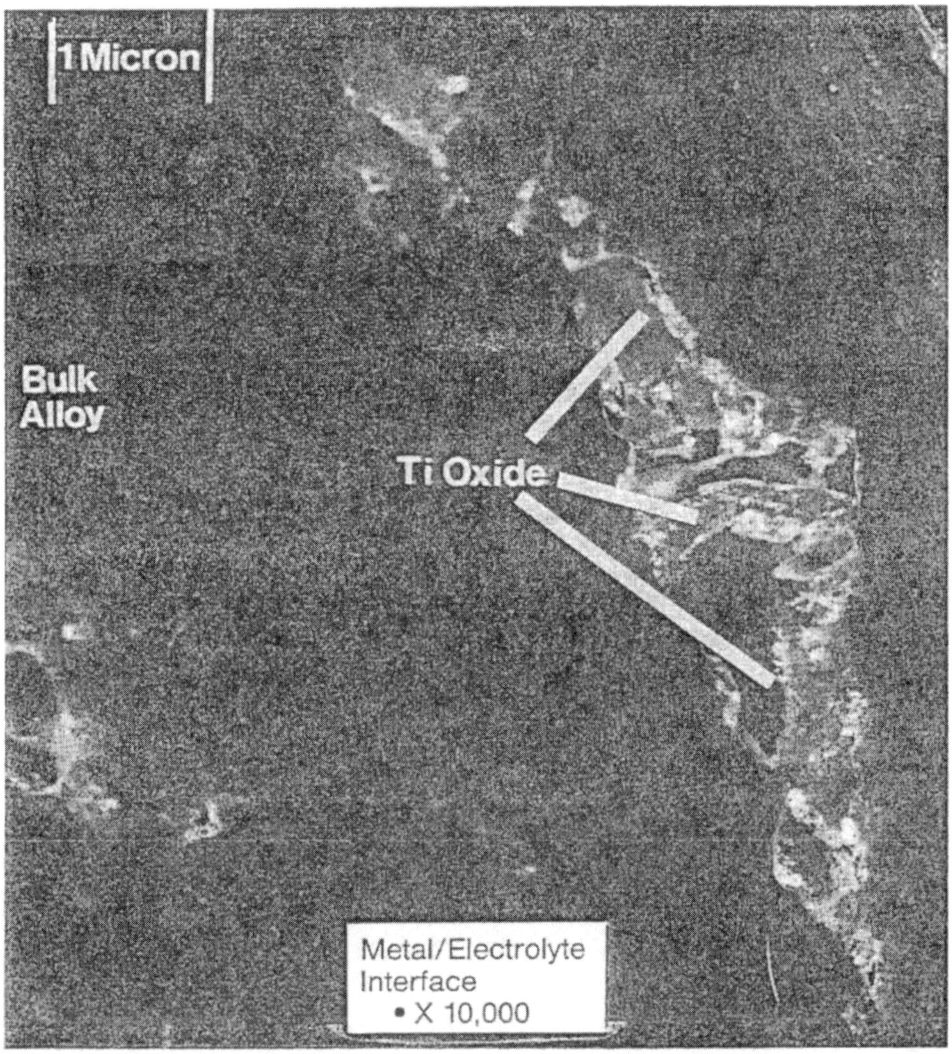

Figure 7. STEM Micrograph of Metal/Electrolyte Interface of poor rate capability sample at 10,000x.

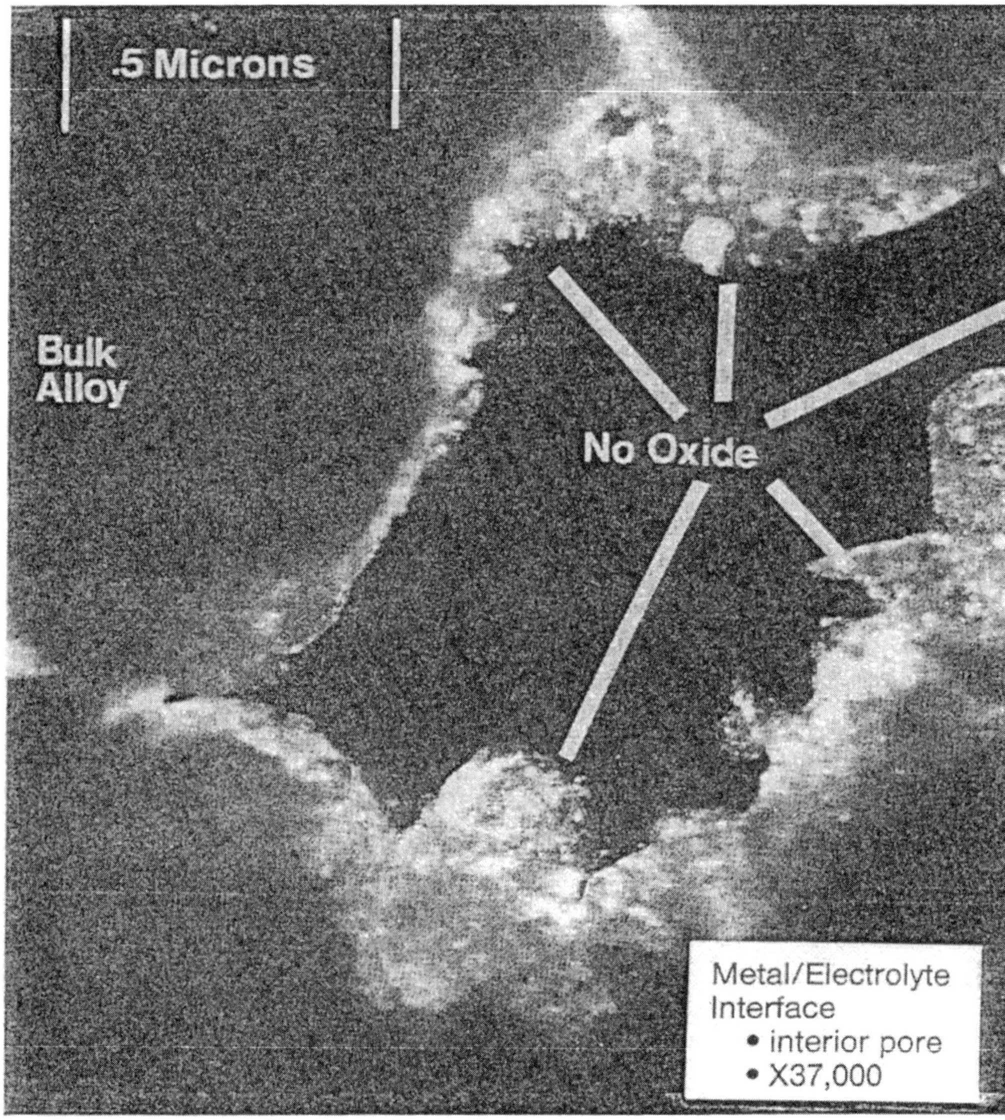

Figure 8. STEM Micrograph of Metal/Electrolyte Interface of good rate capability sample at 37,000x.

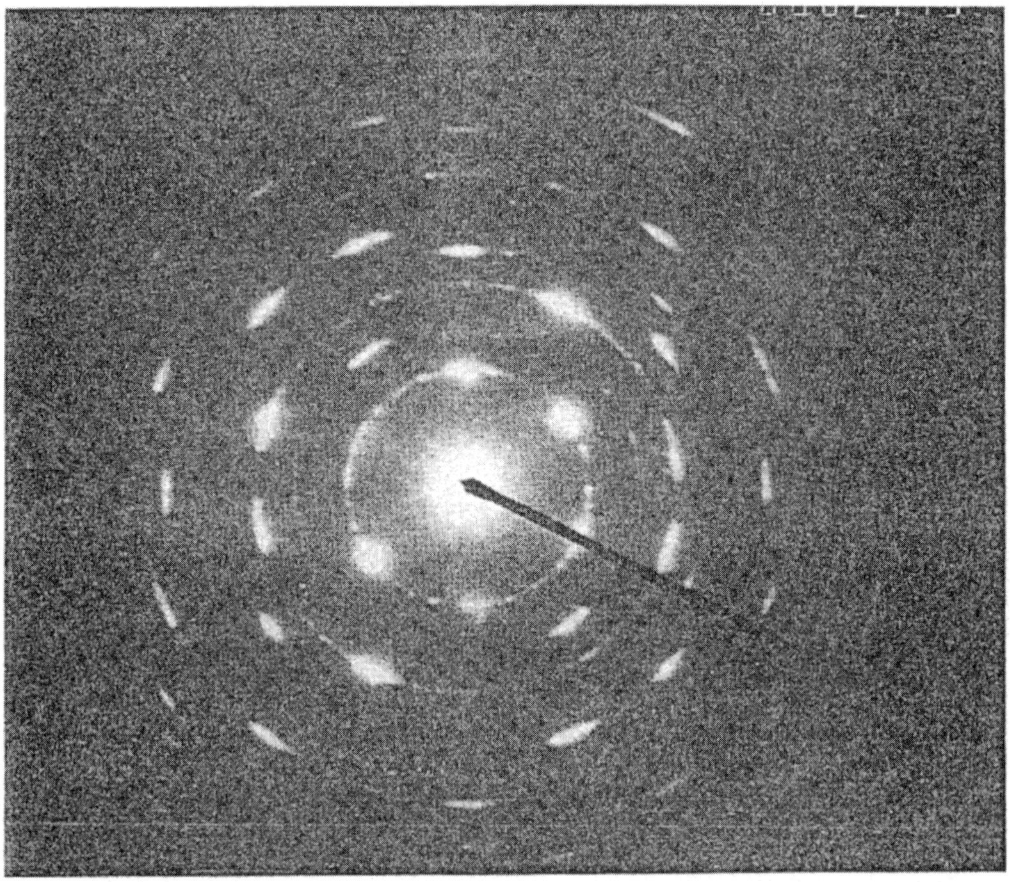

Figure 9. SAED of Bulk Alloy at Electrolyte Interface, with Particular Phase Corresponding to BCC Structure with a_o = 2.95A.

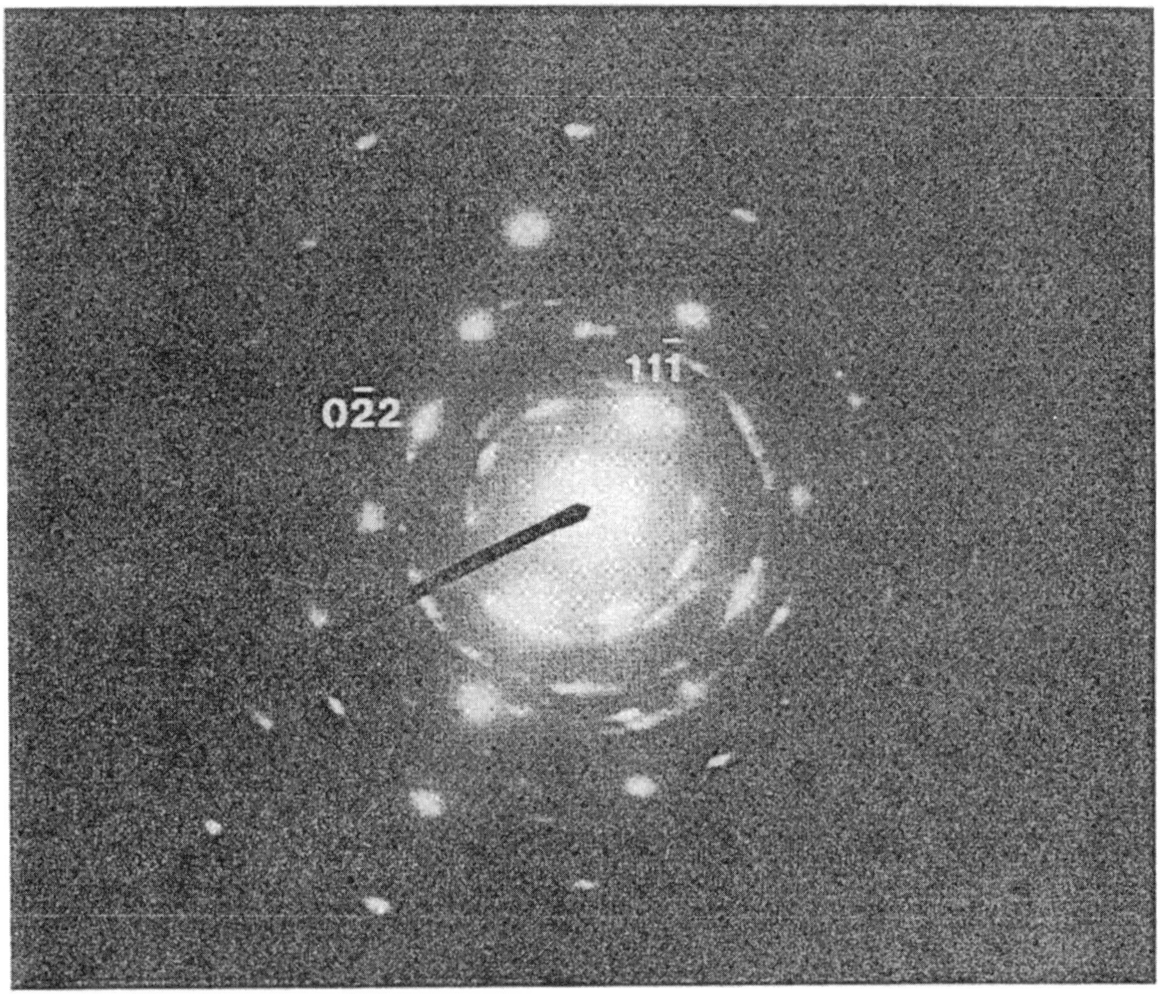

Figure 10. SAED of Surface Oxide, Corresponding to TiO_x with a_o = 4.2A.

Figure 11. STEM micrograph of Metal/Electrolyte interface of good rate capability sample at x330,000.

Reprint Series
9 April 1993, Volume 260, pp. 176–181

SCIENCE

A Nickel Metal Hydride Battery for Electric Vehicles

S. R. Ovshinsky, M. A. Fetcenko, J. Ross

Widespread use of electric vehicles can have significant impact on urban air quality, national energy independence, and international balance of trade. An efficient battery is the key technological element to the development of practical electric vehicles. The science and technology of a nickel metal hydride battery, which stores hydrogen in the solid hydride phase and has high energy density, high power, long life, tolerance to abuse, a wide range of operating temperature, quick-charge capability, and totally sealed maintenance-free operation, is described. A broad range of multi-element metal hydride materials that use structural and compositional disorder on several scales of length has been engineered for use as the negative electrode in this battery. The battery operates at ambient temperature, is made of nontoxic materials, and is recyclable. Demonstration of the manufacturing technology has been achieved.

The interest in electrically powered vehicles extends nearly as far back as interest in vehicles powered by hydrocarbon fuels. Throughout this period, however, there has been a major technological barrier to the development of practical electric vehicles (EVs) that can compete in performance and cost with those that use internal combustion (IC) engines. This barrier has been the lack of an economical battery with sufficient energy density and other essential performance criteria. In this article, we describe the science and technology of a nickel metal hydride (NiMH) battery that will permit future EVs to replace IC-powered vehicles in many applications.

S. R. Ovshinsky and M. A. Fetcenko are at Energy Conversion Devices, Inc., 1675 West Maple Road, Troy, MI 48084. J. Ross is in the Chemistry Department, Stanford University, Palo Alto, CA 94305, and consultant to Energy Conversion Devices, Inc.

Recently, U.S. federal and state governments have been providing an impetus for the development of an EV industry through legislation aimed at increasing national energy independence and reducing the impact of automobile emissions on the environment. California has passed laws that demand that 2% of new cars sold in 1998 be emission-free, and this percentage is slated to grow to 10% by the year 2003; 12 eastern states are planning similar laws. A comprehensive energy bill passed by Congress contains a tax credit for EV buyers. This bill also requires state and federal governments to purchase alternative-fuel fleet vehicles, with the percentage of new, cleaner fuel vehicles growing to 90% by the year 2000. It is expected that EVs will make up an increasing portion of alternative fuel vehicles as the market grows.

There are several important advantages of EVs compared with IC-powered vehicles. First, EVs are emission-free: they produce no pollution during operation. This quality is particularly important in city centers where congested automobile traffic is the primary source of local air pollution. The overall unwanted emissions that result from combustion of fossil fuels for the generation of electricity are also far less per mile of EV travel than the emissions produced directly by a fossil fuel–powered car. This fact, discussed in detail in a study by the Electric Power Research Institute (EPRI) (*1*), results from the sophisticated emissions controls that can be used economically by large, efficient, central power-generation facilities. Second, the EPRI study also details how the primary energy efficiency of electric transportation can exceed the efficiency of gasoline-powered vehicles in many instances. For example, the study shows that electric-powered commercial fleet vans that are used in urban areas have a significant advantage in energy efficiency over their gasoline-powered counterparts, traveling about 1100 miles per barrel of oil consumed at the power plant compared with 620 miles per barrel of oil refined into gasoline. This difference results primarily from the higher energy efficiency of power plant combustion—approximately twice as high as combustion of gasoline in an IC engine in urban traffic. Third, conversion from cars directly powered by fossil fuel to ones powered by electricity can shift the choice of hydrocarbon fuels that are consumed in the United States from oil to coal and gas. This change could possibly reduce the oil imports and, consequently, reduce the U.S. trade imbalance and the strategic vulnerability of its energy supply. Photovoltaic and other renewable energy sources are also increasingly available to generate pollution-free electricity for EVs.

In response to the need to develop a practical battery for EVs, the U.S. federal government authorized the establishment of the U.S. Advanced Battery Consortium (USABC) in 1990. Under the aegis of the Department of Energy, USABC brings together Chrysler, Ford, General Motors (GM), and EPRI to sponsor research and development of EV batteries. Although energy density is one of the most important requirements for an EV battery system, USABC has identified a number of other battery criteria as necessary for the development of economically viable EVs (Table 1 provides the primary midterm goals of USABC). Ovonic Battery Company (OBC), a subsidiary of Energy Conversion Devices, has received the first contract from USABC toward the continued development and fabrication of the company's proprietary NiMH batteries and has agreed to establish EV battery production facilities.

In this article, we describe the science and technology of the Ovonic NiMH battery, with emphasis on the materials science aspects of the metal hydride (MH) electrode and their effect on battery performance (*2*). The MH electrode offers an important opportunity for materials engineering and optimization when compared with negative electrodes for other nickel-based battery systems. In these other systems, the negative electrode (Cd, Zn, or Fe) is typically fabricated from relatively pure elemental metals, and the oxidation-reduction reactions associated with battery charge and discharge convert the electrode back and forth between a metal and a metal oxide that is a poor electric conductor. This type of chemical reaction can be undesirable in a practical battery design because of accompanying changes in the physical properties of the electrode. Changes in the mechanical integrity and surface morphology of the electrode as a result of dissolution and recrystallization and of its reduced electrical conductivity in the oxidized state are sources of many of the performance deficiencies in these systems.

The MH electrode, by contrast, uses a chemical reaction that reversibly incorporates hydrogen into a metal alloy. In this oxidation-reduction reaction both chemical states are metallic, and so electrical conductivity is high in both the charged and discharged states. Furthermore, the small size of the hydrogen atom allows it to enter the metal lattice during formation of the hydride (reduced) state with only about 10% volumetric expansion and without the changes in crystallography associated with oxidation and reduction of the Cd, Zn, or Fe electrodes.

In effect, the MH negative electrode can be regarded as a matrix for the chemical incorporation of the hydrogen atom. In the Ovonic NiMH battery, we have exploited the ability of this matrix to be engineered through the use of multi-element alloys, using compositional and structural disorder to produce materials with desirable battery properties.

Cell Reactions

The NiMH battery, which has a nominal voltage of 1.2 V, stores hydrogen as a reaction product in the solid hydride phase, unlike the nickel-hydrogen battery that stores hydrogen as a high-pressure gas. The negative electrode of a conventional NiMH battery consists of a hydrogen storage material (*3–5*) that can allow electrochemical storage and release of hydrogen during battery charge and discharge processes. The nickel hydroxide positive electrode (*6–9*) is electrochemically reversible between $Ni(OH)_2$ and nickel oxyhydroxide, usually written as NiOOH. At both electrodes, oxidation-reduction reactions take place in an alkaline medium consisting of 30% by weight KOH in water. During charge, the $Ni(OH)_2$ electrode is oxidized and the MH electrode is reduced. As a result, water is separated into hydrogen and hydroxyl ions, with hydrogen reacting with the metal in the negative electrode to form MH. At the positive electrode, the hydroxyl ion reacts with the $Ni(OH)_2$ electrode to form NiOOH. This reaction results in a change in the Ni oxidation state from +2 to +3. The half-cell reactions on charge and discharge of the battery can be written as

$$M + H_2O + e^- \underset{\text{Discharge}}{\overset{\text{Charge}}{\rightleftarrows}} MH + OH^- \quad (1)$$

$$Ni(OH)_2 + OH^- \underset{\text{Discharge}}{\overset{\text{Charge}}{\rightleftarrows}} NiOOH + H_2O + e^- \quad (2)$$

As a consequence of reactions 1 and 2, there is no net change in electrolyte quantity or concentration over the charge-discharge cycle. This result contrasts with other alkaline electrolyte systems such as NiCd where water is generated at both electrodes during charge and consumed at both electrodes during discharge. Although transient electrolyte concentration gradients can occur in the NiMH battery, its constant average concentration has the important consequences of good overall performance in gas recombination, kinetics, high- and low-temperature operation, and resistance to cycle-life limitations produced by corrosion and swelling.

Material Requirements

The MH materials used for an NiMH battery electrode must satisfy an extensive list of requirements. Above all, the amount of hydrogen that the MH material can absorb determines the electrochemical storage capacity of the electrode and, consequently, the energy storage capacity of the battery. It is desirable to have high electrode storage capacity that is electrochemically reversible. To ensure reversibility, an important aspect of the MH design is the range of metal-to-hydrogen bond strengths, which must be about 6 to 12 kcal mol^{-1}. If the bond strength is too weak, hydrogen will not react with the alloys and will be evolved as a gas. If the bond strength is too large, the MH electrode is extensively oxidized and does not store hydrogen reversibly.

Even with an optimally adjusted metal-hydrogen bond strength, the problem of electrode oxidation in the MH battery

Table 1. Primary USABC midterm performance goals for the EV battery and actual performance of the current OBC NiMH battery. DOD, depth of discharge.

Property	USABC	OBC
Specific energy (Wh kg^{-1})	80 (100 desired)	80*
Energy density (Wh per liter)	135	215*
Power density (W per liter)	250	470
Specific power (W kg^{-1}) (80% DOD in 30 s)	150 (>200 desired)	175
Cycle life (cycles) (80% DOD)	600	1000
Life (years)	5	10
Environmental operating temperature	−30° to 65°C	−30° to 60°C
Recharge time	<6 hours	15 min (60%) <1 hour (100%)
Self discharge	<15% in 48 hours	<10% in 48 hours
Ultimate projected price (dollars per kWh) (10,000 units at 40 kWh)	<$150	$200

*A specific energy of 80 Wh kg^{-1} and an energy density of 215 Wh per liter have been achieved in a laboratory prototype, with 50-Ah cells under a discharge rate at which the battery energy capacity is exhausted in 3 hours.

remains. The NiMH battery operates in a strongly oxidizing medium composed of a high-concentration alkaline electrolyte. Because many chemical elements react to form oxides in an alkaline electrolyte, it follows that if these elements are used as electrodes, they will oxidize and fail to store hydrogen reversibly. In addition, MH electrodes are typically designed for use in totally sealed batteries where oxygen recombination occurs at their surfaces. In this aggressively oxidizing environment, oxidation and corrosion resistance of MH electrode materials is critical. Because some oxidation at the metal-electrolyte interface is inevitable and because both passivation and corrosion can have adverse effects on battery performance, these unwanted processes must be controlled in a practical NiMH electrode design.

Another consideration in the use of hydride materials in NiMH batteries relates to electrochemical kinetics and transport processes. The power output of the battery depends critically on these processes. During discharge, hydrogen stored in the bulk metal must be brought to the electrode surface by diffusion. The hydrogen must then react with hydroxyl ions at the metal-electrolyte interface. As a consequence, surface properties such as oxide thickness, electrical conductivity, surface porosity and topology, surface area, and degree of catalytic activity affect the rate at which energy can be stored in and removed from the NiMH battery.

For the battery to operate as a sealed system, it must also tolerate the consequences of chemical reactions that occur during cell overcharge and overdischarge. In overcharge, oxygen gas is generated at the $Ni(OH)_2$-positive electrode and must recombine with hydrogen at the MH electrode to form water. In overdischarge, which occurs when a low-capacity cell in a series-connected string is subjected to reverse polarity, hydrogen is generated at the $Ni(OH)_2$ electrode and must be recombined at the surface of the MH electrode to form water. In a sealed system, these gas recombination reactions must occur at sufficient rates to avoid pressure buildup. This condition requires adequate electrode area, a thin electrolyte film, and, for the hydrogen absorption process, catalytic activity at the MH electrode surface to promote rapid dissociation of hydrogen.

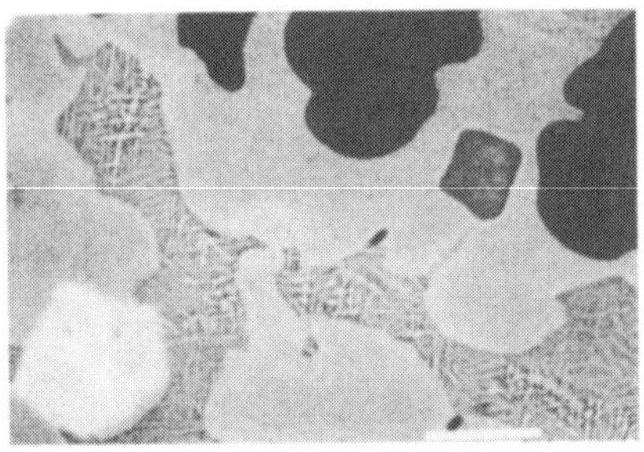

Fig. 1. Scanning electron micrograph of a bulk region of an Ovonic MH battery electrode that shows compositionally and structurally disordered multiphase alloy regions. Scale bar, 10 μm.

Chemical and Structural Disorder in Engineered Materials

The diverse properties required for a superior MH battery electrode can be attained by the engineering of new hydrogen storage materials on the basis of the concepts of structural and compositional disorder (*2, 10–13*). Compositional and structural disorder is designed into the new MH materials on three different length scales through the use of elemental composition and processing techniques of alloys and electrodes. The length scales over which disorder is created can be designated: local (or atomic), which comprises regions with dimensions up to a few nearest-neighbor atomic distances; intermediate range, which comprises regions typically about 10 to 20 nm and extending up to about 100 nm; and long range, which involves regions with a dimension larger than about 100 nm. Disorder on each of these length scales is used to achieve different goals in the engineered alloys.

This approach allows one to consider a range of alloys for electrode materials containing elements that, if used alone, would be unacceptable for thermodynamic reasons, in particular oxidation or corrosion. Among the elements that become available for alloy formation in disordered electrode materials are Li, C, Mg, Al, Si, Ca, Ti, V, Cr, Mn, Fe, Co, Ni, Cu, Y, Zr, Nb, Mo, Sn, La, W, and Re. The list contains elements that can increase the number of hydrogen atoms stored per metal atom (Mg, Ti, V, Zr, Nb, and La). Other elements allow the adjustment of the metal-hydrogen bond strength (V, Mn, and Zr); provide catalytic properties to ensure sufficient charge and discharge reaction rates and gas recombination (Al, Mn, Co, Fe, and Ni); or impart desirable surface properties such as oxidation and corrosion resistance, improved porosity, and electronic and ionic conductivities (Cr, Mo, and W). The wide range of physical and chemical properties that can be produced in these alloys allows the MH battery performance to be optimized.

Compositional and structural disorder on a long-range length scale is used in the bulk of Ovonic MH alloys to give considerably higher hydrogen storage and better kinetics than possible in the conventional MH alloy structures, which are compositionally ordered and crystalline. The processing of disordered alloys can be optimized to produce polycrystalline, compositionally multiphase material. Figure 1 shows a scanning electron micrograph of a representative bulk region of a typical V, Ti, Zr, Ni, and Cr Ovonic MH alloy. Electron backscattering imaging was used to produce visual contrast between regions of the alloy that have different elemental compositions. This material contains five major distinct compositional phases as determined by energy-dispersive x-ray analysis. In addition, it has been determined separately by x-ray diffraction that the alloy contains three crystal structures: body-centered-cubic (bcc), hexagonal, and ^{14}C Laves crystal structures. From synthesis and characterization of separate, individual phases of a particular alloy, we conclude that the bcc structure can react to store large quantities of hydrogen (2.5% by weight) but lacks sufficient catalytic activity to be discharged at the required rates for battery applications. On the other hand, surrounding phases that may store less hydrogen but exhibit greater catalytic activity can effectively "channel" the hydrogen for rapid electrochemical discharge (*12*).

Intermediate-range structural and chemical disorder plays a number of important roles, primarily at interfaces both within the bulk of the MH electrode and at the electrode-electrolyte interface. Formation of the polycrystalline, compositionally multiphase bulk alloy gives rise to a high density of grain boundaries between com-

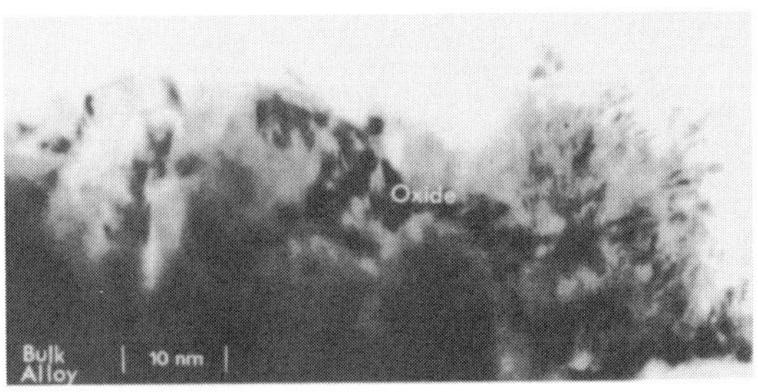

Fig. 2. Scanning transmission electron micrograph of the metal-electrolyte interface of an Ovonic MH battery electrode that shows the structure of the engineered multiphase bulk alloy and surface oxide.

positional and structural phases (for example, Fig. 1). The intermediate-range disorder that occurs at the grain boundaries increases surface area, which can greatly increase the density of catalytic sites. At the electrode-electrolyte interface, disorder on a length scale of approximately 10 to 100 nm is created during electrode processing and activation by the exploitation of chemical properties that are traceable to the elemental constituents of the MH alloy. The high-magnification scanning transmission electron micrograph of the electrode-electrolyte interface in Fig. 2 shows the presence of disorder on all three length scales but is particularly useful in illustrating the intermediate-range structural and compositional disorder that occurs in the engineered oxide layer that forms on the Ovonic MH electrode.

The basic Ovonic MH electrode typically contains elements such as V, Ti, Zr, Ni, and Cr. Although the alloy is a system with many characteristics such as MH bond strength that depend on interactions among the elemental constituents, some alloy properties are influenced by the chemistry of individual components. The primary role of V, Ti, and Zr in the alloy is hydrogen storage. All three elements, rather than just the least expensive (Ti), are used in the alloy for several reasons. Titanium and Zr form thick, dense, passive oxides in alkaline solutions, whereas V forms soluble oxides. These chemical characteristics are used in the preparation of the Ovonic MH electrode during the electrochemical activation step in which soluble oxides are intentionally corroded to produce intermediate-range structural disorder. This change

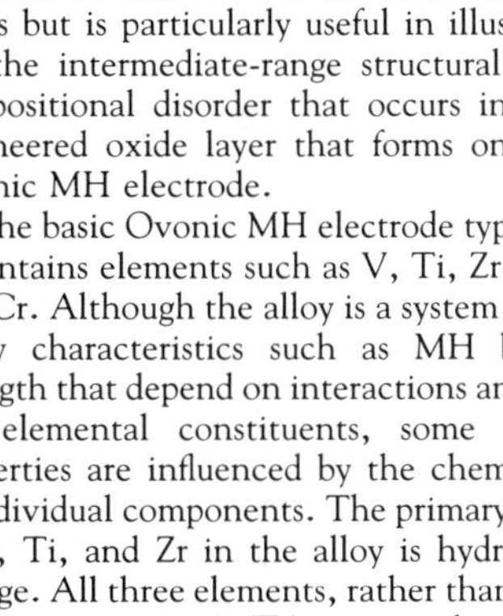

Fig. 3. Equilibrium hydrogen pressure versus hydrogen concentration (percent by weight) at 30°C for a series of Ovonic MH electrode alloys. Data show how variation in alloy composition may be used to control metal-hydrogen bond strength. The MH alloy compositions shown here, expressed as atomic percent, are (**A**) ($V_{21}Ti_{15}Zr_{15}Ni_{31}Cr_6Co_6Fe_6$), (**B**) ($V_{15}Ti_{15}Zr_{21}Ni_{31}Cr_6Co_6Fe_6$), (**C**) ($V_{18}Ti_{15}Zr_{18}Ni_{31}Cr_6Co_6Fe_6$), and (**D**) ($V_{15}Ti_{15}Zr_{20}Ni_{28}Cr_5Co_5Fe_6Mn_6$).

gives rise to increased electrode surface area and microporosity and thereby increases charge acceptance. Chromium is used to limit the unrestrained corrosion of V and to control the alloy microstructure.

Zirconium contributes the important property of controlled hydrogen embrittlement, which leads to high surface area and, hence, to fast cell reactions. However, because excessive embrittlement can produce mechanical disintegration of the electrode that leads to poor electrical conductivity, high polarization, and low charge-recharge cycle life, one must control this property carefully in designing the electrode alloy.

Nickel serves several functions in the alloy. First, NiH has a weak bond strength. The bond strengths of elemental Ti, Zr, and V with hydrogen are too high for electrochemical applications. However, formation of alloys from these elements in various concentrations with Ni allows control of the alloy bond strength as was discussed earlier. Second, Ni is a catalyst for dissociation of H_2 and subsequent absorption of atomic H into the alloy. Third, Ni is resistant to oxidation. The combination of Ni with Zr, V, and Ti makes the alloy more resistant to oxidation and produces oxide films at the electrode-electrolyte interface that contain regions of metallic Ni. These regions help provide the necessary electrical conductivity and catalytic activity in the oxide film. The interface is characterized by a heterogeneous oxide region (Fig. 2) rather than a sharply defined homogeneous oxide film. We believe that this disordered oxide region is microporous and contains electrically conductive Ni regions that can catalyze the electrochemical discharge reaction (*14*).

Compositional disorder on the atomic scale is used to increase hydrogen storage capacity and improve catalytic activity

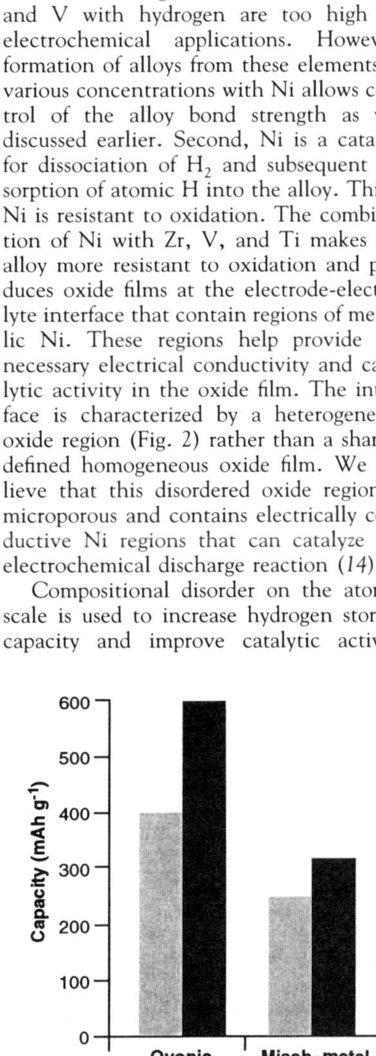

Fig. 4. Comparison of the charge storage capacity of present (shaded bars) and projected (solid bars) Ovonic MH battery and misch-metal ($LaNi_5$) electrodes.

through the incorporation into the MH alloy of elements that generate new chemically active sites. These sites offer an increased variety of hydrogen bonding possibilities and enhanced rates as a result of increased catalysis. Incorporation of elements with multidirectional *d* orbitals increases the range of stereochemical possibilities for bonding hydrogen, as confirmed by the increased amount of hydrogen absorbed and by increased catalytic activity. These effects also occur to a lesser extent with elements containing *f* orbitals that extend in still more directions than *d* orbitals but that are closer to the nucleus of the metal atom and, therefore, are less accessible.

Local compositional disorder is also used to adjust the metal-hydrogen bond strength in the MH alloy. Measurements of equilibrium hydrogen pressure versus MH hydrogen concentration at 30°C are shown in Fig. 3 for a series of multicomponent MH alloys in which the ratio of V to Zr is systematically varied. The equilibrium hydrogen pressure, p, in these measurements is related to the change in Gibbs free energy, ΔG, which occurs for the reaction between gaseous hydrogen and the hydrogen storage alloy to form MH. This value can be written:

$$\Delta G = \Delta H - T\Delta S = RT \ln p \qquad (3)$$

Because the entropy term, $T\Delta S$, is small at room temperature compared with the enthalpy change, ΔH largely determines ΔG. Thus, determination of p provides a measure of ΔH, which is related to the metal-hydrogen bond strength. In Fig. 3, variations in the ratio of V to Zr give rise to the observed changes in p for a given hydrogen concentration in the MH alloy. This result indicates that these compositional variations have changed the metal-hydrogen bond strength.

Status of the Ovonic Battery

The storage capacity of current and future Ovonic MH electrodes is shown in Fig. 4, along with the storage capacity of current and improved conventional $LaNi_5$, or "misch-metal," MH electrodes. The latter materials are frequently referred to as misch metal because they are traditionally made from a mixture of naturally occurring rare-earth elements that can include Ce, La, Nd, and Pr. The data for current electrodes were obtained from electrochemical half-cell measurements of commercial Ovonic and misch-metal battery electrodes, as described in (*11*). Data for projected misch-metal electrodes are from (*15*) and are based on electrochemical half-cell measurements of prototype materials. Data for projected Ovonic MH electrodes are based on electrochemical measurements of advanced laboratory thin-film materials.

The USABC contract contains three main program goals: a scale-up from small portable cells (1 to 5 Ah) to large EV cells (50 to 250 Ah), an increase in energy density from 56 to 80 Wh kg^{-1}, and the construction of series-connected battery modules to produce the voltages required for EV propulsion (180 to 320 V). The performance capabilities of the present Ovonic NiMH technology are shown in the second column of Table 1. To a large extent, prototype batteries of several designs and sizes have demonstrated performance characteristics that satisfy or exceed individual USABC midterm goals shown in the first column. Efforts to optimize all performance characteristics within a particular cell design are an ongoing activity. Another program objective, scale-up of cell size, has also been achieved. The ability to manufacture batteries with this technology has been demonstrated by OBC and several of its licensees who have been producing small portable batteries since 1987. Series-connected battery modules up to 40 V have been constructed by OBC and are under test. A 12-V, 3-kWh module is shown in Fig. 5.

Comparison with Other Candidates for EV Batteries

Battery characteristics have a dominant influence on overall EV performance. For example, battery-specific energy (in watt-hours per kilogram) controls vehicle range. Similarly, battery power (in watts per kilogram) translates into vehicle acceleration. The Bertone Blitz, a high-performance prototype EV sports car (*16*), has achieved an impressive acceleration of 0 to 100 km per hour (0 to 62 miles per hour) in 6 s, in part through the use of Ni-Cd batteries with high peak power. Some candidate battery technologies for EV applications are listed in Fig. 6, which shows a comparison plot of peak power versus depth of discharge. These measurements were made independently at Argonne National Laboratories (*17*). High peak power (>150 W kg^{-1}), as required by the USABC goals shown in Table 1, must be maintained over the entire depth of discharge of the battery for satisfactory vehicle performance. The Ovonic NiMH battery provides the highest peak power and can maintain it over almost the full range of discharge.

Although Ni-Cd is a rechargeable battery technology with high peak power that is extensively used in consumer products such as electronic devices and power tools, its energy density does not meet USABC requirements, it uses toxic materials (Cd), and in the large sizes used for EVs, it is not a totally sealed system.

The Na-S battery has a high energy density, but its low peak power is a significant deficiency. In addition, the operating temperature of the battery is approximately 300°C, which must be maintained at all times because the battery can withstand only a few cycles of cooling and heating. The presence of molten sodium and sulfur is potentially hazardous, and corrosion has limited the reliability and life of prototype batteries.

Of the remaining batteries shown in Fig. 6, only the Pb-acid battery has been tested sufficiently to serve as a practical, immediate candidate for EV applications. However, its typical energy density of 30 Wh kg^{-1} is substantially below USABC requirements, and its limited cycle life would force it to be replaced every 32,000 km (20,000 miles).

Long cycle life is a feature of the Ovonic NiMH battery technology that will have economic consequences for EVs. Over 1000 charge-discharge cycles at 100% depth of discharge have been demonstrated (*13*) with Ovonic batteries. Under conditions of 30% depth of discharge, Ovonic NiMH cells designed for aerospace applications have demonstrated (*18*) a lifetime of more than 10,000 cycles. It is expected that in EV applications, batteries will experience a typical depth of discharge of about 80%. Under these conditions the cycle life of the Ovonic NiMH battery is projected to be 2000 to 3000 cycles, according to a numerical model (*18*).

Battery cycle life can be converted into EV battery-life driving range when the characteristics of the EV are specified. For example, in a GM Impact-type vehicle, replacement of the Pb-acid battery with an Ovonic NiMH system of the same volume increases its range to 480 km (300 miles). For 80% depth of discharge [385 km (240 miles)], even a conservative estimate of 500 cycles for the battery life will give a 200,000-km (120,000-mile) battery-life driving range. The electrical energy necessary to provide the 385-km range per charge costs only $2.32 at $0.08 per kilowatt-hour, compared with approximately $14 worth of gasoline needed to provide the same range for a typical IC-powered vehicle. Lifetime EV maintenance costs will also be smaller than for typical IC-powered vehicles. Therefore, EVs that are powered by batteries with long cycle lives and that meet the USABC initial cost goal of $150 per kilowatt-hour can be economically competitive on a lifetime basis.

Technology Improvements

The range of an EV will depend on many factors besides battery energy density, such as vehicle weight, tire rolling friction, and electric motor efficiency. Information published by GM on its pioneering Impact vehicle can be used to establish a benchmark for conversion of battery energy density to vehicle range (*19*) for EVs of this type. The data from GM show that their vehicle will travel 180 km (113 miles) with a battery that stores 13.5 kWh. For a battery of the same weight, the current Ovonic NiMH technology will, therefore,

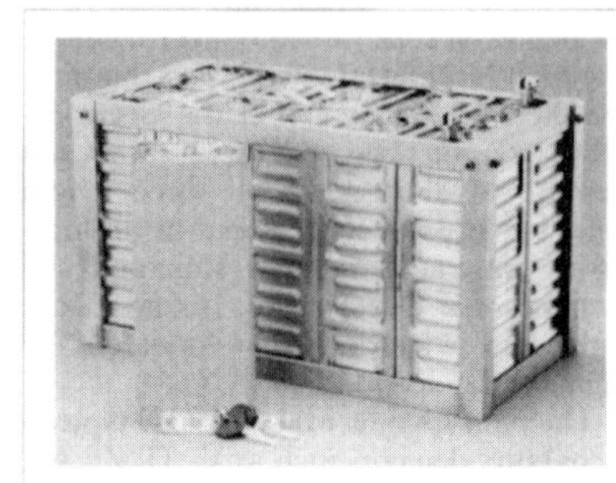

Fig. 5. A 3-kWh, 12-V series-connected Ovonic NiMH battery module and 250-Ah Ovonic NiMH single cell.

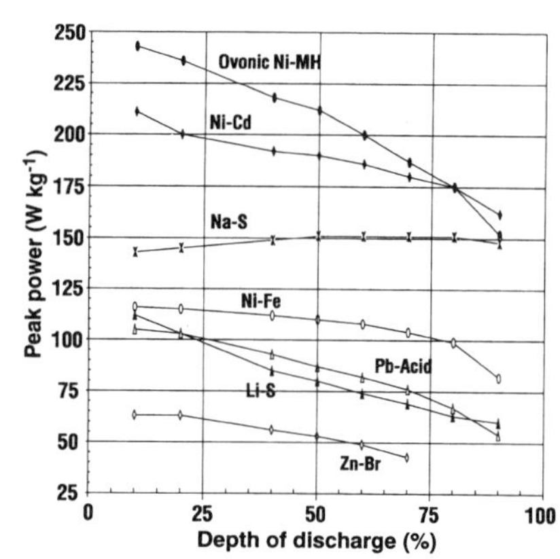

Fig. 6. Peak power versus depth of discharge, as measured (*17*) at Argonne National Laboratories, for a number of candidate EV battery technologies.

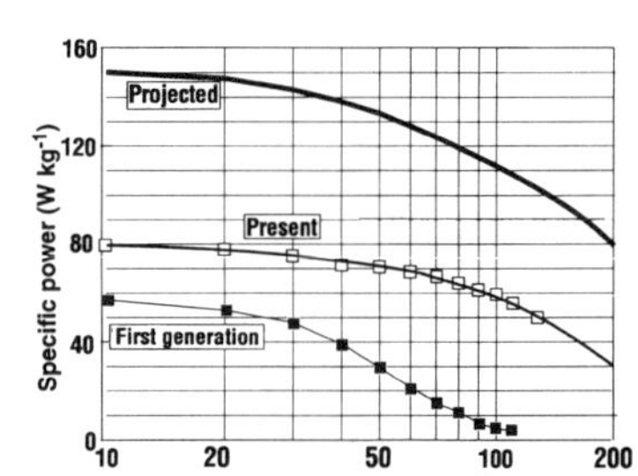

Fig. 7. Specific energy versus specific power for first-generation, present, and projected Ovonic NiMH batteries. Data shown for first-generation devices were obtained from 4-Ah "C" size cells; data for present devices were obtained from 50-Ah prismatic cells.

provide 31.6 kWh of energy storage, which will provide a vehicle range of 415 km (264 miles). For the same battery volume, the Ovonic NiMH battery will increase the range per charge to 480 km (300 miles). The environmental impact of eventual disposal of the Ovonic NiMH battery has also been studied (*20*). Knoll and colleagues concluded that, according to existing Environmental Protection Agency regulations, batteries that use this technology can be safely disposed of in landfills. It has also been shown (*21*) that with existing technology, Ovonic batteries can be recycled into metallurgical additives for cast iron, stainless steel, or new Ovonic NiMH battery electrodes. The commercial viability of each of these technologically feasible recycling programs will depend on process economics.

Future developments of Ovonic NiMH batteries will include improvements through the continued optimization of the MH materials and electrodes as well as improvements to the positive electrode and cell design (*2*). For example, some of the ongoing research at OBC focuses on application of the company's synthetic materials techniques to the development of an improved positive electrode with enhanced storage capacity through the use of engineered valence control. The chemical reaction that occurs during the charge of a conventional $Ni(OH)_2$ electrode involves transfer of one electron per Ni atom. We are developing materials that use the exchange of up to two electrons per atom. In addition, MH alloys with twice the storage capacity of first-generation materials have been measured in the laboratory, and cell designs in which lightweight substrates, current collection components, and containers are used are now being developed. Because the overall energy density of the battery is determined by the entire system, these combined approaches are targeted at the fabrication of batteries with both an energy storage density of 150 Wh kg^{-1} and the characteristics shown in Fig. 7.

Conclusion

In the development of the Ovonic NiMH battery, we have used aspects of physics, chemistry, metallurgy, and materials science. In particular, materials concepts (*10–13*) were focused on structural and compositional disorder to develop an NiMH battery with the characteristics necessary for practical EV use in the near, middle, and distant future.

REFERENCES AND NOTES

1. P. Jaret, *EPRI J.* **4**, 4 (1992).
2. S. R. Ovshinsky and M. A. Fetcenko, U.S. Patent 5 135 589 (1992); 23 additional patents.
3. M. A. Gutjahr, H. Buchner, K. D. Beccu, H. Saufferer, in *Power Sources 4*, D. H. Collins, Ed. (Oriel, Newcastle upon Tyne, United Kingdom, 1973), p. 79.
4. A. Percheron-Guegen, U.S. Patent 4 107 405 (1978).
5. M. H. J. van Rijswick, in *Proceedings of the International Symposium on Hydrides for Energy Storage* (Pergamon, Oxford, 1978), p. 261.
6. G. Halpert, in *Proceedings of the Symposium on Nickel Hydroxide Electrodes*, Electrochemical Society, Hollywood, FL, 16 to 18 October 1989 (Electrochemical Society, Pennington, NJ, 1990), pp. 3–17.
7. E. J. McHenry, *Electrochem. Technol.* **5**, 275 (1967).
8. T. A. Edison, U.S. Patent 1 402 751 (1922).
9. S. U. Falk and A. J. Salkind, *Alkaline Storage Batteries* (Wiley, New York, 1969).
10. S. R. Ovshinsky, *J. Non-Cryst. Solids* **32**, 17 (1979); for additional references, see S. R. Ovshinsky, *Disordered Materials: Science and Technology—Selected Papers*, D. Adler, B. B. Schwartz, M. Silver, Eds. (Institute for Amorphous Studies Series, Plenum, New York, ed. 2, 1991).
11. K. Sapru, B. Reichman, A. Reger, S. R. Ovshinsky, U.S. Patent 4 623 597 (1986).
12. M. A. Fetcenko, S. Venkatesan, K. C. Hong, B. Reichman, in *Proceedings of the 16th International Power Sources Symposium* (International Power Sources Committee, Surrey, United Kingdom, 1988), p. 411.
13. S. R. Ovshinsky, S. Venkatesan, M. Fetcenko, S. Dhar, in *Proceedings of the 24th International Symposium on Automotive Technology and Automation* (Automotive Automation, Croyden, United Kingdom, 1991), p. 29.
14. M. A. Fetcenko, S. Venkatesan, S. R. Ovshinsky, in *Proceedings of the Symposium on Hydrogen Storage Materials, Batteries, and Electrochemistry* (Electrochemical Society, Pennington, NJ, 1992), p. 141.
15. M. Tadokoro, K. Moriwaki, M. Nogami, T. Ise, N. Furakawa, in *ibid.*, p. 92.
16. L. Ciferri, *Autoweek* **1992**, 14 (7 September 1992).
17. W. H. DeLuca, paper presented at the 1991 Annual Automotive Technology Development Contractors Coordination Meeting, Dearborn, MI, 24 October 1991.
18. B. Otzinger, in *Proceedings of the 6th Annual Battery Conference on Applications and Advances* (California State University, Long Beach, 1991).
19. *General Motors Electric Vehicles Progress Report* (summer 1992).
20. C. R. Knoll, S. M. Tuominen, J. R. Peterson, T. R. McQueary, in *Proceedings of Battery Waste Management Seminar*, S. Wolsky, Ed. (Ansum Enterprises, Deerfield Beach, FL, 1990).
21. C. R. Knoll, S. M. Tuominen, R. E. Walsh, J. R. Peterson, in *Proceedings of the 4th International Seminar on Battery Waste Management*, S. Wolsky, Ed. (Ansum Enterprises, Deerfield Beach, FL, 1991).

22. We thank the research staff at Energy Conversion Devices–OBC, particularly S. Venkatesan, for their contributions to the developments described in this article and S. J. Hudgens for his critical comments and helpful suggestions during preparation of the manuscript.

Presented at the 185th Meeting of the Electrochemical Society
San Francisco, California, May 1994

The Electrochemical Society Proceedings Volume 94-21

DISORDERED MATERIALS IN CONSUMER AND ELECTRIC VEHICLE NICKEL METAL HYDRIDE BATTERIES

S.R. Ovshinsky, M.A. Fetcenko, S. Venkatesan, B. Chao
Ovonic Battery Company, Inc.
1707 Northwood, Troy, Michigan 48084

A proprietary nickel metal hydride battery has been developed by the Ovonic Battery Company which stores hydrogen in the solid hydride phase and has high energy density, high power, and a long cycle life. Nickel metal hydride batteries readily tolerate abuse, have a wide range of operating temperature, can be quick-charged, and are totally sealed and maintenance-free. A broad range of multi-element metal hydride materials that use structural and compositional disorder on several scales of length have been engineered for use as the negative electrode in this battery. The battery operates at ambient temperature, is made of nontoxic materials, and is completely recyclable. Metal hydride materials and electrodes are now in high volume production. Small consumer batteries are being produced that have 70-80 Wh/kg specific energy, over 240 Wh/l energy density, and excellent overall characteristics. Larger size batteries up to 52 kWh have been demonstrated in electric vehicles. Recent performance results are presented, as well as projections for next generation product performance.

Introduction

Nickel Metal Hydride (NiMH) batteries are now in commercial production for consumer applications and have been successfully demonstrated as 18-52 kWh packs in electric vehicles. Today, NiMH batteries are rapidly replacing nickel-cadmium in portable applications due to the increasing demand for higher energy density and environmentally acceptable materials. For electric vehicle applications, NiMH technology has emerged as the only viable alternative to lead-acid batteries. NiMH batteries have demonstrated:

- Energy density (80 Wh/kg, 240 Wh/l);
- Power (> 200 W/kg);
- Cycle life (1000 cycles at 100% DOD);
- Operating temperature (-30 to +70°C);
- Quick charge (60% in 15 minutes);
- Sealed, maintenance free operation;
- Tolerance to overcharge, overdischarge;
- Operation at ambient temperature with an aqueous electrolyte;
- Safe, recyclable materials; and
- Easy manufacture.

NiMH batteries were initially regarded as an "interim EV technology" until other more advanced systems could be developed. However, NiMH batteries are now recognized as potentially the best long term EV battery technology based on indications that their energy density can be increased to more than 120 Wh/kg, which will allow a vehicle range of 300-400 miles.

Ovonic Battery Company (OBC), a subsidiary of Energy Conversion Devices, Inc., received the first contract from The United States Advanced Battery Consortium (USABC) for the continued development and fabrication of the company's proprietary NiMH batteries. On March 9, 1994, General Motors and Ovonic Battery Company announced the formation of a partnership to manufacture and commercialize Ovonic NiMH batteries for electric vehicles.

In this paper, we will review some of the significant achievements at ECD/OBC in NiMH battery performance as well as discuss the fundamental material science advances that have brought about these achievements.

Cell Reactions

The NiMH battery has a nominal voltage of 1.2 V. Unlike the nickel-hydrogen battery that stores hydrogen as a high-pressure gas, the NiMH battery stores hydrogen in the solid hydride phase. The negative electrode of the NiMH battery consists of a hydrogen storage material capable of electrochemical storage and release of hydrogen during battery charge and discharge processes. The nickel hydroxide positive electrode is electrochemically reversible between $Ni(OH)_2$ and nickel oxyhydroxide (usually written as NiOOH). At both electrodes, oxidation-reduction reactions take place in an aqueous alkaline medium of 30% by weight KOH. During charge, the $Ni(OH)_2$ positive electrode is oxidized and the metal hydride alloy negative electrode is reduced. As a result, water is separated into hydrogen and hydroxyl ions. At the negative electrode the hydrogen reacts with the metal to form a metal hydride (MH). At the positive electrode, the hydroxyl ion reacts with the $Ni(OH)_2$ electrode to form NiOOH. This reaction results in a change in the Ni oxidation state from +2 to +3. Thus, the half-cell reactions on charge and discharge of the battery can be written as:

$$M + H_2O + e^- \underset{\text{discharge}}{\overset{\text{charge}}{\longleftrightarrow}} M\text{-}H + OH^- \qquad [1]$$

$$Ni(OH)_2 + OH^- \underset{\text{discharge}}{\overset{\text{charge}}{\longleftrightarrow}} NiOOH + H_2O + e^- \qquad [2]$$

As a consequence of reactions [1] and [2], there is no net change in electrolyte quantity or concentration during the charge-discharge cycle. This is in contrast to other alkaline electrolyte systems such as the NiCd system where water is generated at both electrodes during charge and consumed at both electrodes during discharge. Although transient electrolyte concentration gradients can occur in the NiMH battery, the constant average concentration of the electrolyte is an important factor in the NiMH battery's excellent overall performance in gas recombination, kinetics, and high and low temperature operation. In addition, the uniform electrolyte concentration is a factor in the NiMH battery's resistance to cycle life limitations produced by corrosion and swelling.

Metal Hydride Alloy Development

In this article, we describe the science and technology of the Ovonic NiMH battery, with emphasis on the materials science aspects of the MH electrode and its effect on battery performance (1). The MH electrode offers an important opportunity for materials engineering and optimization when compared with negative electrodes for other nickel based battery systems. In these other systems, the negative electrode, commonly Cd, Zn, or Fe, is typically fabricated from relatively pure elemental metals, and the oxidation-reduction reactions associated with battery charge and discharge convert the electrode back and forth between a metal and a metal oxide that is a poor conductor of electricity. This type of chemical reaction is undesirable in practical battery design because of accompanying changes in the physical properties of the electrode. Changes in the mechanical integrity and surface morphology of the electrode as a result of dissolution and recrystallization, and the reduced electrical conductivity of the electrode in the oxidized state are sources of many of the performance deficiencies in these systems.

The MH electrode, in contrast, uses a chemical reaction to reversibly incorporate hydrogen into a metal alloy. In this oxidation-reduction reaction both chemical states are metallic. This means that electrical conductivity is high in both the charged and discharged states. Furthermore, the small size of the hydrogen atom allows it to enter the metal lattice during formation of the hydride state (reduced) with only about 10% volumetric expansion and without the changes in crystallography associated with oxidation and reduction of the Cd, Zn, or Fe electrode, for example.

In effect, the MH negative electrode is a matrix for the chemical incorporation of a hydrogen atom. In the Ovonic NiMH battery, we specifically engineer this matrix through the use of multi-element alloys and compositional and structural disorder to produce negative electrode materials with the most desirable battery properties.

Despite decades of development by companies all over the world, the first commercial NiMH batteries did not appear until 1987. More recently, several manufacturers have started commercial production of NiMH batteries for use in portable consumer products like laptop computers, cellular telephones, and video camcorders.

The greatest initial obstacle to NiMH batteries was the fundamental material limitations of the early metal hydride alloys. These early alloys were based on the $LaNi_5$ and TiNi systems (2, 3). While these systems had only moderate hydrogen storage capacity, the real limitation was instability of the electrode material due to oxidation and corrosion. Such instability results, for example, in poor cycle life, high internal cell pressure, and low power.

By applying ECD's proprietary concepts of disordered materials, we learned that it is possible to design and engineer metal hydride alloys for optimized performance (4, 5). It is notable that multicomponent, multiphase alloys that use these concepts are now the standard in both the $LaNi_5$ and TiNi systems. This has allowed the development of disordered LaCePrNdNiCoMnAl alloys in the $LaNi_5$ family and disordered VTiZrNiCrCoMnAlFe alloys in the TiNi family. It is noteworthy that in both families specific elements may have widely varying percentages or even the elimination of certain elements in some cases.

Material Requirements

The MH materials used for an NiMH battery electrode must satisfy an extensive list of requirements. Above all, the amount of hydrogen that the MH material will absorb from the aqueous electrolyte determines the electrochemical storage capacity of the electrode and, consequently, the energy storage capacity of the battery. It is desirable to have high electrode storage capacity that is electrochemically reversible. To ensure reversibility, an important aspect of the MH design is the range of metal-to-hydrogen bond strengths, which must be about 6 to 12 kcal mol^{-1}. If the bond strength is too low, hydrogen will not react with the alloys and will be evolved as a gas. If the bond strength is too high, the MH electrode will be extensively oxidized and will not store hydrogen reversibly.

Even with an optimally adjusted metal-hydrogen bond strength, the problem of electrode oxidation in the MH battery remains. The NiMH battery operates in a strongly oxidizing medium composed of a high-concentration alkaline electrolyte. Because many chemical elements react to form oxides in an alkaline electrolyte, it follows that if these elements are used as electrodes, they will oxidize and fail to store hydrogen reversibly. In addition, MH electrodes are typically designed for use in totally sealed batteries where oxygen recombination occurs at their surfaces. In this aggressively oxidizing environment, oxidation and corrosion resistance of MH electrode materials is critical. Because some oxidation at the metal-electrolyte interface is inevitable and because both passivation and corrosion can have adverse effects on battery performance, these unwanted processes must be controlled in a practical NiMH electrode design (6).

Another consideration in the use of hydride materials in NiMH batteries relates to electrochemical kinetics and transport processes. The power output of the battery depends critically on these processes. During discharge, hydrogen stored in the bulk metal must be brought to the electrode surface by diffusion. The hydrogen must then react with hydroxyl ions at the metal-electrolyte interface. As a consequence, surface properties (such as oxide thickness, electrical conductivity, surface porosity and topology, surface area, and degree of catalytic activity) affect the rate at which energy can be stored in and removed from the NiMH battery.

For a battery to operate as a sealed system, it must also tolerate the consequences of chemical reactions that occur during cell overcharge and overdischarge. In a NiMH battery at overcharge, oxygen gas is generated at the $Ni(OH)_2$-positive electrode and must recombine with hydrogen at the MH electrode to form water. In overdischarge, which typically occurs when a low-capacity cell in a series-connected string is subjected to reverse polarity, hydrogen is generated at the $Ni(OH)_2$ electrode of a NiMH cell and must be recombined at the surface of the MH electrode to form water. In a sealed system, these gas recombination reactions must occur at sufficient rates to avoid pressure buildup. Adequate gas recombination rates require sufficient electrode area, a thin electrolyte film, and, for the hydrogen absorption process, catalytic activity at the MH electrode surface to promote rapid dissociation of hydrogen.

Disordered Materials

The required properties of metal hydride alloys for electrochemical use are diverse:

- high hydrogen storage capacity;
- good range of metal/hydrogen bond strength;
- oxidation/corrosion resistance;
- good gas recombination rates;
- fast kinetics and transport rates;
- easy manufacturability; and
- reasonable cost.

Multicomponent, multiphase materials allow the engineering of alloys having these required properties by introducing compositional and structural disorder on a local, intermediate and long range scale (7, 8). Local order, having dimensions of a few nearest neighbor atoms, substantially increases the number of chemically active sites by offering an increased variety of hydrogen bonding possibilities to increase capacity and improve catalysis. Intermediate range disorder plays a number of important roles at grain boundaries of the bulk alloy and at the electrode/electrolyte interface. Long range disorder provides increased hydrogen storage and improved kinetics. OBC proprietary disordered materials can be amorphous, microcrystalline, and/or polycrystalline multiphase materials.

The use of multicomponent, multiphase hydride electrode alloys allows one to consider a range of alloys for electrode materials containing elements that, if used alone, would be unacceptable for thermodynamic reasons. Among the elements that become available for alloy formation in disordered electrode materials are Li, C, Mg, Al, Si, Ca, Ti, V, Cr, Mn, Fe, Co, Ni, Cu, Y, Zr, Nb, Mo, Sn, La and other rare earth elements, W, and Re. Some of these elements increase the number of hydrogen atoms stored per metal atom (Mg, Ti, V, Zr, Nb, and La). Other elements allow the adjustment of the metal-hydrogen bond strength (V, Mn, and Zr); or provide catalytic properties to ensure sufficient charge and discharge reaction rates and gas recombination (Al, Mn, Co, Fe, and Ni). Still others impart desirable surface properties such as oxidation and corrosion resistance, improved porosity, and electronic and ionic conductivities (Cr, Mo, and W). The wide range of physical and chemical properties that can be tailored in these alloys allows the MH battery performance to be optimized.

Compositional and structural disorder on a long-range length scale were first used in bulk Ovonic MH alloys. This resulted in considerably higher hydrogen storage capacity and better kinetics than were possible in conventional MH alloy structures of the time which were compositionally ordered and crystalline. The processing of disordered alloys can be optimized to produce polycrystalline, compositionally multiphase material. Figure 1 shows a scanning electron micrograph of a representative bulk region of a typical V, Ti, Zr, Ni, and Cr Ovonic MH alloy. Electron backscattering imaging was used to produce visual contrast between regions of the alloy that have different elemental compositions. This material contains five major distinct compositional phases as determined by energy-dispersive x-ray analysis. In addition, it has been determined separately by x-ray diffraction that the alloy contains three crystal structure: body-centered-cubic (bcc), hexagonal, and C14 Laves crystal structures. From synthesis and characterization of separate, individual phases of a particular alloy, we conclude that the bcc structure can react to store large quantities of hydrogen (2.5% by weight) but lacks sufficient catalytic activity to be discharged at the required rates for battery applications. On the other hand, surrounding phases that may store less hydrogen but exhibit greater catalytic activity can effectively "channel" the hydrogen for rapid electrochemical discharge (12).

Intermediate-range structural and chemical disorder play a number of important roles primarily at interfaces within the bulk of the MH electrode and at the electrode-electrolyte interface. Formation of a disordered polycrystalline, compositionally disordered, multiphase bulk alloy gives rise to a high density of grain boundaries between compositional and structural phases. The intermediate-range disorder that occurs at the grain boundaries increases surface area, which can greatly increase the density of catalytic sites. At the electrode-electrolyte interface, disorder on a length scale of approximately 10 to 100 nm is created during electrode processing and activation as a result of the chemical properties that are traceable to the elemental constituents of the MH alloy. The high magnification transmission electron micrograph (TEM) of the electrode-electrolyte interface in Figure 2 shows the presence of disorder on all three length scales and are particularly useful for illustrating the intermediate-range structural and compositional disorder that occurs in the engineered oxide layer of an Ovonic MH electrode.

The basic Ovonic MH electrode typically contain elements such as V, Ti, Zr, Ni, and Cr, Co, Mn, Fe, Al, Mo, W, Si, Sn, and Zn. Although Ovonic alloys are a system with many characteristics (such as MH bond strength) that depend on interactions among the elemental constituents, some alloy properties are influenced by the chemistry of individual components. It should be noted that concentrations of specific elements can vary widely and that some alloy formulas completely eliminate use of particular elements.

The primary role of V, Ti, and Zr in the alloy is hydrogen storage. All three elements, rather than just the least expensive (Ti), are used in the alloy for several reasons. Ti and Zr form thick, dense, passive oxides in alkaline solutions, whereas V forms soluble oxides. These chemical characteristics are used in the preparation of the Ovonic MH electrode during the electrochemical activation step in which soluble oxides are intentionally corroded to produce intermediate-range structural disorder. This change gives rise to increased electrode surface area and microporosity and thereby increases charge acceptance. Zr also contributes the important property of controlled hydrogen embrittlement, which leads to high surface area and, hence, fast cell reactions. However, because excessive embrittlement can produce mechanical disintegration of the electrode (leading to poor electrical conductivity, high polarization, and low charge-recharge cycle life) one must control this property carefully in designing the electrode alloy.

Cr is used to limit the unrestrained corrosion of V and to control the alloy microstructure.

Ni serves several functions in the alloy. First, NiH has a weak bond strength in contrast to Ti, Zr, and V all of which have bond strengths with hydrogen that are too high for electrochemical applications. However, when Ni is combined with Ti, Zr, and V the resulting materials have nearly ideal bond strengths for the reversible, electrochemical storage of hydrogen. Second, Ni contributes to reversibility because Ni is a catalyst for the dissociation of H_2 and subsequent absorption of atomic H into the alloy. Third, Ni is resistant to oxidation. The combination of Ni with Zr, V, and Ti makes the alloy more resistant to oxidation and produces oxide films at the electrode-electrolyte interface that contain regions of metallic Ni. These regions help provide the necessary electrical conductivity and catalytic activity for the oxide film. The electrode-electrolyte interface in Ovonic alloys is characterized by a heterogeneous oxide region (Fig. 2) rather than a sharply defined homogeneous oxide film. This disordered oxide region is microporous and contains electrically conductive microcrystalline Ni grains and these Ni regions can catalyze the electrochemical discharge reaction (Figures 3, 4).

Co also serves several functions in the alloy, promoting improved capacity, cycle life, and gas recombination. From a metallurgical standpoint, Co inhibits the precipitation of vanadium rich phases during alloy solidification which are susceptible to corrosion in the electrochemical environment. In addition, Co influences the surface oxide, providing microporosity and catalytic activity which allows improved reaction rates during charge, discharge and overcharge. Controlled dissolution of cobalt also provides desirable poisoning resistance for the nickel hydroxide positive electrode, reducing premature oxygen evolution brought about by undesirable corrosion species.

The addition of Mn to metal hydride alloys promotes higher electrochemical capacity and improves overall performance in terms of characteristics such as high rate discharge and low temperature discharge. In the bulk material, Mn will extend the region of electrochemically useful hydrogen storage capacity by broadening the plateau of the PCT curve, a measure of a more uniform metal-hydrogen bond strength as a function of hydrogen content. Mn influences discharge rate capability in several ways. It has been observed through SEM analysis that alloys modified with Mn exhibit less distinct grain boundaries, Figure 5. In some cases, instead of increasing overall hydrogen storage, Mn increases the utilization of available hydrogen by increasing the rate of bulk hydrogen diffusion through the multiphase microstructure. The Mn modification alters the MH bond strength within particular phases, making hydrogen storage sites more accessible. Finally, the presence of Mn influences the quality of the surface oxide, providing increased oxidation and corrosion resistance as well as increased oxygen recombination.

Compositional disorder on the atomic scale is used to increase hydrogen storage capacity and improve catalytic activity through the incorporation into the MH alloy of elements that generate new chemically active sites. These sites offer an increased variety of hydrogen bonding possibilities and enhanced rates (as a result of increased catalysis). For VTiZrNiCrCoMn based alloys local disorder is illustrated at the metal-electrolyte interface in the high magnification TEM micrograph (Figure 6). Incorporation of elements with multidirectional *d* orbitals increases the range of stereochemical possibilities for bonding hydrogen. This is confirmed by an increase in the amount of hydrogen absorbed and by increased catalytic activity. These effects occur to a lesser extent with elements containing *f* orbitals. While *f* orbitals extend in still more directions than *d* orbitals, *f* orbitals are closer to the nucleus of the metal atom and, therefore, are less accessible.

Local compositional disorder is also used to adjust the metal-hydrogen bond strength in the MH alloy. Measurements of equilibrium hydrogen pressure versus MH hydrogen concentration at 30°C are shown in Fig. 7 for a series of multicomponent MH alloys in which the ratio of V to Zr is systematically varied. The equilibrium hydrogen pressure in these measurements is related to the change in Gibbs free energy, ΔG, which occurs for the reaction between gaseous hydrogen and the hydrogen storage alloy to form MH. This value can be written:

$$\Delta G = \Delta H - T\Delta S = RT \ln P \qquad [3]$$

Because the entropy term, $T\Delta S$, is small at room temperature compared with the enthalpy change, ΔH largely determines ΔG. Thus, determination of P provides a measure of ΔH, which is related to the metal-hydrogen bond strength. In Fig. 7, variations in the ratio of V to Zr give rise to the observed changes in P for a given hydrogen concentration in the MH alloy. This result indicates that these compositional variations have changed the metal-hydrogen bond strength.

Portable Battery Performance

NiMH batteries for portable consumer applications such as laptop computers, cellular telephones and other applications are now in commercial production in a variety of sizes, AA and Af being the most common. Compared to first generation NiMH AA cells that had a capacity of 1000 mAh, a specific energy of 56 Wh/kg and an energy density of 180 Wh/L, advanced NiMH AA cells now exhibit a 1200-1500 mAh storage capacity, a specific energy of 70-80 Wh/kg, and an energy density of 240-260 Wh/L. As shown in Figure 8 with a Goldpeak NiMH AA cell, these advanced high energy storage products provide excellent performance even up to a 3C continuous discharge rate. These same cells provide excellent performance under low temperature discharge. At -10°C, these cells provide 93% of room temperature capacity at C/5 discharge and about 50% of room temperature capacity under C rate discharge (9).

We have significantly improved the cycle life of Ovonic NiMH cells under $-\Delta V$ charge termination. As shown in Figure 9, Goldpeak Af cells now provide 500 cycles under C rate charge and discharge to 100% DOD. However, cycling performance is highly dependent on the charge termination method. Using the increasingly popular $-\Delta V$ chargers, cycle life could be dramatically reduced due to premature oxygen evolution and resultant premature heat generation. Product performance under these test conditions has been significantly improved in present Goldpeak production cells where 350-400 cycles is now typical.

As presented previously, we have significantly improved charge retention (10). We reduced self-discharge from an 85% loss to a 15% loss in 30 days at ambient temperature by optimizing the negative alloy, the separator, positive electrode composition, and various processing steps. The alloy, separator, and positive electrode had approximately equal relative contribution to these improvements. In general, the factors contributing to improved charge retention are the level of residual nitrates, the low corrosion of species detrimental to nickel hydroxide stability, the use of a nickel hydroxide active material that is resistant to poisoning and has a low equilibrium hydrogen pressure, and the use of a separator resistant to decomposition upon exposure to alkaline electrolyte and hydrogen gas. As measured by Texas A&M, by combining all aspects of cell design we have achieved a dramatic reduction in self discharge to 20% loss in one month.

Performance - EV Batteries

The excellent overall performance achieved in small cells has been realized in full size EV packs using 25-250 Ah cells (11, 12). A comparison of OBC NiMH battery performance to USABC primary midterm goals is presented in Table 1. EV batteries of 10-52 kWh have been successfully tested in actual road driving conditions. Other significant achievements of OBC's EV battery development program include:

- Specific energy and power density, scaled up from laboratory scale to actual electric vehicles in totally sealed cells;
- Full size EV batteries demonstrated and displayed in the United States, Japan, Korea, Hong Kong, and Singapore;
- An 18.8 kWh battery provided over 150 miles during on-road range test and over 170 miles on a dynamometer test;
- Demonstrated 15 minute recharge (to 60% capacity);
- Tolerance to regenerative braking;
- Excellent thermal characteristics in an operating EV; and
- Independent evaluation of cells, modules and packs.

In addition to battery energy density, the range of an EV will depend on many factors such as vehicle weight, tire rolling friction, air drag, and electric motor efficiency. Information published by GM on its pioneering *Impact* vehicle can be used as a benchmark for conversion of battery energy density to vehicle range (13). Utilizing a 16.7 kWh advanced lead-acid battery, GM forecasts an *Impact* range of 70-100 miles depending on driving conditions (city/highway). For a battery of the same weight, current Ovonic NiMH technology, since OBC NiMH can be cycled at 100% depth of discharge will provide, 80 Wh/kg or 39 kWh of energy storage, for a vehicle range of 200-292 miles.

Battery characteristics have a dominant influence on overall EV performance. For example, battery specific energy (Wh/kg) controls vehicle range. Similarly, battery power (W/kg) translates into vehicle acceleration. In a comparison of peak power versus depth of discharge for candidate EV battery technologies. At Argonne National Laboratory, high peak power (>150 W/kg), as required by the USABC goals shown in Table 1, was maintained over the entire depth of discharge of the battery (14).

Battery cycle life can be converted into EV battery-life driving range when the characteristics of the EV are specified. As stated earlier, in a GM *Impact*-type vehicle, replacing Pb-acid batteries with Ovonic NiMH batteries of the same weight and volume increases range to approximately 250 miles. At 80% depth of discharge (200 miles), even a conservative estimate of 500 cycles for the battery life will give a 100,000-mile battery life driving range. The electrical energy necessary to provide the 200 mile range per charge costs only $2.32 at $0.08 per kilowatt-hour, compared with approximately $14 worth of gasoline needed to provide the same range for a typical internal combustion engine powered (ICE) vehicle. Lifetime EV maintenance costs will also be smaller than for typical ICE-powered vehicles. Therefore, EVs that are powered by batteries with long cycle life can be economically competitive on a lifetime basis. Argonne National Laboratory has verified cycle life at over 500 cycles in ongoing tests of EV modules. At OBC, EV modules have now achieved over 700 cycles.

The environmental impact of the disposal of the Ovonic NiMH battery has also been studied (15). Knoll and colleagues concluded that, according to existing United States Environmental Protection Agency regulations, batteries that use this technology can be safely disposed of in landfills. It has also been shown (16) that with existing technology, Ovonic batteries can be recycled into metallurgical additives for cast iron, stainless steel, or new Ovonic NiMH battery electrodes. The commercial viability of each of these technologically feasible recycling programs will depend on process economics.

Advanced R&D

Despite today's achievement of 80 Wh/kg specific energy in sizes ranging from 1500 mAh AA cells to 250 Ah EV cells, NiMH technology is still in its infancy. We know that Ovonic NiMH can readily achieve a specific energy of 120 Wh/kg or greater based on material improvements that have already been observed in laboratory thin film materials. Ongoing development to scale these advances to bulk levels are underway. In addition, we are researching novel cell designs and configurations as well as overall systems engineering that includes separator, electrode construction, cases, restraint systems, and hardware.

The future development of Ovonic NiMH batteries includes the continued optimization of the MH materials and electrodes as well as improvements to the positive electrode and improved cell designs. For example, ongoing research at OBC is focused on applying the techniques of disorder to the development of an improved positive electrode with enhanced storage capacity through the use of engineered valence control. The chemical reaction that occurs during the charge of a conventional $Ni(OH)_2$ electrode involves the transfer of one electron per Ni atom. We are developing materials capable of exchanging up to two electrons per atom. In addition, MH alloys with twice the storage capacity of first generation materials have been measured in the laboratory, and cell designs in which lightweight substrates, current collection components, and containers are now being developed and evaluated. Because the overall energy density of the battery is determined by the entire system, the combined effects of these developments are targeted at the fabrication of batteries with both an energy storage density of 150 Wh/kg.

While the engineering approaches to achieve 150 Wh/kg have been demonstrated, we also are addressing advanced battery materials and configurations with the potential to achieve up to 500 Wh/kg specific energy. Such a tremendous specific energy would translate to a vehicle range of over 1200 miles, which is obviously more than necessary. Instead, this level of performance could be translated into a range of 300 miles and maintained with an extremely small, light, and inexpensive battery, allowing Electric Vehicle cost to rival that of a gasoline powered car.

Conclusions

The application of disordered multicomponent, multiphase materials has allowed commercial production of consumer NiMH batteries and the demonstration of full size Electric Vehicle batteries. Current NiMH batteries deliver excellent overall performance. Which has resulted in widespread acceptance that NiMH batteries are the best available for EVs. Opportunity for significant improvement has further strengthened the position of Ovonic NiMH as the best overall EV technology, both now and in the future.

References

1. S. R. Ovshinsky, M. A. Fetcenko, J. Ross, *Science*, Volume 260, April 9, 1993, pp. 176-181.

2. M. A. Gutjahr, H. Buchner, K. D. Beccu, H. Saufferer, in *Power Sources 4*, D. H. Collins, Ed. (Oriel, Newcastle upon Tyne, United Kingdom, 1973), p. 79.

3. J. J. G. Willems, (1984), *Thesis*, Technical University, Eindhoven.

4. S. R. Ovshinsky, K. Sapru, B. Reichman and A. Reger, U.S. Patent 4,623,597 (1986).

5. M. A. Fetcenko, S. Venkatesan, S. R. Ovshinsky, in *Proceedings of the Symposium on Hydrogen Storage Materials, Batteries, and Electrochemistry* (Electrochemical Society, Pennington, NJ, 1992), p. 141.

6. M. A. Fetcenko, S. Venkatesan, K. C. Hong, B. Reichman, in *Proceedings of the 16th International Power Sources Symposium* (International Power Sources Committee, Surrey, United Kingdom, 1988), p. 411.

7. S. R. Ovshinsky, *J. Non-Cryst. Solids* **32**, 17 (1979); for additional references, see S. R. Ovshinsky, *Disordered Materials: Science and Technology - Selected Papers*, D. Adler, B. B. Schwartz, M. Silver, Eds. (Institute for Amorphous Studies Series, Plenum, New York, ed. 2, 1991).

8. S. R. Ovshinsky, M. A. Fetcenko, U.S. Patent No. 5,096,667, No. 5,104,617, No. 5,135,589, No. 5,238,756, and No. 5,277,999.

9. S. R. Ovshinsky, S. K. Dhar, M. A. Fetcenko, S. Venkatesan, A. Holland, P. R. Gifford, D. A. Corrigan, in *Proccedings of the Eleventh International Seminar on Primary and Secondary Battery Technology and Application,* S. Wolsky, Ed. (Annsum Enterprises, Deerfield Beach, FL, 1994).

10. M. A. Fetcenko, S. Venkatesan, S. R. Ovshinsky, *Proceedings of The Electrochemical Society*, Electrochemical Society, Pennington, NJ, May 16-21, 1993.

11. S. R. Ovshinsky, S. Venkatesan, M. A. Fetcenko, S. K. Dhar, in *Proceedings of the 24th International Symposium on Automotive Technology and Automation* (Automotive Automation, Croyden, United Kingdom, 1991), p. 29.

12. S. R. Ovshinsky, S. K. Dhar, M. A. Fetcenko, in *Proceedings of Japan Society of Automotive Engineers,* Feb. 17, 1994.

13. *General Motors Electric Vehicles Progress Report* (Summer 1992).

14. W. H. DeLuca, paper presented at the 1991 Annual Automotive Technology Development Contractors Coordination Meeting, Dearborn, MI, 24 October 1991.

15. C. R. Knoll, S. M. Tuominen, J. R. Peterson, T. R. McQueary, in *Proceedings of Battery Waste Management Seminar*, S. Wolsky, Ed. (Ansum Enterprises, Deerfield Beach, FL, 1990).

16. C. R. Knoll, S. M. Tuominen, R. E. Walsh, J. R. Peterson, in *Proceedings of the 4th International Seminar on Battery Waste Management,* S. Wolsky, Ed. (Ansum Enterprises, Deerfield Beach, FL, 1991).

Figure 1. Scanning electron micrograph of a bulk region of an Ovonic MH $V_{22}Ti_{17}Zr_{16}Ni_{39}Cr_7$ battery alloy that shows disordered multiphase regions (X100).

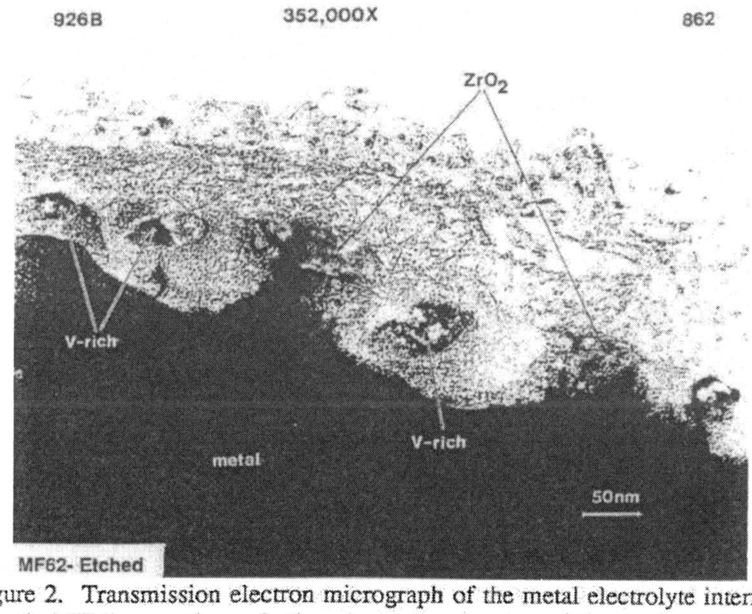

Figure 2. Transmission electron micrograph of the metal electrolyte interface of an Ovonic MH battery electrode that shows the structure of the engineered multiphase bulk alloy and surface oxide (X352,000).

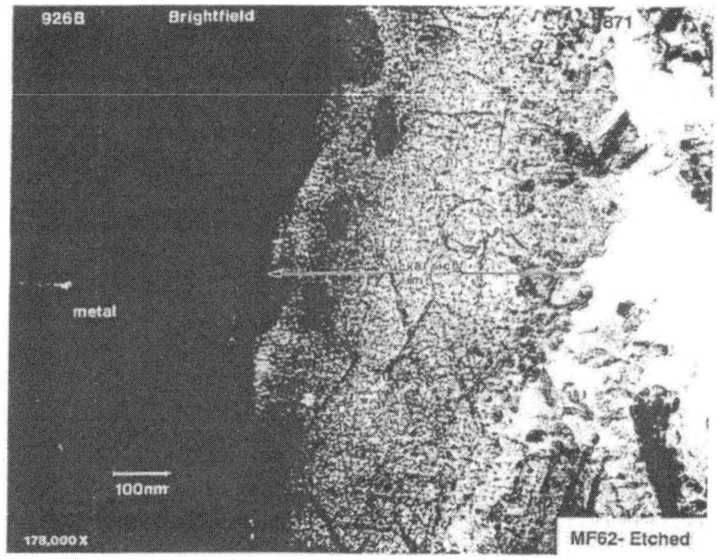

Figure 3. Brightfield transmission electron micrograph of the metal/electrolyte interface of an Ovonic MH battery electrode that shows physical strucutre of the surface oxide.

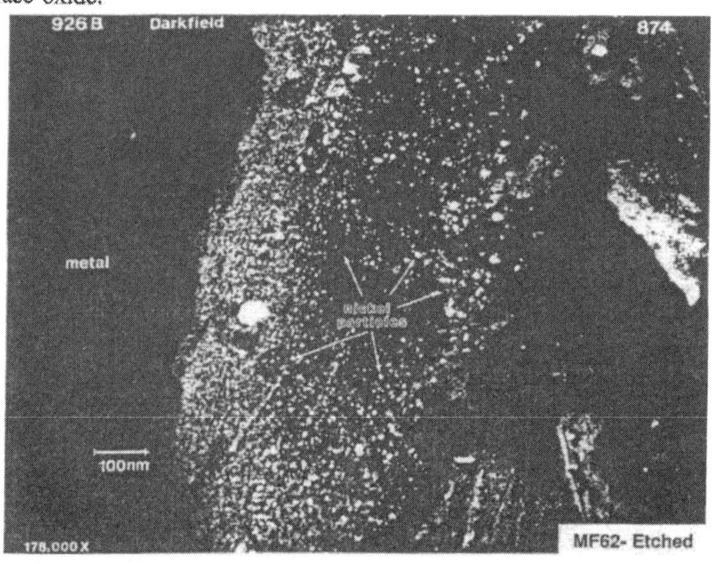

Figure 4. Darkfield transmission electron micrograph of the metal/electrolyte interface of an Ovonic MH battery electrode that shows distribution of metallic nickel regions within the surface oxide.

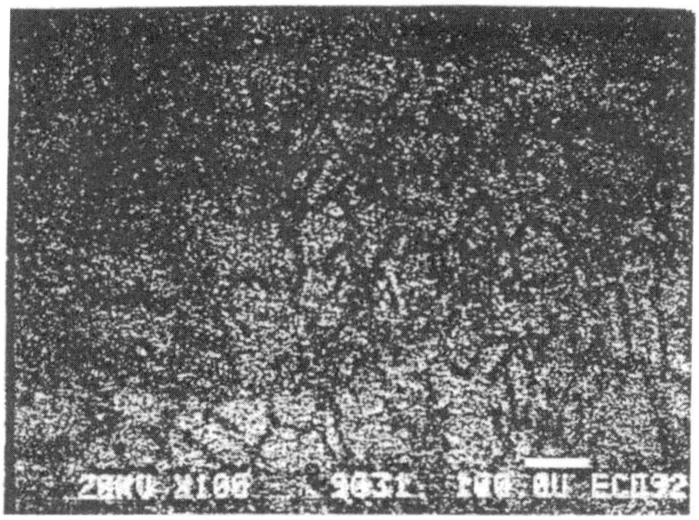

Figure 5. Scanning electron micrograph of a bulk region of an Ovonic MH $V_{18}Ti_{15}Zr_{18}Ni_{29}Cr_5Co_7Mn_8$ battery electrode that shows compositionally and structurally disordered multiphase alloy regions and less distinct phase boundaries (X100).

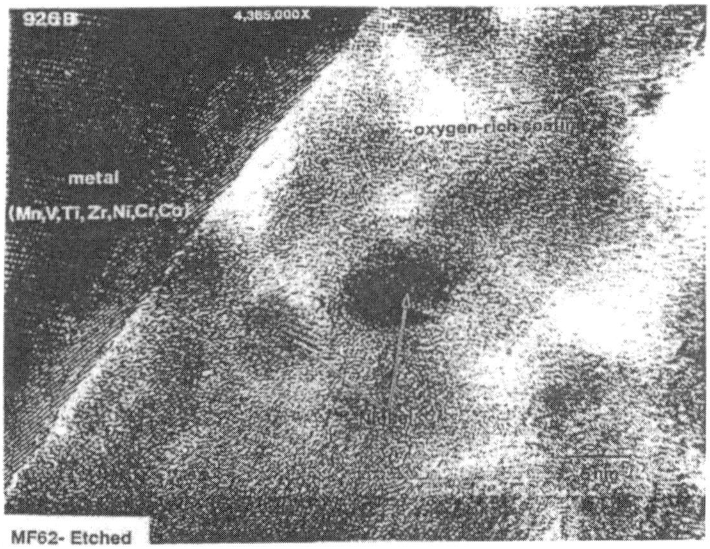

Figure 6. High magnification scanning transmission electron micrograph of the metal electrolyte interface that shows lattice imaging and heterogeneous nature of the bulk alloy and surface oxide.

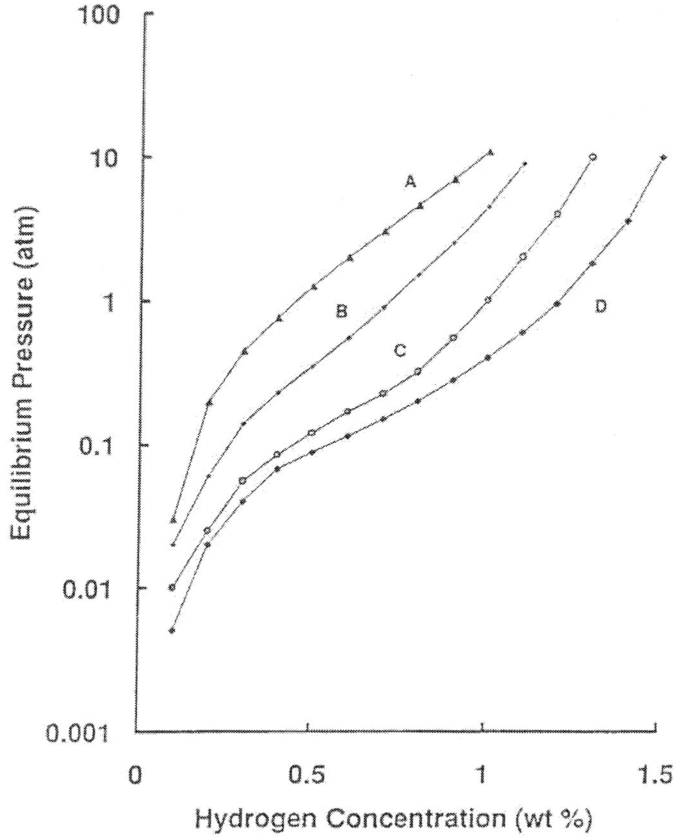

Figure 7. Equilibrium hydrogen pressure versus hydrogen concentration at 30°C for a series of Ovonic MH electrode alloys. Data show how variation in alloy composition may be used to control metal-hydrogen bond strength. The MH alloy compositions shown here, expressed as atomic percent, are (**A**) ($V_{21}Ti_{15}Zr_{15}Ni_{31}Cr_6Co_6Fe_6$), (**B**) ($V_{15}Ti_{15}Zr_{21}Ni_{31}Cr_6Co_6Fe_6$), (**C**) ($V_{18}Ti_{15}Zr_{18}Ni_{31}Cr_6Co_6Fe_6$), and (**D**) ($V_{15}Ti_{15}Zr_{20}Ni_{28}Cr_5Co_5Fe_6Mn_6$).

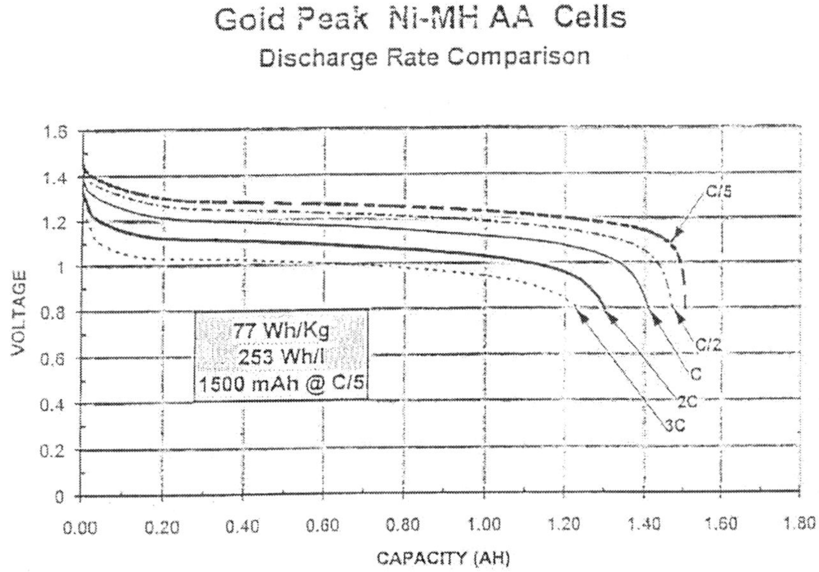

Figure 8. Capacity and rate capability of Goldpeak NiMH AA cells.

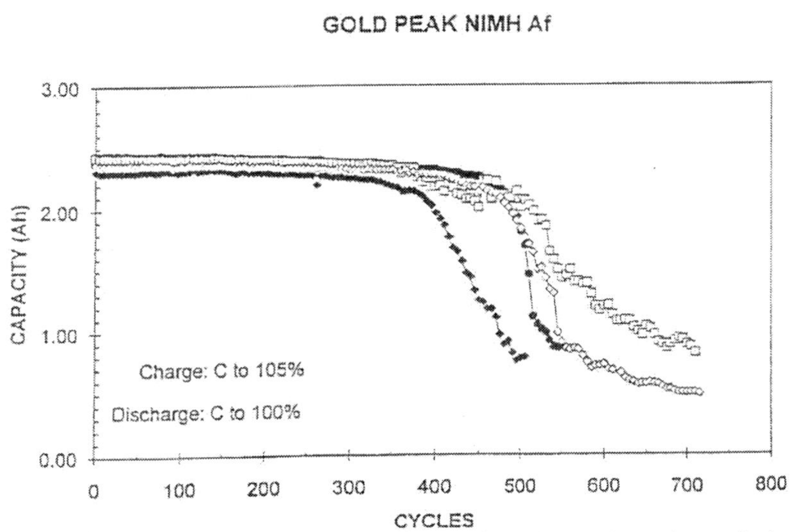

Figure 9. Cycle life of Goldpeak 4/3 Af NiMH cells under one hour charge-discharge to 100% depth of discharge.

USABC PRIMARY CRITERIA	USABC Mid-Term Goals	OBC Current
Specific Energy Wh/kg (C/3 discharge rate)	80 (*100 desired)	80
Energy Density Wh/l (C/3 discharge rate)	135	210
Specific Power W/kg (80% DOD/30 sec)	150 (*200 desired)	175
Power Density W/l	250	475
Life (years)	5	10
Cycle Life (cycles) (80% DOD)	600	1000
Power & Capacity Degredation (% of rated spec)	20%	20%
Ultimate Price ($/kWh) (10,000 units @ 40 kWh) (40,000 units @ 40 kWh)	< $150	$230 - 250 < $200
Operating Environment	-30 to 65°C	-20 to 60°C
Recharge Time	< 6 hours	1 hour
Continuous Discharge in one hour (no failure)	75 % of rated energy capacity	90 % of rated energy capacity

Table 1. Comparison of USABC - midterm primary goals to Ovonic NiMH performance.

ELSEVIER

Journal of Power Sources 65 (1997) 1–7

JOURNAL OF POWER SOURCES

1

Nickel/metal hydride technology for consumer and electric vehicle batteries — a review and up-date

S.K. Dhar, S.R. Ovshinsky, P.R. Gifford *, D.A. Corrigan, M.A. Fetcenko, S. Venkatesan

Ovonic Battery Company, 1707 Northwood Drive, Troy, MI 48084, USA

Received 4 November 1996; accepted 13 November 1996

Abstract

Nickel/metal hydride batteries today represent the fastest growing market segment for rechargeable batteries due to the high energy density and more environmentally acceptable chemistry offered by this technology. The high energy density of nickel/metal hydride batteries coupled with high power density and long cycle life make this battery chemistry a key enabling technology for practical electric vehicles, including cars, vans, trucks, and other forms of transportation such as scooters, bicycles, and three-wheelers. This paper provides a review of Ovonic technology and up-dates recent developments in materials and cell development for both consumer electronic and EV applications, and highlights areas for future development.

Keywords: Metal hydride anode batteries

1. Introduction

The worldwide market for rechargeable batteries for consumer electronic applications is growing at a record pace due to increased consumer demand for portable devices such as cellular phones, lap-top computers, camcorders, and other personal electronic devices. Nickel/metal hydride batteries today represent the fastest growing segment of this rechargeable battery market for consumer electronics due to their higher energy density and more environmentally acceptable chemistry relative to Ni/Cd. Current estimates indicate that the total portable rechargeable battery market is expected to reach $2 billion in 1996, growing to over $5 billion by 2000 [1].

An exciting new battery market is emerging due to the growing interest and demand for emission-free vehicles, which today can only be achieved through the use of battery power. The limitations of current electric vehicle (EV) battery technologies have been well documented in numerous articles decrying the lack of vehicle range offered by today's EVs. The high energy density, excellent power density, and long cycle life of nickel/metal hydride batteries make this the leading technology of choice for the battery power source for EVs. Several leading automotive companies, including Honda, Toyota, Nissan, General Motors, and Hyundai, have recently announced that they will be offering nickel/metal hydride batteries on EVs to provide long range, high power, and long battery life.

Ovonic Battery Company is the recognized world leader in the development of nickel/metal hydride battery technology, and today over 95% of the worldwide major manufacturers of nickel/metal hydride cells for consumer applications are producing under license to Ovonic. Ovonic continues to develop advanced negative electrode and positive electrode materials for continued product improvements. Additionally, Ovonic is actively involved in the development of cells and batteries for EV applications. Through its joint venture company, GM Ovonic, we are actively entering into the commercialization of this battery technology for EVs. Ovonic is also exploring other applications for nickel/metal hydride technology including transportation applications such as hybrid vehicles, bicycles, scooters, etc., as well as remote and stand-by power applications.

Ovonic nickel/metal hydride batteries today demonstrate specific energies over 80 Wh kg^{-1} and energy densities of over 200 Wh dm^{-3} in commercial cells and up to 95 Wh kg^{-1} in advanced cell designs. Continued advances in electrode materials and cell designs are projected to further improve product performance for a wide range of battery applications.

This paper describes the basic chemistry of the Ovonic nickel/metal hydride technology and reviews cell perform-

* Corresponding author.

ance characteristics of these batteries, both for portable electronic and EV applications. EV battery performance at the vehicle level is described for some of the many conversion vehicles and 'ground up' EVs that have employed Ovonic nickel/metal hydride batteries. Current R&D efforts are described as well as areas for future development and the projected impact on battery performance.

2. Nickel/metal hydride cell chemistry

The basic cell reaction for nickel/metal hydride can be written as:

$$MH + NiOOH = M + Ni(OH)_2$$

where M represents an intermetallic alloy capable of forming a metal hydride phase. It is interesting to note that the overall cell reaction consists of transfer of a hydrogen ion from one electrode to the other, in much the same manner as a lithium-ion battery functions through two insertion electrodes for the Li^+ ion. Therefore, one could describe nickel/metal hydride batteries as a 'hydrogen ion' battery, or a 'protonic' battery [2,3]. It is the simplicity of the total cell reaction that provides for the fast kinetics and long cycle life demonstrated by nickel/metal hydride batteries.

An important feature of the nickel/metal hydride chemistry is the cell's ability to tolerate both overcharge and overdischarge through gas recombination reactions that result in no net change in battery electrolyte and prevent a build up of pressure inside the sealed and totally maintenance free cell. This ability to tolerate both overcharge and overdischarge is particularly advantageous for EV applications where system voltages over 300 V are common. Under these conditions, with over 200 cells series connected, individual cells will be subjected to varying degrees of overcharge during charging and individual cells can experience overdischarge and cell reversal, particularly during acceleration or at very low states of battery pack charge. The ability of nickel/metal hydride to accept overcharge and overdischarge eliminates the need for single cell voltage monitoring, simplifying battery management in comparison to other high energy systems such as NaS, lithium-ion, or Li-polymer electrolyte.

3. Metal hydride alloy development

Early development work on hydride alloys for rechargeable batteries focused on traditional hydride materials of well-defined composition and crystal structure. The most widely studied of these materials were alloys of the $CaCu_5$ family, most notably $LaNi_5$ [4]. While early prototype batteries using these hydride materials in negative electrodes displayed high specific energy when coupled with conventional nickel electrodes, these cells suffered in other performance areas such as cycle life, internal cell pressure, and corrosion of the hydride alloy. These performance limitations were largely due to the simple single phase nature of these early hydrides.

Ovshinsky and his team of material scientists at Energy Conversion Devices employed a fundamentally different approach to developing hydride materials specific to a rechargeable battery application [5]. The material requirements for an electrochemical application were defined and engineered alloys specifically developed to meet these demanding and diverse materials requirements. Among the required properties for a negative electrode hydride material are:

(a) high hydrogen storage capacity
(b) proper metal to hydrogen bond strength
(c) oxidation and corrosion resistance
(d) fast gas recombination kinetics
(e) manufacturable at low cost

The engineering of multicomponent, multiphase materials allows for alloys that satisfy these required properties by introducing compositional and structural disorder [6]. By controlling the alloy composition one can control the metal-to-hydrogen bond strength to the desired value. Control of alloy microstructure provides for increased hydrogen storage sites as well as improved kinetics and corrosion resistance. Ovonic and Ovonic licensees have successfully employed this concept of compositionally complex, multiphase materials to develop alloys that in today's commercial nickel/metal hydride batteries provide high energy, high power, and long cycle life. All commercial batteries today incorporate alloys based on the concepts developed at Energy Conversion Devices (our parent company) and Ovonic of multielement composition and controlled microstructure. Today for example, those alloys derived from the $LaNi_5$ family are in actuality complex materials containing generally 6 to 8 elements with complex phase structures. Similarly, the V–Ti–Zr–Ni alloys commonly employed at Ovonic are equally complex and contain local disorder and multiple phases [7].

Ovonic has continued to focus predominately on alloys of the V–Ti–Zr–Ni type due to their intrinsic ability to store very high amounts of hydrogen. While typical rare earth based materials store <300 mAh g^{-1} specific capacity, commercial Ovonic alloys today store up to 400 mAh g^{-1}. Moreover, unlike simple crystalline materials, there is no readily definable theoretical limit for hydrogen storage in these complex materials and Ovonic is actively developing materials that store 550 to 700 mAh g^{-1} hydrogen [2,8]. The specific role of each component in these intermetallic compounds is well understood and has been described in previous publications [6,9].

Ovonic currently operates a production facility capable of producing hydride alloys and negative electrode 'belt' material for both consumer battery and electric vehicle battery applications. Rolls of compacted electrode strip suitable for cutting or profiling into electrodes are produced in a low-cost, continuous, roll-to-roll process. Upon completion of our most recent capital expansion, Ovonic will have the capability of producing over 800 000 linear feet of electrode material per month to meet the ever-growing demand for hydride battery electrodes.

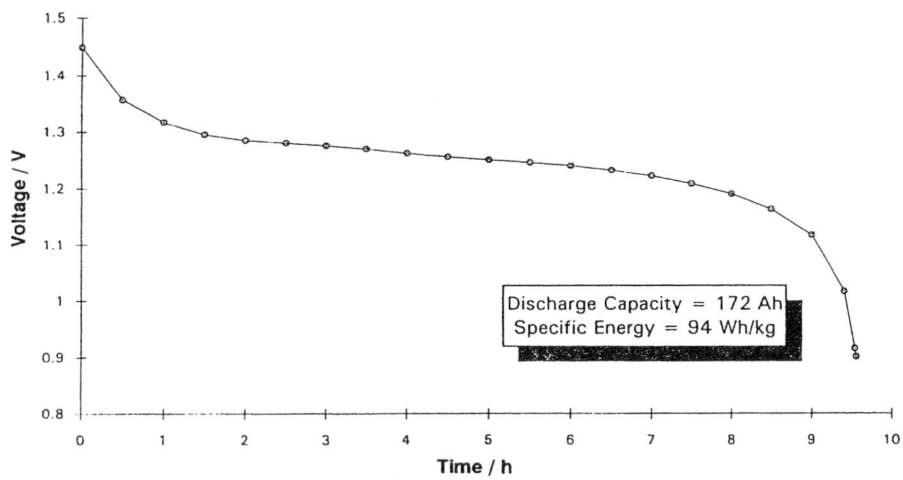

Fig. 1. Discharge curve for advanced prototype EV cell demonstrating a specific energy of approximately 95 Wh kg^{-1}.

4. Positive electrode development

During the early stages of product development, Ovonic developed a high capacity sintered, chemically impregnated ('CI') $Ni(OH)_2$ electrode that surpassed the Ni battery industry standard at that time. Through careful control of sintering parameters and development of optimized impregnation conditions, Ovonic was able to produce CI electrodes with specific energies of 150 mAh g^{-1} total electrode weight (>550 mAh cm^{-3}).

However, even with these high energies, we recognized that the high specific energies required for today's portable products and, in particular, for EV applications, could not be achieved with the decades old CI electrode technology. To solve this problem, Ovonic undertook development of newer, mechanically impregnated ('MI') electrodes using the new high porosity, low weight Ni foam and felt structures. These MI electrodes have the benefit of higher specific energy and energy density while maintaining high power capability and long cycle life. MI electrodes also utilize a simpler manufacturing process requiring less capital investment than sintered, CI electrodes, providing for opportunities in cost reduction of Ni electrodes, particularly for the high volume manufacturing required for EV batteries.

By proper selection of active material composition together with proper types and levels of active material additives, one can achieve high utilizations of $Ni(OH)_2$ together with high electrode loadings. For good loading and utilization, a high density, spherical $Ni(OH)_2$ active material is used which also incorporates other transition metal ions as co-precipitates. Other electrode additives such as cobalt compounds are typically added to improve active material utilization and cycle life [10]. In addition to optimizing the electrode slurry composition, Ovonic has developed a mechanical impregnation manufacturing process that provides for high electrode loading. This process, together with active material utilizations of 90%, yields positive electrodes with specific capacities of up to 200 mAh g^{-1} based on finished electrode weight.

While this level of energy represents a significant advance over the CI electrode used in previous cell technology, Ovonic is continuing to pursue further advances in positive electrode cost and energy. In particular, while the Ni electrode charge/discharge reaction is commonly written as a simple one-electron transfer reaction, it is well known that this reaction is much more complex [11]. In fact, up to 1.6 electron transfers are theoretically possible for this reaction. Energy Conversion Devices and Ovonic researchers have been working on advanced $Ni(OH)_2$ materials incorporating the same principles of compositional and structural disorder as those demonstrated so successfully in hydride materials. These proprietary materials have demonstrated up to 1.5 electron transfers in thin-film electrodes and up to 1.3 electrons per Ni atom in battery electrodes containing bulk spherical powders produced through a proprietary process [12,13]. Scale-up of these materials and their incorporation into full cell designs will result in energy densities of 95 Wh kg^{-1} and greater in commercial consumer and EV batteries, as has already been demonstrated in prototype EV cells employing these advanced positive active materials (Fig. 1).

5. Cell development — portable batteries

Ovonic has provided regular up-dates on the progress of nickel/metal hydride technology for wound cells developed for portable consumer electronic applications [2,14–16]. The history of these cell development activities has clearly been one of continuous improvement. Ovonic nickel/metal hydride batteries today meet the strictest performance requirements for the '3C' applications of communication, computing and camcorder. Recent advances in cell performance include higher capacity designs, longer cycle life, improved operating temperature range and lower self-discharge rates [16].

Early Ovonic batteries exhibited specific energies in the 55 to 60 Wh kg^{-1} range, corresponding to an energy density

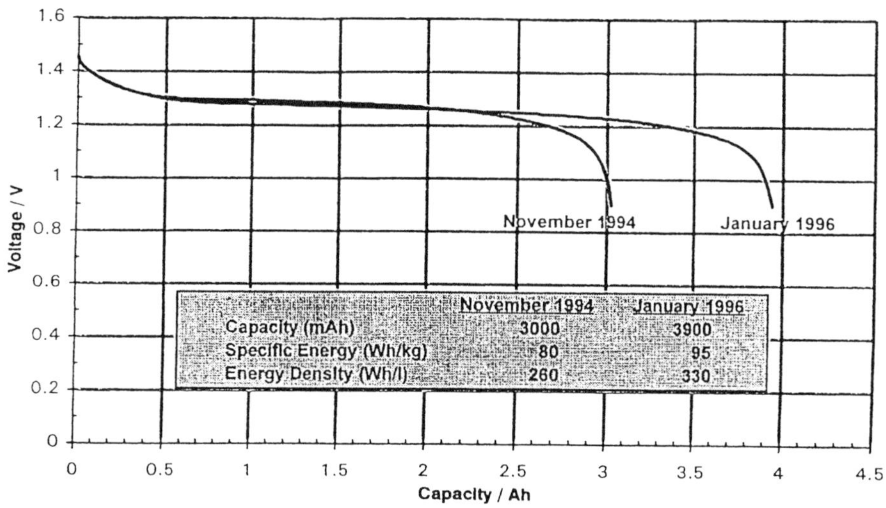

Fig. 2. Discharge curves for commercial and advanced nickel/metal hydride cells, 17.5 mm diameter × 67.5 mm long (type 7/5 Af, as produced by Gold Peak).

of 180 Wh dm^{-3}. While this represented a significant improvement over other rechargeable batteries, continued product development and advances in battery materials have allowed for continuous increases in cell capacity. One Ovonic licensee, Gold Peak, first announced achievement of 80 Wh kg^{-1} in 1994 [15]. Since then GP has continued to improve product design and manufacturing, together with advanced Ovonic materials, resulting in achievement of 95 Wh kg^{-1} in prototype 7/5 Af cells (17.5 mm diameter × 67.5 mm long; weight 46 g) having a rated capacity of 3900 mAh (Fig. 2) [16,17]. Perhaps more importantly for portable devices is battery size rather than weight. Energy density for these prototype 7/5 Af cells is an impressive 330 Wh dm^{-3}, greater than existing lithium-ion batteries [18].

In addition to the traditional OEM products requiring rechargeables, improvements in cell capacity and charger technology make nickel/metal hydride batteries an attractive option to consumers to replace primary alkaline batteries. Traditionally consumers have shied away from rechargeables due to unacceptable device run times and long charging cycles. The recent market introduction of 're-usable' alkaline manganese batteries indicates that the consumer is willing to consider rechargeables if he can maintain the same or nearly identical device performance. However, these batteries exhibit poor cycle life, offering a limited incentive to the consumer. High capacity Ovonic nickel/metal hydride batteries now offer the same, or greater, run-time than a primary battery, can be recharged in 1–3 h, and provide many hundreds of cycles. Due to the bobbin construction of alkaline cells, rate capability for these primary batteries is quite poor. This is true even for the re-useable alkaline manganese technology. The wound construction and low impedance of nickel/metal hydride batteries provides high capacity over a wide range of currents, exceeding primary battery capacity even at relatively moderate drain rates (Fig. 3). The replacement of literally hundreds of primary batteries with a single nickel/metal hydride cell offers a strong economic incentive to the properly educated consumer. Additionally, the environmental impact of disposing of billions of primary batteries annually is significant and can be alleviated by the use of nickel/metal hydride rechargeables.

6. Cell and battery development — electric vehicles

It has long been acknowledged that the key enabling technology for practical EVs has been the development of advanced batteries that would provide the requisite range and performance. While considerable sums of money were spent in the 1970s and 1980s to develop such technologies, EV developers continued to have little choice of battery beyond those technologies that had been employed in the earliest EVs. While advances have certainly been achieved in lead–acid and Ni/Cd batteries, these technologies continue to fall short of providing the desired performance level for a market-driven electric vehicle.

In recognition of this fundamental limitation in battery technology, the Big Three automotive companies, together with the US Department of Energy, with cooperation from the Energy Power Research Institute (EPRI) established the United States Advanced Battery Consortium (USABC) for the express goal of accelerating development of batteries for EVs for the late 1990s time frame and beyond. USABC established a set of performance criteria for mid-term and long-term battery technologies and solicited proposals for the development of technologies capable of achieving these goals.

USABC awarded its first battery development contract in May of 1992 to Ovonic Battery Company for a three phase programme to scale up nickel/metal hydride technology for EV applications and to continue development of the technology to meet USABC 'mid-term' criteria. Within weeks of

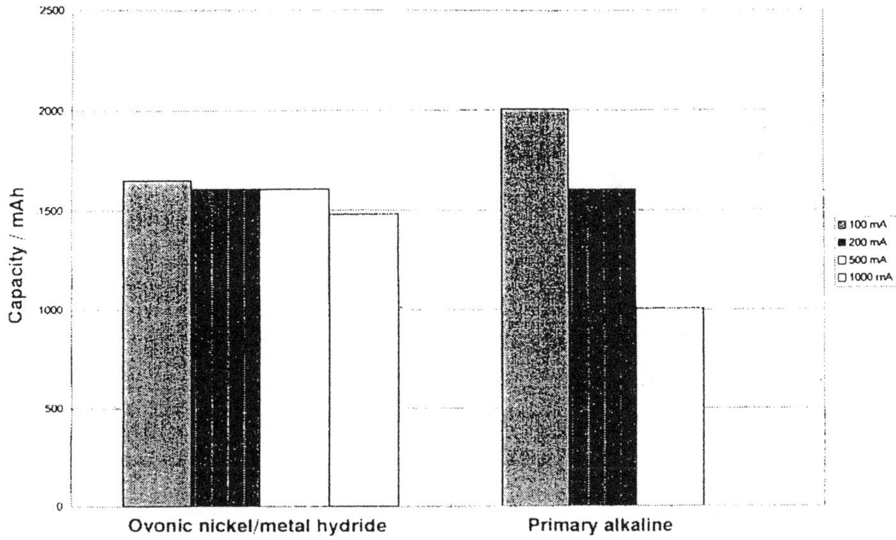

Fig. 3. Comparison of discharge capacity vs. rate for primary alkaline and Ovonic nickel/metal hydride 'AA'-size cells.

receiving this contract, Ovonic delivered its first prismatic cells to USABC, demonstrating successful scale-up from small wound cells to EV design cells with no loss in energy density or performance. Continued development led to increases in energy density, specific energy, power, life, and charge retention [19,20].

Based on the early success of this program, USABC placed an order for prototype batteries in April 1993, almost two years ahead of the original program plan. The first vehicle demonstration of a nickel/metal hydride battery was achieved in August 1993 when a Chrysler TEVan soundlessly rolled down the streets of Troy, MI. Even this early prototype battery provided 28 kW h energy with over 60 Wh kg^{-1} demonstrated at the vehicle pack level.

Since this early vehicle demonstration, Ovonic has continued to improve its technology and battery design expertise, achieving improvements in vehicle range and power [21,22]. Some of the more notable achievements include:

- March 1994 — Ovonic-powered Solectria Force wins APS 500 electric stock car race, with an average speed of greater than 65 m.p.h. (104 k.p.h.) over the 125 mile (200 km) event;
- May 1994 — Ovonic-powered Solectria wins Tour de Sol road rally with a record 214 mile range on a single charge;
- May 1995 — Ovonic-powered Solectria Sunrise wins Tour de Sol road rally with a record 238 mile range on a single charge;
- May 1996 — Ovonic-powered Solectria Sunrise wins Tour de Sol road rally with an impressive 373 mile range on a single charge with a 33 kW h battery pack employing the latest Ovonic technology.

Today Ovonic, together with its EV battery business partner GM Ovonic, has successfully completed dozens of vehicle conversions of all types, including conversion sedans, ground-up EVs, vans, and light trucks.

Hyundai Motor Company, a licensee of Ovonic technology, has established a prototype battery manufacturing facility and has designed and built nickel/metal hydride batteries for a number of conversion vehicles, including the Hyundai Sonata, Grace van, Elantra, and Accent vehicles [22]. Performance of these vehicles has far exceeded that obtained with lead–acid or Ni/Cd batteries. A converted Accent vehicle equipped with nickel/metal hydride batteries, for example, is capable of travelling 390 km (242 miles) on a single battery charge.

An Ovonic nickel/metal hydride battery pack was first tested in the state-of-the-art GM Impact vehicle in January 1994. With this first battery pack design, a range of over 200 miles was demonstrated, with acceleration from 0 to 60 m.p.h. in under 8 s [23]. GM has since announced that it will be producing for sale in 1996 the EV1, a commercial vehicle based on the Impact concept car. GM has announced that this vehicle, while initially equipped with VRLA lead–acid batteries, will be introduced with nickel/metal hydride batteries within two years [24].

These impressive vehicle demonstrations are possible because of the high level of performance that has been achieved with the Ovonic battery. The current status of battery performance relative to the USABC mid-term criteria is shown in Fig. 4. The specifications for the initial GM Ovonic production design nickel/metal hydride EV battery module are given in Table 1.

Advanced battery designs presently being provided to customers for evaluation exceed the current specifications for energy and power, demonstrating 80 Wh kg^{-1} and over 220 W kg^{-1} at 80% depth of discharge. Ovonic batteries have been tested over a wide range of ambient temperatures, from -20°C to over 50°C, including testing under actual vehicle conditions. By proper thermal management of the battery pack using forced air cooling, battery performance is maintained over this temperature range (Fig. 5).

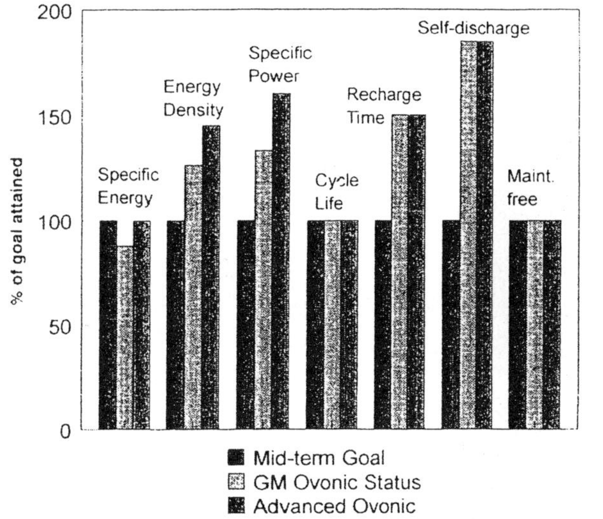

Fig. 4. Ovonic EV battery performance vs. USABC mid-term technical goals.

Table 1
Product specification for nickel/metal hydride EV battery module

Product code	Gen 1 design, model 13-EV-90
No. of cells	eleven
Nominal voltage	13.4
Capacity	90 Ah (at $C/3$ rate)
Energy	1.2 kWh
Specific energy	70 Wh/kg
Energy density	170 Wh/dm^3
Specific power	200 Wh/kg (30 s at 80% DOD)
Power density	485 Wh/dm^3
Charge retention	93% at 26°C (80°F) after 48 h
	85% at 38°C (100°F) after 48 h
Operating temperature	<43°C (110°F) for maximum life
	<65°C (150°F) to avoid damage
Cycle life	600 D.S.T. cycles to 80% DOD
Other attributes	maintenance free
	no spillable liquids
	recyclable

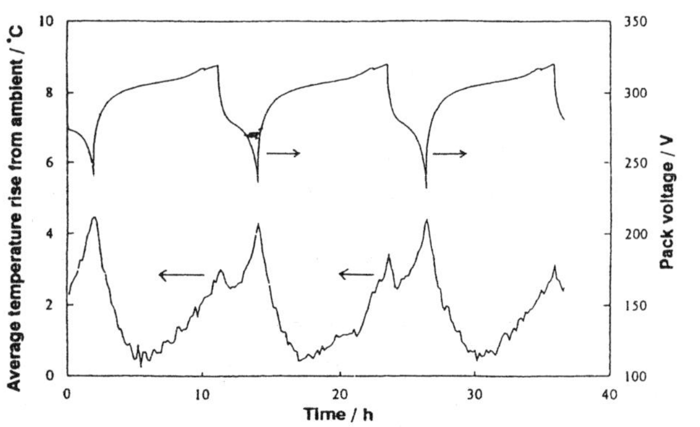

Fig. 5. Thermal performance of air-cooled Ovonic nickel/metal hydride EV battery pack vs. pack voltage.

We have previously described the recombination chemistry that allows for overcharge and overdischarge of individual cells. Additionally, these batteries are resistant to physical abuse. Batteries have been subjected to a wide range of 'abuse' tests including short-circuit, prolonged high-rate overcharge and overdischarge, puncture, crush, drop, etc. In all instances, no safety issues were encountered with the battery module or cell beyond those normally encountered in traditional sealed cells such as Ni/Cd or VRLA.

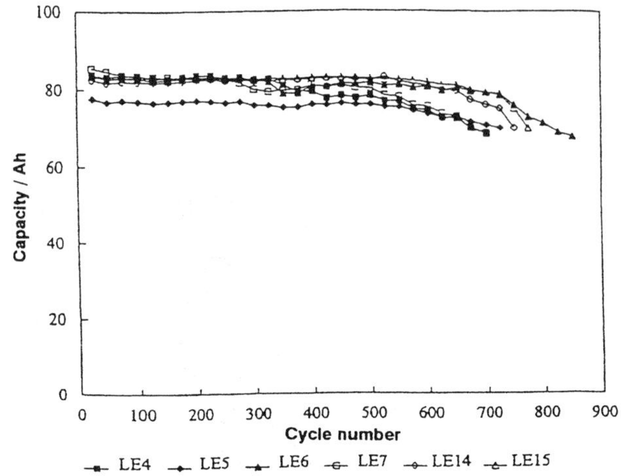

Fig. 6. Cycle life performance of Ovonic nickel/metal hydride EV cells, discharged at $C/3$ rate to 80% depth-of-discharge.

A key advantage of Ovonic nickel/metal hydride batteries is long cycle life (Fig. 6). EV cells and modules cycled to 80% DOD under the demanding 'DST' (Dynamic Stress Test) simulated driving profile have achieved over 600 cycles. For an average vehicle range of 200 miles, this would correspond to 96 000 miles of driving. Battery cycle life improves at lower states of charge, so that under actual driving conditions the driver could expect even longer battery life.

Other battery performance goals include the ability for rapid recharge and charge retention. Ovonic modules have been subjected to rapid recharging and meet the USABC goals of recharge from a 40% SOC to 80% SOC in 15 min, with a charge acceptance >99%. Charge retention for current modules of >90% for 48 h exceeds the USABC goal of 85% charge retention.

Perhaps equally important as battery performance is that these batteries are manufacturable using already established manufacturing processes. Energy Conversion Devices and General Motors have joined together in a joint venture, GM Ovonic, to commercialize and manufacture these batteries for EV customers [25]. Prototype batteries are currently being constructed at our Troy, MI, sample build facility for customer vehicle applications. GM Ovonic is validating the manufacturing processes and implementing the quality systems needed for high reliability at this facility. Particular focus today is on reducing the material, labor, and processing costs for this battery. Advanced cell designs optimized for manufacturability, together with advances in battery active materials, are projected to lead to significant cost savings as production volumes increase. GM Ovonic is projecting that in commercial scale production, batteries can be manufac tured for $200 to $250 per kW h. Further advances in battery technology are currently under evaluation at Ovonic that will result in higher specific energy and reduced costs to meet the aggressive cost targets required for a market driven EV.

7. Summary and future work

Today, Ovonic nickel/metal hydride batteries for consumer electronic applications are in high volume commercial production around the world by our growing network of licensees. Ovonic batteries for EVs are being demonstrated in a wide variety of vehicle types in the US, Europe, and Asia. GM Ovonic has established prototype manufacturing and will be scaling-up battery production to meet market demand while actively pursuing development programs to reduce battery cost.

Ovonic batteries today offer high specific energy and power with long cycle life, low self-discharge, wide operating temperature range, and quick recharge. Next generation products will offer even higher performance, outperforming other advanced batteries such as lithium-ion at lower cost without the need for complex protective circuits required for safe operation of lithium-ion batteries.

Researchers at Ovonic, and our parent company Energy Conversion Devices, continue to develop advanced materials for negative and positive electrodes that will provide for higher specific energy and lower cost batteries. The goal of 95 Wh kg^{-1} has already been achieved and it is believed that specific energies approaching 120 Wh kg^{-1} will be reached in the near future.

Ovonic is continuing to look at new markets and applications for this battery technology. We have already demonstrated the application of this technology for such forms of personal transportation as scooters and bicycles and several licensees are developing commercial products for these applications. Stored energy for off-grid power or emergency power is another exciting application for the nickel/metal hydride battery. As technology advances and costs are reduced, other applications and markets are certain to open up.

References

[1] *Business Week*, 7 Oct. 1996, p. 142.

[2] S.R. Ovshinsky, M.A. Fetcenko, S. Venkatesan, S.K. Dhar, A. Holland, R. Young, P.R. Gifford and D.A. Corrigan, *13th Int. Seminar Primary and Secondary Battery Technology and Application, Deerfield Beach, FL, Mar. 1996.*

[3] M. Yamashita, Y. Wataru and T. Ohzuku, *Meet. Abstr., 190th Society Meet.*, Vol. 96-2, The Electrochemical Society, Inc., Pennington, NJ, 1996, Abstr. No. 880.

[4] M.H.J. van Rijswick, in A.F. Andresen and A.J. Maeland (eds.), *Hydrides for Energy Storage*, Pergamon, Elmsford, NY, 1978, pp. 261–272.

[5] S.R. Ovshinsky, K. Sapru, B. Reichman and A. Reger, *US Patent No. 4 623 597* (Nov. 1986).

[6] S.R. Ovshinsky, M.A. Fetcenko and J. Ross, *Science, 260* (1993) 176–181.

[7] S.R. Ovshinsky and M.A. Fetcenko, *185th Meet. Electrochemical Society, San Francisco, CA, May 1994.*

[8] S.R. Ovshinsky, R.C. Stempel, S. Dhar, M.A. Fetcenko, P.R. Gifford, S. Venkatesan, D.A. Corrigan and R. Young, *29th Int. Symp. Automotive Technology and Automation, Florence, Italy, June 1996.*

[9] M.A. Fetcenko, S. Venkatesan and S.R. Ovshinsky, *180th Meet. Electrochemical Society, Phoenix, AZ, Oct. 1991.*

[10] M. Oshitani, H. Yufu, K. Takashima, S. Tsuji and Y. Matsumaru, *J. Electrochem. Soc., 136* (1989) 1590.

[11] J.L. Weininger, in R.G. Gunther and S. Gross (eds.), *Proc. Symp. Nickel Electrode*, Vol. 82-4, The Electrochemical Society, Inc., 1982, pp. 1–18.

[12] S.R. Ovshinsky, D. Corrigan, S. Venkatesan, R. Young, C. Fierro and M.A. Fetcenko, *US Patent No. 5 384 822* (Sept. 1994).

[13] S.R. Ovshinsky, M.A. Fetcenko, C. Fierro, P.R. Gifford, D.A. Corrigan, P. Benson and F.J. Martin, *US Patent No. 5 523 182* (June 1996).

[14] M.A. Fetcenko, S. Venkatesan, S.K. Dhar and S.R. Ovshinsky, *3rd Int. Rechargeable Battery Seminar, Deerfield Beach, FL, Mar. 1992.*

[15] S.R. Ovshinsky, S.K. Dhar, M.A. Fetcenko, S. Venkatesan, A. Holland, P.R. Gifford and D.A. Corrigan, *11th Int. Seminar Primary and Secondary Battery Technology and Application, Deerfield Beach, FL, Mar. 1994.*

[16] M.A. Fetcenko, S.K. Dhar, S. Venkatesan, A. Holland, P.R. Gifford, D.A. Corrigan and S.R. Ovshinsky, *12th Int. Seminar Primary and Secondary Battery Technology and Application, Deerfield Beach, FL, Mar. 1995.*

[17] A. Ng and S.R. Ovshinsky, *Gold Peak*, Ovonic Press Release, 17 Jan. 1996.

[18] N. Furukawa, *Nikkei Electronics Asia*, May 1996, pp. 80–83.

[19] S.R. Ovshinsky, S.K. Dhar, S. Venkatesan, M.A. Fetcenko, P.R. Gifford and D.A. Corrigan, *11th Int. Electric Vehicle Symp., Florence, Italy, Sept. 1992.*

[20] P.R. Gifford, M.A. Fetcenko, S. Venkatesan, D.A. Corrigan, A. Holland, S.K. Dhar and S.R. Ovshinsky, *186th Meet. Electrochemical Society, Miami, FL, Oct. 1994.*

[21] D.A. Corrigan, S. Venkatesan, P.R. Gifford, M.A. Fetcenko, S.K. Dhar and S.R. Ovshinsky, *12th Int. Electric Vehicle Symp., Anaheim, CA, Dec. 1994.*

[22] *The Korea Herald*, Sat. 14 May 1994.

[23] M. Shnayerson, *The Car That Could*, Random House, New York, 1996, pp. 179–181.

[24] *Indianapolis Star*, Sun. 6 Oct. 1996.

[25] *Wall Street Journal*, 2 Dec. 1994.

special report

With their impressive energy performance and their outstanding power capability, NiMH batteries are equally suitable for use in pure and hybrid electric vehicles

Nickel-metal hydride: ready to serve

ROBERT C. STEMPEL,
STANFORD R. OVSHINSKY,
PAUL R. GIFFORD, &
DENNIS A. CORRIGAN

Ovonic Battery Co.

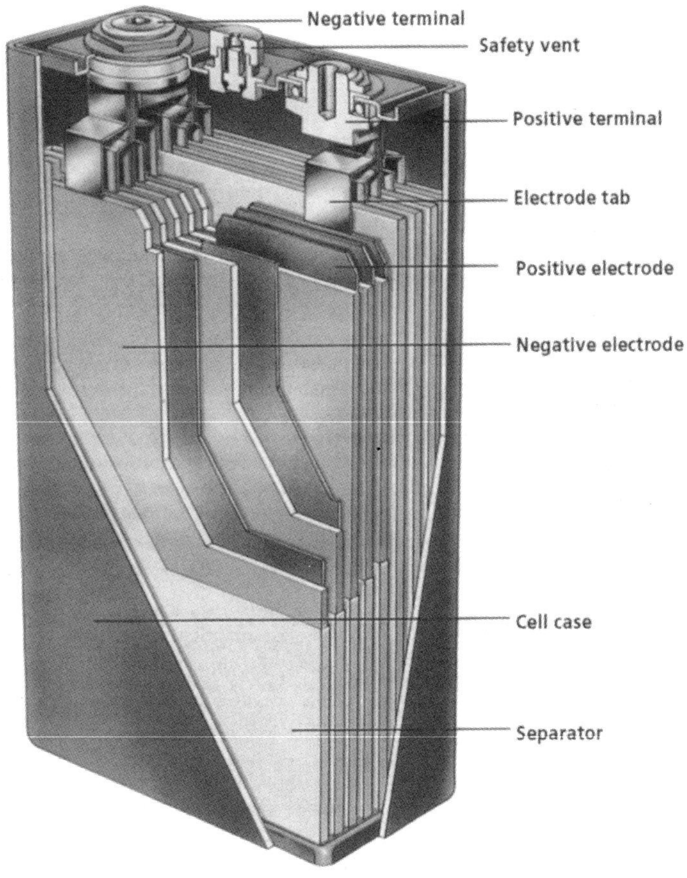

[1] NiMH batteries like this prismatic unit from Ovonic Battery deliver and accept charge at very high rates.

The leading choice for electrically powered vehicles today is the nickel-metal hydride battery [see "The great battery search," pp. 21–28]. Leading car makers have not been shy about jumping on the NiMH bandwagon.

Over the last year, the Honda EV Plus and the Toyota RAV-4 EV—both powered by NiMH batteries—were introduced for lease in California. General Motors is now offering the Chevrolet S-10 Electric Pickup truck powered by NiMH batteries for lease throughout the United States, and GM will offer NiMH versions of the EV1 later this year. Chrysler and Ford have also announced they will be introducing EVs equipped with NiMH batteries.

Further, NiMH batteries are perhaps even more suited to hybrid electric vehicles (HEVs) because their excellent power performance allows them to both deliver and accept charge at very high rates for short periods [Fig. 1]. NiMH batteries already power the Toyota Prius hybrid EV, the first mass-produced vehicle with electric propulsion, which is now being sold in Japan.

What is it that gives NiMH batteries the edge over earlier technologies like nickel-cadmium? The answer is rooted

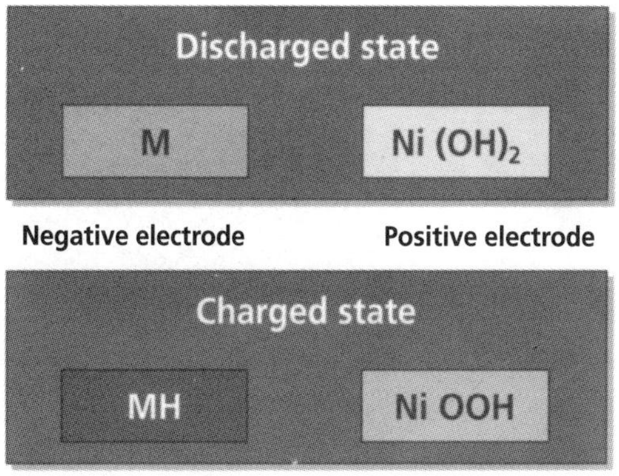

[2] When a NiMH cell is charged, hydrogen generated by reaction with the cell electrolyte is stored in the metal alloy (M) in the negative electrode. Meanwhile, at the positive electrode, which consists of nickel hydroxide loaded into a nickel foam substrate, a hydrogen ion is ejected and the nickel is oxidized to a higher valency. On discharge, the reactions reverse .

in the basic nickel-metal hydride technology. For one thing, hydrogen can be stored in metal-hydride alloys at very high volumetric densities, comparable to the density in liquid hydrogen, leading to the superior energy density of metal-hydride batteries. In an NiMH cell, a nickel hydroxide positive electrode is coupled with a metal hydride negative electrode [Fig. 2]. In the latter, a metal hydride alloy is compacted onto a conducting substrate. In the positive electrode, nickel hydroxide is loaded into a nickel foam substrate.

During charge, hydrogen is generated by reaction with the electrolyte and stored in the metal alloy in the negative electrode. At the positive electrode, a hydrogen ion is ejected while the nickel is oxidized to a higher valency within the brucite structure of nickel hydroxide. Both reactions are fully reversible.

Both simple...

This charge storage reaction makes the NiMH unit a simple hydrogen-transfer battery, in which hydrogen is transferred back and forth between the nickel hydroxide and the metal hydride without soluble intermediates or complex phase changes. For this reason, the nickel hydroxide battery is known as a "rocking chair" or "swing" battery. Soluble intermediates in everyday rechargeable batteries, such as lead-acid and nickel-cadmium types, have irreversible side reactions whose kinetics can compromise power and cycle life. It is the simplicity of the NiMH battery that makes for its high power and exceptionally long intrinsic cycle life.

Another virtue of the NiMH battery is its intrinsic toleration of electrical abuse. If conventional batteries are charged beyond the fully charged state, irreversible reactions occur that can damage the battery or create safety concerns. With the NiMH battery, overcharging generates oxygen at the positive electrode, which can be easily recombined at the negative electrode, providing that the gas is not generated at too high a rate.

Similarly, overdischarge in conventional batteries generally results in deleterious irreversible reactions. But overdischarging the NiMH battery generates hydrogen at the positive electrode, which can be easily recombined at the negative electrode, again provided that the rate of gas generation does not exceed the rate at which it can be recombined. The result is benign: no net chemical reaction on overcharge or overdischarge.

Because of these oxygen cycle and hydrogen cycle recombination reactions, moderate overcharge and overdischarge can be tolerated without damage to the battery or safety concerns. This toleration for electrical abuse is particularly advantageous for EV applications where high-voltage systems are common. With large numbers of cells in series connection (as is necessary to develop high voltages), all of the cells will not always be at exactly the same state of charge, resulting in overcharge and/or overdischarge of some of the cells during normal operation. The recombination reactions of the NiMH battery do away with the costly need to monitor and balance the cells individually, thereby simplifying battery management.

...and chaotic

But in contrast to the simplicity of its operation, the active materials of the nickel-metal hydride battery are complex. By and large, rechargeable batteries have been designed around single-component, single-phase active materials. Lead, lead dioxide, lead sulfate, cadmium, cadmium oxide, and nickel hydroxide spring to mind. Similarly, in the early history of metal hydride battery development, the emphasis was on single-phase hydrogen storage materials.

A new approach utilized the principles of disordered materials. Proprietary hydrogen storage materials were developed that consist of disordered, multicomponent alloys, and these form the basis of NiMH battery technology today [see To Probe Further, p. 34]. This technology is being used by battery manufacturers worldwide to produce over 500 million NiMH consumer cells per year in addition to batteries for EVs.

At Ovonic Battery Co., Troy, Mich., for example, the hydrogen storage alloys used in the negative electrodes of NiMH EV batteries are patented, highly catalytic, alloys of vanadium, titanium, zirconium, and nickel. Alloy materials currently under development will more than double the specific capacity of these materials. Materials engineering of disordered nickel hydroxide positive electrode materials, also based on disordered-material concepts, has led to the development of a new, patented class of high-capacity nickel electrodes, which have proven amenable to low-cost, pasted-electrode manufacturing processes.

What the demands are

Designing a practical EV battery is a tall order. Besides having high energy and power densities (both gravimetric and volumetric), they must be long-lived, tolerant of electrical abuse, capable of maintenance-free operation over a wide temperature range, and—above all—safe. And they must be manufacturable at an acceptable cost.

Although numerous EV batteries have been developed in this century, none so far has met these criteria as well as NiMH batteries. Significant performance improvements are still under way, yet NiMH batteries already exceed the mid-term performance goals of the U.S. Advanced Battery Consortium

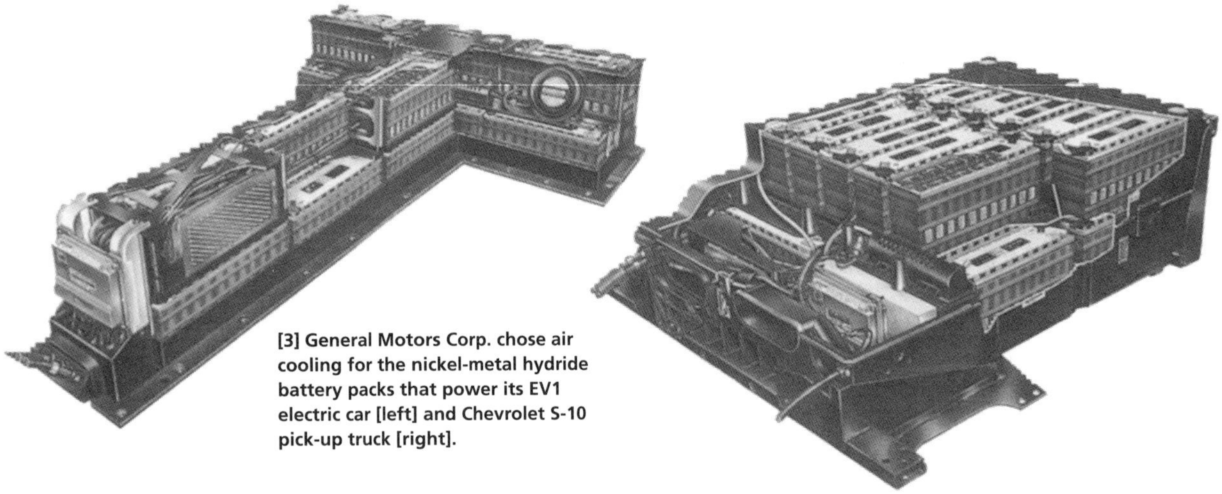

[3] General Motors Corp. chose air cooling for the nickel-metal hydride battery packs that power its EV1 electric car [left] and Chevrolet S-10 pick-up truck [right].

(USABC) as demonstrated by the delivery of 150-Ah battery modules by Ovonic Battery to the U.S. Department of Energy (DOE) last year for testing.

The 13-V batteries exhibited a specific energy of 80 Wh/kg and an energy density of 215 Wh/L, in comparison with the consortium's goals of 80 Wh/kg and 135 Wh/L, respectively. Power levels of 250 W/kg and 675 W/L were achieved, well in excess of the USABC goals of 150 W/kg and 250 W/L. What's more, the USABC cycle life goal of 600 dynamic stress test (DST) simulated driving cycles was exceeded. To top things off, NiMH batteries are environmentally friendly and can be totally and profitably recycled. Even prior to recycling, there are second-use options for NiMH EV batteries for load-leveling and other applications.

NiMH EV batteries go to market

At the 14th International EV Symposium (EVS14) held last year, several battery manufacturers from around the world displayed NiMH batteries designed for EVs. Some of the batteries were clearly still in the developmental stage, but others—notably, those belonging to Panasonic and GM Ovonic—were advertised as being currently in production and available for commercial sale, albeit in very limited quantities.

According to product brochures, the Panasonic battery is a 1.1-kWh module that delivers 95 Ah at 12 V. The specific energy is 63 Wh/kg and the energy density is 150 Wh/L. The specific power is given as 200 W/kg. Another company, Saft America, Cockeysville, Md., offers two versions of its monoblock design, one at 12 V and one at 24 V nominal voltage, both with 93-Ah rated capacity. The design incorporates accommodation for liquid cooling. Both designs are described as having a specific energy of 64 Wh/kg and specific power of 150 W/kg. The energy density is 125 Wh/L.

At the present time, GM Ovonic offers a 1.2-kWh module that delivers 90 Ah at 13.2 V (a module has eleven 90-Ah cells connected in series). This first-generation product, GMO1, has impressive power performance and the best volumetric energy density available for EV applications. The specific energy is 70 Wh/kg and the volumetric energy density is 170 Wh/L. The specific power is 200 W/kg. [See Table 1, which gives some of this battery's key specifications alongside those of another Ovonic NiMH battery optimized for use in hybrid electric vehicles.]

Future NiMH EV battery products are expected to improve markedly in performance. For example, GM Ovonic's GMO2 design, which will be introduced in 1999, will have specific energy and energy density ratings of at least 80 Wh/kg and 200 Wh/L, respectively. Concurrent with GMO2 development, Ovonic Battery Co. (OBC) is developing active materials that offer higher specific capacity. Introduction of these materials into the GMO3 design will offer still further improvements in specific energy and lower cost. The GMO3 battery will provide 95-Wh/kg performance with greater power than the current production battery.

Construction and manufacture

All NiMH batteries contain positive and negative electrodes, separator, electrolyte, and cell container [Fig. 1, again]. Stacking enough positive and negative plates in parallel achieves the desired cell capacity. The plates are electrically isolated by a nonwoven separator,

1. Batteries for electric and hybrid electric vehicles

Characteristic	Units	13-EV-90	12-HEV-20
Weight	kg	18	5.2
Volume	L	7.1	2.3
Nominal voltage	V	13.2	12.0
Rated capacity	Ah	90	20
Specific energy	Wh/kg	70	50
Energy density	Wh/L	170	110
Specific power	W/kg	200	550
Power density	W/L	485	1300

which also retains the aqueous alkaline electrolyte.

For the negative electrode, active material is manufactured by vacuum induction melting of the raw materials to produce the intermetallic alloy. The resulting alloy ingot is reduced to a powder either by mechanical grinding or by a hydride/dehydride process. Final grinding and classifying follow. Electrode strip can be produced in either of two ways: either by wet pasting a slurry of powder and binder onto perforated metal or expanded metal or, alternatively, by a roll compaction process. Ovonic Battery employs a dry powder compaction process to yield a finished electrode that has exceptional volumetric energy density in the absence of binders or other additives.

To produce the positive electrodes, generally a mechanical impregnation process is used. In this case, a highly porous nickel substrate, such as nickel foam or felt, is filled with an active material slurry composed of nickel hydroxide, performance-enhancing additives (generally cobalt compounds), and a binder. The active material is loaded into the substrate, and then dried, compressed to final thickness, and profiled into finished plates.

The nickel-hydroxide active material is a special battery grade of high-density, spherical particles, and generally includes other metal ion additives through coprecipitation. Ovonic Battery and its parent company, Energy Conversion Devices, Troy, Mich., have developed proprietary nickel-hydroxide materials based on the same concepts of disorder applied successfully to hydride alloys. The result is a greater specific capacity than is the case with standard battery-grade materials.

To construct a cell, alternating negative and positive plates are stacked and enclosed in separators. The stack is inserted into a cell container of metal or plastic construction, and the electrode tabs welded to the terminals. GM Ovonic uses metal cell cans for strength and heat transfer, as do Varta and Daug, whereas Panasonic and others employ plastic cases.

Each cell has a nominal voltage of 1.2 V. To build a battery with the desired voltage, enough cells are connected in series and combined into a final battery module. The battery designer thus has flexibility in system voltage and system packaging constraints. While the standard GM Ovonic battery consists of 11 cells, module designs of 8 to 14 cells in series have been constructed, depending on customer requirements.

Module construction constrains individual cells between two end plates with metal bands holding everything together. Saft, however, employs a plastic case monoblock design, as opposed to single cells. After assembly into the final module, the battery undergoes a series of repeated electrical charge/discharge cycles that fully activate it.

The final vehicle battery pack is constructed from the battery modules, with the pack voltage determined by whatever the vehicle drive train needs. Today's ac propulsion drives generally operate at voltages in excess of 300 V, requiring 24 or more batteries in series connection. Proper thermal management of the pack is needed to maintain acceptable temperature levels during both charging and driving. Since the early market for EVs was in such hot locations as Los Angeles and Phoenix, ambient operating temperatures can be well in excess of 40 °C. To offset that heat, either water-cooling or air-cooling strategies have been employed on NiMH battery packs. GM uses active air cooling for the S-10 and EV1 vehicle packs [Fig. 3], as does Toyota for its RAV4 vehicle. Honda has chosen water cooling for the battery pack on its Honda EV Plus.

The mass production of NiMH batteries for EVs has started, as several companies enter low-volume manufacture throughout the world. In addition to GM Ovonic, which was formed in June 1994, manufacturers include Panasonic EV Energy Co., a 1996 partnership of Toyota and Matsushita, and most recently Saft announced the commissioning of a pilot plant in Bordeaux, France, with an annual capacity of 500 batteries. Hyundai Motor Co. has established a pilot manufacturing facility in Korea to produce hydride alloys and NiMH EV batteries. Other leading battery companies such as Varta, GS Saft Yuasa, and Furukawa, have announced development activities and showcased prototype batteries, but to date have made no public announcements of production plans.

On the road

The annals of nickel-metal hydride batteries are impressive. Since they first "turned wheels" in 1993, hundreds of NiMH battery packs have powered electrically propelled vehicles around the world. These included electric cars, vans, trucks, scooters, bicycles, and military vehicles. EVs with NiMH battery packs have been driven year-round in sun, rain, and snow, demonstrating successful operation over ambient temperatures ranging from –25 °C to +50 °C. In total, more than 1.5 million kilometers of real world testing has been completed.

Thanks to their higher energy density, NiMH batteries have provided big improvements in range over lead acid. Typically, the range of a lead-acid–powered EV can be more than doubled with a NiMH pack of equivalent size and weight. For example, General Motors has announced that the EV1 achieves a real world range of 250 km with a GM Ovonic NiMH battery pack, as opposed to a range of about 100 km with the current lead-acid battery. Similarly for the relatively less efficient S-10 conversion truck, a range of 110–130 km is obtained with no tradeoff in vehicle payload. Toyota has announced a range of 200 km for its RAV4 conversion EV, as has Honda for its EV Plus.

Since much of the testing of NiMH batteries in EVs has been performed at various automotive companies around the world, not all results are publicly available. One publicly sponsored event for EVs that has gained increasing popularity and support is the annual Tour de Sol in the United States, an EV road rally competition held over public roads sponsored by the Northeast Sustainable Energy Association (NESEA), Greenfield, Mass. Ovonic Battery has participated in this event for the last five years, supplying battery packs to participants to gain real world driving experience for its NiMH batteries. In particular, the company's collaboration with Solectria Corp., Wilmington, Mass., in this event has demonstrated that the range of EVs equipped with NiMH batteries can easily meet the driving needs for most drivers and can even approach the range for gasoline-powered vehicles. In May 1996, NiMH batteries from Ovonic Battery powered the Solectria Sunrise to a world record 600 km. The Sunrise [Fig. 4] is a purpose-built EV with about the same amount of passenger and trunk space as a Chevrolet Lumina. (As a matter of possible interest, in October of 1997, a Sunrise traveled from Boston to New York on a single charge with this article's editor, Mike Riezenman, riding shotgun to certify that the car was indeed not charged during the trip [EV Watch, *IEEE Spectrum*, December 1997, pp. 68–70].)

This year, in conjunction with the 10th annual Tour de Sol, NESEA performed head-to-head energy-efficiency measurements on the Solectria Force EV, powered by Ovonic batteries, and a gasoline-powered Geo Metro. This comparison is especially meaningful because the Solectria Force is, in fact, a Geo Metro converted into an electric vehicle.

Under stop-and-go New York City driving, the Geo Metro fuel economy was about 24 L/100 km compared with 7.69 kWh/100 km (equivalent to just 2.7 L/100 km) achieved by the Solectria Force. Under these conditions, not only the energy efficiency, but even the range of the EV was a good deal higher than that of its internal-combustion counterpart. The range of the Geo Metro with 45 L of fuel was limited to about 200 km. The range of the Solectria Force with 27 kWh of NiMH batteries was 350 km under these conditions.

Back-to-back victories at the feature event of the 1996 and 1997 Arizona Public Service EV road race demonstrated the excellent range and power performance of NiMH batteries in a converted Saturn coupe entered by AeroVironment Inc., Monrovia, Calif. Equipped with a 24-kWh pack of GM Ovonic production batteries, this EV set new records in the 80-km sprint race both years, reaching speeds over 120 km/h. This high power capability was further demonstrated in a production Chevy S-10 EV truck, setting a new EV course record in the 1997 Pike's Peak hill climb. The S-10 EV truck traveled over a 20-km course winding through 156 turns and climbing 1435 meters to the summit in just 15 minutes and 32 seconds.

Hybrid electric vehicle batteries

NiMH batteries are fast becoming a leading contender in another area as well. Because of their high-power capability, they are as well suited to applications in HEVs as in pure EVs. As indicated in Table 1 of the preceding article, there are three main differences between EV and HEV battery requirements:

- HEVs require higher power capabilities—on both charge and discharge.
- HEVs have less in the way of energy storage demands.
- HEV batteries need to withstand many more charge-discharge cycles.

Development of NiMH batteries for HEVs is under way at several battery companies, including Ovonic, Panasonic, and Varta. Work at Ovonic Battery has proceeded to the stage of several types of prototype battery modules that meet the basic HEV performance requirements.

The high power performance of the Ovonic 13-HEV-60 is delivered over a wide range of state-of-charge. Over 500 W/kg and 1000 W/L are delivered from full charge, zero percent depth of discharge, down to 90 percent depth of discharge. The ability to accept charge also exceeds 500 W/kg and 1000 W/L over a similar range of state-of-charge.

A particular advantage of NiMH batteries in HEVs is their ability to accept regenerative braking energy at high rates. Figure 5 shows a plot of voltage-current measurements for an Ovonic 13-HEV-60 battery module during an aggressive simulated HEV driving cycle.

Results for a high-power lead-acid battery, shown for comparison, exhibit a bent V-I plot. The slope, proportional to resistance, is a good deal higher for charge than discharge, indicating a higher resistance, and thus lower regenerative power-acceptance capability. The cause is the complex chemistry of the lead-acid battery, leading to a change in reaction mechanism between charge and discharge. For the NiMH battery, on the other hand, results show a constant lower slope, providing lower resistance, and excellent ability to accept regenerative braking energy. This is a consequence of the relatively simple proton-transfer chemistry of the NiMH battery.

The high efficiency of the NiMH HEV battery is due to its low internal resistance. Under aggressive HEV driving schedules, efficiencies of 80–90 percent have been measured. This high efficiency is maintained for tens of thousands of cycles at temperatures ranging up to 60 °C. In on-going HEV cycle life testing, over 90 000 charge-discharge cycles have been achieved, corresponding to nearly 160 000 km.

The 13-HEV-60 module shows great promise for HEV applications due to its high power capability. The high power levels, however, are achieved with little tradeoff in specific energy, which for this model is 70 Wh/kg. This unique combination of high power and high energy provides for excellent HEV performance and is coupled with a fairly high range in the purely electric (zero-emissions) mode. A high zero-emissions range (ZEV) for HEVs will

[4] The Solectria Sunrise has an extremely efficient powertrain (92 percent overall efficiency from its battery to its wheels) and a lightweight composite body—a combination that let it cover more than 600 km on a single charge of its Ovonic high-capacity NiMH battery pack.

PORTER GIFFORD/LIAISON AGENCY

become important if a requirement develops for purely electric operation in the center of large cities, as seems likely at least in Europe. In addition, a high ZEV range is favored by regulators in the United States, including the California Air Resources Board, to help minimize vehicular emissions on hybrid vehicles.

Still, the 13-HEV-60 battery module is not suitable for the full range of HEV applications. It is excellent for range-extender HEV applications where large batteries are needed. However, power-assist HEVs need only small batteries, and dual-mode HEVs need intermediate sizes. For these reasons, Ovonic Battery has developed several HEV cell designs to match this variety of HEV applications. Prototype batteries are now available with 20, 28, and 60 Ah capacity. The batteries have comparable performance, including power and regenerative capability of about 550 W/kg and 1200 W/L. Prismatic

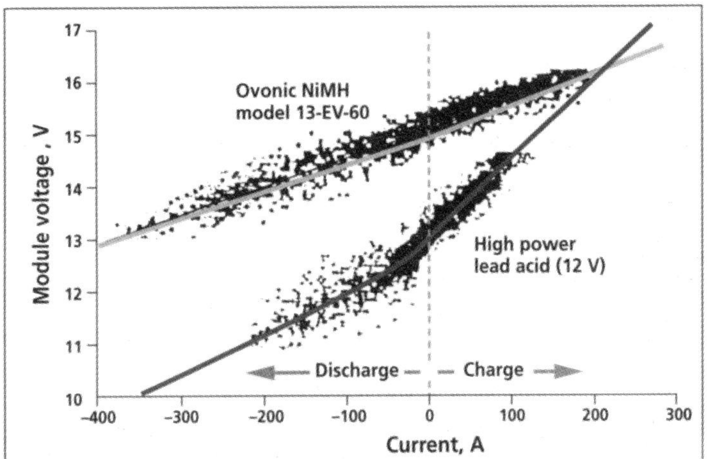

[5] These voltage-current plots for an Ovonic 13-HEV-60 NiMH battery [yellow] and a high-power lead-acid battery [red] highlight the former's superior ability to accept charge at very high rates. The slopes of the curves are proportional to resistance. Note that the slope of the red curve is bent and is considerably higher for charge than discharge; the reason is a higher internal resistance that keeps the lead-acid battery from accepting charge as quickly as the NiMH unit.

(parallel-plate rectangular) battery modules with a capacity of 12 Ah are under development. Prototype Ovonic NiMH cylindrical cells have achieved a specific power of over 700 W/kg.

At Panasonic and Varta, NiMH HEV battery development has apparently focused on low-capacity cylindrical cells aimed at power-assist HEV applications. Panasonic has announced a 6.5-Ah D-cell with a specific power of about 500 W/kg, but with a specific energy of only 43 Wh/kg. Varta has reported a 10-Ah cylindrical cell with a specific power rating of around 800 W/kg, but with a specific energy of only 30 Wh/kg. In these cases, power performance was apparently maximized at the expense of specific energy and energy density. However, results at Ovonic Battery have shown that higher power levels can be achieved, even in larger cells. Power levels of over 1000 W/kg and 3000 W/L have been achieved in a 30-Ah prismatic cell, and engineering analysis shows power levels of 2000 W/kg and 5000 W/L are reasonable development targets.

Since NiMH technology is still in its early stages, continued improvement in battery technology is expected. Development cells aimed at EV applications have already achieved substantial improvements in specific energy to over 95 Wh/kg and energy density to 300 Wh/L. Even higher performance is foreseen for the future. Similarly, power levels of 1000 W/kg and 3000 W/L have been attained in developmental HEV cells—and double those levels are achievable. While low-volume battery cost is a concern, as technology matures and production volumes increase, NiMH batteries will become much less expensive, expanding today's EV market as drivers experience the comfort and enjoyment of driving modern EVs. In fact, say automotive engineers in the EV field, EVs will ultimately cost less than gasoline-powered vehicles. ◆

To probe further

A scientific review of the chemistry and physics of the nickel-metal hydride battery can be found in "A Nickel-Metal Hydride Battery for Electric Vehicles," by S.R. Ovshinsky, et al., in *Science,* Vol. 260,1993, pp. 176-181.

The key role played by the NiMH battery in the development of General Motors' revolutionary electric car prototype, the Impact, is given in M. Shnayerson's *The Car That Could* (Random House, New York, 1996).

Updates of the state of development of NiMH batteries for electric vehicle (EV) and hybrid EV applications are given in the *Proceedings of the 14th International Electric Vehicle Symposium,* Orlando, Fla., December 1997, as well as in the proceedings of the 15th symposium in 1998, both available from the Electric Vehicle Association of Americas.

Further details of the multicomponent, multiphase alloys on which NiMH batteries are based may be found in U.S. Patents 4 623 597, 18 Nov. 1986, and 5 536 591, 16 July 1996. Details on Ovonic's proprietary positive electrodes are given in U.S. Patents 5 348 822, 20 Sept. 1994 and 5 637 423, 10 June 1997.

About the authors

Robert C. Stempel is chairman of Energy Conversion Devices Inc., an energy and information company in Troy, Mich., and of its subsidiary there, Ovonic Battery Co. He is on the board of managers of GM Ovonic LLC, a joint manufacturing venture between General Motors Corp., Detroit, and Ovonic Battery. Stempel is a former chairman and chief executive officer of General Motors.

Stanford R. Ovshinsky is the founder and chief executive officer (CEO) of Energy Conversion Devices Inc., CEO of Ovonic Battery, and on the GM Ovonic board. He is a pioneer in the field of disordered materials, developing engineered materials for energy conversion applications, including storage (batteries) and generation (photovoltaics), and for information systems (switches and memory). He holds 200 U.S. patents, including fundamental ones for nickel-metal hydride batteries.

Paul R. Gifford is vice president, business development, at Ovonic Battery and technology manager for GM Ovonic. He's had extensive experience in R&D of battery systems, with over 15 years spent at AlliedSignal, Gates Energy Products, and Duracell.

Dennis A. Corrigan is vice president of electric vehicle battery systems at Ovonic Battery, in charge of developing nickel-metal hydride battery packs for electric and hybrid vehicles. He joined Ovonic in 1992 after 12 years in EV R&D on EV batteries at General Motors Research Laboratories.

Spectrum editor: Michael J. Riezenman

Appl. Phys. A 72, 239–244 (2001) / Digital Object Identifier (DOI) 10.1007/s003390100776

Applied Physics A
Materials
Science & Processing

Development of high catalytic activity disordered hydrogen-storage alloys for electrochemical application in nickel–metal hydride batterie

S.R. Ovshinsky, M.A. Fetcenko*

Energy Conversion Devices/Ovonic Battery Company, 1675 W. Maple, Troy 48084, MI, USA

Received: 14 August 2000/Accepted: 6 November 2000/Published online: 9 February 2001 – © Springer-Verlag 2001

Abstract. Multi-element, multiphase disordered metal hydride alloys have enabled the widespread commercialization of nickel–metal hydride (NiMH) batteries by allowing high capacity and good kinetics while overcoming the crucial barrier of unstable oxidation/corrosion behavior to obtain long cycle life. Alloy-formula optimization and advanced materials processing have been used to promote a high concentration of active hydrogen-storage sites vital for raising NiMH specific energy. New commercial applications demand fundamentally higher specific power and discharge-rate kinetics. Disorder at the metal/electrolyte interface has enabled a surface oxide with less than 70 Å metallic nickel alloy inclusions suspended within the oxide, which provide exceptional catalytic activity to the metal hydride electrode surface.

PACS: 71.55.Jv; 61.46.+w; 68.37.Lp; 68.47.De

Nickel–metal hydride (NiMH) batteries are in high-volume commercial production for small portable battery applications, beginning in 1989 and achieving over 900 million annual worldwide cell production in 1999 [1]. The driving force for the rapid growth of NiMH is both technical and environmental, with energy and performance advantages over nickel–cadmium fueling the explosive growth of portable electronic devices such as communication equipment and laptop computers.

NiMH batteries have become the dominant advanced battery technology for electric vehicle (EV) and hybrid electric vehicle (HEV) applications by having the best overall performance in the wide-ranging requirements set by automotive companies. In addition to the essential performance targets of energy, power, cycle life and operating temperature, the following features of NiMH have established the technology pre-eminence:

- Flexible cell sizes from 60 mA h – 250 mA h;
- Safe operation at high voltage (320+ V);
- Excellent volumetric energy and power, flexible vehicle packaging;
- Easy application to series and series/parallel strings;
- Choice of cylindrical or prismatic cells;
- Safety in charge and discharge, including tolerance to abusive overcharge and overdischarge;
- Maintenance-free;
- Excellent thermal properties;
- Capability to utilize regenerative braking energy;
- Simple and inexpensive charging and electronic control circuits; and
- Environmentally acceptable and recyclable materials.

Recent development activity in NiMH batteries has focused on further improvements in peak power for HEV and portable power-tool applications. As the licensor of essentially all commercial NiMH batteries, the Ovonic Battery Company continues to develop advanced NiMH technology and is uniquely positioned to comment on development goals from an applied industrial perspective. In this paper, we will report our results in raising power and high-rate discharge capability, with particular emphasis on the metal hydride electrode surface catalytic activity at the metal/electrolyte oxide interface.

1 Experimental

Metal hydride alloys having formulas typified by $V_5Ti_9Zr_{26.2}\cdot Ni_{38}Cr_{3.5}Co_{1.5}Mn_{15.6}Al_{0.4}Sn_{0.8}$ and $Ti_8Zr_{29}Ni_{29}Cr_8Co_{11}Mn_{15}$ were prepared by vacuum induction melting to have multiphase C14 and C15 structures. Alloy powder was produced by a single hydride/dehydride cycle and pulverization to below 75 μm. Electrodes were prepared by compacting or pasting the powder onto a copper expanded-metal substrate using 0.5% PTFE binder.

C-size cylindrical cells were assembled using a pasted-style positive electrode consisting of NiCoZnCaMg-formula active nickel hydroxide with high conductivity additives consisting of nickel metal fibers, cobalt and cobalt oxide [2].

The C cells were tested for power by comparing cell voltage and resistance rat the C rate and 10C rate discharge using 10-s pulses as

$$P_{\text{peak}} = 0.5V_{\text{oc}} \times 0.5I_{\text{max}}.$$

*Corresponding author. (E-mail: MFetcenko@MCImail.com)

AC-impedance studies of the metal hydride electrode (~ 1.5 cm^2, ~ 200 mg) were conducted in a flooded electrolyte, half-cell configuration with respect to an Hg/HgO reference electrode. AC-impedance measurements were performed using an EG&G 263 A potentiostat and a Solatron SI1250 frequency-response analyzer.

Analysis of the metal hydride surface was conducted by disassembly of discharged C cells in an argon glove box, with the metal hydride electrode rinsed and dried of KOH. Samples were prepared for examination by progressive polishing followed by dimple grinding to a specimen thickness of about 50 μm. The samples were then ion-milled at cryogenic temperature using 3–6 k eV argon ions to a specimen thickness of about 500 Å. Investigation of the MH surface oxide was performed on a JEOL-2010 scanning transmission electron microscope (STEM) using bright-field and dark-field imaging. Crystal structures were studied using selected-area electron diffraction (SAED). Quantitative energy-dispersive spectroscopy (EDS) analyses were obtained using a Cliff–Lorimer thin-film correction procedure in conjunction with experimental X-ray correction factors (*k*-factors) obtained from thin-film standards.

2 Results and discussion

2.1 Alloy-design concept

NiMH batteries are an unusual battery technology in that the metal hydride active material is an engineered alloy made up of many different elements and that the MH alloy formulas vary to a significant degree [3]. Following the principles we have developed [4], disordered metal hydride materials based on transition, rare-earth and magnesium alloys have been produced. The active material in the negative electrode is of either the disordered AB_5 (LaCePrNdNiCoMnAl) or the disordered AB_2 (VTiZrNiCrCoMnAlSn) type, where the "AB_x" designation refers to the ratio of the A-type elements (LaCePrNd or TiZr) to that of the B-type elements (VNiCrCoMnAlSn). Disorder takes into account that both 3*d* and 5*f* element orbitals can be used to produce multiphase materials having a spectrum of hydrogen binding energy with either type of orbital serving as a host matrix. In either case, the materials have complex microstructures that allow the hydrogen-storage alloys to operate in the aggressive environment (30% KOH electrolyte, oxygen gas in overcharge recombining at the MH surface) within the battery where most of the metals are thermodynamically more stable as oxides. AB_5-type alloys are more common, despite significantly lower hydrogen-storage capacity as compared to AB_2 (320 vs. 440 mA h/g). Significant ongoing development has improved the properties of AB_2 materials such as cycle life, charge retention, and power to take advantage of the inherently higher energy that is especially important to reduce cost [5, 6].

Electrochemical utilization of metal hydride materials as anodes in NiMH batteries requires meeting a demanding list of performance attributes including hydrogen-storage capacity, suitable metal-to-hydrogen bond strength, acceptable catalytic activity and discharge kinetics and sufficient oxidation/corrosion resistance to allow for long cycle life. Multi-element, multiphase, disordered alloys of the $LaNi_5$ and VTiZrNiCr types are attractive development candidates for atomic engineering due to a broad range of elemental addition and substitution, availability of alternate crystallographic phases which form the matrix for chemical modification and a tolerance for non-stoichiometric formulas. Through the introduction of modifier elements, ease of activation and formation has been achieved. Special processing steps have been developed, such as alloy melting and size reduction suitable for these metallurgically challenging materials in terms of molten reactivity, degree of alloying and high hardness.

The metal hydride active material and electrode construction also have similar special design options. The active materials may be adjusted to influence one or more of capacity, power and/or cycle life. Disorder permits the extra degree of freedom that allows for a high level of chemical substitution. For the AB_5 system, a typical formula is $La_{5.7}Ce_{8.0}Pr_{0.8}Nd_{2.3}Ni_{59.2}Co_{12.2}Mn_{6.8}Al_{5.0}$ (atomic percent a/o). While the capacity of various AB_5 alloys is usually around 290–320 mA h/g, other overall performance attributes can be greatly influenced [7]. It is common for the ratio of La/Ce to be reversed to emphasize cycle life and power. The total amount of Co, Mn and Al significantly affect ease of activation and formation but increased cobalt has cost implications. After production of the AB_5 alloy ingot, it is common to further refine the microstructure of the material by a post-anneal treatment of perhaps 1000 °C for 10 h. The annealing treatment can have a significant effect on capacity, discharge rate and cycle life by adjusting crystallite size and grain boundaries, as well as eliminating unwanted phases precipitated during ingot melting and casting [8, 9]. Commercial AB_5 alloys have a predominantly $CaCu_5$ crystallographic structure. However, within that structure, there is a range of lattice constants brought about by compositional disorder within the material that are important to catalysis, storage capacity and stability to the alkaline environment and embrittlement [10]. These materials also precipitate a nickel–cobalt phase that is important to high rate discharge [11].

AB_2 alloys also have formula and processing choices. Popular AB_2 alloy formulas (a/o) include $V_{18}Ti_{15}Zr_{18}Ni_{29}\cdot Cr_5Co_7Mn_8$ and $V_5Ti_9Zr_{26.2}Ni_{38}Cr_{3.5}Co_{1.5}Mn_{15.6}Al_{0.4}Sn_{0.8}$.

Alloy capacity may range from 385 to 450 mA h/g. High vanadium content alloys may suffer from higher rates of self-discharge due to the solubility of vanadium oxide and its consequent ability to form a special type of redox shuttle [12]. The concentrations of Co, Mn, Al and Sn are important for easy activation and formation and long cycle life. The ratio of hexagonal C14 to cubic C15 phase is important to emphasize capacity or power [13, 14].

2.2 Alloy and battery performance

NiMH specific energy can vary from 42–100 W h/kg depending on the particular application requirements. For laptop computers where run time is paramount, NiMH batteries need not have high power capability or even ultra-long cycle life. On the other hand, for extremely high power charge and discharge, extra current collection, high N/P ratios (proportion of excess negative electrode capacity to positive electrode capacity) and other cell design and construction decisions can additively affect specific energy. Figure 1 presents

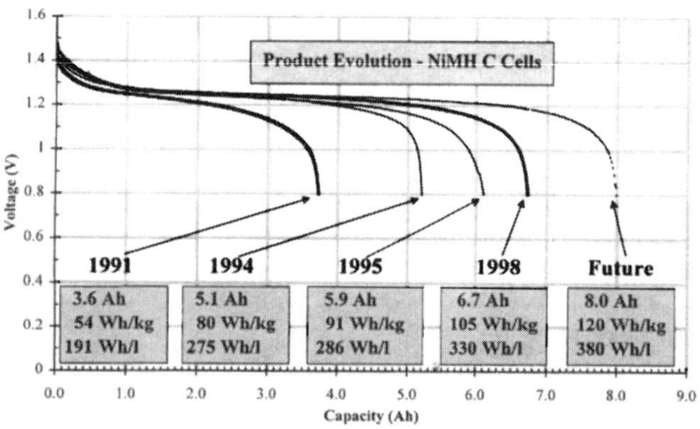

Fig. 1. Advances in NiMH technology specific energy

NiMH specific energy improvements over the last 10 years in consumer (portable) cylindrical cells. For the most common small consumer NiMH batteries, specific energy is usually about 75–95 W h/kg, for EV batteries usually about 65–80 W h/kg and for HEV batteries and other high-power applications about 45–60 W h/kg [15]. While gravimetric energy usually receives the attention for advanced battery technologies, in many cases volumetric energy density in watt–hours per liter is actually more important. NiMH has exceptionally good energy density, achieving up to 320 W h/liter. Cost reduction is at the forefront of NiMH development. High-volume consumer battery production has seen NiMH cost come at or below the \$/W h cost of NiCd, previously thought to be unobtainable. NiMH cost reduction begins with the recognition that the technology cost is primarily materials-intensive. Effort to raise specific energy involves development of metal hydride alloys with higher hydrogen-storage capacity (from 320–385 mA h/g active material to 450 mA h/g) and higher utilization nickel hydroxide (from 240 mA h/g active material to 280–300 mA h/g). Each of these higher utilization active materials involves innovative materials research involving highly modified alloy formulas and advanced processing techniques.

While electric vehicles may use a battery of about 200 W/kg, hybrid electric vehicles are now becoming the prominent environmentally conscious vehicle approach. HEV applications require battery specific power in excess of 500 W/kg and preferably 1000 W/kg. In addition, traditional lead–acid starting–lighting–ignition (SLI) batteries are problematic in size and weight when automotive requirements call for 42-V electrical systems. NiMH batteries are considered the leading advanced battery for high-voltage SLI applications due to the following advantages compared to lead–acid: higher gravimetric and volumetric energy and longer cycle life, especially under high depth of discharge. Ongoing work will include improving low-temperature power at −30 °C and initial cost at low to moderate production volumes.

2.3 MH *surface oxide and catalysis*

A critical design factor within the metal hydride surface oxide is to achieve a balance between surface oxide passivation and corrosion. Porosity within the oxide is important to allow ionic access to the metallic catalysts and therefore promote high rate discharge. While passivation of the oxide is problematic for high rate discharge and cycle life, unrestrained corrosion is equally destructive. Oxidation and corrosion of the anode metals consumes electrolyte, changes the state of charge balance by release of hydrogen in the sealed cell, and creates corrosion products which are capable of poisoning the positive electrode by causing premature oxygen evolution. Establishing a balance between passivation and corrosion for stability is a primary function of compositional and structural disorder.

For both AB_5 and AB_2 metal hydride alloys, the metal/electrolyte surface oxide interface is a crucial factor in discharge-rate capability and cycle-life stability [16–18]. Original $LaNi_5$ and TiNi alloys extensively studied in the 1970s and 1980s for NiMH battery applications were never commercialized due to poor discharge-rate and cycle-life capability [19–21]. Lack of catalytic activity at the surface oxide limits high rate discharge and lack of sufficient oxidation/corrosion resistance is a critical obstacle to long cycle life. The complicated chemical formulas and microstructures of present disordered AB_5 and AB_2 alloys extend to the surface oxide. At the oxide, important factors include thickness, microporosity and catalytic activity. In particular, the oxide interface between the metal hydride and the electrolyte has been identified as essential for low voltage loss under pulse discharge. Of crucial importance to discharge rate was the use of ultra-fine metallic nickel alloy particles having a size less than 70 Å dispersed within the oxide. These particles are excellent catalysts for the reaction of hydrogen and hydroxyl ions, and are especially important for reducing activation polarization [22].

The ultra-fine metallic catalysts are created by preferential corrosion within the multi-element hydrogen-storage alloy. In particular, the dissolution and precipitation of the less noble vanadium, titanium and zirconium allow nickel in the presence of cobalt, manganese and aluminum to form the metallic clusters. This requires careful design of the surface composition and structure based on the lack of such proximity of nickel atoms in the unit cell. The keys to success of the catalysts are size, number, density and topology of the metallic particles. Nickel in the metallic state is electrically conductive and catalytically active. By further alloying the catalysts with cobalt (∼ 20% a/o), the relative size of the catalysts can be reduced from about on average 50–70 Å to about 20–50 Å. Surface segregation via enrichment and selective oxidation of La in $LaNi_5$ to lower surface energy and the formation of surface nickel precipitates have been previously reported. However, there was no disclosure of size, composition, proximity or stability of such precipitates and the resultant surface formation of $La(OH)_3$ and $Ni(OH)_2$ upon electrochemical cycling was discussed [23].

Disorder allows such finely divided metallic catalysts ("surface sites") by taking into account the entire spectrum of local order effects such as porosity, topology, catalyst size and proximity. While a tremendous amount of research investigates development of bulk nanocrystalline materials, it is important to note that such materials exist in today's 900 million per year NiMH batteries, albeit on a small volume fraction at the surface interface. The catalyst size is important in that at less than 70 Å size, there are an approximately equivalent amount of surface atoms and bulk atoms. The close proximity of the metallic catalysts, typically about 100-Å apart, may also be referred to as "density of sites". It is the interaction of the local chemical electronic bonds with the reacting H^+ and OH^- that makes these catalysts so effective, and it is their metallic nature, size and number which provide excellent poisoning resistance.

The investigation of the MH surface oxide was conducted using a scanning transmission electron microscope having the capability for EDS, SAED and EELS (electron energy loss spectroscopy) [24]. Figures 2 and 3 present high-magnification bright-field and dark-field TEM images of the surface oxide exhibiting high catalytic activity. The bright inclusions under dark-field imaging are useful for determining the size of the catalyst and are indicative of very strong electron diffraction from nanocrystals having high crystallinity. Wide-area EDS shows an enriched nickel surface to be expected from the preferential corrosion of the other base-alloy constituents. Fine-beam EDS of the inclusion shows almost no oxygen relative to a high nickel or nickel–cobalt peak intensity while the area surrounding the nickel inclusion is rich in oxygen. To provide further corroboration, SAED is presented in Fig. 4. Indexing of this ring pattern is consistent with metallic nickel rather than nickel oxide. Final support for

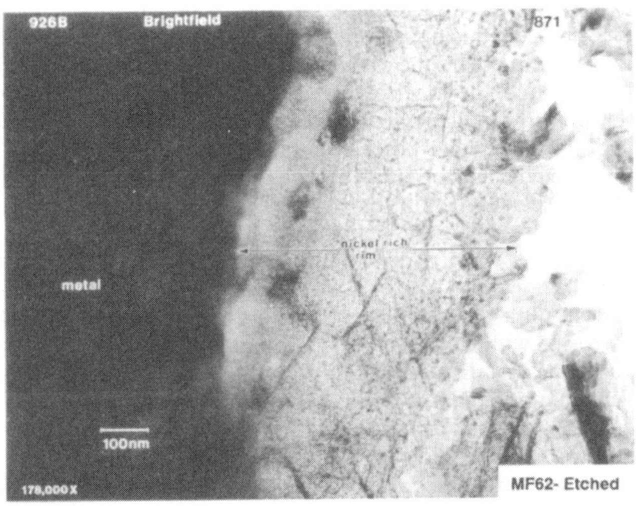

Fig. 2. TEM micrograph of VTiZrNiCrCoMnAlSn alloy surface oxide having high catalytic activity (brightfield imaging)

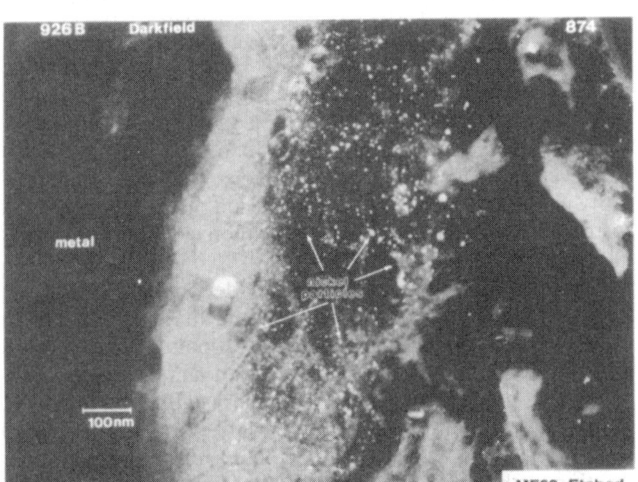

Fig. 3. TEM micrograph of VTiZrNiCrCoMnAlSn alloy surface oxide having high catalytic activity (darkfield imaging)

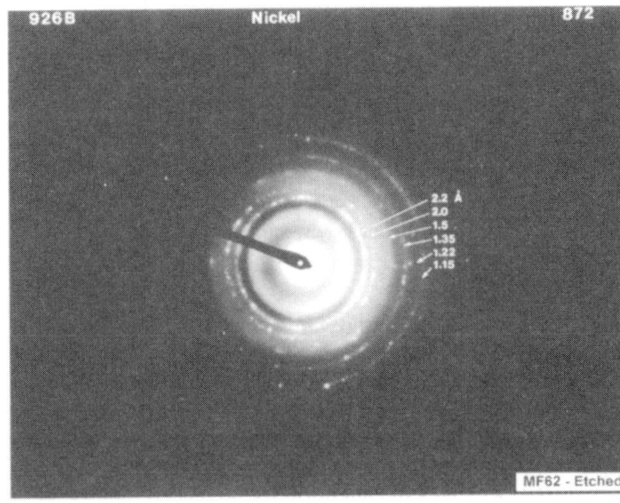

Fig. 4. Select area electron diffraction (SAED) pattern of VTiZrNiCrCoMn-AlSn alloy surface oxide inclusions (Ni regions in Fig. 3)

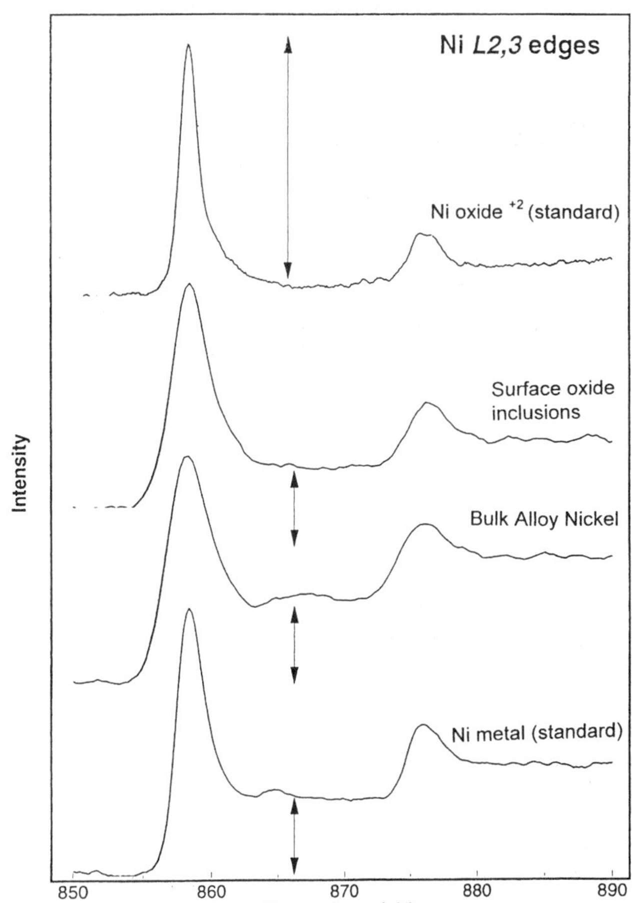

Fig. 5. Electron energy loss spectroscopy (EELS) of VTiZrNiCrCoMnAlSn alloy surface oxide having high catalytic activity

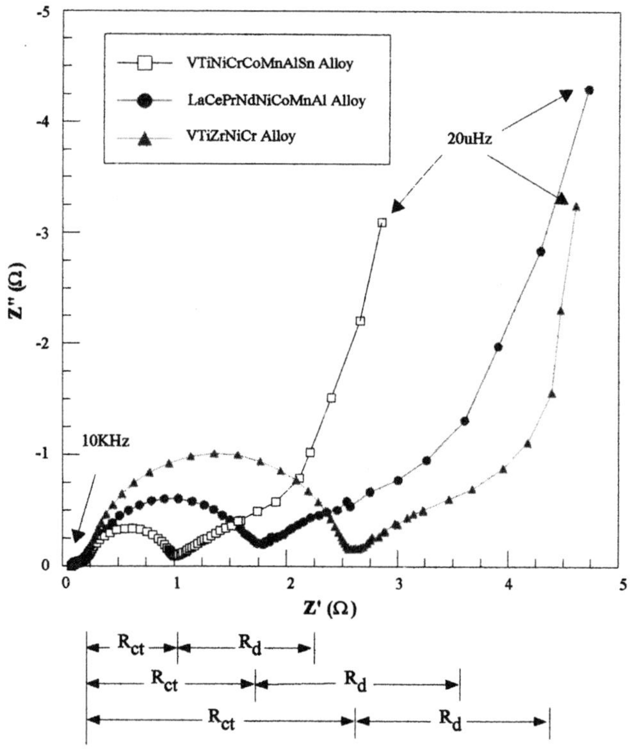

Fig. 6. AC impedance Nyquist analysis of metal hybride alloy catalytic activity (R_{ct} charge transfer resistance, R_d diffusion resistance)

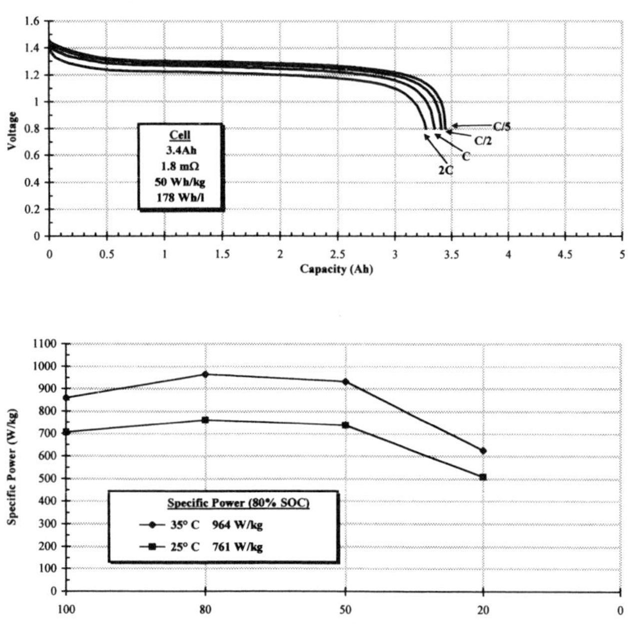

Fig. 7. NiMH C cell capacity and voltage under varying rate discharge (*top*) and specific power measured at C rate, 10 C rate discharge using 10 seconds pulses (*bottom*)

our finding that the improved alloy catalytic activity is the result of the MH surface oxide interface including the presence of the ultra-fine metallic nickel inclusions was provided by EELS, where electron-energy loss shows the inclusions to be at a Ni^{+0} oxidation state (Fig. 5).

Comparable tests on earlier low catalytic activity MH materials showed the absence of inclusions within the oxide under dark-field imaging, a high oxygen signal throughout the oxide under EDS, SAED patterns characteristic of nickel oxide and EELS spectra indicative of nickel in the +2 oxidation state.

AC-impedance measurements of the highly catalytic MH materials on a half-cell basis are presented in Fig. 6, where low-frequency resistance measurements indicate a higher rate of hydrogen diffusion for hydride alloys having the surface oxide catalytic sites and the charge-transfer resistance was two to three times smaller [25, 26]. The resulting higher exchange current density of the surface-modified materials reflects the higher catalytic activity of the metal oxide surface interface with the electrolyte.

NiMH C-cell and HEV batteries were assembled using the MH materials exhibiting the improved surface oxide. C-cell specific power of 1057 W/kg at 350 °C was achieved under the HEV required test conditions of 10-s pulses at a current of 10-C rate (40 A for a 4-A h C cell), as presented in Fig. 7. Typical commercial high-power NiMH cylindrical cells have a specific power in the 600-W/kg range.

3 Conclusions

Specific power and high rate discharge of disordered hydrogen-storage alloys for electrochemical application in NiMH batteries were significantly improved by combining physics and chemistry in the design of the surface oxide on the metal hydride materials. By introducing extremely small diameter metallic nickel alloy inclusions throughout the oxide through the method of preferential corrosion, catalytic activity through reduced charge-transfer resistance was significantly increased and a NiMH battery specific power of 1057 W/kg was attained.

References

1. S.R. Ovshinsky, S.K. Dhar, M.A. Fetcenko, K. Young, B. Reichman, C. Fierro, J. Koch, F. Martin, W. Mays, B. Sommers, T. Ouchi, A. Zallen, R. Young: 17th Int. Seminar & Exhibit on Primary and Secondary Batteries, Ft. Lauderdale, Florida, 6–9 March 2000
2. M. Oshitani, H. Yufu, K. Takashima, S. Tsuji, Y. Matsumaru: J. Electrochem. Soc. **136**(6), 1590–1593 (1989)
3. S.R. Ovshinsky, M. Fetcenko, J. Ross: Science **260**, 176 (1993)
4. S.R. Ovshinsky: *In Disordered Materials: Science and Technology*, ed. by D. Adler, B. Schwartz, M. Silver (Institute for Amorphous Studies Series) (Plenum, New York 1991)
5. H. Nakano, S. Wakao: J. Alloys Compd. **231**, 587 (1995)
6. T. Gamo, Y. Tsuji, Y. Moriwaki: Electrochem. Soc. Proc. **94–27**, 155 (1994)
7. M. Latroche, A. Percheron-Guegan, Y. Chabre, J. Bouet, J. Pannetier, E. Ressouche: J. Alloys Compd. **231**, 537 (1995)
8. T. Sakai, H. Miyamura, N. Kuriyama, H. Ishikawa, I. Uehara: J. Alloys Compd. **192**, 155 (1993)
9. R. Mishima, H. Miyamura, T. Sakai, N. Kuriyama, H. Ishikawa, I. Uehara: J. Alloys Compd. **192**, 176 (1993)
10. T. Weizhong, S. Guangfei: J. Alloys Compd. **203**, 195 (1994)
11. P.H.L. Notten, J.L.C. Daams, R.E.F. Einerhand: Ber. Bunsenges. Phys. Chem. **96**(5), 656–667 (1992)
12. M.A. Fetcenko, S. Venkatesan, S. Ovshinsky: In Proc. Symp. Hydrogen Storage Materials, Batteries, and Electrochemistry (Electrochemical Society, Pennington, NJ 1992) p. 141

13. H. Nakano, S. Wakao, T. Shimizu: J. Alloys Compd. **253–254**, 609 (1997)
14. M. Bououdina, D.L. Sun, H. Enoki, E. Akiba: J. Alloys Compd. **288**, 229 (1999)
15. R.C. Stempel, S.R. Ovshinsky, D.A. Corrigan: IEEE Spectrum **35**(11), 29–34 (1998)
16. A. Zuttel, F. Meli, L. Schlapbach: J. Alloys Compd. **200**, 157 (1993)
17. A. Zuttel, F. Meli, L. Schlapbach: J. Alloys Compd. **231**, 157 (1995)
18. J.W. Kim, S.M. Lee, H.H. Lee, D.M. Kim, J.Y. Lee: J. Alloys Compd. **255**, 248 (1997)
19. K.D. Beccu: United States Patent 3 669 745 (1972)
20. J.R. van Beek, H.C. Donkersloot, J.J.G. Willems: Proc. 14th Int. Power Sources Symposium (1984)
21. M.A. Gutjahr, H. Buchner, K.D. Beccu, H. Saufferer: Power Sources 4, ed. by D.H. Collins (Oriel, Newcastle upon Tyne, UK 1973) p. 79
22. M.A. Fetcenko, S.R. Ovshinsky, B. Chao, B. Reichman: United States Patent 5 536 591 (1996)
23. L. Schlapbach: Surface properties and activation, In *Hydrogen in Intermetallic Compounds II*, ed. by L. Schlapbach (Top. Appl. Phys. 67) (1992) pp. 42–44, 72–75
24. J.P. Bradley, D.A. Brooks: *Analytical Electron Microscopy in Materials Science* (American Laboratory, **20**, No. 6 Shelton, Conneticut, USA 1988)
25. B. Reichman, W. Mays, M.A. Fetcenko, S.R. Ovshinsky: Electrochem. Soc. Proc. **97–16**, 236–246 (1997)
26. B. Reichman, W. Mays, K. Young, M.A. Fetcenko, S.R. Ovshinsky, T. Ouchi: Electrochem. Soc. Proc. **98–15**, 111–119 (1998)

Battery Publications

Alloy Effects on Cycle Life of Ni-MH Batteries (with M.A. Fetcenko, S. Venkatesan, K. Kajita, M. Hirota and H. Kidou), 17th Power Sources Symp. 13 (1991) 149.

Selection of Metal Hydride Alloys for Electrochemical Applications (with M.A. Fetcenko and S. Venkatesan), Electrochem. Soc. Proc. 92-5 (1992) 141.

A Nickel Metal Hydride Battery for Electric Vehicles (with M.A. Fetcenko and J. Ross), Science 260 (1993) 176.

Advances in Ovonic Nickel Metal Hydride Batteries for Electric and Hybrid Vehicles (with P.R. Gifford, M.A. Fetcenko, S. Venkatesan, D.A. Corrigan, A. Holland and S.K. Dhar) in Hydrogen and Metal Hydride Batteries, eds. P.D. Bennet and T. Sakai (Miami Beach, FL, 1994)

Disordered Materials in Consumer and Electric Vehicle Nickel Metal Hydride Batteries (with M.A. Fetcenko, S. Venkatesan and B. Chao), Electrochem. Soc. Proc. 94-21 (1994) 344.

Product Development Through Advances in Materials Science at ECD/OBC (with M.A. Fetcenko and S.J. Hudgens), Daido Journal 4 (1995) 123.

Ovonic Nickel-Metal Hydride Batteries Making Electric Vehicles Practical (with R.C. Stempel), Proc. of the Japanese Society of Electric Vehicles,Tokyo, Japan (February 1997).

Nickel/Metal Hydride Technology for Consumer and Electric Vehicle Batteries—A Review and Up-Date, (with S.K. Dhar, P.R. Gifford, D.A. Corrigan, M.A. Fetcenko and S. Ventakesan), J. of Power Sources 65 (1997) 1.

Nickel Metal Hydride Batteries: The Enabling Technology for Electric and Hybrid Electric Vehicles (With R.C. Stempel, P.R. Gifford and D.A. Corrigan), IEEE Spectrum 35 (1998) 29.

Advancing Batteries (with R.C. Stempel, S.K. Dhar and P.R. Gifford), Electric & Hybrid Vehicle Technology "98" (1998) p. 80.

High Conductivity Negative Electrode Substrates for EV and HEV Ovonic NiMH Batteries (with S. Venkatesan, B. Prasad, B. Aladjov, D. Corrigan and S. Dhar), ibid. p. 263.

Development of Advanced Ni/MH Batteries for Electric and Hybrid Vehicles (with P. Gifford, J. Adams, D. Corrigan and S. Venkatesan), J. of Power Sources 80 (1999) 157.

High Temperature Charge Acceptability Improvements in Ovonic Nickel Metal Hydride Batteries (with S. Venkatesan, B. Aladjov, K. Fok, T. Hopper, B. Prasad, L. Taylor, J. Strebe, M. Arno and S.K. Dhar) 39th Power Sources Conf. (2000) 278.

Technology's Tortoise and Hare – The sociological dynamics are now right for the electric car to eclipse its rival, book review, Nature (November 16, 2000) p.289.

Effect of Alloy Composition on the Structure of Zr Based Metal Alloys (with B.S. Chao, R.C. Young, D.A. Pawlik, B. Huang, J.S. Im and B.C. Chakoumakos), Mat. Res. Soc. Symp. Proc. 575 (2000) 193.

High Performance Zr-based Metal Hydride Alloys for Nickel Metal Hydride Batteries (with R.C. Young, B. Huang, B.S. Chao and Y. Li), Mat. Res. Soc. Symp. Proc. 575 (2000) 187.

Development of High Catalytic Activity Disordered Hydrogen-Storage Alloys for Electrochemical Applications in Nickel- Metal Hydride Batteries (with M.A. Fetcenko), Applied Physics A 72 (2001) 239.

US patents - battery

Electrolytic cell anode
4537674 08/27/1985

Catalytic electrolytic electrode
4544473 10/01/1985

Rechargeable battery and electrode used therein
4623597 11/18/1986

Coatings for electrochemical electrodes and methods of making the same
4624902 11/25/1986

Catalytic hydrogen storage electrode materials for use in electrochemical cells and electrochemical cells incorporating the materials
5104617 04/14/1992

Metastable hydrogen storage alloy material and electrochemical cells incorporating same
5135589 08/04/1992

Electrode alloy having decreased hydrogen over pressure and/ or low self-discharge
5238756 08/24/1993

Electrochemical hydrogen storage alloys and batteries fabricated from these alloys having significantly improved performance characteristics
5277999 01/11/1994

Metal hydride cells having improved cycle life and charge retention
5330861 07/19/1994

Optimized positive electrode for alkaline cells
5344728 09/06/1994

Disordered nickel hydroxide positive electrode material
5348822 09/20/1994

Electrochemical hydrogen storage alloys and batteries fabricated from these alloys having significantly improved capacity
5407761 04/18/1995

Apparatus for deposition of thin-film, solid state batteries
5411592 05/02/1995

Electrochemical hydrogen storage alloys and batteries fabricated from mg containing base alloys
5506069 04/09/1996

A solid state, electrically insulating, ion conducting electrolyte material and a thin-film, solid state battery employing same
5512387 04/30/1996

Enhanced nickel metal hydroxide positive electrode materials for alkaline rechargeable electrochemical cells
5523182 06/04/1996

Improved electrochemical hydrogen storage alloys for nickel metal hydride batteries
5536591 07/16/1996

A solid state, electrically insulating, ion conducting electrolyte material and a thin-film, solid state battery employing same
5552242 09/03/1996

Electrochemical hydrogen storage storage and batteries containing heterogeneous powder particles
5554456 09/10/1996

Optimized cell pack for large sealed nickel metal hydride batteries
5558950 09/24/1996

A nickel metal hydride battery containing a modified disordered multiphase nickel aluminum based positive electrode
5567549 10/22/1996

A nickel metal hydride battery containing a modified disordered multiphase nickel hydroxide positive electrode
5569563 10/29/1996

Electrochemical hydrogen storage alloys and batteries fabricated from mg containing base alloys
5616432 04/01/1997

Compositionally and structurally disordered multiphase nickel hydroxide positive electrode for alkaline rechargeable electrochemical cells
5637423 06/10/1997

Nickel-metal hydride batteries having high power electrodes and low-resistance electrode connections
5851698 12/22/1998

High power nickel-metal hydride batteries and high power electrodes for use therein
5856047 01/05/1999

Nickel battery electrode with multiple composition nickel hydroxide active materials
5861225 01/19/1999

Mechanical and thermal improvements in metal hydride batteries, battery modules and battery packs
5879831 03/09/1999

Beta to gamma phase cycle able electrochemically active nickel hydroxide material
5905003 05/18/1999

Optimized positive electrode for alkaline cells
5948564 09/07/1999

A solid state battery having a disordered hydrogenated carbon negative electrode
5985485 11/16/1999

Nickel positive electrode having high temperature capacity
6017655 01/25/2000

Active nickel hydroxide material having controlled water content
6019955 02/01/2000

Structurally modified nickel hydroxide and method for making same
6086843 07/11/2000

Method of synthetically engineering alloys formed of high melting point and high vapor pressure materials
6143373 11/07/2000

Nickel positive electrode material comprising rare earth minerals
6150054 11/21/2000

Composite positive electrode material and method for making same
6177213 01/23/2001

High storage capacity alloys enabling a hydrogen-based eco system
6193929 02/27/2001

Improved hydrogen storage alloys and methods and improved nickel metal hydride electrodes and batteries using same
6210498 04/03/2001

Active electrode compositions comprising raney based catalysts and materials
6218047 04/17/2001

Nickel hydroxide positive electrode material exhibiting improved conductivity and engineered activation energy
6228535 05/08/2001

Monoblock battery assembly
6255015 07/03/2001

Modified electrochemical hydrogen storage alloy having increased capacity, rate capability and catalytic activity
6270719 08/072001

Hydrogen-based ecosystem
6305442 10/23/2001

A hybrid electric vehicle incorporating an integrated propulsion system
6330925 12/18/2001

Composite positive electrode material and method for making same
6348285 02/19/2002

Mechanical and thermal improvements in metal hydride batteries, battery modules and battery packs
6372377 04/16/2002

High power nickel-metal hydride batteries and high power alloys/electrodes for use therein
6413670 07/02/2002

Nickel hydroxide electrode material and method for making the same
6416903 07/09/2002

Nickel hydroxide positive electrode material exhibiting improved conductivity and engineered activation energy
6447953 09/10/2002

Electrochemically stabilized cani5 alloys and electrodes
6524745 02/25/2003

Nickel positive electrode material with misch metal additives
6537700 03/25/2003

Composite positive electrode material and method for making same
6548209 04/15/2003

Composite positive electrode material and method for making same
6569566 05/27/2003

Monoblock battery assembly with cross-flow cooling
6689510 02/10/2004

Finely divided metal catalyst and method for making same
6841512 01/11/2005

Monoblock battery assembly with cross-flow cooling
6864013 03/08/2005

Mechanical and thermal improvements in metal hydride batteries, battery modules and battery packs
6878485 04/12/2005

Method of making a catalyst
7045484 05/16/2006

Performance enhancing additive material for the nickelhydroxide positive electrode in rechargeable alkaline cells
7172710 02/06/2007
7201857 04/10/2007

Mechanical and thermal improvements inmetal hydride batteries, battery modules and battery packs
7217473 05/15/2007

Active electrode composition with conductive polymeric binder
7238446 07/03/2007

US patents - electric/ hybrid vehicles

A hybrid electric vehicle incorporating an integrated propulsion system
6330925 12/18/2001

Hybrid electric vehicle
6557655 05/06/2003

A hybrid electric vehicle incorporating an integrated propulsion system
6565836 05/20/2003

A very low emission hybrid electric vehicle incorporating an integrated propulsion system including a hydrogen powered internal combustion engine and a high power NI-MH
6759034 07/06/2004

Method and system for hydrogen powered internal combustion engine
6820706 11/23/2004

A hybrid electric vehicle incorporating an integrated propulsion system
6837321 01/04/2005

Onboard hydrogen storage unit with heat transfer system for use in a hydrogen powered vehicle
6860923 03/01/2005

A hydrogen powered scooter
6918382 07/19/2005

Onboard hydrogen storage unit with heat transfer system for use in a hydrogen powered vehicle
6918430 07/19/2005

Chapter VI: Hydrogen Storage, Fuel Cells and the Hydrogen Energy Loop

Our energy sources have changed radically during the past 150 years. One can observe an encouraging trend as more and more nations consider the transition from a carbon economy toward a hydrogen economy. Our major source of energy shifted over the decades from carbon rich wood to coal and more recently to carbon poor natural gas and oil. The energy we use is derived from burning (oxidizing) carbon to carbon dioxide and hydrogen to water. With the world's population and economy expanding even the relativity carbon poor fuels are causing unacceptable atmospheric concentrations of carbon dioxide which warm the earth and cause climate changes. Fuel for transportation alone accounts for one third of the greenhouse gas CO_2 emission in the United States. This climate changing pollution as well as our dependence on imported fossil fuels can be halted by taking the next step in fuel technology to a clean hydrogen economy.

Our concerns with pollutants and political problems related to fossil fuel energy sources would be forever over if we could control and hence use the fusion process of hydrogen nuclei into helium nuclei which is the main source of energy of our sun and the stars in the Universe. Even though the control of this fusion reaction on Earth is an object of intense research, we must unfortunately write this attractive option off as a distant technology. We instead propose to make use of the chemical reaction of hydrogen with oxygen as a clean energy source. The reaction product is just clean water. 1kg of hydrogen has the energy equivalent of 1 gallon of gasoline. Hydrogen as a fuel can directly power combustion engines, be used for portable and stand-by power generation and produce electricity by combining with oxygen in fuel cells. While hydrogen has a large potential as a fuel, a major drawback in its widespread utilization has been the lack of safe, lightweight hydrogen storage media. Hydrogen compressed to 3000 psi in a heavy metal cylinder contains less energy than 1 gallon of gasoline and requires furthermore an expensive compressor. Liquid hydrogen while denser than the compressed gas, unfortunately requires cryogenic refrigeration to a temperature below -253 degree C where hydrogen becomes a liquid. This is expensive and the liquid is highly volatile when spilled. Liquefaction of hydrogen is therefore not a practical solution. Ovshinsky developed a much better and safer hydrogen storage system.

In reading the previous chapters on Ovshinsky's NiMH battery developments one should realize, as Ovshinsky did, that the negative electrode of the NiMH battery is close to an ideal hydrogen storage medium. The material of the electrode can be filled with hydrogen from an electrolyte or directly from hydrogen gas because its surface layer contains, in the form of nickel-alloy nano-particles, the catalytic sites that break molecular hydrogen into atoms which then diffuse into the electrode material. This hydrogen fills the spaces between the atoms in such material to an amazing density exceeding that of liquid hydrogen. Yet, the hydrogen storage material filled with hydrogen can be handled at room temperature and shipped without danger of fire or combustion. The hydrogen can be released in a controlled manner only upon heating the storage material. When empty the material is ready to be filled again with hydrogen.

By replacing the gasoline tank of a Toyota hybrid car with his hydrogen storage material, Ovshinsky demonstrated safe and efficient storage of hydrogen and the use of hydrogen as a fuel in the Toyota's combustion engine. With a 3 kg hydrogen storage capacity, the modified Toyota proved to have a 150 miles driving range with zero emission. The detailed engineering and test results are discussed in the publications which follow. Ovshinsky and his team truly established that the pollution free hydrogen economy is not a future dream but ready for practical uses today.

Even though hydrogen can fuel a car's combustion engine, a commercially viable fuel cell is a key for opening another door to the hydrogen economy because it converts the reaction of hydrogen with oxygen directly into electricity. The hydrogen fuel cell was invented by Sir William Grove about 160 years ago. Modern versions of fuel cells have played a crucial role already by providing electricity in all manned spacecraft of the US space program. The alkaline fuel cells used by NASA resemble batteries in that the electricity is generated electrochemically. It therefore comes as no surprise that Ovshinsky and his team became major players in the field of fuel cells.

With small modifications, the NiMH negative electrode material and the related hydrogen storage materials become the anode of Ovshinsky's alkaline fuel cell. These materials have excellent catalytic activity for converting hydrogen molecules, the primary energy source of fuel cells, into hydrogen ions. These ions diffuse through the alkaline electrolyte to the cathode. Meanwhile electrons pass from the anode through the external electrical circuit to the cathode where they produce hydroxyl ions. Near the cathode-electrolyte interface the hydroxyl ions react with the hydrogen ions forming water. As long as hydrogen gas is available at the anode and oxygen at the cathode, the electrical current flowing through the external circuit is a source of electrical energy. The new part of the fuel cell is the cathode in which catalytic activity has to transform oxygen gas molecules into oxygen atoms and, with the help of the electrons coming through the electrical circuit, into hydroxyl ions. There are a number of material engineering challenges which have to be met to remove the water from the cell to avoid diluting the electrolyte as well as designing a separator between anode and cathode which transmits hydrogen ions but blocks electrons. NASA used very pure oxygen for their fuel cells in space. For practical terrestrial use one wishes to use air, whose carbon dioxide contaminant can cause undesired reactions which must be avoided. One further unique aspect of Ovshinsky's fuel cell is its regenerative feature. If the electrical current is reversed, as can occur when one steps on the car's brake, the electrical energy flows into the Ovshinsky fuel cell. The cell acts like a battery which is being charged. This energy is immediately available for use, thus eliminating any start-up delay during the conventional fuel cell operation.

The American Chemical Society honored these achievements by naming Stan Ovshinsky and his wife Iris "Heroes of Chemistry 2000" for "advances in electrochemical, energy storage and generation, including the development of Ovonic nickel metal hydride (NiMH) rechargeable batteries, regenerative fuel cells, solid hydrogen storage systems, and amorphous silicon photovoltaics" and having "made significant and lasting contributions to global human welfare." Moreover, Ovshinsky's achievements were profiled in a book published by the MIT Press, in association with the Lemelson-MIT Program for Invention and Innovation, as one of 35 American inventors over the past century "who helped to shape the modern world." (*Inventing Modern America: From the Microwave to the Mouse*, MIT Press, 2001).

Ovshinsky's innovations are cornerstones for the development of a sustainable hydrogen energy economy. Hydrogen is the simplest element consisting of one proton and one electron. It is by far the most abundant element in the universe and the source of stellar energy, by fusing into helium. On Earth, however, it is not available as a primary energy source like coal or oil. Hydrogen is chemically bound in form of water and carbohydrates. Water can be split into hydrogen and oxygen by passing an electric current through it. This electrolysis is, in a way, the reverse reaction of a fuel cell in which hydrogen and oxygen combine to water and produce a current. Splitting water by electrolysis unfortunately is an energy intensive process because of the strong chemical binding force which holds the water molecule together. Yet, an abundance of solar radiation hitting Earth can be converted to electricity by photovoltaic panels. This together with wind power and off-peak power generation can make hydrogen production economically feasible and avoids the burning of carbon fuels at every step of the process. Hydrogen stored in Ovshinsky's hydrogen storage alloys can serve two important functions. Locally it can fulfill the indispensable energy storage function of smoothing out the unavoidable fluctuations of solar, wind, and other power sources. In addition the solid hydrogen storage systems allow the use of hydrogen energy together with Ovshinsky's NiMH batteries in gas/electric hybrid cars, trucks and buses.

Early in 2007, The New York Times invited 10 companies to demonstrate their hydrogen powered vehicles at the spot where the Hindenburg, the largest zeppelin aircraft, crashed on May 6, 1937 ("Hydrogen's Second Coming on the Road, Hope for a Zero-Pollution Car", Don Sherman, NYT 4/29/07). The purpose was to see what the cars of the future hydrogen economy may look like. The HydroGen3 of General Motors carried 6.8 pounds of hydrogen compressed to 10.000 psi for a driving range of 168 miles. The Ford E-450, a large van of Ford Motor carried 66 pounds of hydrogen compressed to 5,000 psi in six tanks and drove 150 miles. The winner was clearly Ovshinsky's converted Toyota Prius which drove 190 miles with its 7.9 pounds of hydrogen safely contained in two metal alloy hydrogen storage tanks. It is exciting to see in the following publications how Ovshinsky and his team have solved the technical problems of a future that no longer is dependent on fossil fuel.

Effect of Alloy Composition on the Structure of Zr Based Metal Alloys

B.S. CHAO, R.C. YOUNG, S.R. OVSHINSKY, D.A. PAWLIK, B. HUANG, J.S. IM, and B.C. CHAKOUMAKOS*
Energy Conversion Devices, Inc., 1675 West Maple Road, Troy, MI 48084
*Neutron Scattering Section, Oak Ridge National Lab., Oak Ridge, TN 37831

Abstract

The structures of Ovonic multi-element, multi-phase Zr-based transition metal alloys for hydrogen storage are studied. The alloys are designed to be multi-phase materials. The hexagonal and diamond cubic structures, known as C14 and C15 Laves structures respectively, are the two major hydrogen storage phases in the alloys. In both C14 and C15 structures, Zr and Ti are the elements typically occupying the hydride former (A) sites and V, Cr, Mn, Fe, Co and Ni the catalytic (B) sites. Based on the application of Ovshinsky's design principles for disordered materials, both A and B sites are compositionally disordered by the corresponding elements in cubic (C15) structure. However, the hexagonal (C14) structure has two distinct B sites, B(I) and B(II) with a 1:3 ratio. Preliminary neutron diffraction studies indicate that the B(I) sites are predominantly occupied by Ni and the B(II) sites are randomly mixed with V, Cr, Mn, Fe, and Ni. It is then proposed that the formula of the hexagonal structure should become $A_2B(I)_1B(II)_3$ in the current multi-element Zr-based alloys. Minor phases close to the structures of Zr_7Ni_{10}, Zr_9Ni_{11}, and ZrO_2 are also present in some alloys. The average electron concentration factor (e/a) derived from the alloy composition dictates the alloy structures. The alloys with higher electron concentration factor (> 7.1) favor the diamond cubic structure. On the other hand, the hexagonal structure is associated with the alloys with lower electron concentration factor (< 6.5). The alloys having electron concentration factors in between are mixtures of the diamond cubic and hexagonal structures.

Introduction

Energy Conversion Devices, Inc. (ECD) and its subsidiary Ovonic Battery Company, Inc. (OBC), have pioneered the development and commercialization of nickel metal hydride secondary batteries. The introduction of the Ovonic alloys, based upon the multi-element, multi-phase, compositional disorder design principles [1], led the way to the successful commercialization of the nickel metal hydride technology of both Zr-based and mischmetal-based systems. The Ovonic nickel metal hydride battery has become the battery of choice for electric vehicle application due to its superior properties [2]. Among the nickel metal hydride alloys, it has been known that the Zr based AB_2 Laves structure alloys have higher capacity [mAh/g] than the La based AB_5 alloys [3]. Recently, Young et al. have demonstrated that the advanced Ovonic multi-element Zr-Ti-V-Cr-Mn-Co-Ni alloys can deliver a capacity of 465 mAh/g [3]. Total capacity and the high rate discharge capability of the Ovonic alloys are superior to the mischmetal based AB_5 alloys currently used for commercial electronic products. $ZrMn_2$ and ZrV_2 binary compounds have the C14 and C15 type Laves structures and can absorb hydrogen contents up to $ZrMn_2H_{3.0}$ and $ZrV_2H_{5.2}$, respectively [4,5]. It is known, however, that these two compounds can not be used for battery applications due to the lack of appropriate thermodynamic properties. It is also known that the stability of the transition metal AB_2 Laves

Mat. Res. Soc. Symp. Proc. Vol. 575

structures are generally influenced by the size-factor principles and the electron concentration factor *(e/a)* [6-8]. The former, or the atomic size ratio (R_A/R_B), is ideally 1.225 with a range of 1.05-1.68. Laves and Witte's [6,7] investigation found that for the Mg based binary alloy systems, with increasing valence electron concentration, the structure moved from $MgCu_2$ (C15 type) to $MgNi_2$ (C36 type) to $MgZn_2$ (C14 type). The works by Bardos et al. [9] and Watson and Bennett [10] were successfully extended to correlate the average electron concentration and the d-band hole number to the crystal structure in the transition metal Laves structures. In this study, we report that the Ovonic multi-element Zr based AB_2 alloys are multiphase materials. Zr and Ti are the elements used for the A sites. B site elements are V, Cr, Mn, Co and Ni. The R_A/R_B ratio is typically at about 1.15 – 1.23. In general, the alloys consist of a mixture of C14 and C15 structures along with minor phases similar to Zr_7Ni_{10} and Zr_9Ni_{11}. The C14 and C15 structures are the major host matrices for hydrogen storage. In addition, a ZrO_2 impurity phase is also present in the alloys. We found that there is a correlation between the electron concentration factor *(e/a)* of the alloy and the formation of the C14 and C15 structures. The electron concentration factor is defined as the average number of electrons per atom outside the closed electron shells of the component atoms [8]. It is directly derived from the alloy composition. The alloys having *e/a* of 7.1 or larger favor the C15 structure. On the other hand, the C14 structure appears in the alloys with *e/a* of 6.5 or smaller. The *e/a* between 6.5-7.1 results in alloy mixtures of C14 and C15.

Experimental

Arc melt casting, induction melt casting and rapid quench by melt spinning are the three techniques used to fabricate the Ovonic metal alloys in this report. The raw materials used for alloy synthesis were 99% pure metals. Arc melting was performed under an argon atmosphere. The mixtures of the elements were melted in a water-cooled copper crucible. The ingot was remelted several times to ensure its homogeneity. For the induction melt casting, a BN crucible was used to melt and then to pour the molten alloy into either a stainless steel or graphite pan. For the melt spun alloys, the starting materials were melted in an induction melting furnace and ejected from the bottom of the BN crucible through a precisely machined ejection nozzle to the surface of a vertically spinning copper-beryllium wheel. The alloys were examined by XRD for structure, micro-morphology and phase composition by SEM/EDS, and overall average alloy composition by ICP. X-ray diffraction was performed on a Philips Θ - 2Θ XRD powder diffractometer equipped with a sample spinner, fixed optics, and diffracted beam monochromator. Using Cu Kα radiation (λ_{Cu} = 0.1542 nm), typical run parameters were $15s/0.01^{o}$ to $60s/0.005^{o}$ on side drifted fine powder samples. Diffraction analyses were accomplished through use of "Materials Data, Inc." software: "Jade+" for basic structural phase analyses and rough least squares fits, "Shadow" for full profile lattice refinements, and "Riqas" for Rietveld method crystal structure refinement. Micro-morphology examination was carried out in a JEOL model 35C SEM microscope equipped with a Kevex EDS detector. Average composition was performed on a Varian Liberty 100 ICP-AES system.

Results

Figure 1(a) displays the XRD trace recorded from the as-prepared Zr151 alloy made by arc melt (nominal composition of $Zr_{25}Ti_{8.5}V_8Cr_{20}Mn_{13}Ni_{25.5}$). The alloy is dominated by the C14 structure along with minor Zr_7Ni_{10}, Zr_9Ni_{11}, and ZrO_2 phases.

Figure 1. XRD patterns of Zr151 alloy of a) as-prepared and b) hydrided. The insert represents the high resolution scan of C14 (213) peak.

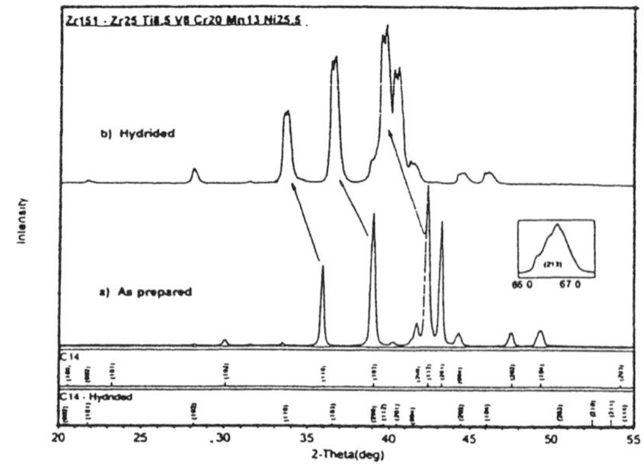

The lattice constants of the C14 structure are centered at about $a_0 \sim 0.499$ nm and $c_0 \sim 0.816$ nm. The XRD trace of the hydrided Zr151 alloy is shown in figure 1(b). All peaks related to the C14 structure have been shifted to lower 2Θ angles. This corresponds to a lattice expansion of the host material due to hydrogen absorption. The expanded lattice constants of the hydride Zr151 alloy are centered around $a_0 \sim 0.532$ nm and $c_0 \sim 0.871$ nm. It represents a 6.6% lattice increase in both a and c axes, equivalent to about 21% volume expansion. It is also noted that there are multiple peaks for each reflection of the hydride alloy. What is the origin of the multiple peaks? The as-prepared alloy was carefully examined by the high resolution XRD scans. It is found that the peak profile of each reflection is not a single peak. For example, the insert is the peak profile of the C14 (213) reflection. The (213) reflection actually consists of at least ***three*** peaks. It implies that there are at least three sets of C14 lattice constants in the Zr151 alloy. The multiple sets of lattice constants in the alloy are associated with the multiple local bonding environments, that is, differences in composition providing multiple lattice constants. The amount of stored hydrogen is slightly varied among the multiple sets of lattice constants. Greater stored hydrogen content corresponds with the larger expansion in the lattice constant (larger peak shift to lower 2Θ values in the XRD trace). Less stored hydrogen then correlates to slightly smaller lattice expansion, smaller shift in peak position. The compositional disorder of the alloy is the origin of the multiple sets of lattice constants, i.e., the origin of the structural disorder. Neutron diffraction result presented in the later section will provide the evidence of the compositional disorder of the alloys.

The MF139Z alloy ($Zr_{27}Ti_8V_5Cr_5Mn_{16}Ni_{38}$) is one of the OBC current production alloys. Figure 2(a) is the XRD pattern of the as-prepared MF139Z made by the induction melt method. In the 2Θ range of 27-53° shown in figure 2(a), the (220), (311) and (222) peaks of the C15 structure totally overlap to the (110), (112), and (004) peaks of the C14 structure, respectively. However, the C14 structure has more allowed peaks in the same range, namely the (102), (103), (200), (201), (202) and (104) peaks. It is noticed that the peak intensity relationship of the MF139Z is very different than that of the Zr151 displayed in figure 1(a). The result of Rietveld crystal structural refinement indicates that the MF139Z alloy is a mixture of C14 and C15 structures. The volume fractions are ~63% C14 and ~37% C15. The lattice constants are $a_o \sim 0.498$ nm/ $c_o \sim 0.812$ nm for C14 and

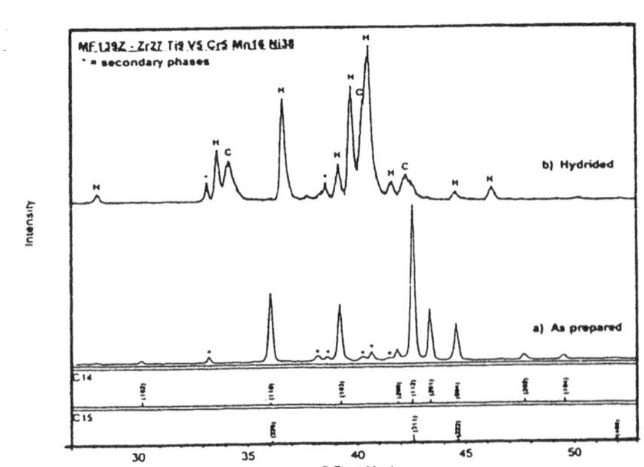

Figure 2. XRD patterns of MF139Z alloy of a) as-prepared, and b) hydrided. H: hydrided C14, C: hydrided C15, and (*): minor phases.

$a_o \sim 0.703$ nm for C15. The peaks marked with the (*) are due to the minor Zr_7Ni_{10}, Zr_9Ni_{11}, and ZrO_2 phases. The characteristics of a C14/C15 mixture is much more evident in the hydride state, see figure 2(b). All major peaks are observed to shift to lower 2Θ values. The full width at half maximum of the peaks is also slightly increased. Peaks marked by H are related to the C14 structure in the hydride state. The expanded lattice constants of the C14 structure are centered at $a_o \sim 0.532$ nm and $c_o \sim 0.868$ nm, equivalent to a 22% expansion in volume from the as-prepared alloy. Peaks labeled with C are due to the hydrided C15 structure. Its expanded lattice constant is centered at $a_o \sim 0.746$ nm (19% volume expansion).

Table I
Electron concentration factor (e/a) and atomic size of the alloying elements

	Zr	Ti	V	Cr	Mn	Fe	Co	Ni
e/a	4	4	5	6	7	8	9	10
Goldschmidt radius (nm)	0.161	0.145	0.136	0.128	0.131	0.127	0.126	0.125

Table I lists the electron concentration factor (e/a) [11], and Goldschmidt atomic radii [12] of the elements used in the Ovonic alloys presented in this paper. The average *e/a* of the alloys is derived from the following equation:

$$(e/a)_{alloy} = \Sigma_i \, [(e/a)_i \times (at.\%)_i] \qquad (1)$$

where i is the ith element. In applying equation (1), *e/a* of the Zr151 and MF139Z alloys is 6.40 and 6.91, respectively.

Melt spinning is one of the rapid quenching techniques currently used in industry to fabricate the material. Ovonic metal alloys prepared by this non-equilibrium process have been found to reveal a unique TEM microstructure [13]. Figure 3 represents the XRD curves, shown in the 2Θ range of $28^o - 52^o$, of three melt spun alloys: (a) MS139X173 ($Zr_{26}Ti_9V_4Cr_4Mn_{15}Ni_{42}$), (b) MS139-326 ($Zr_{26}Ti_9V_5Cr_6Mn_{17}Ni_{37}$), and (c) MS139X202 ($Zr_{25}Ti_7V_8Cr_{10}Mn_{22}Ni_{28}$). These three XRD patterns are distinctly different. For the MS139X173 alloy in curve (a), all three peaks are associated with the (220), (311), and (222) reflections of the cubic C15 structure with a lattice constant of 0.704 nm. Several weak features located at the 2Θs of $\sim 30^o$, $\sim 40^o$, $\sim 43.5^o$ are due to the minor C14 structure in the alloy. Curve (c) is very similar to the XRD pattern shown in figure 1(a). It indicates that the MS139X202 alloy is also dominated by the C14 structure. The peaks marked by arrows in curve (c) are reflections that are only related to the C14 structure. Their reflection indices are shown on the bottom of the figure. The XRD pattern of the MS139-326 alloy, curve (b), is very similar to that of the MF139Z alloy shown in figure 2(a). The structure of the alloy

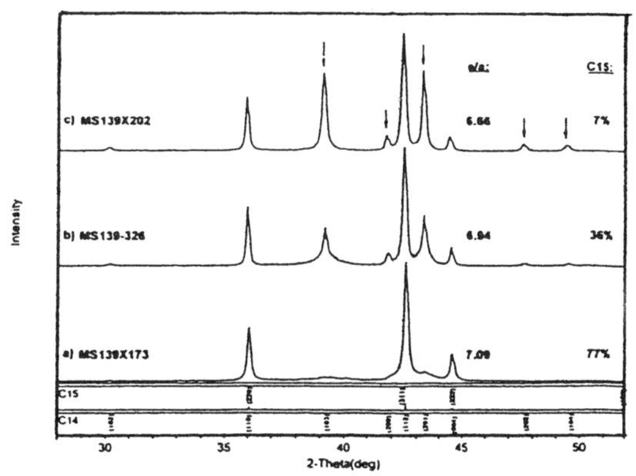

Figure 3. XRD patterns of three alloys, a) MS139X173, b) MS139-326, and c) MS139X202. The arrows indicate the peaks solely attributed to the C14 structure.

MS139-326 is also a mixture of C14 and C15 structures. The peaks related to the minors in the MF139Z alloy are all absent in the MS139-326. Thus, rapid solidification inhibits gross segregation and the development of the minor phases [13]. The *e/a* ratios of the three alloys are (a) 7.09, (b) 6.94, and (c) 6.66. Profile refinement analysis indicates that the volume fractions of the C15 are approximately 77%, 36% and 7% for alloys in (a), (b), and (c) respectively.

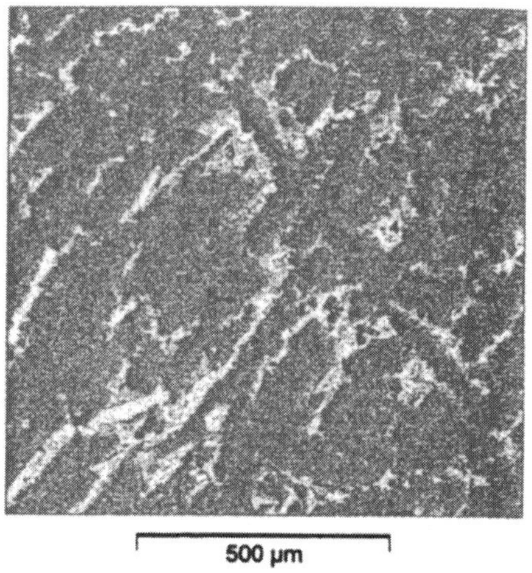

Figure 4. SEM-BEI picture taken from the as-prepared MF139Z alloy. Five image areas are shown. #1 and #2 are the areas for hydrogen storage. #3-#5 are the minors.

Figure 4 is a typical BEI (backscattered electron image)-SEM photograph of the as-prepared MF139Z alloy. It shows five electron image contrasts resulting from differences in atomic compositions. The gray color region is the primary matrix and is responsible for the hydrogen storage. It actually consists of two areas labeled by #1 (light gray) and #2 (dark gray) where the average composition is $Zr_{25.3}Ti_{8.4}V_{5.1}Cr_{3.8}Mn_{17.8}Ni_{39.6}$ and $Zr_{25.3}Ti_{6.7}V_{7.6}Cr_{9.1}Mn_{21.8}Ni_{29.5}$, respectively. The e/a of #1 is 7.04, close to the 7.09 of MS139X173. It is reasonable to suggest that the former one is related to the C15 structure. A similar argument can also be applied to #2, its e/a is 6.68, closely related to the C14 structure. The regions labeled #3 - #5 are the minors. The compositions of the regions #3 and #4 are $Zr_{38.4}Ti_{4.6}Cr_{0.2}Mn_{4.1}Ni_{41.2}Sn_{11.5}$ and $Zr_{33.7}Ti_{10.4}Cr_{0.2}Mn_{1.6}Ni_{52.3}Sn_{11.5}$. The presence of Sn in these regions is a consequence of the zircalloy used as the starting raw material for Zr. Zircalloy contains ~0.5 at.% Sn which segregates out to the minors during the melt. These two regions have phases similar to Zr_7Ni_{10} and Zr_9Ni_{11}. Region #5 is due to the ZrO_2 phase.

The schematics of the C14 and C15 Laves structures are illustrated in Figure 5. In the C14 and C15 structures, each A atom is coordinated with 12 B atoms in a predominantly icosahedral configuration, and surrounded tetrahedrally by 4 A atoms. In transition metal alloys, Zr and Ti, the hydride formers, are the typical A site elements, and V, Cr, Mn, Co, Ni are selected for B sites. In the case of C15 cubic structure, B atoms form the tetrahedral unit B_4 that are linked together by sharing a vertex to form a three-dimension network. For the C14 hexagonal structure tetrahedral units are built up from two crsytallographically distinct B(I) and B(II) atoms and are connected by B(I) vertex sharing and by triangle B(II)-B(II)-B(II) face sharing. The ratio of the B(I) and B(II) sites is 1 to 3. As pointed out in figure 2, the higher *e/a* alloy (curve 2(a)) favors the C15 structure over the C14 structure. It seems to suggest that the vertex-sharing B(I) sites of the C14 structure would have higher electron concentration than the face-sharing B(II) sites. Therefore, we expected that the later transition elements would preferentially go to the B(I) sites. In order to confirm this hypothesis, neutron diffraction was performed on the MF139Z alloy. Rietveld crystal structural refinement indicates that in the C14 structure, Zr and Ti are randomly mixed at the A sites, Ni atoms prefer to occupy the B(I) sites [14] and V, Cr, Mn and the remaining Ni are randomly distributed in the B(II) sites. For the C15 cubic structure, Zr and Ti randomly occupy A sites and all B sites are randomly mixed by the V, Cr, Mn, and Ni elements [15].

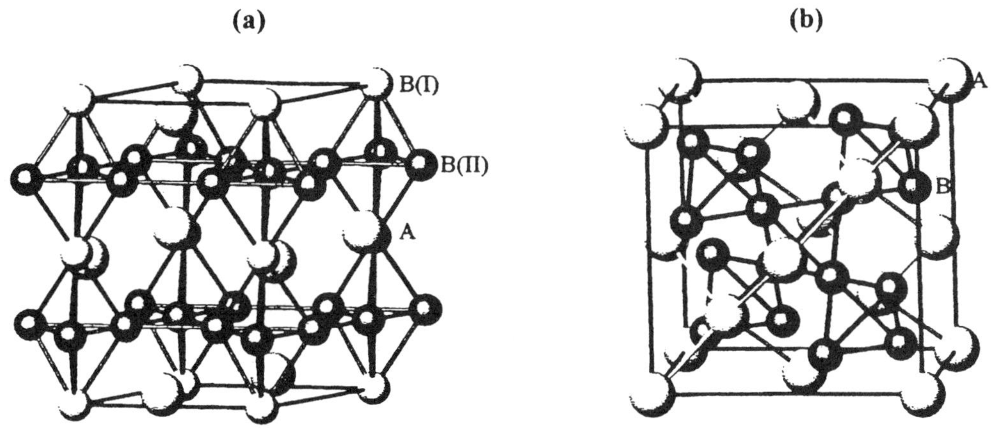

Figure 5. The schematic diagrams of a) C14 and b) C15 Laves AB_2 structures.

Conclusions

The Ovonic alloys specially engineered by the principles of multi-element, compositional disorder are multi-phased. The C14 and C15 AB_2 structures are the two major hydrogen storage phases in the alloys. Neutron diffraction studies confirm the compositional disorder in the alloys. In the C15 cubic structure, both A and B sites are randomly distributed by the corresponding elements. For the C14 structure, Zr and Ti are randomly distributed in A sites, Ni atoms are preferentially occupied in B(I) sites and V, Cr, Mn, and Ni are randomly mixed in B(II) sites. The alloys with higher *e/a* favor the C15 structure.

References

1. K.Sapru, B. Reichman, A. Reger, S.R. Ovshinsky, U.S. Patent 4 623 597 (1986).
2. S.R. Ovshinsky, M.A. Fetcenko, J. Ross, Science **260**, 176 (1993).
3. R.T. Young, B. Huang, B.S. Chao and S.R. Ovshinsky, Mater. Res. Soc. Proc. Spring Symp. CC, (1999).
4. J.J. Didisheim, K. Yvon, D. Shaltiel, P. Fisher, Solid State Comm. **31**, 47 (1979).
5. D. Fruchart, A. Rouault, C.B. Shoemaker, D.P. Shoemaker, J. Less-Comm. Met. **73**, 363 (1980).
6. F. Laves, H. Witte, Metallwirschaft **14**, 645 (1935).
7. F. Laves, H. Witte, Metallwirschaft **15**, 840 (1936).
8. J.H. Zhu, P.K. Liaw, C.T. Liu, Mater. Sci. and Eng. **A239-240**, 260 (1997).
9. D.I. Bardos, K.P. Gupta, P.A. Beck, Trans. Met. Soc. AIME **221**, 1087 (1961).
10. R.E. Watson, L.H. Bennett, Acta Metall. **32**, 477 and 491 (1984).
11. Periodic Table of the Elements from the Sargent Welch (1998).
12. "Smithell's Metal Reference Book" edited by E.A. Brandes and G.B. Brook, 7th Edition, Butterworth Heinemann, p. 4-41 (1998).
13. B.S. Chao et al., will be published .
14. M. Bououdina, J.L. Soubeyroux, D. Fruchar, P. de Rango, J. Alloys Comp **257**, 82 (1997).
15. M. Yoshida, E. Akiba, J. Alloys Comp **224**, 121 (1995).

2004-01-0060

Hydrogen-Fueled Hybrid: Pathway to a Hydrogen Economy

Richard Geiss, Bruce Webster, S. R. Ovshinsky, Robert Stempel
Energy Conversion Devices, Inc.

Rosa Chiang Young, Yang Li, Vitaliy Myasnikov
Texaco Ovonic Hydrogen Systems, LLC

Bruce Falls, Alwin Lutz
Quantum Technologies, Inc.

ABSTRACT

A production hybrid-electric vehicle with gasoline internal combustion engine (ICE) has been modified to operate on hydrogen fuel and to demonstrate an advanced metal hydride fuel storage system. Fuel economy, range, performance, and emissions of the modified vehicle are similar to levels achieved with current fuel cell vehicles.

This project was conceived and executed by Energy Conversion Devices, Inc. All vehicle modifications, engine and chassis dynamometer tests, and road tests were performed by Quantum Technologies, Inc. at their Lake Forest, CA facility. Project activities and test results are chronicled in this paper.

INTRODUCTION

The development timeline for a Texaco Ovonic Metal Hydride (TOMH) onboard fuel storage program specified demonstration of a TOMH system in a roadable vehicle by June of 2003. A fuel cell vehicle could not be obtained to support that timeframe. It was decided to commission a new project that would modify a production gasoline vehicle to operate on hydrogen fuel. This vehicle would then be used to demonstrate the TOMH storage system.

The specific project objective was to develop and demonstrate a hydrogen-fueled, ICE-powered (H2ICE) vehicle with overall operating characteristics similar to those of current prototype fuel cell vehicles.

If this could be achieved, the H2ICE vehicle would be proposed to fleet operators as a means to increase near-term utilization of their hydrogen fueling stations (or to justify installation of additional stations). Deployment of H2ICE vehicles might thus support and/or accelerate expansion of the hydrogen-fueling infrastructure.

PROJECT ACTIVITIES

<u>Establishing "Vehicle Fuel Efficiency" Objectives</u>

The initial project task was to establish quantifiable performance criteria for current fuel cell vehicles. These would become project vehicle objectives.

Honda's 4-passenger FCX is the first fuel cell vehicle certified by the Environmental Protection Agency (EPA) [1]. The FCX achieved EPA city economy of 51 miles per kilogram of hydrogen (MPK) and highway economy of 48 MPK. On an energy basis, one MPK is equivalent to one mile per gallon of gasoline (MPG). The reported FCX driving range is 170 miles (273 km). These data (with range adjusted to reflect vehicle fuel storage capacity) were adopted as project vehicle objectives.

<u>Assessing Project Feasibility</u>

Fuel Economy and Driving Range

Calibrated lean, the H2ICE was projected to improve EPA fuel economy by 10-15% vs. a baseline gasoline ICE (due to reduced pumping losses and operation at mean-best-torque spark). The TOMH storage tank and adaptation add approximately 200kg to base vehicle weight, increasing fuel consumption by an estimated 3 to 5%. With regenerative braking (as used on the FCV hybrid vehicle), this penalty might be reduced. Achieving project vehicle economy objectives dictated selection of a base gasoline vehicle capable of at least 45 MPG.

The FCX stores 3.75 kg of hydrogen to achieve its 170-mile driving range. The prototype TOMH tank stores 3.0 kg of hydrogen. Based on proportional fuel storage, the project vehicle must achieve a 136-mile range to "match" FCX vehicle fuel efficiency. A 140-mile (225 km) driving range was adopted as the project vehicle objective.

Vehicle Acceleration and Hill Climbing Capabilities

Significant reductions in power and torque occur when conventional ICEs employ gaseous fuels. Peak power is generated at a stoichiometric air-fuel ratio (AFR). With AFR at stoichiometry, a 28% reduction in H2ICE peak power (compared to gasoline) occurs when gaseous hydrogen in the intake charge reduces the amount of air available for combustion. (A direct injection strategy that might overcome this disadvantage was deemed beyond the scope of this project.) The charge cooling effect attributable to gasoline vaporization is also lost. At the much-leaner-than-stoichiometric AFR required to avoid "backflash" and to minimize NOx emissions, significant additional power reductions are encountered [2]. Under these conditions, the reserve power available for vehicle acceleration and hill climbing is severely compromised. This reserve power shortfall made it apparent that a power-boosting strategy was required.

To maintain extended hill-climbing capability with the additional weight of the OMH system, peak power of the H2ICE must exceed gasoline peak power. This was projected to be achievable using an appropriate turbocharger and aggressive intercooling [3]. Satisfying the acceleration objective with a turbo configuration would be more difficult (due to reduced boost capability at low engine speeds and to "turbo lag" effects.) The appeal of a hybrid-electric base vehicle became apparent. Electric motor torque would supplement engine torque during vehicle launch and acceleration.

Emissions

Honda's FCX has received ZEV emission status [4], and produces no carbon dioxide (CO_2) emissions. For the project vehicle, PZEV compliance with "zero" CO_2 was adopted as the emission objective. Satisfying the PZEV NOx standard of 0.02 g/mi would present a significant challenge. Two distinct calibration approaches were considered: (1) lean operation with AFR set to approximately 0.55 equivalence ratio (ER) without an exhaust catalyst, and (2) operation with variable AFR using a 3-way or lean NOx catalyst. The first approach was preferred, but both would ultimately require testing.

Project Vehicle Selection and Finalization of Objectives

Requirements for 45+ MPG and seating for four people led to selection of a Toyota Prius as the project demonstration vehicle*. A 2002 model year Prius was purchased from a local Toyota dealership. Emissions, acceleration, hill-climbing capability, and ICE output of the Prius were documented. On the basis that the Prius represents a "commercially acceptable" vehicle, Prius levels of acceleration and gradeability were adopted as project vehicle objectives (Fig. 1).

*Toyota did not participate in this project.

Fuel Economy 51 MPK City (FTP) 48 MPK Hwy (FEC)	**ICE Output @ 4000 RPM*** 67 BHP (50 kw) 82 lb-ft (111 Nm)
Driving Range 140 miles (225 km)	**Gradeability* (calculated)** 13.4% @ 55 MPH
Acceleration* 0 to 40 MPH - 6.9 sec. 0 to 60 MPH - 14.1 sec.	**PZEV Emissions** 0.02 g/mi NOx
	"Zero" CO2 Emissions

* Prius gasoline data

Figure 1 - Project Vehicle Objectives

Adapting Engine and Controls for Hydrogen Fueling

An "aftermarket conversion" approach was selected. The gasoline fuel rail was replaced with a fabricated rail incorporating an ignition-activated fuel cutoff valve plus fuel pressure and temperature sensors. The four gasoline port injectors were replaced with four high-flow Quantum gaseous fuel injectors. Cooler, non-Pt spark plugs were installed.

No other base engine upgrades or modifications were required. The production 13.5:1 mechanical compression ratio (9.5:1 Atkinson cycle) was deemed appropriate for this H2ICE application. Valve timing and camshaft positioning were reviewed and left unaltered.

A supplementary H2ICE controller containing appropriate injector drivers was installed. Signals from key engine sensors were intercepted and used by the controller to calculate and command injector pulse width and spark timing. An exhaust AFR sensor was added to enable closed loop control at a selectable AFR.

The production Prius controllers retain command of all vehicle functions other than those specifically associated with the hydrogen fueling system (e.g. hydrogen pressure regulation, etc). This includes camshaft positioning, the electronic throttle actuator, idle shutdown and restart strategy, transmission ratio control, all aspects of electric motor and battery charging control, and all onboard diagnostic (OBD) routines.

Adapting the Vehicle for Hydrogen Fueling

The Prius gasoline tank and filler assembly, fuel supply lines, evaporative control system components, and exhaust catalysts were removed.

Pending availability of the prototype TOMH fuel storage system, a conventional Quantum compressed hydrogen storage system was temporarily installed. The trunk-mounted 5000-psi tank provided capacity to store 0.78 kg of hydrogen. Filtered hydrogen was supplied to the H2ICE fuel rail at a regulated pressure of 30 psi. This system was used for initial vehicle testing and evaluation prior to installation of the TOMH system.

The TOMH system consists of a MH containment tank (Figs. 2a and 2b) with integral heat exchanger HE1, and a tank coolant circulation system (Fig. 3). The trunk-mounted OMH tank is bolted into the vehicle structure via a cradle mount and two tank retention straps.

Figure 2a - Prototype TOMH Hydrogen Storage Tank

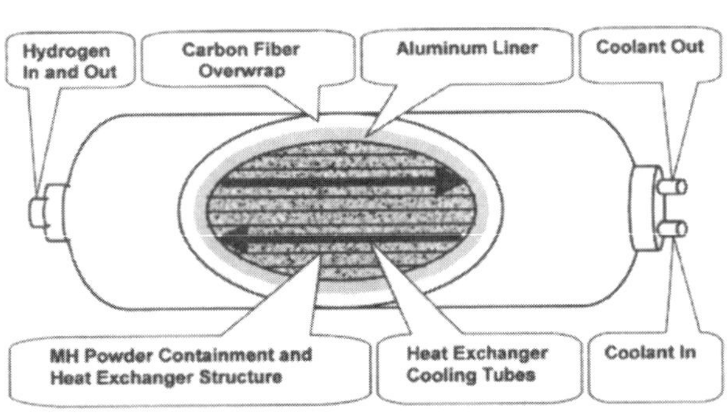

Figure 2b - TOMH Tank Construction Schematic

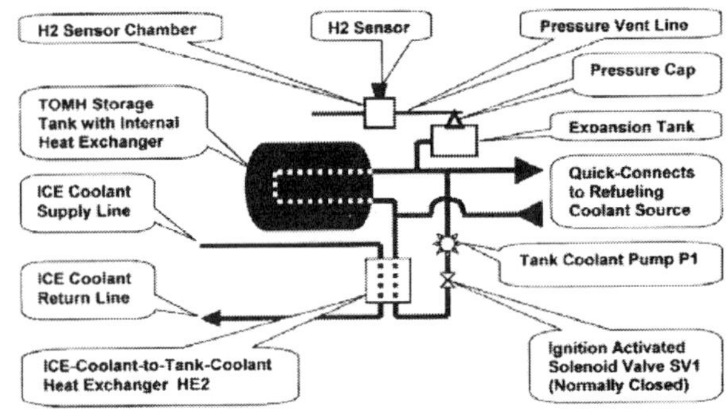

Figure 3 - TOMH Tank Coolant System Schematic

An engine-coolant-to-tank-coolant heat exchanger HE2 (Fig. 4) is inserted into the ICE coolant bypass circuit downstream of the vehicle heater core. During vehicle operation, tank coolant is continuously circulated through HE2 and HE1 by a 12V electric pump P1 (Fig. 5). Devices P1 and HE2 are mounted underbody in the area formerly occupied by the gasoline tank.

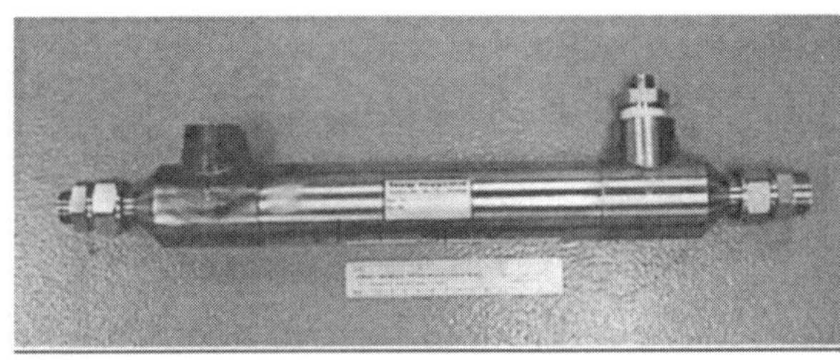

Figure 4 - Coolant Heat Exchanger HE2 (ICE to Tank)

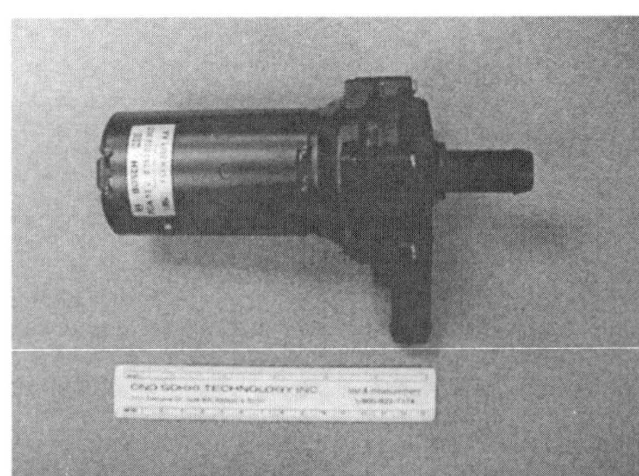

Figure 5 - Tank Coolant Circulation Pump P1

During vehicle refueling, a coolant solenoid valve SV1 isolates P1 and HE2 from the tank coolant circuit. An off-vehicle coolant supply system is connected to the vehicle and circulates coolant through HE1. Inlet and outlet connections for the off-vehicle coolant supply system are located adjacent to the hydrogen fueling connector behind the refueling access panel (Figs. 6a and 6b).

Fast-filling with "high" pressure hydrogen (1500 psi maximum) requires about 10 minutes for a 90% fill. For fast filling, coolant flow requirement is about 5 gal/min (19 liters/min) with coolant temperature at 15-20 deg-C.

Slow-filling to about 80% (maximum) tank capacity can be accomplished by supplying hydrogen at 200 psi (e.g. directly from an electrolyser).

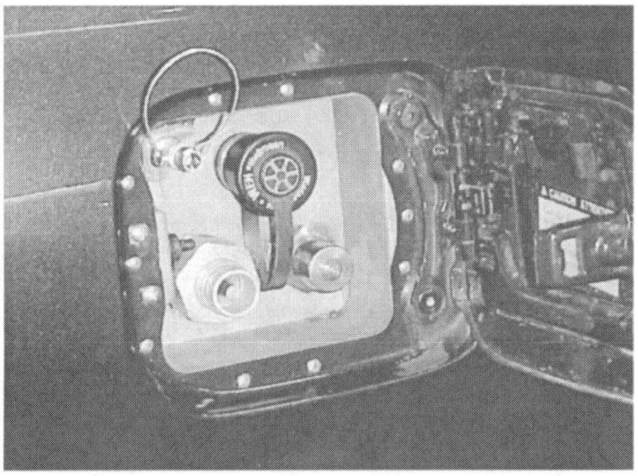

Figure 6a - Refueling Access Panel (open) showing Hydrogen fueling receptacle (top, with plastic dust cap), inlet and outlet connectors for external tank coolant, and fuel dispenser electronics receptacle (upper left).

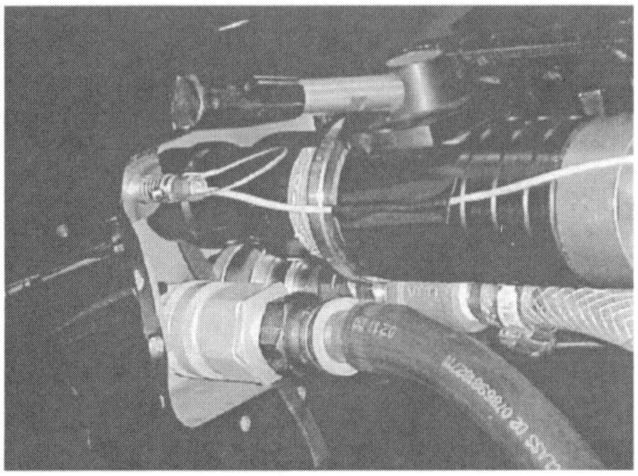

Figure 6b - Refueling lines attached

Other Vehicle Modifications

Battery Booster Pack (Fig. 7)

To supplement electric motor torque, 5 additional battery modules containing 30 1.2V NiMH cells were installed in series with the Prius battery. This increased the total number of cells to 258, elevating system voltage by 13% (309V vs. 273V) and providing a corresponding increase in motor torque available for vehicle acceleration.

Fuel Consumption Monitor Display Unit (Fig. 8)

To measure fuel usage with the TOMH fuel system, a multi-function display unit was calibrated and installed on the passenger compartment "tunnel." This unit determines fuel consumption using accumulated injector pulsewidth to calculate fuel flow. The unit is manually reset to display "100.0" and "3.000" after each vehicle refueling, representing the "percent full" (100%) and "fuel remaining" (3.0 kg). During vehicle operation, cumulative fuel consumption is tracked to continuously decrement these values. The unit also uses the vehicle odometer signal to calculate and display instantaneous and average fuel economies (MPK) and uses the "average MPK" value to calculate "distance to empty" (miles). Engine RPM is also displayed.

Figure 7 - Battery Booster Pack (shown with 6 modules)

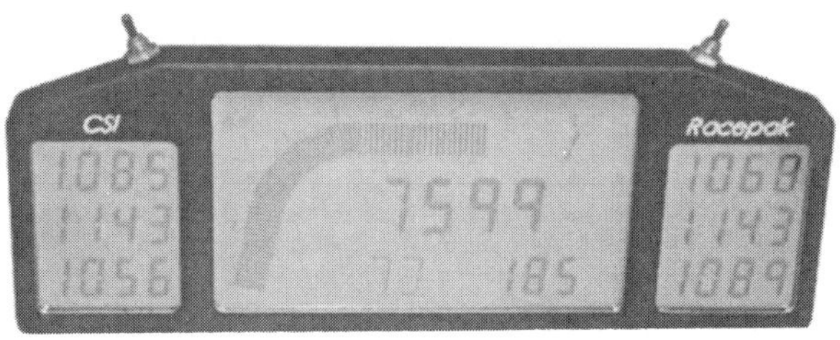

Figure 8 - Fuel Consumption Monitor Display Unit

Rear Suspension

Retaining a "level" vehicle attitude required modification of rear spring perches to compensate for additional weight of the TOMH fuel storage system (Fig. 9.)

Hydrogen Safety Features

The hydrogen storage tank is contained within a metal enclosure inside the vehicle trunk (Fig. 10.) Air is permitted to enter the enclosure via a fuel line access port that communicates with the vehicle underbody, and to exit via a top-mounted vent line that terminates in the trunk seal channel (outside the seal). The enclosure prevents any hydrogen that might escape the tank or tank fittings from entering the passenger compartment.

With the TOMH system, an FMEA identified a possibility of fuel leakage into the tank coolant circuit. That circuit contains a trunk-mounted expansion tank (Fig. 11) fitted with a vented pressure cap. The cap vent line is routed through a chamber containing a hydrogen sensor. Any hydrogen present in the coolant will cause the pressure cap to vent and subsequently activate that sensor. The chamber containing the sensor is externally vented.

To detect any hydrogen that might inadvertently escape the tank enclosure or the expansion tank, a second hydrogen sensor is installed at the top of the trunk compartment (Fig. 12). Activation of either hydrogen sensor causes the fuel consumption monitor display digits to "pulse." Should this occur, the vehicle operator is advised to open the vehicle windows and proceed at once to a service facility, or, alternatively, to park the vehicle; open the trunk, hood, and windows; and call for assistance.

Figure 9 Ovonic Hydrogen Prius (level attitude)

Figure 10 - TOMH Tank Enclosure

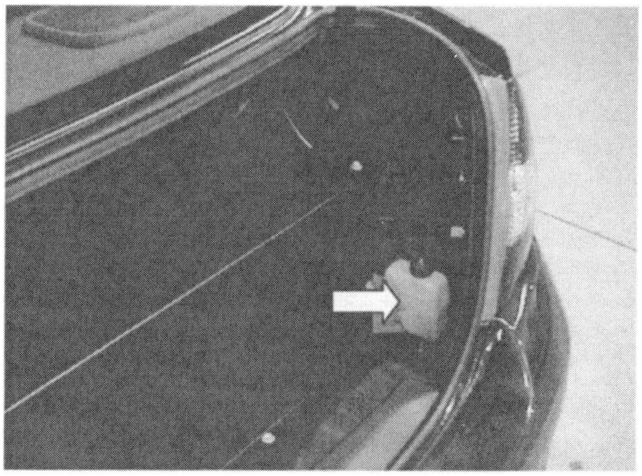

Figure 11 - Coolant Expansion Tank

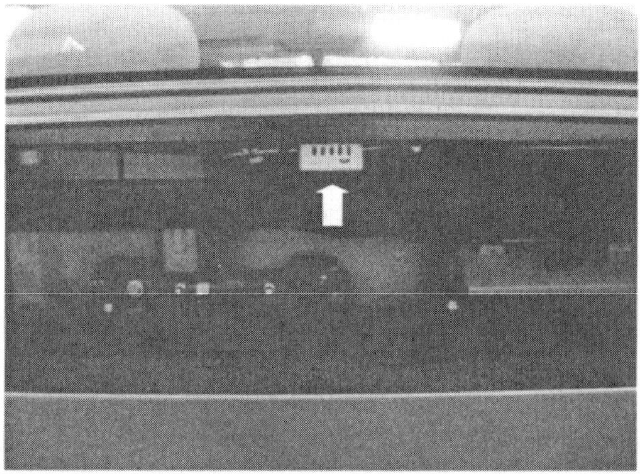

Figure 12 - Trunk Hydrogen Sensor

<u>Engine Dynamometer Testing Activities</u>

TOMH System Testing

A MH fuel storage system offers superior fuel storage density compared to systems utilizing compressed hydrogen (Fig. 13), but with an attendant weight penalty and with additional vehicle integration requirements. [5] To maintain constant temperature in the TOMH storage tank, 14 kJ of heat energy must be supplied for each gram of hydrogen released. H2ICE coolant is the source of this heat energy. Testing was conducted to confirm that the TOMH system heat exchangers and coolant flow rates were properly specified and capable of maintaining an adequate supply of hydrogen to support worst-case H2ICE fuel demand following cold starts.

Figure 13 - TOMH tank with 3.0 kg reversible capacity @ 1500 psi (left) and compressed Hydrogen storage tank with 0.78 kg capacity @ 5000 psi (right)

Engine Power Development

A Prius ICE was procured specifically for dynamometer usage. It was initially tested on gasoline to establish wide-open-throttle (WOT) power and torque baselines that became objectives for the H2ICE.

Testing was then conducted on three unique H2ICE configurations: naturally aspirated (NA), supercharged (SC), and turbocharged with intercooling (TCI). The NA version retained the production Prius air induction system. The SC version employed an integrated Rootes-type supercharger and intake manifold unit (Fig. 14) designed for an aftermarket Toyota racing application. The SC unit was crankshaft driven using dedicated pulleys and drive belt. Extensive modification of the unit was required to accommodate vehicle underhood packaging constraints. This configuration did not permit packaging an intercooler, and intake charge temperature became a factor limiting maximum allowable boost. The TCI version retained the production Prius intake manifold. A new exhaust manifold was fabricated to accommodate the turbocharger (Fig. 15 and 16), a production wastegated unit from a European diesel engine application. A liquid-to-air intercooler (Fig. 17) was installed between the TC and throttle body. Coolant at 25 degree C was supplied to the intercooler.

Figure 14 - Integrated Supercharger-Intake Manifold (minus rotors and driveshaft assembly)

Figure 15 - Turbocharger Assembly

Figure 16 - H2ICE showing Turbocharger

Figure 17 - H2ICE showing Intercooler

NOx Emissions

H2ICE NOx production was measured vs. AFR and spark timing to investigate calibration alternatives. A brief evaluation of water injection (as a means of reducing NOx) was conducted. A small production 3-way catalyst was evaluated at rich and lean AFRs.

Vehicle Testing Activities

Acceleration Capability

Vehicle acceleration was measured with each of the three H2ICE configurations (NA, SC, TCI). Output from the vehicle distance sensor ("speed sensor") was recorded vs. time on a series of standing-start WOT accelerations. Prior to each acceleration run, battery state-of-charge (SOC) was normalized to remove the influence of that variable on vehicle acceleration. All tests were conducted using the battery booster pack.

Fuel Economy and Vehicle Emissions

Economy and emissions were measured at a California Air Resources Board (CARB) certified, SULEV capable test facility operated by Quantum Technologies, Inc. Chassis dynamometer absorber settings duplicated those specified for Prius certification tests.

Hydrogen consumption was determined by one of three techniques. With compressed fuel storage, the change in tank pressure between the start and completion of each test was used to calculate fuel consumption for that test. With the TOMH system, a MicroMotion fuel flow sensor with a flow totalizer function was employed for chassis dyno testing. This sensor was also used to calibrate the fuel consumption monitor, and that device was used to determine fuel consumption during vehicle road testing.

As a final confirmation and tank capacity check, a "run out of fuel" test was conducted. A standard 10-minute refueling was performed. The vehicle was then driven on the chassis dynamometer at 55 MPH until that speed could not be maintained.

Additional Activities

The development process identified several opportunities for vehicle and fuel system optimization that were not implemented because of project budget and timing constraints. Several of these items are recommended for future investigation, and are briefly discussed below.

Modification of the current Prius transmission ratio selection strategy would allow H2ICE operating characteristics to be better matched to vehicle operating requirements. Specifically, vehicle acceleration and fuel economy might benefit from a revised ratio schedule and/or a broader ratio capability. Throttle positioning strategy could also be modified to compensate for reduced output of the H2ICE. (Both of these strategies are embedded in the Prius controllers and are not accessible for simple recalibration.)

Several ideas were proposed to capitalize on the heat absorption characteristics of the TOMH tank - either to reduce vehicle cooling system capacity, cost, and weight; or to enhance intercooler performance during hot weather by employing the tank as a "chiller". The considerable thermal inertia of the tank also suggested the possibility of employing it as a "heat battery" to accelerate ICE and passenger compartment warmup in colder climates. Implementation of these opportunities is contingent upon developing an overall tank temperature control and tank coolant pump operating strategy. (In the current vehicle, the pump operates continuously whenever the H2ICE is running.)

The possibility of utilizing an "on demand" electrically operated boost device [6] was briefly investigated, but a suitable device was not identified. Several suppliers are actively pursuing devices that might provide increased boost at low engine airflow (compared to the current turbocharger). The high voltage electrical systems used in hybrid applications might make such devices practical.

To reduce H2ICE pumping losses and resultant fuel consumption, a strategy to eliminate the H2ICE throttle and to control H2ICE output by modulating AFR was considered. This approach was evaluated briefly on the dynamometer. NOx emissions increase dramatically if AFR is richened beyond 0.55 ER. For this project, the small incremental benefit in fuel economy did not warrant the effort required to fully implement this strategy. However, this opportunity warrants future investigation.

An all-in-one fueling "nozzle" that combines hydrogen, coolant, and electrical connections would simplify and "fool-proof" the refueling process.

TEST RESULTS

<u>H2ICE Dynamometer Tests</u>

Power, Torque, and BSFC

Wide-open-throttle (WOT) power and torque of the three H2ICE configurations (calibrated at 0.55 ER) and of the production gasoline baseline (production calibration) are shown on Figs. 18 and 19.

With the TCI configuration, H2ICE peak power exceeded gasoline power, satisfying this project objective. The turbocharger wastegate controlled boost to 0.85 Bar (12.5 psi) at engine speeds above 3200 RPM. At speeds below 2400 RPM, the turbocharger was unable to generate positive boost, an indication that the turbine section is oversized for this application. The subject turbine is the smallest available in this production series of turbochargers. A purpose-built unit might enable improved low speed boost capability.

With the SC configuration, lower than anticipated power is attributed to two factors: boost limitations imposed by high intake charge temperature (an intercooler could not be packaged) and an imperfect "match" between the H2ICE and available SC unit (which had excess capacity for this application.) These factors also contributed to unacceptable penalties (Fig.20) in brake specific fuel consumption (BSFC). It is possible that rotor clearances were affected by welding operations during the SC unit adaptation, exacerbating the high unit friction and producing unacceptably noisy operation.

In the NA configuration, WOT power was also measured with AFR calibrated to 0.65 ER. This 18% increase in fuel flow generated a net power increase of 30%. If this relationship between additional fuel flow and increased power is maintained, the 50 kW produced by the gasoline ICE would be matched by the H2ICE if fuel flow was increased by 83%. This was achieved in the H2ICE TCI configuration with boost at 0.83 Bar (12.2 psi).

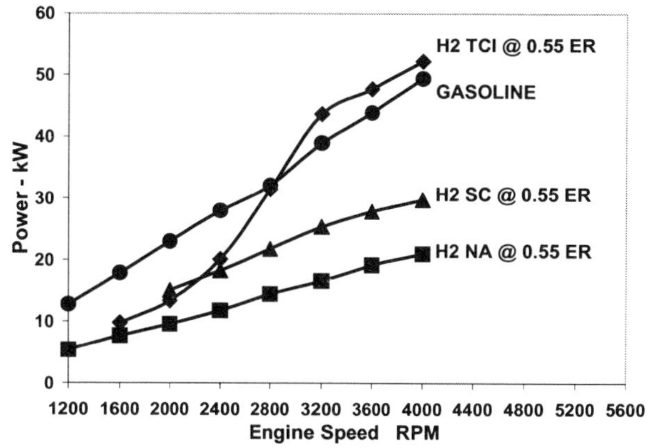

Figure 18 - Engine **Power** at WOT

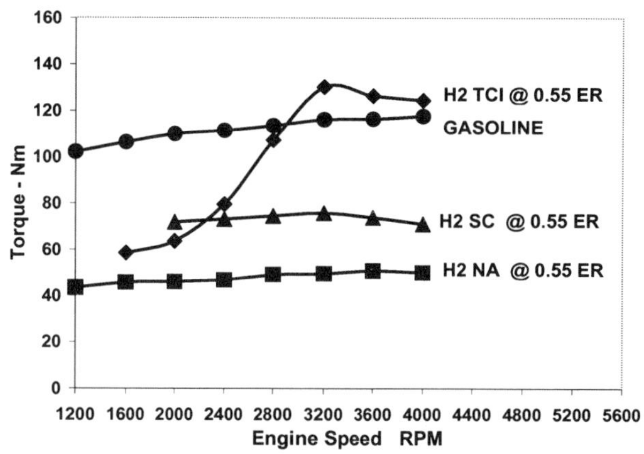

Figure 19 - Engine **Torque** at WOT

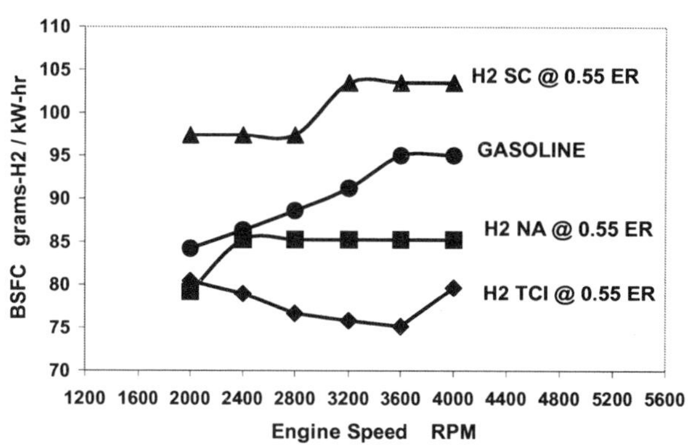

Figure 20 - Engine **BSFC** at WOT

NOx Emissions

A detailed study of NOx production vs. AFR was not performed, but spot checks were performed with AFR set to either side of 0.55 ER. Results confirmed previously reported findings that show a sharp upturn in NOx production with AFR richer than 0.55 ER [2]. To address the "no catalyst" objective, 0.55 ER was tentatively selected as the target AFR for this application.

In the NA configuration, exhaust NOx concentrations (Fig. 21) were measured upstream and downstream of a small production 3-way catalyst (3WC) installed specifically for this test series.

MAF	ER	NOx IN	NOx OUT	NOx EFF
g/sec		ppm	ppm	%
13.0	1.09	1438	1.7	99.90
7.9	1.09	985	0.9	99.90
7.5	1.08	1090	0.9	99.90
13.5	0.55	7	1.7	75.70
20.2	0.55	8	4.6	42.50
30.0	0.55	16	15.5	3.10

Figure 21 3-Way Catalyst Efficiency Data

It was projected that the 3WC would have no effect on NOx at the "lean" (0.55 ER) calibration, because no reducing agent was anticipated in the exhaust at this condition. At higher mass airflow (MAF), the data confirm this projection. But significant NOx activity was recorded with MAF below 20 g/sec. This indicates the presence of some reducing agent - most likely unburned hydrogen. It was postulated that cylinder-to-cylinder and/or cycle-to-cycle fueling variability was resulting in "partial burns" as injector pulse width reached the non-linear threshold during lean operation at low MAF. Implementation of a "skip-fire" (alternate cylinder fueling) routine to improve this condition was considered, but not implemented because of project timing constraints. The finding did suggest that a catalyst should be considered for vehicle usage, even with the "lean" calibration.

NOx testing was also performed at a "rich" (1.10 ER) calibration, and catalyst efficiency approached 100% at this condition. Tailpipe NOx concentrations were equal to or below levels measured during lean operation.

Two observations were made: (1) a calibration strategy employing a 3WC with stoichiometric operation at low MAF and lean operation (to avoid backflash) during high MAF "power demand" might provide best vehicle tailpipe NOx emissions, and would also improve engine low-range torque; (2) this calibration would compromise EPA fuel economy and range vs. a "lean" approach.

A brief attempt was made to evaluate water injection as a means of reducing NOx production. A single water injector was mounted above the throttle blade and delivered atomized water into the intake charge. NOx was reduced by up to 50%, but the quantity of water required was excessive. The sumped Prius intake manifold is not suitable for "wet" operation, as evidenced by water pooling in the sump area. A multi-point (port) water injection system was considered but, due to project timing constraints, was not implemented.

TOMH Fuel Storage System Testing (Fig. 22)

In the NA configuration, testing was conducted to confirm that coolant flow through the vehicle heater circuit (the ICE bypass circuit) provided sufficient heat energy to maintain adequate desorption of hydrogen to satisfy worst-case H2ICE fuel demand. The TOMH tank was first cooled to a uniform internal temperature of zero degrees C by permitting the release of hydrogen to atmosphere. The H2ICE (at room temperature) was then started and accelerated to a condition simulating 70 MPH road load. TOMH tank internal temperature dropped an additional one degree C during H2ICE warmup before stabilizing after four minutes of operation and rising from that point onward. Fuel supply pressure remained steady at an acceptable level throughout the test. Additional data gathered during testing projected capability to sustain H2ICE operation to at least a minus 20 degree C ambient condition. It was determined that tank coolant flow rate could be reduced from the initial design level, permitting a reduction in coolant pump electrical draw. An alternate control strategy was developed to modulate tank coolant pump operation based on TOMH tank temperature. This control strategy was not implemented on the project vehicle, but it will be considered for any subsequent vehicle builds.

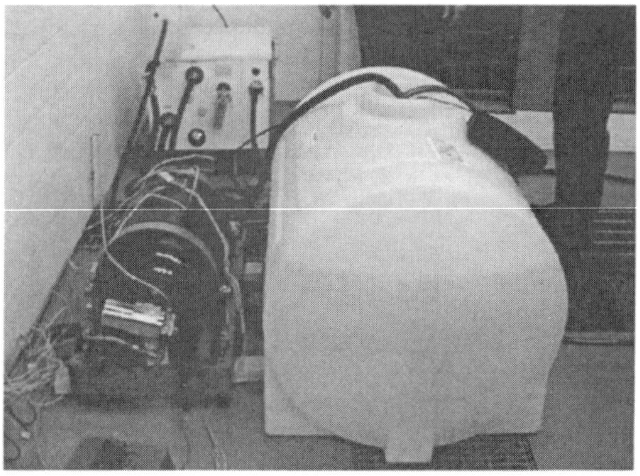

Figure 22 Dyno Testing Setup for TOMH Fuel System. Located adjacent to engine dyno cell and connected to dyno cell fuel supply via lines in the "trench" connecting rooms. Large tank contains coolant circulated through the TOMH tank heat exchanger during tank refueling. Device at upper left is an auxiliary coolant heater made available to pre-adjust internal TOMH tank temperature.

Vehicle Test Results

The "final" project vehicle incorporated the H2ICE TCI configuration, AFR calibration set to a nominal 0.50 ER increasing to 0.55 ER at higher MAF (>50 g/sec), the production Prius catalyst system, and the TOMH fuel storage system (see Figures 29 - 32 on page 12).

Acceleration

Results with the "final" configuration are plotted on Figs. 23 and 24. Acceleration tests were initiated with battery state-of-charge at a "normal" level of 55 - 60%. H2ICE results include benefit of the battery booster pack.

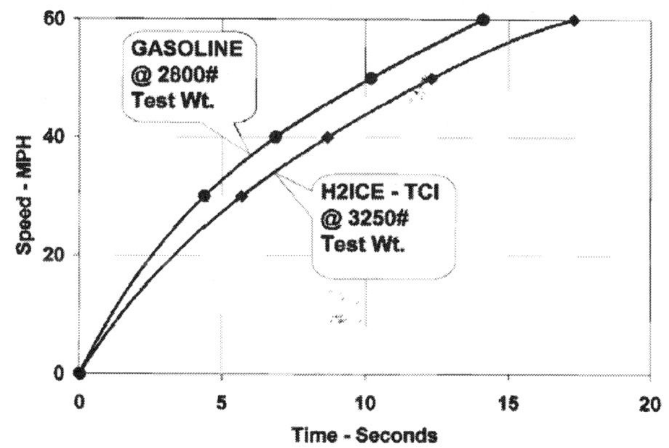

Figure 23 - Speed vs. Time @ WOT

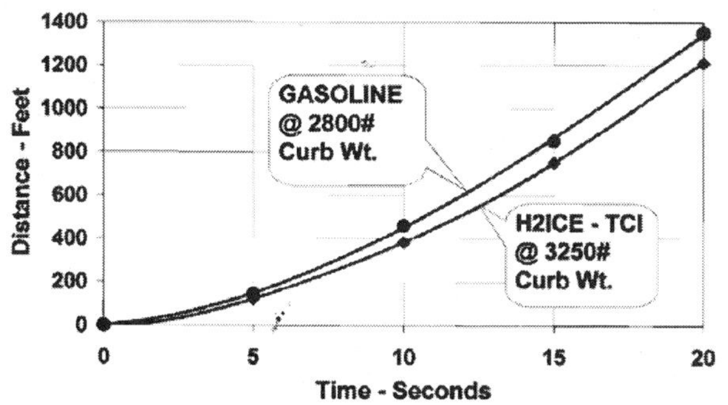

Figure 24 - Distance vs. Time @ WOT

Data for all configurations tested are summarized on Fig. 25. The data in column 5 represent the effect of a "simple" vehicle conversion using compressed H2 storage. Working to the left, column 4 shows the performance effect of adding the SC. Column 3 reflects the effect of three coincident changes: (1) installation of the TOMH fuel storage system, (2) a 259% increase in onboard fuel capacity, and (3) a 204kg increase in vehicle weight. On an equal fuel storage basis, TOMH system weight and the resultant vehicle performance penalty would be substantially reduced.

	Gasoline	H2ICE TCI	H2ICE SC	H2ICE SC	H2ICE NA
Curb weight (lbs.)	2800	3250	3250	2800	2800
(kg)	1273	1477	1477	1273	1273
Fuel storage system	gasoline	TOMH	TOMH	5Kpsi	5Kpsi
ICE Peak BHP (kW)	67 (50)	70 (52)	40 (30)	40 (30)	28 (21)
0-40 MPH (sec.)	6.9	8.7	10.2	8.2	9.8
0-60 MPH (sec.)	14.1	17.3	22.0	17.3	20.4
Distance (ft.)					
at 5 seconds	145	120	95	110	100
at 10 seconds	460	385	340	410	380
at 20 seconds	1360	1200	1075	1220	1150
Gradability (calc.*)					
% at 55 MPH	13.4	11.2	5.9	6.8	3.9

* @ curb wt. + 350 lbs.

Figure 25 Data Summary - Vehicle WOT Acceleration

Emissions

Results are summarized on Fig. 26. Emission levels reflect the results of an abbreviated calibration effort and limited access to testing facilities. Trace levels of carbon emissions (NMOG, CO, and CO_2) are attributed to combustion of engine lubricating oil introduced to the intake charge via the positive crankcase ventilation (PCV) system and/or past the piston rings and valve stem seals and/or past the turbo bearing seal.

NOx emissions satisfied the 0.02 g/mi SULEV / PZEV standard without an exhaust catalyst, but significant spark retard (from MBT timing) was required to achieve this level of compliance. NMOG emissions (resulting from oil consumption) required an exhaust catalyst to satisfy the PZEV standard. For these tests, the standard Prius catalyst system was retained. However, a much smaller and less expensive catalyst would probably satisfy the minimal requirements for NMOG conversion.

A potential NOx reduction option that was not pursued is exhaust gas recirculation (EGR). This approach might reduce the fuel economy penalties that were observed when spark was retarded to control NOx.

	Tailpipe Emissions, g/mi			
	NMOG	CO	NOx	CO2
Hydrogen Fueled:				
Engine Out	0.032	0.041	0.013	2.5
Tailpipe (w. Catalyst)	0.001	0.001	0.018	3.2
Gasoline B/L (w. Catalyst)	0.010	0.386	0.004	222.8
SULEV/PZEV Standard	0.010	1.000	0.020	N/A

Figure 26 - Vehicle C75 Emissions Summary, g/mi

Fuel Economy (Figure 27) and Driving Range

In the "final" H2ICE-TCI-TOMH system configuration, hydrogen consumption was determined (using fuel flowmeter data) on EPA chassis dynamometer test cycles. Hydrogen fuel economy was essentially equal to measured baseline gasoline fuel economy.

	H2ICE			Gasoline	
	TCI	SC	NA	Test	(Cert.)
City	**42.3**	**45.2***	**49.0***	**42.4**	**52**
Hwy	**46.1**	**54.7***	**n/a**	**n/a**	**45**
Comb.	**43.9**	**49.0***			**48.6**
55 MPH	**50.9**				

Figure 27 - Fuel Economy Summary, MPK and MPG

The reader will note that measured gasoline city fuel economy is significantly less than city fuel economy achieved by the gasoline Prius on certification tests. There is an explanation for this difference.

Emissions and fuel economy tests were conducted on a single-rolls chassis dynamometer. With only the front wheels turning, Prius antilock braking system (ABS) diagnostics detected that the rear wheels were "locked", and imposed a speed limit on the front wheels that precluded operating the vehicle on EPA test cycles.

To overcome this issue, tests were conducted with the rear wheel ABS speed sensors disconnected - "standard procedure" when testing vehicles with ABS. In this situation, a conventional vehicle will simply set an ABS fault code and disable ABS operation.

On the Prius, the effects are more far reaching. When ABS is disabled, regenerative braking is also disabled. Elimination of regen requires that more battery charging "work" must be provided by the ICE, which in turn allows less frequent ICE shutdowns at closed throttle. The net result is a significant increase in city fuel consumption, and (potentially) increased vehicle emissions.

After all project testing was completed, it was learned that a special "diagnostic test mode" can be activated on the Prius to overcome the issue described above. With this "test mode" activated, the Prius can be operated on a chassis dynamometer without the need to disconnect rear wheel speed sensors, and regenerative braking will remain active. This "test mode" is employed during gasoline certification testing. The effect is a 10+ MPG improvement in city fuel economy. This was confirmed on tests with a gasoline Prius. It is theorized that a similar improvement would be seen with hydrogen fueling, but a confirming test has not been conducted.

Highway cycle fuel consumption is less affected by this procedural difference. Measured hydrogen highway fuel economy matches the gasoline Prius certification data.

After filling the TOMH tank to 90% of capacity, a "run out of fuel" test was conducted. At 55 MPH, the vehicle accumulated 136 miles before 55 MPH could no longer be maintained. Fuel consumed on this test totaled 2.67 kg, confirming the 2.7 kg "90% fill" capacity. On this test, average fuel economy was 50.9 MPK.

In the SC and NA configurations, high-pressure fuel storage was in use, and hydrogen consumption was determined using a tank pressure differential calculation. The higher calculated fuel economy results (shown with asterisks) also reflect calibration differences - these tests were run before spark was retarded to control NOx.

Measured city and highway fuel economy data and EPA's 55/45 city/highway driving assumption produce a combined EPA economy of 43.9 MPK and a calculated EPA driving range (with 3.0 kg of onboard Hydrogen) of 131.7 miles (212 km). This does not include several additional miles of "electric-only" operating range.

If it is assumed that "certification" city economy would be 10 MPK higher (52 MPK), combined economy would be 49.2 MPK yielding a range of 147.6 miles (237 km).

Vehicle Refueling

Refueling a compressed hydrogen storage system is accomplished by attaching the refueling station "nozzle" to a mating receptacle on the vehicle and allowing fuel pressure in the storage tank to equilibrate with supply pressure. To achieve a final "settled" tank pressure of 5000 psi, it is typical to fuel to an equilibrium pressure above 6000 psi to allow for post-fill fuel cooling. During vehicle operation, tank pressure decreases in a near-linear relationship with fuel usage (Fig. 28).

TOMH system refueling can be accomplished in similar fashion. For "fast filling", it is necessary to "force cool" the MH material by circulating coolant through the TOMH tank internal heat exchanger. Using this procedure, refueling to 90% of tank capacity is accomplished in 10 minutes using a fuel supply pressure of 1500 psi. During vehicle operation, tank pressure quickly decreases to the MH equilibrium pressure (typically below 250 psi) and remains at that level until fuel is depleted (Fig. 28).

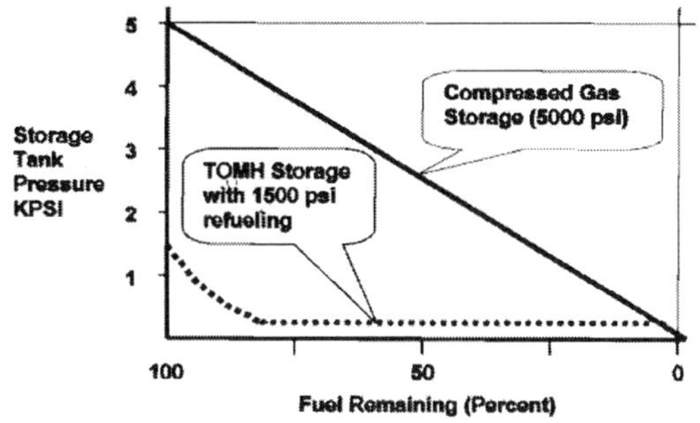

Figure 28 - Storage Pressure vs. Fuel Remaining

CONCLUSIONS

The Ovonic Hydrogen Prius satisfies all project objectives except WOT acceleration and gradeability. These are diminished by 15-20% due to increased vehicle weight resulting from the hydrogen conversion. Part-throttle vehicle performance remains subjectively acceptable. PZEV tailpipe emission requirements are satisfied, and over 99% of greenhouse gas emissions are eliminated. City and highway fuel economies are similar to levels achieved by the gasoline Prius and the Honda FCX. Driving range with the enabling TOMH onboard fuel storage system matches the FCX on a proportional fuel storage basis.

In the same physical space, a current TOMH system stores about 3x as much hydrogen as a 5000 psi system. Alloys under development might improve this advantage. A current TOMH system provides higher storage density than either 10,000 psi or liquid hydrogen systems.

A complete TOMH storage system weighs about 150% more than an equal-capacity 5000 psi system. TOMH system weight and its effects on vehicle payload and performance are significant. On many vehicles (like this Prius), providing space for adequate hydrogen storage presents equal if not greater vehicle design challenges.

Fig. 29 - Ovonic Hydrogen Prius Underhood

Fig. 30 - Intercooler Radiator (front bumper above)

Fig. 31 - TOMH Tank Installation (enclosure removed)

Fig. 32 - Ovonic Hydrogen Prius

ACKNOWLEDGMENTS

The authors thank Sid Jeffe, Ian Kinoshta, Phil Schnell, David Swan, Tom Asmus, and the many involved personnel at ECD, TOHS, and Quantum for their support, enthusiasm, and hard work.

REFERENCES

1."EPA Technical Highlights", Office of Transportation and Air Quality, EPA420-F-02-055, Feb. 2003

2. Tang, X., et al, "Ford P2000 Hydrogen Engine Dynamometer Development", SAE 2002-01-0242

3. Natkin, R. J. et al, "Hydrogen IC Engine Boosting Performance and NOx Study", SAE 2003-01-0631

4. "City of Los Angeles Becomes First Customer for Fuel Cell Car", http://www.prnewswire.com/comp/372013.html

5. Young, R. C., et al, "A Hydrogen ICE Vehicle Powered by Ovonic Metal Hydride Storage", SAE 2004-01-0699

6. Carney, D., "Forced Induction", Automotive Engineering International, September 2003

04ANNUAL-606

A Hydrogen ICE Vehicle Powered by Ovonic Metal Hydride Storage

R. C. Young, B. Chao, Y. Li, V. Myasnikov, B. Huang and S. R. Ovshinsky

Texaco Ovonic Hydrogen Systems

ABSTRACT

Among the various alternative fuels, hydrogen is the only fuel, which is clean and sustainable. More importantly, if hydrogen is produced from renewable resources, virtually no CO_2 emissions are produced.

Texaco Ovonic Hydrogen Systems (TOHS) has converted a gasoline internal combustion engine (ICE) hybrid vehicle to a hydrogen vehicle using its advanced metal hydride storage. Compared to its gasoline counterpart, the converted vehicle demonstrated a reduction in CO_2 tailpipe emission of 220 grams per mile.

Ovonic metal hydride hydrogen storage systems consist of three critical elements: (1) an advanced metal hydride alloy, (2) an efficient heat exchanger and (3) a lightweight fiber wrapped pressure container. A 50 liter Ovonic metal hydride storage vessel installed in the vehicle provides a storage capacity of 3 kg hydrogen and a driving range of 130-150 miles.

BACKGROUND

Because of continuing growth in transportation energy demand, growing awareness of energy security, climate change, air pollution, and dwindling fossil fuel reserves, it is critical to identify a transitional path of adopting hydrogen as the transportation fuel of the future. Hydrogen can be generated from a variety of sources such as from coal, natural gas and from wind, solar, hydroelectric, etc. Although it would be most cost effective to generate hydrogen using fossil fuel sources in the short term, upstream greenhouse gas emissions can only be minimized if hydrogen is generated using renewable energy sources.

It is generally believed that it is no longer a question of *if* a hydrogen economy will come, but rather *when*. To pave the transition, a well-thought, well-planned hydrogen program targeted simultaneously to both hydrogen fueled vehicles and its supporting infrastructure will be crucial to a seamless transition. Attention should also be paid to minimizing well-to-wheels greenhouse gas emission. Ultimately, a cost effective renewable hydrogen generation and delivery system will have to be developed and implemented. In this paper, we will address the opportunity and challenge of adopting hydrogen as the transportation fuel and its impact on the tailpipe greenhouse gas emissions. The analysis of upstream greenhouse gas emissions is not included in the scope of this paper.

While hydrogen powered fuel cell vehicles would be the most desirable end goal, they are not quite ready to be a near term affordable solution. Hydrogen powered ICE vehicles could serve as a viable bridging technology option. These hydrogen vehicles, whether ICE or fuel cell, will share the same refueling infrastructure, adopt the same codes and standards, and provide the consumer with the same hydrogen safety awareness. A successful near term hydrogen ICE vehicle and supporting refueling demonstration can provide a positive impact in adopting hydrogen as a transportation fuel. It may even shorten the time frame for the commercialization of fuel cell vehicles, and the availability of a cost effective renewable hydrogen production.

In vehicle demonstrations, the safety aspect of fuel storage, the driving range and a user friendly refueling systems are of primary concerns. None of the hydrogen storage approaches employed in current OEM demonstration vehicles will meet these requirements. Each hydrogen storage approach has at least one major shortcoming: High-pressure (5000 psi) gaseous storage provides the limited driving range. Liquid hydrogen storage requires the handling of a cryogenic liquid. Use-once chemical storage (e.g. NaBH4) presents waste recycling challenges. Metal hydride onboard storage systems, even though it is heavier, are better able to satisfy these concerns than the aforementioned hydrogen storage methods.

Metal hydride onboard storage, due to its compactness, low pressure and safe operation, received much attention in the 70s and early 80s. (1). The first SAE paper with the concept of using metal hydrides as a fuel source for vehicle propulsion was presented by K.C. Hoffman et al as early as 1969. (2). However, many design challenges remain unresolved. For example the hydriding/dehydriding process is accompanied by the volume expansion and contraction, which not only decrepitates the coarse particles into fine powder, but also causes powder settling and vessel swelling. (3). Furthermore, because FeTi type of alloys were predominantly used in the early systems, material issues, such as limited pressure tailorability, capacity degradation with cycles, were encountered. (4). To be practical, much more development was needed.

Unfortunately, the quest of using hydrogen as a transportation fuel and the research required to develop a better metal hydride onboard system declined and eventually reached a low in the early 80s. However, upon entering into the 21st century, the urgency of energy security and the awareness of climate change resulted in new national and worldwide hydrogen initiative once again. (5).

To pave the transition, ChevronTexaco and Energy Conversion Devices (ECD Ovonics) formed a joint venture in 2000 called Texaco Ovonic Hydrogen Systems, LLC (TOHS). Its mission is to develop and to advance solid hydrogen storage systems for portable power, stationary and onboard applications. In this paper, we report the advancement of the metal hydride onboard vessels made by TOHS. A vessel has been integrated into a converted hydrogen ICE hybrid vehicle. The heat integration and vehicle fuel economy and tailpipe emissions are measured. As anticipated, the greenhouse emission is virtually eliminated.

ONBOARD STORAGE CHALLENGES

Hydrogen has the highest heat of combustion among all fuels; 33.3 kWh per kg of H_2 vs 12.2 kWh per kg of gasoline. However, hydrogen, the lightest gaseous fuel, is very difficult to compactly contain. Under ambient temperature (20 °C) and 1 atm pressure, one gallon (~ 3.8 liters) of hydrogen provides 0.011 kWh of energy; whereas 1 gallon of gasoline provides 33.6 kWh of energy. (Note: the energy content of 1 kg H_2 is equivalent to 1 gallon of gasoline).

To store sufficient amounts of hydrogen to provide a driving range of over 300 miles within the space and weight limitations of a passenger vehicle remains a major technological challenge. In the near term, it will be very difficult for hydrogen powered vehicles to compete with gasoline ICE vehicles in terms of convenience in refueling, driving range, and affordability. Adopting hydrogen as a transportation fuel will require a strong resolve and leadership not only from federal, state, and local governments, but also from automotive OEMs and energy companies.

SOLID HYDROGEN STORAGE

To date, compressed hydrogen has been used in most hydrogen demonstration vehicles, even though it is generally believed that solid hydrogen storage holds the key to the long-term solution. In this section, we will first give a brief review of solid hydrogen storage. We will then discuss various solid hydrogen onboard storage options. Finally, we will discuss the requirements of solid hydrogen storage systems.

SOLID HYDROGEN STORAGE. As the name implies, solid hydrogen storage is hydrogen stored in solid materials, in which hydrogen can be either physically adsorbed or chemisorbed to the solid.

Physical Adsorption. Hydrogen stored in activated carbon is a good example of physical adsorption. Here, molecular hydrogen is adsorbed to the high surface area of carbon by the weakly bonded Van der Waal force. Reducing temperature or increasing pressure dramatically increases the storage capacity of physical adsorption. It is also called cryoadsorption because H_2 is in a condensed form at cryogenic temperatures (77 °K). (6). Because of the high surface area required for storage, the volumetric energy density is not much better than compressed hydrogen storage. Activated carbon hydrogen storage has been known for many years. No practical onboard storage system has been demonstrated.

More recently, many investigators have focused on hydrogen in carbon nano-tubes or other nano forms of carbons. (7). However, complicated processing techniques and small quantities of inconsistent samples have resulted in a wide spectra of non-reproducible results. (8).

Chemical Absorption. Hydrides, in which atomic hydrogen is chemically bonded to the host elements, are typical examples of chemical absorption. Depending on the nature of chemical bonding and its bond strength, the hydride reaction can be either reversible or irreversible. Generally speaking, metallic and some of the ionic bonded hydrides are reversible, while covalently bonded substances are irreversible. For example, the ionic bonded MgH_2 and the metallic bonded $FeTiH_2$, $LaNi_5H_6$ are reversible; whereas the covalently bonded substances such as CH_4 (Methane), C_3H_8 (Gasoline), and most of the chemical hydrides such as $NaBH_4$, $LiAlH_4$ are irreversible.

Irreversible fossil fuels such as methane and gasoline leave no by-product in the fuel tank as they are depleted. Instead, the by-products (CO, CO_2, NO_x, etc.) are released to the air causing air pollution and climate change. Chemical hydrides, such as $NaBH_4$ require by-product removal ($NaBO_2$) at each refilling, as well as, replenishing the fuel tank with new chemicals. Not only are the waste recycling and/or disposal processes very costly, but perhaps more importantly, chemical hydrides are not considered to be a sustainable fuel. Unless, an abundant, low cost chemical hydride can be reprocessed cost effectively, it would be unlikely to be developed as a realistic onboard storage solution.

SOLID HYDROGEN ONBOARD STORAGE SYSTEM. Ultimately, a solid hydrogen onboard system should be reversible and have refilling characteristics similar to today's gasoline. Among the reversible solid hydrogen storage options, metal hydrides are the only solid hydrogen storage medium, which can be realistically demonstrated in a vehicle today. In this sub-section, we will give a brief description of metal hydride onboard storage.

Metal hydrides are a broad class of materials that undergo a reversible reaction with hydrogen. The reaction is written as:

$$M_{solid} + x/2\ H_{2\ gas} \Leftrightarrow MH_{x\ solid}\ + \textit{heat}$$

where M is the hydridable metal alloy, MH_x is the metal hydride and *heat* is the enthalpy of the reaction. The absorption reaction is exothermic - where heat is liberated; whereas the desorption reaction is endothermic - where heat is absorbed. A metal hydride system requires a delicate balance between heat transfer and hydrogen mass transfer. It is a self-regulated system. During absorption, if heat is not removed, absorption will automatically stop. During desorption, the hydrogen flow will stop if the heat supply is cut off. This self-regulating mechanism is another safety feature associated with metal hydrides.

Figure 1 shows a pressure, composition, temperature (PCT) diagram that represents a typical metal hydride system.

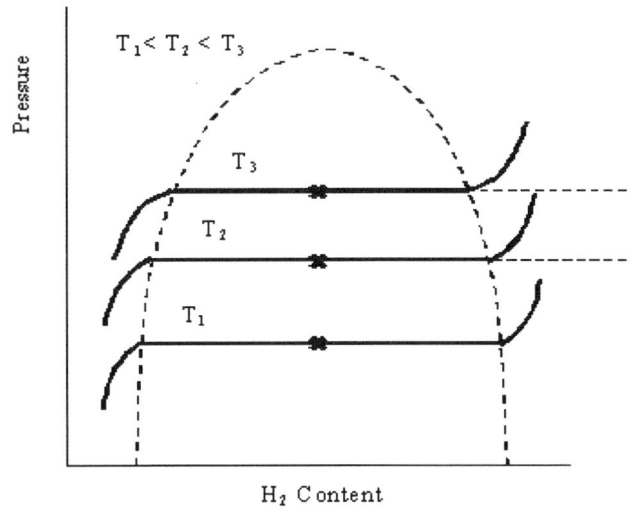

Figure 1. PCT Diagram of a Reversible Metal Hydride System.

Figure 1 depicts isotherms at three temperatures and shows the variation of the equilibrium pressure with the concentration of hydrogen in the metal alloy. The heat of formation and the entropy of a metal hydride can be derived from the PCT curves by means of the van't Hoff equation:

ln P = $\Delta H/RT - \Delta S/R$; where ΔH is the enthalpy of the reaction or heat of formation, ΔS is the change of entropy and R is the gas constant.

At a given temperature, molecular hydrogen will first dissociate on the metal surface into atomic hydrogen, then dissolve in the metal alloy to form a solute solution. This is usually designated as the α phase. Hydride phase or the β phase, starts to grow, when the hydrogen reaches its solubility limit. The β phase formation is a nucleation and growth mechanism. During the β phase formation, the hydrogen pressure remains a flat plateau as more hydrogen is added. The α & β phases coexist in the plateau pressure region. The hydrogen pressure will rise sharply upon the completion of β phase formation. This phenomenon illustrated in Figure 2.

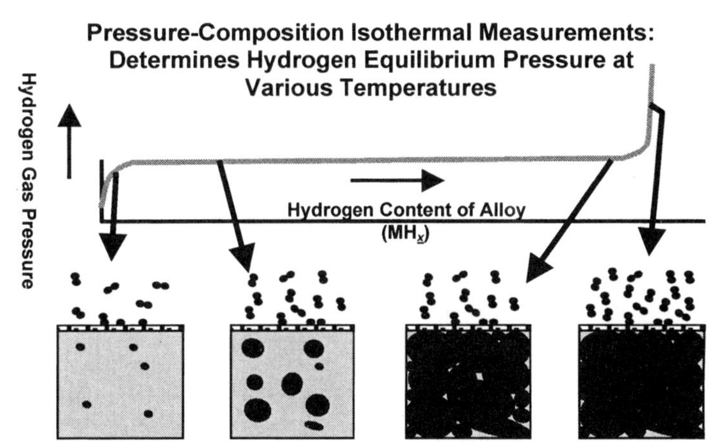

Figure 2. The Nucleation and Growth Process of β Phase Formation.

Typically, at a given temperature and pressure, hydrogen content in the flat plateau region represents the usable hydrogen. In a metal hydride system, quite often, there is a slope in the plateau. Hydrogen stored in the low-pressure part of the plateau may not be able to release because of the insufficient amount of onboard waste heat. Therefore, a large difference between the total stored hydrogen and the usable hydrogen is observed, if the material is improperly designed.

Moreover, in the design of a metal hydride hydrogen storage system, special attention needs to be paid to issues such as, the volume expansion/contraction associated with the hydriding/dehydriding process, powder settling from powder decrepitation, insufficient heat transfer due to the poor thermal conductivity of hydride powders, etc.

In addition to the alloy and design challenges for onboard fuel storage, the system needs to satisfy the following criteria: lightweight, fast refueling, onboard waste heat for desorption, cycle stability, cold temperature start up, and durability.

TOHS METAL HYDRIDE ONBOARD SYSTEM

To satisfy these requirements, our efforts have focused on alloy design and selection, heat exchanger design, powder packaging, and lightweight pressure vessel selection.

ALLOY SELECTION. Over the years, Ovonic has developed a family of metal hydride alloys, from high temperature 7 wt% Mg alloys to low temperature multi-element intermetallic alloys.

In the selection of a particular alloy for onboard vehicle application, we first defined the type of waste heat available from the vehicle. For an ICE vehicle, there are two types of waste heat: the high temperature heat from the exhaust and the low temperature heat from engine-cooling loop. For a PEM fuel cell vehicle, there is no exhaust, the only available waste heat is from the fuel cell cooling loop. The heat of the cooling loop either from the engine or from the fuel cell stack is similar in temperatures, in the range of ~50-85 °C, only varies in its flow rate. To be versatile, our onboard vessel is designed to integrate with heat only from the cooling loop. In this case, low temperature metal hydrides (LTMH) were selected for onboard fuel storage.

Primary criteria in the selection of an appropriate LTMH are: (1) The alloy should be able to provide the highest usable hydrogen or to minimize the trapped hydrogen in the tank, and (2) The alloy would be able to release hydrogen to start the engine in cold weather without the engine heat supply.

Figure 3 shows the pressure-temperature tailorability of Ovonic LTMH. The appropriate alloy to be used in a particular vessel will depend on the required deliverable pressure to the fuel rail, the cold temperature start up requirement, and the maximum allowable filling pressure.

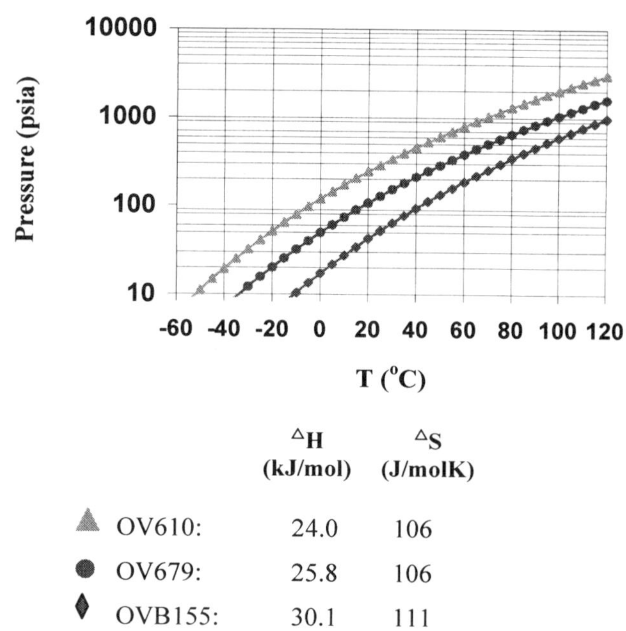

Figure 3. Pressure vs. Temperature of Ovonic alloys.

Figure 4 is the reversible capacity of the alloys as a function of the filling pressure. The measurement was done in a specially designed alloy-screening vessel with an active liquid heat exchanger. The sample size of the measurement was ~ 2 kg. The desorption was done with the water temperature of 85 °C.

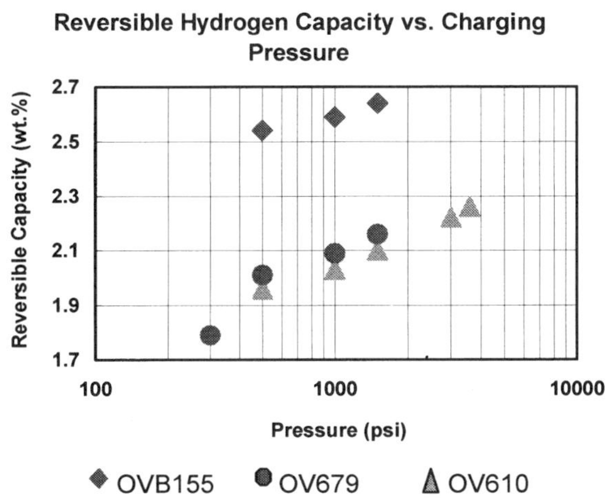

Figure 4. Reversible Hydrogen Capacity vs. Filling Pressure.

As can be seen from Figure 4, OVB155 has the highest reversible capacity of 2.6 wt% at 1500 psig . However, it will require 20 °C to provide the initial fuel to the engine at 30 psig. (See Figure 3). Due to its inability to provide cold temperature start up and other considerations, OVB155 was not chosen for this prototype vehicle demonstration.

HEAT EXCHANGER. One of the most challenging issues associated with MH onboard storage is satisfying the fast refilling requirement. In order to accommodate fast refueling, a very efficient onboard heat exchanger as well as an off-board heat dissipation facility in the filling station are required. Figure 5 shows the average heat removal rate in kW as a function of refueling time for a 6 kg H_2 metal hydride vessel for an alloy with a heat of formation of 25 kJ/mol of H_2.

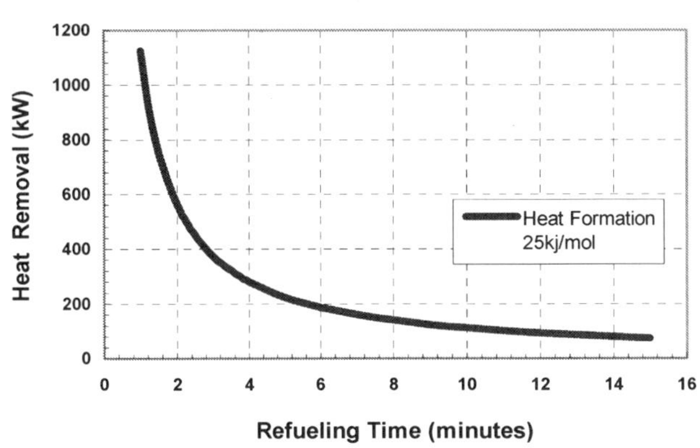

Figure 5. Filling Time vs. Heat Removal Rate (based on 90% of 6 kg hydrogen).

Because of the limited ΔT between the hydride bed and the cooling liquid, to facilitate the fast refueling, it is desirable to use a higher filling pressure. For example, with 20 °C cooling liquid, filling OV610 @ 500 psig, the hydride bed temperature will be about 40 °C (See Figure 3) which results a ΔT of 20 °C, whereas the ΔT will increase to 80 °C if 1500 psig is used for the refueling, which enhance the heat removal rate. For a given filling pressure, OV679, the lower plateau pressure alloy, will have a larger ΔT, than OV610, the higher plateau pressure alloy. However, other attributes of OV610 discussed below makes it a favorable alloy for vehicle applications.

To accommodate the fast refueling, we use a computer model to guide and to optimize the heat exchanger design. A mesh-less analytical model was developed to analyze the non-linear behavior of the metal hydride thermo-physical characteristics during hydrogen absorption. Based on the experimental PCT data and its heat of formation of a given alloy, this model calculates the heat generation and dissipation rate for a given heat exchanger geometry as a function of the filling pressure and coolant temperature. The initial conditions and boundary conditions were refined by comparison of the calculated and experimental kinetics data. Once these conditions were established, for a given heat exchanger geometry and coolant temperature, the simulation provided us with information of the filling time as a function of filling pressure.

Figure 6 is the hydrogen filling time (90% of the total capacity) as a function of the filling pressure of OV679 and OV610. The testing was done using a high-pressure stainless steel vessel with the same heat exchanger design as our current fiber wrapped onboard vessel. The cooling water was kept at 10 °C during the testing. Because of the higher bed temperature of OV679 compared to OV610 at a given filling pressure, OV679 provides a more efficient heat transfer and as a result, a faster refilling. Figure 6 shows the experimental data and the simulation. However, even though OV679 has a faster refilling time, OV610 has some desirable attributes from a vehicle integration and performance perspective. For example, the OV610 has a better cold temperature start up capability, can provide a sustainable hydrogen flow at a pressure higher than 150 psi, which is a condition required by some of the fuel cell vehicles. OV610 will also provide a better driving performance when the vessel is in the near empty condition. To further improve the refilling time of OV610, we need to either increase the filling pressure or improve the heat exchanger. An improved heat exchanger is currently under design, in which we will rearrange the geometry of the heat exchanger and double the surface area without compromising the system weight and volume. Based on the simulation, we predict that the refilling time of OV610 would be able to reduce from 14 minutes to about 6 minutes @1500psig filling pressure. We would like to point out that these experiments and simulations were conducted without the constraint of temperature rise during refilling.

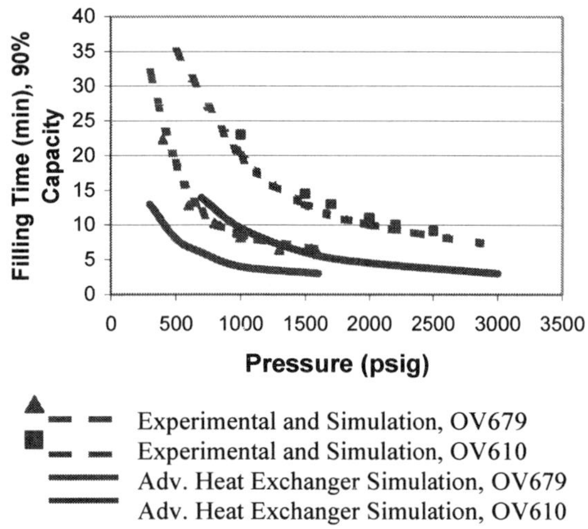

Figure 6. Refilling Time vs. Pressure. Coolant temperature 10 °C.

THE LIGHTWEIGHT PRESSURE VESSEL. In order to reduce the weight of the vessel, we developed a proprietary technology to combine a lightweight fiber wrapped pressure vessel with our heat exchanger and liquid manifold. The fiber wrapped vessel was provided by Dynetek. The vessel with the heat exchanger and liquid manifold was hydrostatically pressure cycled between 290 psi and 4500 psi for 15,000 cycles without leakage. It was then hydrostatically burst. The burst pressure was 13,801 psi, which is above the minimum required burst pressure of 10,800 or 3 times the 3600 psi service pressure.

Figure 7 is an Ovonic prototype metal hydride onboard vessel. One end contains a hydrogen solenoid valve; the other end is the liquid inlet and outlet. The vessel has an internal volume of 50 liters with an external dimension of 32.8 cm OD and 84 cm in length.

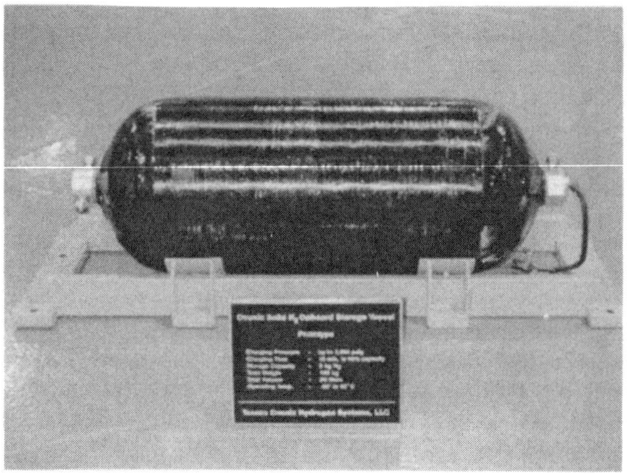

Figure 7. Ovonic Prototype Metal Hydride Onboard Vessel.

Table 1 is a comparison chart of the metal hydride and compressed hydrogen using the same fiber wrapped vessel with an internal volume of 50 liters.

Table 1

	Metal Hydride@ 1500 psig	Compressed H_2 @ 3,600 psig
H_2 Capacity (kg)	3.0	0.88
Total weight (kg)	190	25.2
Weight %	1.58	3.49
Volume (g / liter)	60	17.6

THE PERFORMANCE OF AN OVONIC METAL HYDRIDE SYSTEM. The Testing Facility. A testing facility at ChevronTexaco's Richmond Technology Center (RTC) was designed and constructed to conduct performance and cycle life tests for Ovonic hydrogen storage vessels. Two test stands are in operation, each test stand is capable of controlling two vessels. Both systems are remotely controlled and operated.

Hydrogen flow rates of up to 15000 slpm and 1000 slpm can be monitored during absorption and desorption experiments, respectively. Other features include the recording of thermocouples and strain gauges attached to the vessels under investigation. Each of the test stands can provide up to 25 gal/min heating/cooling liquid (water-propylene glycol) between 8°C and 95°C. Figure 8 is a picture of the test stand.

Figure 8. The Test Stand.

A diaphragm compressor is available to supply hydrogen at pressures of up to 5500 psi, in addition to the RTC hydrogen supply at 3200 psi. The high-pressure storage capacity installed at the compressor is approximately 19 kg of hydrogen at 5500 psi. This storage unit is also used to recover and re-use hydrogen during long-term cycle life experiments.

The equipment was installed in accordance with RTC safety regulations and procedures. All pressurized parts are equipped with safety relief devices and pressure switches for emergency shut-down. Hydrogen and UV/IR detectors were installed in the test cells to trigger a safe shut off and to turn on the water sprinkler system in case of emergency.

The vessel was instrumented with ten strain gauges and eight thermocouples. The strain gauges were mounted on the aluminum liner underneath the carbon composite outer wrapping. Thermocouples were placed at coolant inlet and outlet locations, also at locations embedded in the metal hydride alloy and on the aluminum liner surface. The instrumented vessel was subjected to various absorption and desorption experiments. Because the glass transition temperature of the organic resin of the fiber wrapped vessel is about 110 °C, an e-stop was set at 85 °C on the liner during absorption testing.

Absorption Testing. Figure 9 shows an example of a hydrogen absorption experiment. The figure shows hydrogen delivery pressure, absorption capacity, metal hydride bed temperature (TC1) and aluminum liner surface temperature (TC6,) as functions of time. The flow rate of cooling liquid is about 20 gallons per minute (gpm) with the coolant temperature maintained at 8°C. The hydrogen delivery pressure was initially set at 1000 psig then ramped up to 1700 psig. The experiment ran for 20 minutes with a total storage capacity of 3 kg H_2, of which 2.7 kg or 90 % of the capacity was absorbed in 10 minutes.

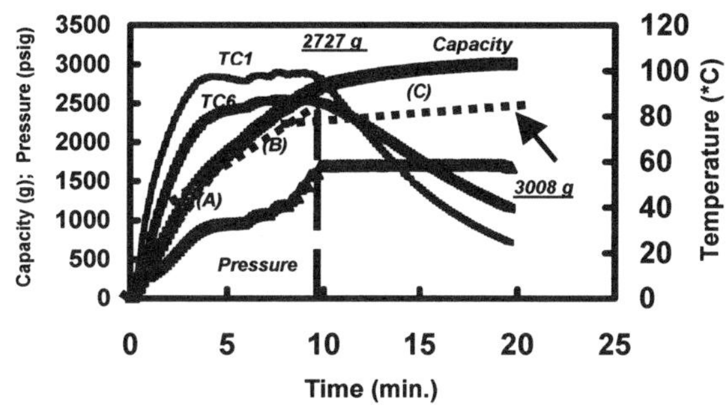

Figure 9. Absorption Experiment.

The capacity curve indicates that the filling process involves three stages, as marked (A), (B) and (C). The first stage (A) is characterized by a rapid hydrogen uptake, in which 1.5 kg of hydrogen is absorbed in the first 3 minutes. This rapid absorption is primarily due to the presence of a large heat sink from the heat capacity of the MH alloy. As can be seen from Figure 9, the rapid reaction during this stage is also accompanied by a rapid rise in both temperature and pressure.

During the second stage (B), an additional 1.2 kg of hydrogen is absorbed in about 7 minutes. The temperatures are kept fairly constant so that the heat generation and heat removal rate remain in balance. The duration of this stage is largely dictated by the effectiveness of the heat exchanger. In this particular experiment, a total of 2.7 kg H2 is absorbed by in the end of the second stage (approximately 10 minutes).

The third stage (C) is the end of the process, where the hydride vessel pressure approaches the delivery pressure. As a result, the hydrogen flow rate and the hydride growth rate are dramatically reduced. It takes an additional 10 minutes to absorb the remaining 300 g of hydrogen . During this stage, the vessel temperature starts to decline because the reaction slows.

Figure 10 illustrates the relationship between the strain and the absorption capacity. The data shows that at 90% of the total absorbed capacity or 2.7 kg H_2, the strain is ~ 1200 $\mu\varepsilon$, which is 33% of the allowable strain. The figure also shows that the strain rapidly increases as hydrogen is further absorbed into the vessel. The strain value is more than double for the last 0.3 kg H_2. Nevertheless, the 3000 $\mu\varepsilon$ measured at the completion of absorption remains below the maximum allowable working limit of 3600 $\mu\varepsilon$. The initial negative strain and the sharp rise of strain value observed in the last 10% of the absorption are currently under more detailed investigation.

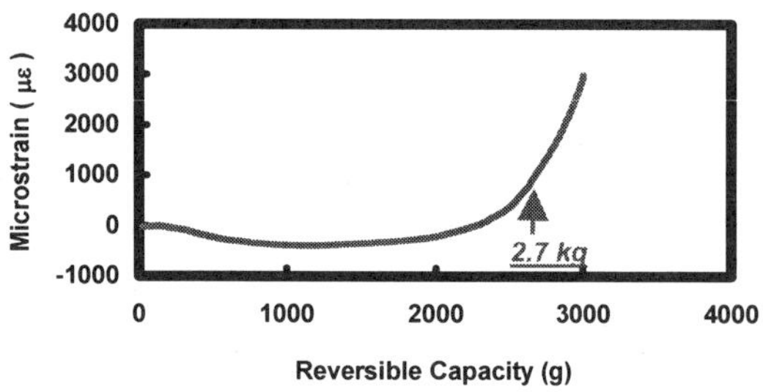

Figure 10. Microstrain During Absorption.

Desorption Testing. Figure 11 is an example of desorption experiment, which includes hydrogen flow rate, desorption capacity , internal vessel pressure, liquid inlet temperature (TC1) as functions of discharge time. The liquid temperature (TC1) was maintained at 75°C during the desorption experiment. The desorption flow rate is controlled @ 350 slpm (31.25 g/min) for approximately 90 minutes or 2.8 kg hydrogen. For a hydrogen ICE hybrid vehicle, a hydrogen flow in the range 17-300 slpm at 30 psi pressure is required. In this case, the vehicle can continue to run until the vessel is almost empty.

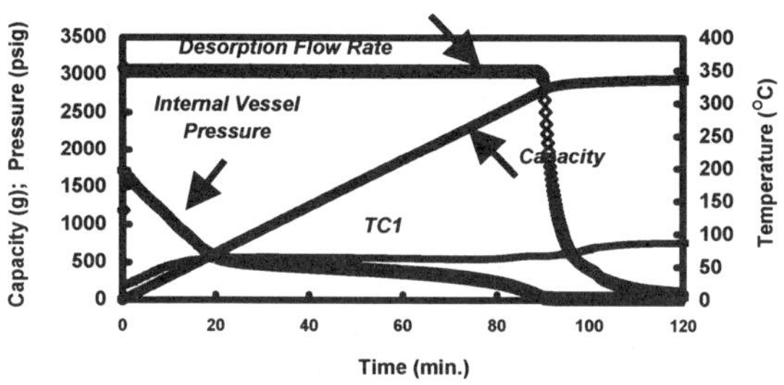

Figure 11. Desorption Data Recorded from an Ovonic MH On-board Hydrogen Storage Vessel.

One of the advantages of metal hydride hydrogen storage is the low pressure operation. Even though, the filling pressure is about 1500-1700 psi, the vessel operates at below 200 psi most of the time. Figure 12 shows the internal pressure of the metal hydride vessel at two different liquid inlet temperatures as a function of % of H_2 remaining in the vessel. For comparison, a 5000 psi compressed H_2 vessel is included in the graph. As can be seen from the graph, the pressure of the metal hydride vessel quickly decreases from the initial refilling pressure to below 200 psig, whereas, the pressure of the compressed H_2 vessel remains above 200 psi until almost empty.

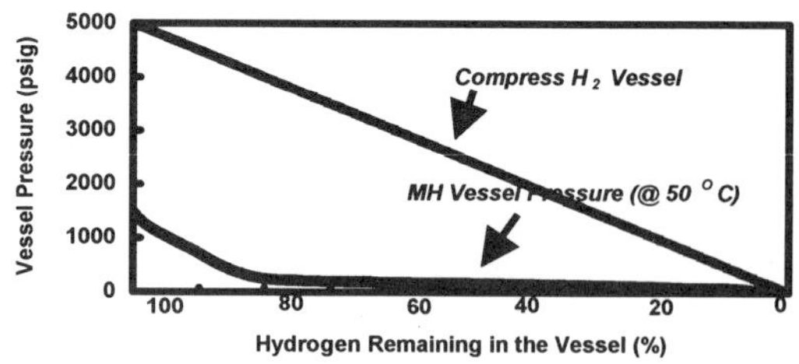

Figure 12. Vessel Pressures, MH vs. 5000 psi Compressed H_2 Vessel, as a Function of % of H_2 Remaining.

FROM LABORATORY TO REAL WORLD

Figure 13 shows a fully instrumented onboard vessel under laboratory testing, in which a chiller is used for cooling liquid for absorption and heat is provided from an electrical liquid heater for desorption.

Figure 13. Fully Instrumented Metal Hydride Onboard Hydrogen Storage Vessel.

Figure 14 shows the vessel installed into the trunk of an ICE hybrid vehicle (2002 Toyota Prius). The waste heat from the engine cooling loop through a liquid-liquid heat exchanger provides the heat for releasing hydrogen and feeding into the engine. Re-circulated water from an off-board water tank is used for heat removal during refilling. Details of vehicle integration and its performance data will be presented in a separate paper at this conference (9).

Figure 14. Metal Hydride Onboard Hydrogen Storage Vessel Installed in Converted H_2 ICE Hybrid Vehicle.

An onboard data logging system is installed in the vehicle, which automatically records data when the engine starts. In order to study transient response, data is recorded in one-second intervals. The data logging system records the vessel pressure, liquid inlet temperatures to the vessel from the engine cooling loop, the liquid outlet temperature from the vessel to the circulation pump and the hydride bed temperature.

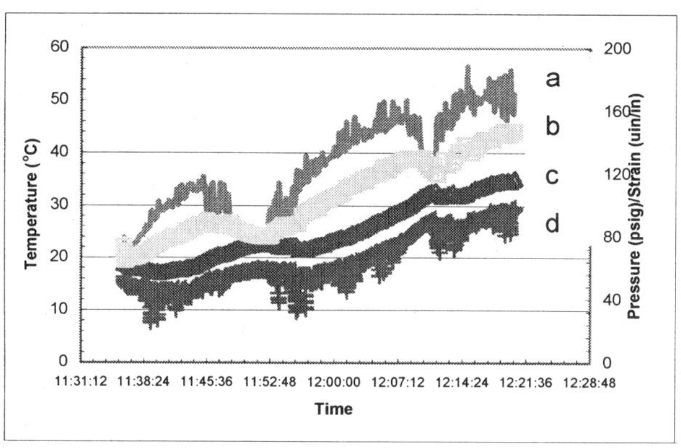

Figure 15. Sample Onboard Data Logged During a 45-Minute Drive.

Figure 15 shows a graph of the data during a 45 minute drive. When the engine starts, the hydride bed temperature is 18 °C, the inlet (a) and outlet (b) liquid temperatures are both about 20 °C, the pressure (d) is about 50 psi, and the vessel is about 1/3 full. At the end of the drive, the hydride bed temperature (c) has reached 34 °C, the liquid inlet/outlet temperatures are 52 °C and 44 °C respectively. The pressure rose to 110 psi and the liquid flow rate through the circulation pump is about 2.5 GPM.

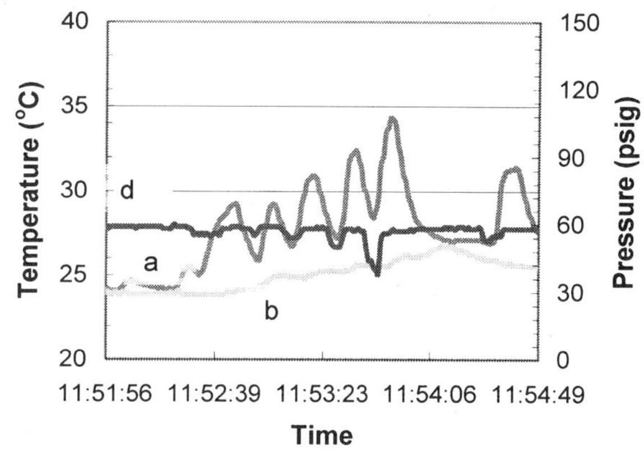

Figure 16. Onboard Data Logged During a 45-Minute Drive.

The spikes in the inlet temperature curve and the vessel pressure curve are an indication of the transient responses. An expanded section between the dotted lines of the figure is shown in Figure 16. The spikes of the inlet temperatures (a) are in good correlation with the pressure drop (d) with an approximate 10-second delay. The pressure drop in the vessel is caused by the acceleration of the engine. Since there is no temperature spike in the outlet (b), the extra engine heat is consumed by the metal hydride to provide a higher hydrogen flow into the engine. It is a self-regulating mechanism.

Figure 17 shows that the vehicle is driven with the vessel near empty. The hydride bed temperature (c) is maintained about 55 °C while the pressure steadily decreases (d). This indicates that the hydrogen in the metal hydride has reached the tail part of the PCT curve. Yet, the vehicle still functions well with good response.

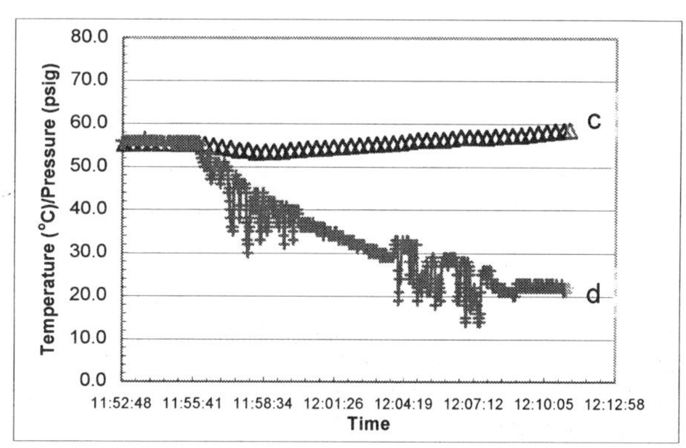

Figure 17. Onboard Vessel Data Logged with a Near Empty Vessel.

A driving range of 137 miles was demonstrated at 55 MPH constant speed with the vessel filled @1200 psi to about 2.7 kg H_2, which has the fuel economy similar to that of a fuel cell vehicle (9).

Table II shows the emission data of the hydrogen Prius vs. gasoline Prius. As can be seen from the table there is about 220 gram/mi CO_2 reduction and 0.385 gram/mi CO reduction in the hydrogen vehicle. The NO_x emission also meets the SULEV standard.

Tailpine Emissions, g/mi				
	NMOG	CO	NOx	CO_2
Hydrogen Prius With Prius Catalyst	0.001	0.001	0.018	3.2
Gasoline Prius	0.010	0.386	0.004	222.8
SULEV / PZEV standard	0.010	1.0	0.020	na

Table II. Emissions data.

Figure 18 shows the fueling port. The top connector is the hydrogen filling nozzle. The two bottom connectors are the cooling inlet and outlet. The small connector on the top left corner is a six pin electrical connector which is connected to the refilling dispenser. The electrical connector provides data such as bed temperature, tank pressure, strain, and hydrogen sensor readings during refilling for this prototype demonstration vehicle.

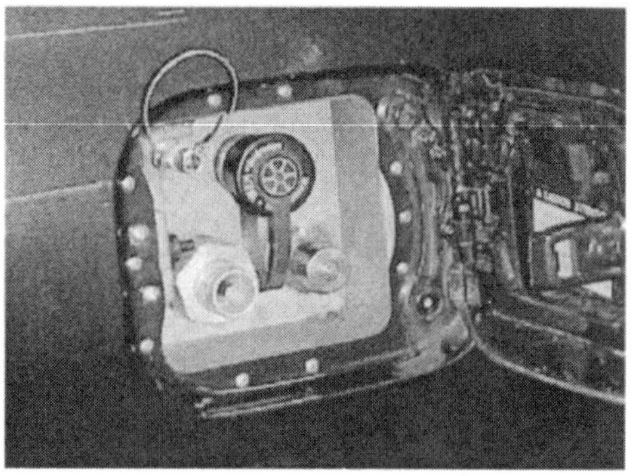

Figure 18. The Fueling Port.

Figure 19 is a portable refilling dispenser, which is designed to be connected to existing 5000 psi compressed hydrogen refilling stations. The dispenser, designed and constructed by TOHS to be user-friendly, will regulate the pressure down to the specified pressure for metal hydride refilling with a pre-programmed refilling algorithm.

Figure 19. Portable Low-Pressure Refilling Dispenser.

Because 5000 psi compressed H_2 vessels will not be able to provide vehicles with sufficient driving range, the DOE and some automotive OEMs are pushing toward the use of 10,000 psi vessels. To provide a vehicle with a driving range of 300 miles (assuming 50 miles per kg H_2), 6 kg H_2 will be needed. Table III is a comparison of a Ovonic metal hydride onboard vessel and a 10,000 psi vessel.

6 Kg H_2 Vessel		
	MH	**10,000 psi compressed**
Internal Vol (l)	100	153
Weight (kg)	375	171
Filling pressure (psi)	1500	12,500

Table III. Comparison of a Ovonic Metal Hydride Onboard Vessel and a 10,000 psi Vessel.

While a 6 kg H_2 storage system using an Ovonic metal hydride vessel will be 200 kg heavier in weight, it is 50 % less in volume. More importantly, its refilling pressure is 8 times less and its operational pressure is about 50-100 times less. Further, because low-pressure gaseous hydrogen is used, instead of liquid H_2 or 10,000 psi compressed H_2, a metal hydride system will also benefit from substantially simpler and lower fuel cost.

CONCLUSION

In this paper, we demonstrated that hydrogen hybrid ICE vehicle powered by metal hydride fuel storage system can virtually eliminate the CO_2 tailpipe emission. The NO_x level can also meet the SULUV standard. It would be an ideal bridging technology in the adapting of hydrogen as the transportation fuel.

Hydrogen onboard storage, however, presents the major technological challenge. The storage system needs to be light in weight, compact in volume, safe in operation, low in fuel cost, fast and user friendly in its refilling operations.

While none of the current storage systems can meet all these requirements, with today's technology, an Ovonic metal hydride system's benefits such as long range, low fuel cost, safe low pressure operation etc. surpass the weight disadvantage. Our strategy is to work on and perfect today's technology while brainstorming and continuing the research on a much improved solid hydrogen storage system which will lead to a long-term hydrogen economy solution.

ACKNOWLEDGMENTS

We acknowledge Franz Gingl and Ming Wang for performing the vessel testing; Isaak Fidel, Nestor Kropelnyckyj, and Alexander Gerasimov for vessel design and assembly; Dick Geiss, Bruce Falls, and Alwin Lutz for vehicle conversion; Bob Stempel, Gene Nemanich and Greg Vesey for encouraging and supporting the project; and Angela Goddard for editing the manuscript.

REFERENCES

1. See for example, HYDROGEN POWER, by L.O. Williams, Pergamom Press Inc. New York, 1980. Turillon, P.P. " Design of Hydride Containers for Hydrogen Storages" in Proceedings of the 4th World Hydrogen Energy Conference, CA, USA. 13-17 June 1982, p. 1289.
2. Hoffman, K. C., Wische, W.E., Wiswall, R.H. , Reilly, J.J., Sheehan, T.V. and Waide, C.H. " Metal Hydrides as a Source of Fuel for Vehicular Propusion". SAE paper 690232 presented at the International Automotive Engineering Conference, Jan, 13-17, 1969, Detroit, USA.
3. Lynch, F.E. and Snape, E., "The Role of Metal Hydrides in Hydrogen Storage and Utilization" in Alternative Energy Sources, Vol. 3, p 1479. Veziorglu, T.N. Ed., Hemisphere Publishing, Washington D.C. 1978.
4. Strickland, G., "State-of-the-Art Summary of the Technical Problems Involved in the Storage of Hydrogen via Metal Hydrides" in Alternative Energy Sources, Vol. 8, p 3699. Veziorglu, T.N. Ed., Hemisphere Publishing, Washington D.C. 1978.
5. See for example; the DOE web site: www.eere.energy.gov/hydrogenandfuelcells/.
6. Noh, J. S. Agarwal, R. K. and Schwarz, J. A. " Hydrogen Storage System Using Activated Carbon" Int. J. Hydrogen Energy. Vol 12, p 693, 1987.
7. Dillon, A.C. etal. Hydrogen Storage in single-wall Carbon Nanotubes, 14th World Hydrogen Energy Conference, Montreal, Canada (June of 2002).
8. Schlapbach, L., Züttel, A. Hydrogen Storage Materials for Mobile Applications, Nature, Vol 414, 353 (2001).
9. Geiss, R., Webster, B., Ovshinsky, S.R., Stempel, R. Young, Li, Y., Myasnikov, V., Falls, B., Lutz, A., "Hydrogen-Fueled Hybrid: Pathway to a Hydrogen Economy". This conference.

CONTACT

Rosa C Young received her PhD in physics from Rensselaer Polytechnic Institute, Troy, New York. Prior to joining Energy Conversion Devices (ECD Ovonics) in 1984, she was a research staff member in the Solid State Division of Oak Ridge National Laboratory. Currently, she is the Vice President of Technology of Texaco Ovonic Hydrogen Systems; a joint venture between ChevronTexaco and ECD Ovonics. The venture's web address is: www.txohydrogen.com, her e-mail address is: ryoung@ovonic.com.

Mat. Res. Soc. Symp. Proc. Vol. 801 © 2004 Materials Research Society BB1.2

New Science and Technology the Basis of the Hydrogen Economy

Stanford R. Ovshinsky
Energy Conversion Devices, Inc., Rochester Hills, MI 48309, U.S.A.

ABSTRACT

Hydrogen is called the "ultimate fuel." It is also the ultimate element. It was born in the Big Bang and almost all of known matter is composed of it. Its condensation into a star, our sun, through fusion, provides us the energy and the photons which power our earth and which we can utilize in the form of photovoltaics to break apart water and generate hydrogen as an energy source on earth.

The hydrogen economy is here. It has been initiated by the electric and hybrid vehicles which depend upon it for their operation through nickel metal hydride batteries and hydrogen as a fuel for the internal combustion engine and by outwitting the Carnot cycle for use as a fuel cell.

I will discuss the complete system needed for the hydrogen economy from generation to storage to infrastructure and use. Any one part of this loop is necessary but not sufficient.

Our global economy is based upon energy and the societal needs for a nonpolluting, non-climate change fuel which does not require strategic military defense as does oil. The transition from fossil fuels to hydrogen is of revolutionary import not only for its societal impact but also for the new materials science that it absolutely requires in all of its aspects. New science and new technologies build much needed new industries, which provide not only jobs but also feedback on our educational system.

Recall that the ages of civilization are known by their materials. Truly, the present age will be known by the materials that make up the twin pillars of our global economy – energy and information. Therefore, I will address the new science, technology and atomic engineering of the materials so necessary to make positive, realistic and productive this revolutionary transition of energy from its fossil fuel beginnings to the present.

PERSPECTIVE

The hydrogen economy was not predicted to occur until approximately 50 years from now as shown in Fig. 1.

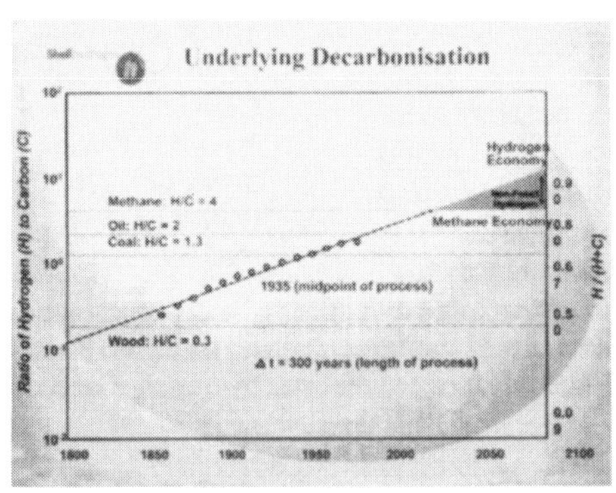

Fig. 1

What I will discuss and illustrate is that we have changed this linear progression from millions of years of dependence on fossil fuel by hominids and humans to hydrogen, the ultimate fuel. To put this in perspective, we are decoupling from fossil fuel and coupling to the first particles of the universe, protons, electrons, and photons, born in the Big Bang.

Rather than a discussion of what is needed, what this paper seeks to show is that there are actual concrete solutions to the problems that were barriers to the hydrogen economy.

Fig. 2 shows that science and technology created a giant step forward when the electric light bulb was invented by Edison just one year before this cartoon. While the fuels and technology which made steam the driving force of the prior industrial era were revolutionary in nature, the new electrical age still depended for its energy on fossil fuels and increasingly on oil.

This drawing is from an 1881 issue of *Punch* magazine, showing King Steam and King Coal anxiously observing the infant "electricity" and asking: What will he grow to? Many of today's industries, ranging from manufacturing to publishing, are similarly seeking answers regarding the growth of computer technology in the next century. (from *Engineers & Electrons*. IEEE Press. 1984.)

Fig 2

In any case, new science and technology were needed for the achievement of the hydrogen economy which had been put off far into the future by extrapolation.

This was particularly important since petroleum was symbiotic with the developing transportation industry and its internal combustion engine.

What was required was new science and technology, especially expressed in materials, for generation, usage, transportation and storage. The conventional approaches of high pressure and liquid hydrogen were well known but could not attack the basic problems.

By storing hydrogen reversibly in solids, and solving the problems of storage, kinetics, that is the speed of getting hydrogen in and out, and cycle life, we have been able to achieve a family of hydrides capable of real world applications.

This has been accomplished by our work in disordered and amorphous materials in which the materials are a system composed of a spectrum and density of catalytic sites and storage sites. This required atomic engineering permitted by the degrees of freedom of disorder where we could use multi-elements and multi-phases and could work across the periodic table, not by throwing darts at it (guesswork), but by following the rules and the scientific base that we have established. In particular, we utilized d- and f-orbital materials for many applications and lighter weight materials modified by d- and f-orbitals to achieve the sites required.

It is not my intention to go more deeply into the science since this paper, as a keynote, is intended to show the results of this approach and therefore the solution of the problems that has allowed us to enable and advance the realistic hydrogen economy.

We have done this by taking a systems approach not only for a family of materials, but for a complete infrastructure, well to wheels, so to speak.
One product, for example, a fuel cell, is necessary but not sufficient, it must be a seamless loop.

The problems brought about by the increased use of fossil fuel energy, particularly oil, in our social, economic and political institutions were putting pressure on our global society in terms of pollution, climate change[1&2] and the dependence on the military to protect the precious stream of oil.

Fortunately, switching to hydrogen, the ultimate and cleanest fuel can break the great dependence on geographical location of fuel since our hydrogen-fusing sun provides us the photons to break up water. Therefore, the hydrogen economy starts with the sun.

We will show that the science and technology of amorphous materials have resulted in the ability to make thin-film, continuous web, multi-junction material devices that can use the entire spectrum of sunlight resulting in the energy necessary to break up water to generate hydrogen.

BACKGROUND

Figs. 3 and 4 show Iris and me planning such a systems loop for hydrogen in 1960 in a storefront laboratory in Detroit. The problem was to build such a loop for one needed to generate, transport, store and utilize hydrogen. In other words, one needed not only to invent the materials, the products and the production technology but to provide a realistic infrastructure that could interface the present one for petroleum.

Fig. 3

Fig. 4

Iris and I founded our company to utilize science and technology to solve serious societal problems and build new industries. The most important barrier we had and still have to overcome was not scientific or technological but being classified as a disruptive technology by the very people to whom we were trying to be constructive and a resource to.

HYDRIDES

Let us see how we have succeeded. First of all, by inventing the nickel metal hydride battery and introducing solid hydrides in a realistic and cost effective manner, we jump started the hydrogen economy. Almost all electric and hybrid vehicles are enabled by our Ovonic batteries and well over a billion consumer nickel metal hydride batteries were sold last year.

The degrees of design freedom of atomically engineered materials permits us to have not only a simple couple such as lead acid, nickel cadmium, etc. but materials that can be designed and improved both in energy density and in power. The key elements for the batteries are d- and f-orbital materials. Some electrodes have as high as eleven different elements. We do not only atomic engineering but orbital engineering.[3&4]

In brief, our materials approach is a systems one. That is in the same multi-elemental, multi-phase material we have designed new types of highly efficient and non-noble metal catalysts and a spectrum of hydrogen storage sites that can be tailored for energy density and for power.

MAKING THE HYDROGEN ECONOMY REALISTIC

Please note in Fig. 5 that the Hydrogen Loop is now shown by actual products.

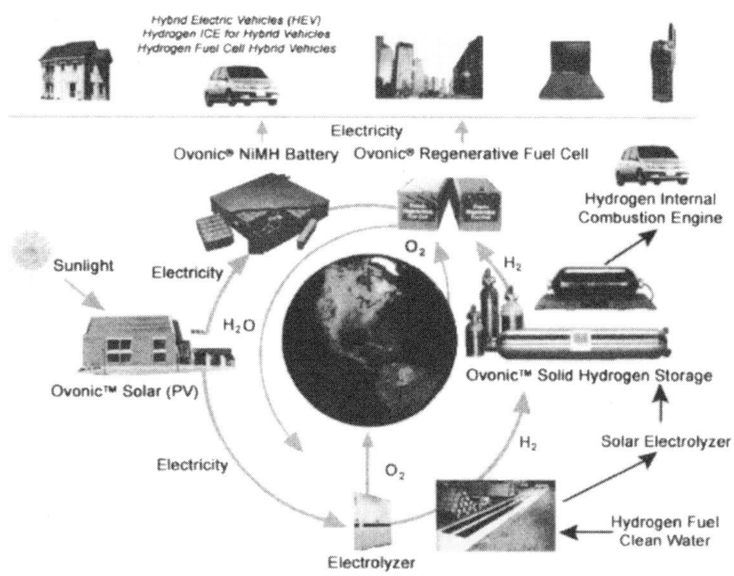

Fig. 5. ECD: Making the Hydrogen Economy Possible
The Only Company with the Complete Hydrogen Loop

INFRASTRUCTURE

Now for the reality of the infrastructure shown in Figure 6. Note how it blends with the existing one without disruption. We provide a transition from fossil fuels to hydrogen with this approach.

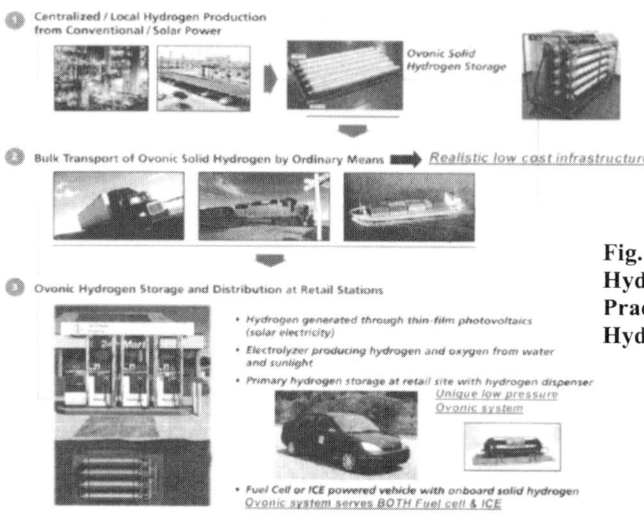

Fig. 6. Ovonic Solid Hydrogen Systems – Practical Solutions for the Hydrogen Infrastructure

Fig. 7 is an old slide showing the electric and hybrid electric vehicles that are powered by our nickel metal hydride (NiMH) batteries.

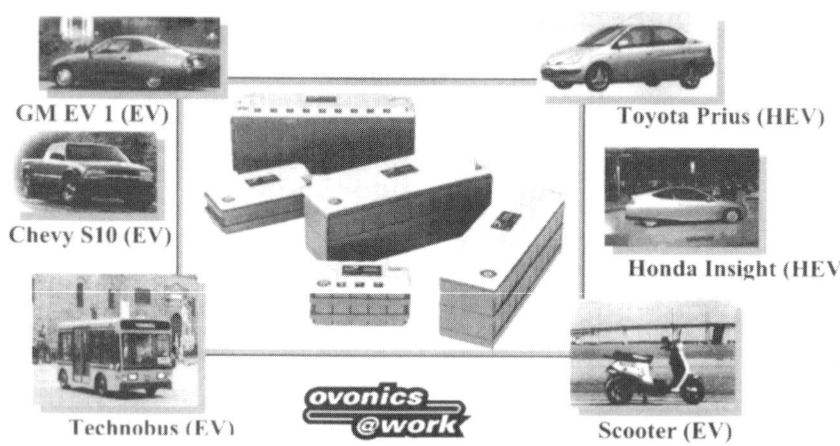

Fig. 7

Fig. 8

Fig. 8 is ECD Ovonic's photovoltaic hydrogen generation and Ovonic all solid hydrogen hybrid ICE vehicle. The Ovonic Uni-Solar thin-film photovoltaic installation on the shelter can be used to generate hydrogen from water.

Fig. 9 depicts the Ovonic renewable hydrogen systems. The same system that fuels the ICE vehicle can also fuel the fuel cell.

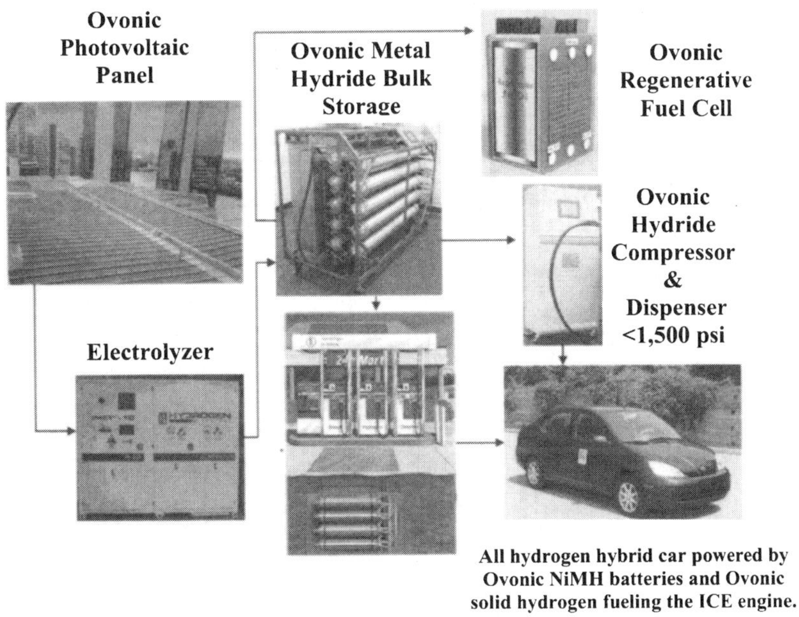

Fig. 9

Fig. 10 shows the advantages of our Ovonic Fuel Cell.

Advantages of Ovonic™ Regenerative Fuel Cells
• Unique proprietary technology
• Atomically engineered, non-noble metal, low cost, and poison-resistant catalyst electrodes
• Instant start capability
• Absorbs regenerative braking energy
• Wide temperature range (-20°C to 100°C)
• Higher efficiency (60%)
• Simple and robust design
• Low cost
• Satisfies the market

Fig. 10

Fig. 11 shows that an Ovonic solid hydrogen ICE hybrid can match a gasoline hybrid in its important parameters. The automotive industry has been committed to the internal combustion engine. Therefore, having a hydrogen burning internal combustion engine, especially in a hybrid form, can provide the transition to the utilization of fuel cells. Indeed, it will probably extend the future for internal combustion engines since there will be a whole family of various kinds of electric/hybrid vehicles. Our approach shows that internal combustion engine burning hydrogen provided by our solid hydrogen storage tank, together with our nickel metal hydride batteries has equality with any other kind of system being considered. This approach shows an equality to utilizing fuel cells and its advantages, therefore, should become part of the various kinds of electric/hybrid electric vehicles that will come on line in the near future.

Ovonic Hydrogen ICE Vehicle
EMISSIONS and FUEL ECONOMY-Dynamometer Test

Code: **BOLD**: **Hydrogen Ovonic vehicle** *ITALICS: Gasoline vehicle*	**Tailpipe Emissions, g/mi**			
	Hydro-carbons HC	**Carbon Monoxide CO**	**Nitrogen Oxides NOx**	**Carbon Dioxide CO_2**
Gasoline (production configuration)	*0.010*	*0.386*	*0.004*	*222.8*
Calibration, with Current Catalyst	**0.001**	**0.001**	**0.018**	**2.5**
SULEV/PZEV Standard	**0.010**	**1.000**	**0.020**	**N/A**

EPA City Economy	*42.4 MPG Gasoline*
	42.3 MPK Hydrogen Ovonic
EPA Hwy Economy	*45.0 MPG Gasoline*
	46.1 MPK Hydrogen Ovonic
EPA Combined	**43.9 MPK Hydrogen Ovonic**
55 MPH steady speed	**50.9 MPK Hydrogen Ovonic**
Range @ 55 MPH	**137 miles with 2.69 Kg (90%) fill.**

NOTE: 1 gallon gasoline has same energy as 1 kg Hydrogen)

Next generation: 6 Kg ➡ ~300 mi

Ovonic Hydrogen vehicle achieves performance similar to those of a fuel cell vehicle

Fig. 11

Most authorities consider that the goal of 250 miles and $3 per kilogram of hydrogen can be reached in 5 years. Some authorities consider as can be seen in Fig. 12 that such a vehicle "isn't likely for many years". The data that I have presented in this paper shows that we can achieve that goal <u>now.</u>

"One of the defining problems that indicates how far research has to progress, said workshop associate chair George Crabtree of Argonne National Laboratory, 'is a system that would store enough hydrogen in a vehicle for a 300-mile trip and be fast enough to allow for acceptable acceleration of the vehicle.'

Such a vehicle isn't likely for many years, . . . Crabtree said..."

Excerpt from article in *Physics Today*, October 2003: Hydrogen-Based Energy Merits Research

Fig. 12

Our approach is to have a family of hydrides for all occasions. They not only operate at room temperature but can be developed for other temperature ranges; for example, Fig. 13 shows how our materials operating at 300 degrees centigrade (use symbols) with 7% storage have solved the problems of storage, kinetics and lifetime.

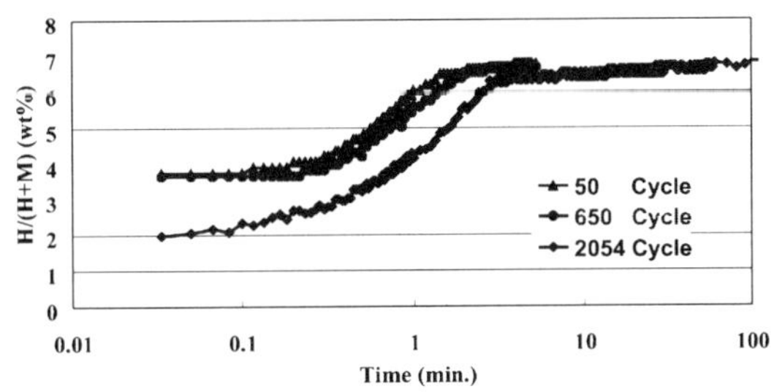

Ideal for onboard storage, provides: 300+ miles, 600,000+ miles life

Fig. 13. High Capacity Metal Hydride Long Cycle Life

Fig. 14 illustrates the special features and benefits of our approach.

Special Features	**Benefit**
• Reversible • Safe • Compact • Tailorable Pressure • Fast Refilling • Low Pressure Operation • Cold Temperature Start-up • Packaging Flexibility • Waste Heat for Desorption	Vehicle: • Long Range • Low Fuel Cost • Safe Operation Stationary: • Small Footprint • Directly Refilled from an Electrolyzer or Fuel Processor Portable: • Compact • Low Pressure Operation • Transportable

Fig. 14. Ovonic Metal Hydride Storage Technology

Fig. 15 shows low pressure operation and its great advantage in that one can charge at less than 1,500 psi and run the vehicle at approximately 150 psi as against the "normal" 5,000 psi gaseous hydrogen operation that are being used in demonstration vehicles. Low-pressure hydrogen in a solid answers the concerns as to leakage into the atmosphere which can occur when high-pressure gaseous or liquid hydrogen is used.

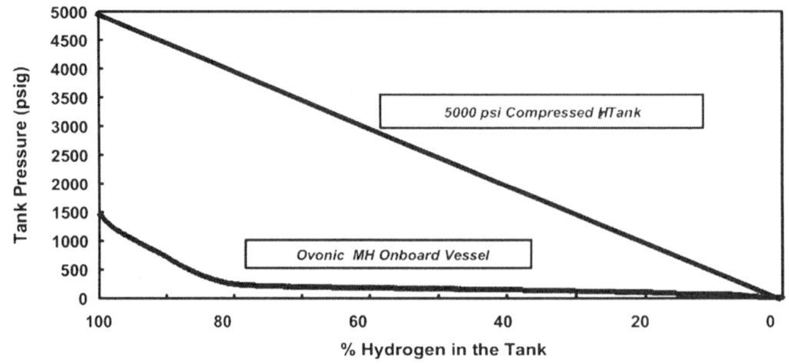

Fig. 15. Low Pressure Operation

We started with our hydrogen burning sun (fusion) and we would now like to show how our amorphous triple-junction photovoltaic production is basic to the hydrogen economy and our complete systems loop approach.

PHOTOVOLTAICS

Fig. 16 is a picture of our 30MW annual capacity continuous roll-to-roll triple junction solar cell machine which is capable of producing 9 miles and 6 tons of thin film photovoltaics in a single run.

Fig. 16

The active material is less than one half micron and has eleven layers with three different bandgaps composed of amorphous silicon and amorphous silicon-germanium. The alloys contain both hydrogen and fluorine. Therefore, the full spectrum of the sun can be utilized. Some of these layers are only 100 angstroms and yet we have very high yields. This machine is our 8th generation machine that we have built since the late 1970s. The early machines in the 1980s worked round the clock, seven days a week with yields close to 100%. Nanostructures are something that we have been doing since our first amorphous devices in the 1950s.

Fig. 17 shows the rolls ready for processing as well as a finished product.

Fig. 17

We would like to cover many of the roofs of the world. That would mean that the hydrogen economy would solve the basic energy problems that bedevil us. Fig. 18 is a typical installation and has aesthetic beauty. Fig. 19 shows our photovoltaic shingles being installed on a school roof. We have them on our own home with our nickel metal hydride batteries for energy storage being charged from our roof.

Fig. 18. Jarecki Center–Acquinas College

Fig. 19. Cass Tech Roof – Detroit, MI

Fig. 20 shows a roof of an office building in Australia. Figures 23 and 24 show the ease of installation of a membrane roof with adhesive backing.

Fig. 20

Fig. 21

Fig. 22

There cannot be a global economy without energy. There are two billion people in the world without electricity and many more beneath the poverty level. We feel that we have shown that these are not just something to be mourned but that we can use our hydrogen loop to be the means of providing the energy necessary to make a livable world.

CONCLUSION

I hope that I have shown that the ultimate element, hydrogen, should be the source of energy for the new age that will ultimately replace the over a million years of fossil fuel dominance; that the hydrogen age has already started and the people, the global economy and the peace of the world can benefit from this uncoupling of energy from the earth to our universe and that this is being achieved by new science and technology associated with amorphous and disordered materials.

As I said in my 1999 MRS paper, "I believe that I have shown that science and technology can be utilized to build new industries that are responsive to societal problems and needs, providing jobs, educational opportunities and the chance to express the creative urge that has driven humankind since time immemorial. Fig. 23 shows a young woman climbing a mountain barefooted with her future on her back, our photovoltaics, and her future in front of her, her child, bringing our photovoltaics to a village that does not have electricity. There can be no civilization without energy and without knowledge (information). We take this picture as inspiration to continue our work."

Fig. 23

ACKNOWLEDGEMENTS

I would like to acknowledge all of my colleagues and collaborators through the years and particularly in the area of Ovonic batteries, the superb leadership of Subhash Dhar and the exceptional talents of Mike Fetcenko, in hydrogen storage, that force of nature, Rosa Young and the hydrogen team, the all-star machine building team of photovoltaics, important contributions of Subhendu Guha, Jeff Yang and the PV team, especially the operating group, my close collaborators on the Ovonic fuel cell, Subhash Dhar, Srini Venkatesen and Zdravko Menjak. I could go on and on and still not do justice to all those who have contributed their wonderful talents, dedication and commitment to help in our making ECD with its unique culture an example of what human society can be. All of us owe gratitude to Bob Stempel, our partner, the ultimate management person with his knowledge of science and tremendous engineering talents. I, and ECD are greatly indebted to Iris who has been my colleague, collaborator, inspiration and the strong bond that has held our company together.

REFERENCES

1. "Disordered Materials: Science and Technology – Selected Papers by Stanford R. Ovshinsky," 2nd Edition, edited by David Adler, Brian B. Schwartz and Marvin Silver, *Institute for Amorphous Studies Series* (Plenum Press, New York, 1991)
2. Stanford R. Ovshinsky, Amorphous and Disordered Materials – The Basis of New Industries, Presented at Materials Research Society (MRS), Boston, MA (November 30 – December 4, 1998); *Mat. Res. Soc. Symp. Proc.* 554, 399 (1999); *Bulk Metallic Glasses,* William L. Johnson, Arihisa Inoue and C.T. Liu (Eds.).
3. H. Akimoto, Global Air Quality and Pollution, Science, 302, 1716-1719, December 5, 2003
4. T.R. Karl and K.E. Trenberth, Modern Global Climate Change, Science 302, 1719-1723, December 5, 2003.

Presented at the National Hydrogen Association Annual Conference 2005
March 29-April 1, 2005
Washington DC, USA

OVONIC INSTANT START FUEL CELLS FOR UPS AND EMERGENCY POWER APPLICATIONS

K. Fok[1], S. Venkatesan[1], D. A. Corrigan[1], S. R. Ovshinsky[1]

ABSTRACT

Ovonic Fuel Cells with novel hydrogen storage fuel electrodes provide instant start capabilities especially useful for UPS and emergency power applications. Unique intrinsic energy storage properties of the fuel cell stack can enable system start up even without an auxiliary battery. This robust technology also features excellent ambient and low temperature operation. Non-noble metal catalysts and simple components provide a manufacturable, scalable, and lower cost approach for a variety of standby power applications requiring extended run time.

1. Introduction

In the UPS and telecommunications industries, there has been significant dissatisfaction with existing battery backup power solutions due to issues such as reliability, life, maintenance, safety, and cost [1-3]. Recent natural disasters, unprecedented acts of terrorism, and power outages from the failure of aging power grids have reinforced the need for reliable back up power. Prolonged outages have emphasized a need for extended run times, especially for critical operations. Run time extensions to more than a few hours are not as practical for traditional battery solutions due to size, safety, and cost issues. Other options including generator sets have their own disadvantages including noise and exhaust emissions problems.

Recently, there has been a consideration and selected introduction of evolving hydrogen and fuel cell technologies. While operational success with PEM fuel cells has been demonstrated [4-7], the high cost of fuel cells is still a barrier to large-scale commercial introduction [8]. Other disadvantages include poor low temperature performance and the need to still include a separate battery for start-up and transient issues. Here, we discuss a new technology option for the UPS/emergency power application, a new type of fuel cell that also can function as a battery with inherently lower materials costs than the PEM fuel cells that are predominant in the fuel cell industry today.

[1] Ovonic Fuel Cell Company LLC, a subsidiary of Energy Conversion Devices, Inc.

2. Metal Hydride Materials

The technological basis for metal hydride fuel cells is the metal hydride materials developed for battery and other applications. Novel concepts of compositional and structural disorder developed by S.R. Ovshinsky at our parent company Energy Conversion Devices, Inc. (ECD Ovonics, see www.ovonic.com) were fundamental to the development of metal hydride materials [9] and their subsequent commercialization into Nickel Metal-Hydride (NiMH) batteries [10] and solid state hydrogen storage devices [11]. NiMH consumer batteries are now a billion dollar a year business with billions of cells manufactured and sold annually under ECD licenses. For the emerging electric and hybrid vehicle industries, the chosen technology is NiMH batteries provided by ECD licensees and joint ventures. Our recent fuel cell advances also have origins in a corporate vision and commitment to the evolving hydrogen economy dating back to the formation of ECD in 1960.

3. Metal Hydride Fuel Cells

The Ovonic Metal Hydride Fuel Cell is a patented technology [12-13] that incorporates metal hydrides into the fuel cell hydrogen electrode, where it serves both as an anodic catalyst for the oxidation of hydrogen and as a hydrogen storage medium. The metal hydride imparts a charge storage or battery functionality to the hydrogen fuel cell electrode providing this fuel cell with a unique intrinsic energy storage capability.

The metal hydride fuel cell is simple in design as shown in Fig. 1. The anode contains metal hydride as the anodic catalyst together with carbon and PTFE materials with a nickel screen current collector. The cathode contains metal oxides as the cathodic catalysts with carbon and graphite materials again with a nickel screen current collector. The electrolyte is potassium hydroxide with compositions similar to those in alkaline rechargeable batteries. The separator is an inexpensive polypropylene screen. The simple design and inexpensive materials provide for a manufacturable and cost-effective product.

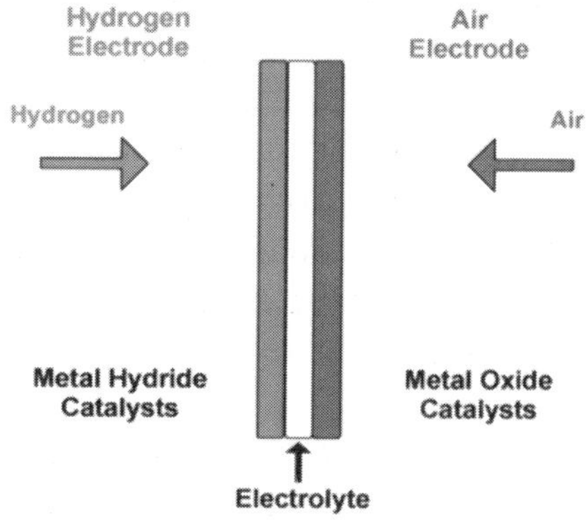

Figure 1: Schematic diagram of metal hydride fuel cell

4. Intrinsic Energy Storage Capability

The Ovonic Metal Hydride Fuel Cell has a unique energy storage capability within the fuel cell stack. This fuel cell can be charged and discharged like a battery. Pulse charge-discharge capability in an early prototype is illustrated in Fig. 2. During a ten second charge pulse, the voltage increases from 0.9 V on open circuit to around 1.15 V. Electrolysis does not occur since the voltage is even below the reversible hydrogen oxygen voltage of 1.23 V. During the subsequent ten second discharge, power is delivered at about 0.8 V. The energy storage or battery functionality of this fuel cell also provides other unique and useful features especially useful for UPS and emergency power applications such as instant start and excellent low temperature performance.

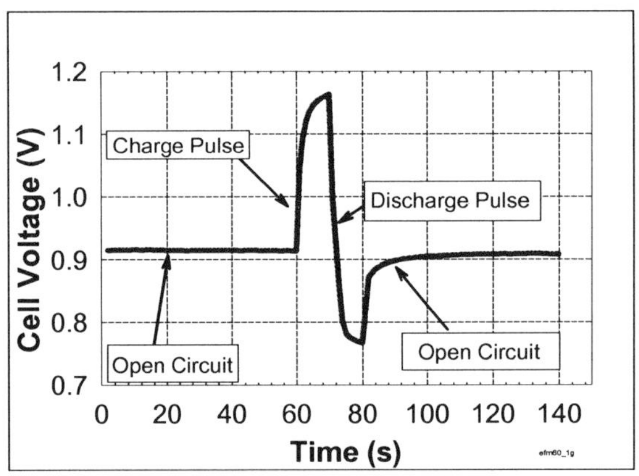

Figure 2: Intrinsic energy storage of metal hydride fuel cell

5. Instant Start

Conventional fuel cells require rather long start-up times. High temperature fuel cell systems typically require several hours. Ambient temperature systems such as conventional PEM fuel cells require many minutes to reach temperatures needed to achieve rated power performance. At least a few seconds may be required for hydrogen to reach the fuel cell stack once the hydrogen is turned on. For many UPS and emergency power applications, fast start-up times on the order of microseconds are often required, something conventional fuel cells cannot provide. A solution for this problem is supplemental batteries or supercapacitors paralleled at the systems level. While providing for instant start, batteries add cost, weight, and complexity and additionally, the traditional battery maintenance and reliability issues. Supercapacitors are even more costly and provide less energy per unit weight and volume.

By contrast, metal hydride fuel cells with inherent battery functionality provide instant start in the fuel cell stack itself on the order of microseconds. This is illustrated in Fig. 3 showing power generation in a few microseconds. Instant power generation is provided even at low temperatures down to -20°C.

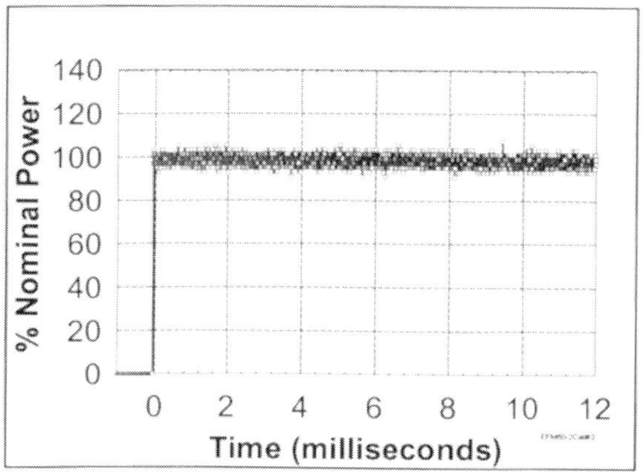

Figure 3: Instant start operation

The instant start feature illustrated in Fig. 3 has a robust fail-safe nature in that power can be provided even in the absence of hydrogen fuel flowing to the fuel cell. The hydrogen stored in the metal hydride anodes can provide power for several minutes at peak power levels, even if the fuel cell is not supplied with hydrogen gas fuel. This unique feature is illustrated in Fig. 4. The intrinsic energy storage capability is thus significant, on the order of 10 Wh/kg with current prototype designs. The 10 Wh/kg intrinsic energy storage density, which already exceeds the level of supercapacitor energy storage devices, can be substantially increased by designs with higher metal hydride contents. Of course, the total system energy density utilizing the fuel cell in combination with hydrogen sources is typically much higher than the intrinsic energy storage density. Total system energy densities on the order of 200-1000 Wh/kg or more are possible.

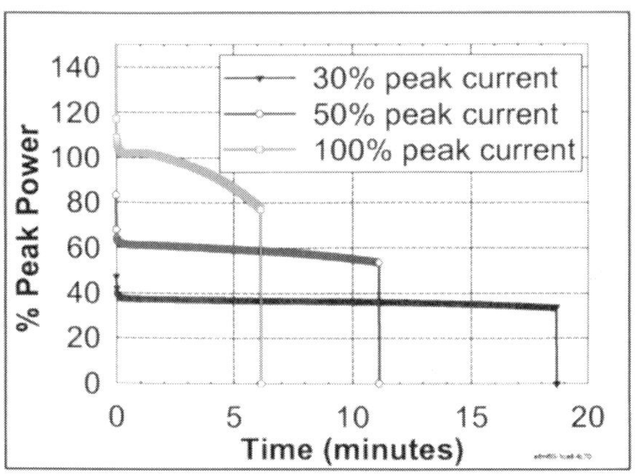

Figure 4: Instant start operation without hydrogen gas input

6. Power Performance

The power performance as a function of current density for a prototype metal hydride fuel cell is shown in Fig. 5. Peak current densities exceeding 250 mA/cm^2 have been demonstrated, which is an excellent result for a fuel cell without noble metal catalysts. Prototype devices have been built and tested demonstrating a specific power of around 100 W/kg and a power density of around 100 W/L. We are currently engineering new prototypes aimed at around 200 W/kg and 200 W/L which will be more than sufficient to build suitable systems for UPS and emergency power applications.

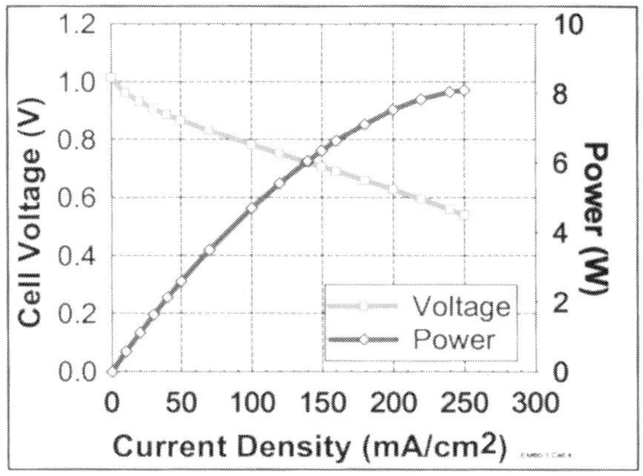

Figure 5: Fuel cell power performance in recent prototype

7. Low Temperature Performance

A significant advantage to this new technology is the operational and storage temperature range. The storage temperature extends to about -40°C, below which the electrolyte freezes. The operational temperature ranges from -20°C to 80°C. The dependence of power on temperature is less than that of conventional fuel cells leading to superior power performance at low temperatures as shown in Fig. 6. Over 75% of the peak rated power is available at room temperature and about 50% is available at 0°C.

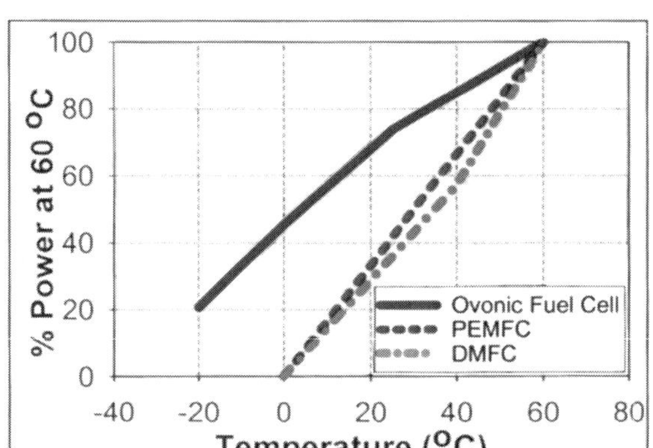

Figure 6: Effect of temperature on power performance

8. Operational and Calendar Life

In life testing of early prototypes, over 1000 hours of operation near peak power have been demonstrated with multi-cell stacks. Up to 5000 hours have been demonstrated in single cell tests of electrodes. The operational life can be extended by the use of carbon dioxide scrubbers at the air inlet to mitigate carbonate formation. There are now a variety of conventional and regenerative carbon dioxide scrubber technologies available. However, the liquid electrolyte design of the metal hydride fuel cell is much more tolerant to carbon dioxide than the traditional immobilized matrix electrolyte design (starved electrolyte type) of alkaline fuel cells used for space applications. Even totally without scrubbers, an operational life of several hundred hours has been demonstrated, more than adequate for UPS and emergency power applications. The calendar life is under study. We expect it to be comparable to that of the 10+ year calendar life of nickel metal hydride batteries.

9. Cost and Manufacturability

A major advantage of the metal hydride fuel cell is the utilization of lower cost materials. Conventional ambient temperature fuel cells such as PEM fuel cells utilize noble metal catalysts and other expensive components. For example, PEM fuel cells use platinum catalysts as well as expensive proton exchange membranes that typically comprise even a larger fraction of the fuel cell cost than the platinum. A comparable fraction is also typically allotted to bipolar conductive plates between cells that are comprised of special materials and require special techniques to machine or form. Furthermore, special care and equipment are required for the manufacturing of high-tech PEM membrane electrode assemblies. Very detailed tolerance levels are required to insure proper sealing and compression.

The metal hydride fuel cell, by contrast, is made from relatively inexpensive materials. The anodic active materials are metal hydrides, composed of common transition metal components. The cathodic active materials are non-noble metal oxides. Other electrode components include conductive graphite and carbon powders, PTFE materials, plastic meshes, and a conductive nickel screen and tab. The most expensive component is the nickel metal screen and tab. Production processing and assembly are also expected to be simpler and less expensive than for PEM fuel cells. Conventional roll-to-roll battery electrode processing methods and equipment can be used in the manufacture of fuel cell electrodes.

Typical metal hydride fuel cell parts and components are illustrated in Fig. 7. Working prototypes are shown in Fig. 8. Electrode processing equipment requires nothing more sophisticated than roll mills such as shown in Fig. 9. Engineering and prototyping activities underway to accelerate product development and demonstration are illustrated in Fig. 10.

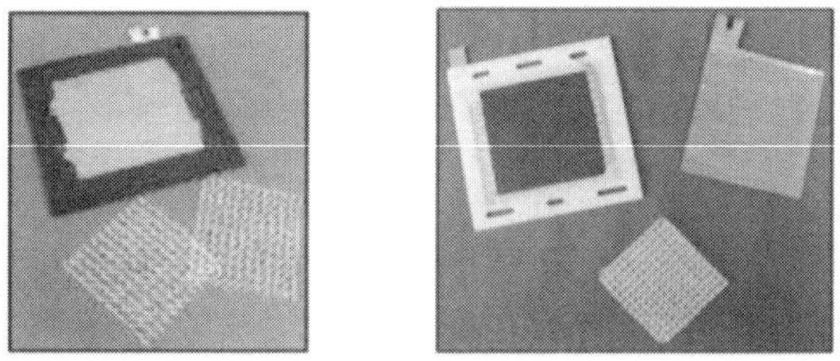

Figure 7: Air electrode, hydrogen electrode, plastic mesh, and nickel screen components

Figure 8: Prototype metal hydride fuel cell stacks

Figure 9: Conventional roll mill for electrode fabrication

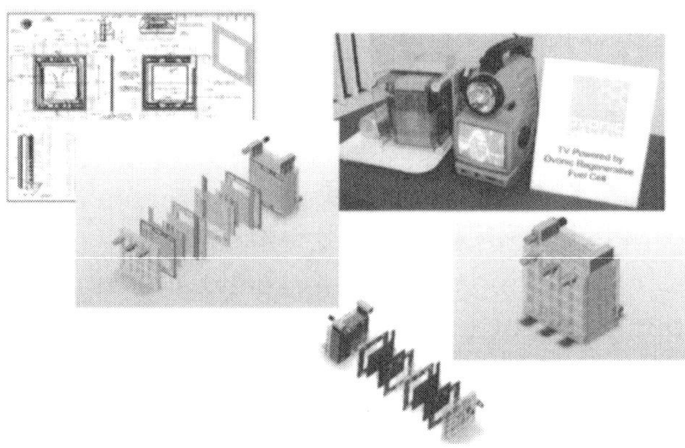

Figure 10: Engineering drawings and early prototype system

10. Summary

Metal hydride fuel cells offer useful new features for UPS/emergency power applications including instant start on the order of microseconds and excellent low temperature performance. Low cost materials and a simple design provide for a lower cost and highly manufacturable fuel cell solution for extended run time applications.

11. References

1. R. Wickham, "*Not Quite Making the Grade,*" Wireless Review, PRIMEDIA, April 15, 1998.
2. M. Corcoran, "*Battery Battle,*" Wireless Review, PRIMEDIA, September 1, 2001.
3. A. Van Sciver, "*Battery Experts Separate Fact from Fiction,*" Power Quality, PRIMEDIA, September 1, 2001.
4. M. Stansberry, "*Hydrogen Fuel Cells: Unlimited UPS,*" Today's Facility Manager, Group C Communications, September 2004.
5. D. DeVries, "*Military Fuel Cells: The Next 5 Years,*" Las Vegas, NV, June 28-29, 2004.
6. D. Ceci, Ballard Power Systems, "*Fuel Cells for Extended Run Backup Power Applications,*" Fuel Cell 2004, Denver, CO., June 8-9, 2004.
7. P. Christensen, "*ReliOn,*" Fuel Cell 2004, Denver, CO. , June 8-9, 2004
8. Frost and Sullivan, "*Global Uninterruptible Power Supply (UPS) Markets,*" A012-27, 2002.
9. S.R. Ovshinsky, M.A. Fetcenko, and J. Ross, Science, 260, 176 (1993).
10. R.C. Stempel, S.R. Ovshinsky, P.R. Gifford, and D.A. Corrigan, IEEE Spectrum, 35, 29 (November 1998).
11. S.R. Ovshinsky, Mat. Res. Soc. Symp. Proc. Vol. 801, G. Nazri, M. Nazri, R. Young, and P. Chen, Eds., p. 3 (2004).
12. S.R. Ovshinsky, S. Venkatesan, B. Aladjov, R. Young, and T. Hopper, U.S. Pat. 6,447,942, Sep 10, 2002.
13. S.R. Ovshinsky, S. Venkatesan, and D.A. Corrigan, "*The Ovonic Regenerative Fuel Cell, A Fundamentally New Approach,*" Hydrogen and Fuel Cells 2004 Conference and Trade Show, Toronto, Canada, September 2004.

Hydrogen Storage and Fuel Cell Publications

Development of a Small Scale Hydrogen Production Storage System for Hydrogen Applications (with K. Sapru, N.T. Stetson, J. Yang and G. Fritz), IECEC (1997).

The Road to Decarbonized Energy – Speeding towards a hydrogen economy – and the obstacles along the way, Book Review, Nature (August 3, 2000) p. 457.

Hydrogen-Fueled Hybrid: Pathway to a Hydrogen Economy (with R. Geiss, B. Webster, R. Stempel, R.C. Young, Y. Li, V. Myasnikov, B. Falls and A. Lutz), Soc. for Automotive Engineers 2004 World Congress, Detroit, MI (2004) p.1.

New Science and Technology - The Basis of the Hydrogen Economy, Mat. Res. Soc. Symp. Proc. 801 (2004) 3.

A Hydrogen ICE Vehicle Powered by Ovonic Metal Hydride Storage (with R. C. Young, B. Chao, Y. Li, V. Myasnikov and B. Huang), Soc. For Automotive Engineers 2004 World Congress, Detroit, MI (2004).

The Ovonic Regenerative Fuel Cell, A Fundamentally New Approach (with S. Venkatesan and D.A. Corrigan), The Hydrogen and Fuel Cells 2004 Conference and Trade Show, Toronto, Canada, (2004).

Ovonic Instant Start Fuel Cells for UPS and Emergency Power Applications (with K. Fok, S. Venkatesan, and D.A. Corrigan), The National Hydrogen Association Annual Conference (2005) p.1.

The Hydrogen Loop – The Means for Making the Hydrogen Economy Realistic, Intl. J. of Nuclear Hydrogen Production and Applications Vol. 1, No. 2. (2005).

Metal Hydride Fuel Cells for UPS and Emergency Power Applications (with K. Fok, S. Venkatesan and D. Corrigan), BATTCON 2005, Miami, FL (2005).

Metal Hydride Fuel Cells, A New Approach (with D.A. Corrigan), Fuel Cell Magazine, June/July 2005, p. 25.

US patents - hydrogen storage alloys

Hydrogen storage materials and method of making same
4431561 02/14/1984

Hydrogen storage materials having a high density of non-conventional useable hydrogen storing sites
5840440 11/24/1998

High storage capacity alloys enabling a hydrogen-based eco system
6193929 02/27/2001

Hydrogen cooled hydrogen storage unit having maximized cooling efficiency
6318453 11/20/2001

Modified magnesium based hydrogen storage alloys
6328821 12/11/2001

Hydrogen cooled hydrogen storage unit having a high packing density of storage alloy and encapsulation
6378601 04/30/2002

Hydrogen storage powder and process for preparing same
6461766 10/08/2002

Method for making hydrogen storage alloy
6478844 11/12/2002

High storage capacity, fast kinetics, long cycle-life, hydrogen storage alloys
6491866 12/10/2002

Atomically engineered hydrogen storage alloys having extended storage capacity at high pressures and high pressure hydrogen storage units containing variable amounts
6536487 03/25/2003

A hydrogen infrastructure, a combined bulk hydrogen storage/single stage metal hydride hydrogen compressor therefore and alloys for use therein
6591616 07/15/2003

Method of activating hydrogen storage alloy electrode
6605375 08/12/2003

High capacity transition metal based hydrogen storage materials for the reversible storage of hydrogen
6616891 11/09/2003

Safe, economical transport of hydrogen in pelletized form
6627148 11/30/2003

A honeycomb hydrogen storage structure with restrictive neck
6708546 03/23/2004

A honeycomb hydrogen storage structure
6709497 03/23/2004

High storage capacity alloys having excellent kinetics and a long cycle life
6726783 04/27/2004

Non-pyrophoric hydrogen storage alloy
6737194 05/18/2004

Modified electrochemical hydrogen storage alloy having increased capacity, rate capability and catalytic activity
6740448 05/25/2004

High storage capacity, fast kinetics, long cycle-life, hydrogen storage alloys
6746645 06/08/2004

Hydrogen storage powder and process for preparing the same
6789757 09/14/2004

Hydrogen storage alloys having a high porosity surface layer
6830725 12/14/2004

Hydrogen storage bed system including an integrated thermal management system
6833118 12/21/2004

Hydrogen storage bed system including an integrated thermal management system
6878353 04/12/2005

Method for producing and transporting hydrogen
6991719 01/31/2006

High capacity hydrogen storage material based on catalyzed alanates
7029600 04/18/2006

Hydrogen storage alloys providing for the reversible storage of hydrogen at low temperatures
7108757 09/19/2006

Method and apparatus for electrorefining impure hydrogen
7175751 02/13/2007

Mg-Ni hydrogen stotage composite having high storage capacity and excellent room temperature kinetics
7211541 05/01/2007

US patents - fuel cell

Fuel cell cathode
4430391 02/07/1984

Fuel cell anode
4487818 12/11/1984

Novel alkaline fuel cell
6447942 09/10/2002

Electrochemical cell having reduced cell pressure
6492057 12/10/2002

Layered metal hydride electrode providing reduced cell pressure
6503659 01/07/2003

Non-pyrophoric hydrogen storage alloy
6517970 02/11/2003

A hydrogen-based ecosystem
6519951 02/18/2003

Method of fuel cell activation
6589686 07/08/2003

Active material for fuel cell anodes incorporating an additive for precharging/activation thereof
6613471 09/02/2003

Novel fuel cell cathodes and their fuel cells
6620539 09/16/2003

Fuel cell hydrogen supply systems using secondary fuel to release stored hydrogen
6627340 09/30/2003

Fuel cell cathode utilizing multiple redox couples
6703156 03/09/2004

Fuel cell cathode with novel redox couple
6777125 08/17/2004

Fuel cell cathode with redox couple
6783891 08/31/2004

Modified redox couple fuel cell cathodes and fuel cells employing same
6790551 09/14/2004

Fuel cell with framed electrodes
6828057 12/07/2004

Double layer oxygen electrode and method of making
6835489 12/28/2004

Catalytic hydrogen storage composite material and fuel cell employing same
6875536 04/05/005

Fuel cell with encapsulated electrodes
6926986 08/09/2005

Parallel flow fuel cell
6933072 08/23/2005

Electrode utilizing fluorinated carbon
6960406 11/01/2005

Fuel cell with overmolded electrode assemblies
6989216 01/24/2006

A hybrid fuel cell
6998184 02/14/006

A drive system incorporating a hybrid fuel cell
7008706 03/07/2006

Regenerative bipolar fuel cell
7014953 03/21/2006

Fuel cell cathode with redox couple
7018740 03/28/2006

Novel fuel cell cathodes and their fuel cells
7033699 04/25/2006

Multi-layered oxygen electrode with peroxide decomposition catalyst
7070878 07/04/2006

Catalyst for fuel cell oxygen electrodes
7097933 08/29/2006

Fuel cell
7132193 11/08/2006

A very low emission hybrid electric vehicle incorporating an integrated propulsion system including a fuel cell
7226675 06/05/2007

Chapter VII: Superconductivity

Nothing comparable has ever happened at an annual meeting of the American Physical Society: the 2000 seat ball room of the New York Hilton was overflowing with researchers, science reporters, radio and TV representatives until the early morning hours to hear about new discoveries. This unusual and historical event occurred in March 1987 and quickly got named the *Woodstock of Physics.*

What stirred these scientists who usually are intellectual conservatives and view news in their field with critical skepticism? It was the discovery of superconductivity at unexpectedly high temperatures in materials which hardly any scientist had investigated before. Superconductivity is a fascinating phenomenon of many metals, a state in which the metal conducts electricity without any resistance or loss. It could revolutionize and cheapen all electrical appliances and power transmissions which suffer often more than 30 percent loss due to the resistances of the wircs. It could do so if the magic phenomenon of superconductivity would exist at room temperature. This unfortunately is not the case. The conventional superconductors require expensive refrigeration to very low temperatures so that their use is normally restricted to cases were cost is of no concern such as medical magnetic resonance imaging systems or particle accelerators. All this seemed to change radically with the discoveries under discussion in March 1987 in the New York Hilton.

J. Georg Bednorz and K. Alexander Mueller of the IBM research laboratory in Zurich, Switzerland, announced in the late 1986 the observation of superconductivity in a lanthanum-barium-copper-oxide material at a temperature 20 degrees higher than what was seen before. Paul Chu of the University of Houston soon topped the record temperature by another 50 degrees in another ceramic material yttrium-barium-copper-oxide. The future promise of practical applications was not however what motivated 2000 scientists to sit until the early hours discussing these results in the ballroom of the New York Hilton. It was the realization that their theoretical understanding of the phenomenon of superconductivity based on the cherished 1957 theory of Bardeen, Cooper and Schrieffer was wrong, and if not wrong, then seriously incomplete. This BCS theory essentially predicted that superconductivity cannot exist at the "high" temperatures discovered by Bednorz, Mueller and Chu. The BCS theory was so successful and its predictions so convincing that any search for new superconducting materials had basically stopped many years prior to the breakthrough of 1986.

The challenge of understanding an essentially new phenomenon of superconductivity in an almost unknown family of materials created an unprecedented rush into the virgin territory. The high temperature superconductors, as they were called, still require refrigeration but with less expensive refrigerants such as liquid air. Hence the hope for finding superconductors for many technological applications certainly heightened the excitement.

These exciting and revolutionary developments in materials sciences fit well with Ovshinsky's approach towards the preparation of new and even higher temperature superconductors. Through his understanding of the crucial role played by d- and f- orbitals of transition metal elements and the strong effect of substituting new elements, Ovshinsky prepared ceramic oxide compounds with the addition of Fluorine. His results indicated the possibility of superconductivity at temperatures as high as 155K (i.e. minus 118 C). As presented in publications included in this book, Ovshinsky and his team verified the occurrence of 155K superconductivity in samples of Yttrium-Barium-Copper-Oxide which they fluorinated by two entirely different methods. Yet, since only a minor fraction of the inhomogeneous fluorinated material became superconducting, attempts to identify the crucial composition unfortunately failed. See the Physical Review Letter paper in this chapter and the Review of Solid State Science paper at the end of the book. The highest superconducting transition temperature of an identified material at present is 138K in Mercury-Thallium-Barium-Calcium-Copper-Oxide.

In other papers Ovshinsky developed structural models for the role of the copper planes in the high temperature superconducting materials. Holes, which are the charge carriers in these materials, are in d orbitals of Cu^{III} shared and hybridized with lone pair p orbitals of adjacent oxygen atoms, thus forming a mixed valence state of Cu^{II} and Cu^{III}. Ovshinsky recognized that the O-Cu-O bond angles differ for the two valence states. He suggests that holes form spin zero pairs because the gain of configurational relaxation is larger for a pair of holes than for two isolated holes. Such spin zero pairs of holes form already above the superconducting transition temperature which explains the observed anomalous properties of these materials. At the transition temperature the spin zero hole pairs undergo a Bose condensation which yields the superconducting quantum state.

Even two decades after the discovery of high temperature superconductivity in the ceramic cuprate compounds, this phenomenon continues to puzzle materials scientists. As of today, no single theory has won out and no understanding gives guidance to new breakthrough materials. It would seem that an informed, intuitive, experimental materials science approach such as that of Ovshinsky and his insight into the local stereochemistry of the essential elements are likely to discover materials with even higher superconductive temperatures.

Superconductivity at 155 K

S. R. Ovshinsky, R. T. Young, D. D. Allred, G. DeMaggio, and G. A. Van der Leeden

Energy Conversion Devices, Inc., Troy, Michigan 48084

(Received 22 May 1987)

Transition to a superconducting zero-resistance state at 155 K is observed for the first time in bulk material. A new five-element compound has been synthesized with nominal composition $Y_1Ba_2Cu_3F_2O_y$. Fluorine plays a critical role in achieving this effect. X-ray diffraction and electron microprobe analysis indicate that the samples are multiphasic. Evidence is presented that the samples contain superconducting phases with onset temperatures considerably above 155 K. Magnetic measurements suggest a flux-trapping effect below 260 K, and diamagnetic deviations from Curie-Weiss behavior in the range 250 K $\geq T \geq$ 100 K indicate a Meissner effect in a small superconducting volume fraction.

PACS numbers: 74.10.+v, 74.70.Ya

The accomplishment of high-temperature superconductivity is of immense scientific and technological importance. Several critical transition-temperature barriers have recently been breached since the long-standing record temperature of 23.2 K for Nb_3Ge was exceeded. The most important milestones were the announcement of $T_c \approx 30$ K in lanthanum barium copper oxide by Bednorz and Müller,[1] whose work was based upon materials developed by Michel and Raveau,[2] and the work of Chu, Wu, and others,[3] based upon the replacement of lanthanum by yttrium, which resulted in superconductivity at temperatures of approximately 95 K.

Indirect measurement techniques[4,5] have been used[6] previously to infer the existence of regions of high-T_c phases in Y-Ba-Cu-O systems. We report here for the first time the direct measurement of a zero-resistance superconducting state at 155 K. This result was observed in a new chemical system, Y-Ba-Cu-F-O.

In the quest for even higher transition temperatures, variations and replacements in the metallic portion of the compounds have not been fruitful. It was found that yttrium could be replaced by most of the rare-earth metals with achievement of approximately the same T_c.[7] Our approach has been to synthesize a new five-element compound, which has resulted in the achievement of considerably higher T_c's. In one sample, the transition to zero resistance was observed at temperatures as high as 155 K. In another sample, an abnormally rapid decline of resistivity was observed starting at room temperature, which suggests the existence of phases exhibiting superconducting onset above room temperature. We have also observed anomalies in the magnetic properties, as well as weak flux trapping below 260 K. This material contains at least four presently identified structural phases. Work is continuing to identify unambiguously the structure of the high-T_c phases. In previous papers[8] we have discussed the significant role that fluorine plays in affecting the electronic and structural properties in other multielemental materials, particularly in affecting orbital interactions. These factors[9] motivated us to synthesize the fluorinated Y-Ba-Cu-O compounds reported here.

Samples with nominal compositions $Y_1Ba_2Cu_3F_xO_y$ (x=0, 1, 2, 3, and 4) were prepared from two master compositions, $Y_1Ba_2Cu_3O_{6.5}$ and $Y_1Ba_2Cu_3F_4O_{4.5}$, which define the compositional extremes (x=0 and x=4). The starting reagents used to prepare the oxide and fluoroxide were (Y_2O_3, $BaCO_3$, and CuO) and (Y_2O_3, BaF_2, and CuO), respectively. Each master composition was prepared by a mixing of the sieved reagent powders, and grinding and firing of them in air in a Pt crucible at 950 °C for 8 h. After the initial firing the master compositions were reground. Samples were prepared from mixtures of the two master compositions which were pressed into $\frac{1}{2}$-in. pellets and sintered at 950 °C in flowing O_2 for 48 h, and then cooled to 200 °C over a period of 6 h. Samples were examined by x-ray powder diffraction and microprobe to determine their structure, phases, and composition. Table I summarizes the various phases and

TABLE I. Various phases and compositions of $Y_1Ba_2Cu_3F_xO_y$ from x-ray diffraction and microprobe analysis.

x	$Y_1Ba_2Cu_3O_{7-\Delta}$	$Y_1Ba_2Cu_3O_y$:F	$Y_2Ba_1Cu_1O_y$:F	$Y_1Cu_2O_y$:F	BaF_2	CuO
x=0	×					
x=1		×	×		×	×
x=2		×	×		×	×
x=3			×	×	×	×
x=4				×	×	×

compositions which have been identified.

Electrical resistance was measured by means of a standard four-probe method on samples with silver-paint contacts. Rectangular bars $10\times2\times1$ mm^3 were cut from pellets for the measurements. The applied constant current ranged from 100 μA to 10 mA depending on the sample resistance. The dc magnetic susceptibility and flux trapping were studied with a SQUID magnetometer.

Of the five nominal compositions studied, only those with $x=0$, 1, and 2 show superconductivity, as measured by resistivity and magnetic susceptibility measurements. The room-temperature resistivity of samples with $x=3$ and $x=4$ is greater than 20 MΩ. The samples with $x=0$ show resistance-temperature behavior similar to that reported by others[7] for typical $YBa_2Cu_3O_{7-\Delta}$, with a modest decrease of resistivity from room temperature to the onset temperature of 95 K, reaching the zero-resistance state at 90–92 K. The $x=1$ sample showed a behavior essentially identical to that of the $x=0$ sample. Dramatically different and complex behavior was observed, however, in samples with $x=2$.

The temperature dependence of the resistance of a portion of the $x=2$ pellet is shown in Fig. 1. During the initial cooling, the sample (sample 1) completely lost its resistance at ≈ 168 K. After warming and measurement during second cooling, an increase of resistance was observed and the zero-resistance state was not reached until 148 K. An appreciable change in resistance was again observed during the second warming. The zero-resistance state remained (within our instrument noise level of $\approx 10^{-8}$ V) until 155 K. Several higher-temperature resistance transitions were also observed. The resistance anomalies seen upon warming and cooling are remarkable and may be connected with filamentary conduction.

Figure 2 shows a plot of the temperature dependence of the average resistivity for another sample (sample 2) with the $x=2$ composition. The average resistivity calculated with the assumption of uniform current density falls dramatically with temperature for this sample beginning at room temperature or above, and achieves resistivity values 5 times lower than that of single-crystal copper before the zero-resistance state is achieved at ≈ 91 K. We were able to fit the resistivity to a good approximation over the entire range by using a $T^{8.3}$ law. Since the resistivity became so low with decreasing temperature, a large current of 10 mA, or a current density of 0.5 A/cm^2, had to be applied during measurement. With the assumed multiphase filamentary type of conduction, this applied current density might have washed out more sharply defined high-T_c transitions. For comparison, the resistivity-temperature plot of pure copper[10] is also included in Fig. 2. The actual resistivity of the current-carrying phases will certainly be much less than

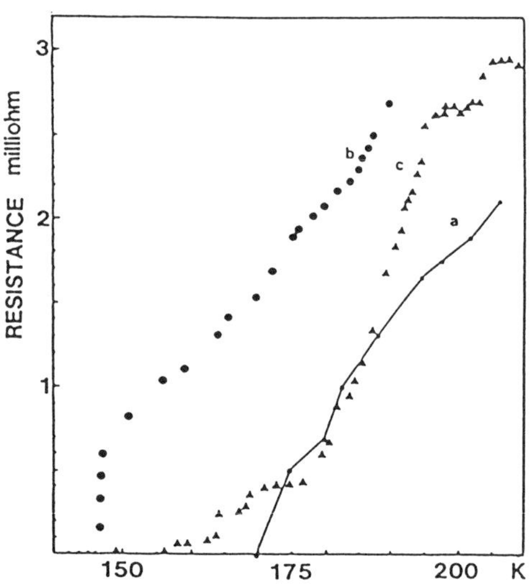

FIG. 1. Resistance vs temperature measured at constant current of 1 mA for sample 1 with nominal composition $YBa_2Cu_3F_2O_y$. Curve a shows the resistance upon initial cooling (line drawn to guide the eyes); curve b, data obtained upon second cooling; and curve c, data from warming after second cooling. In the superconducting state, voltage is less than 10^{-8} V, the noise level of the voltmeter. Sample dimensions are area $=1\times2$ mm^2, length $=10$ mm. This gives an average resistivity estimate of less than 2×10^{-7} Ω cm in the superconducting state.

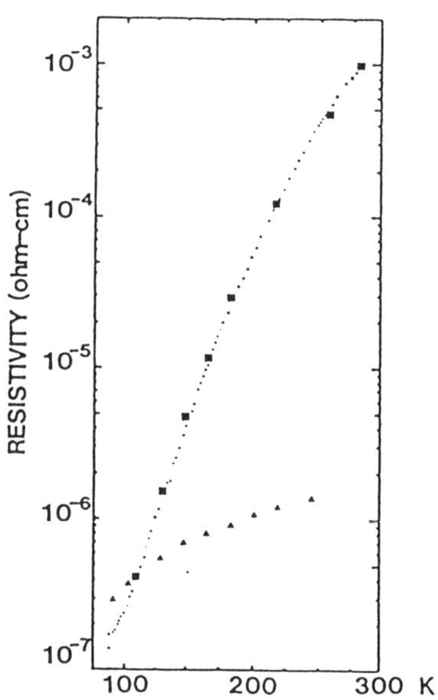

FIG. 2. The logarithm of the average resistivity of sample 2 of $YBa_2Cu_3F_2O_y$. Average resistivity was calculated on the assumption of uniform current density. Resistivity was found to follow T^n, where $n\simeq 8.3$; these points are indicated by squares. The ideal resistivity of pure copper (Ref. 10) is also plotted (triangles).

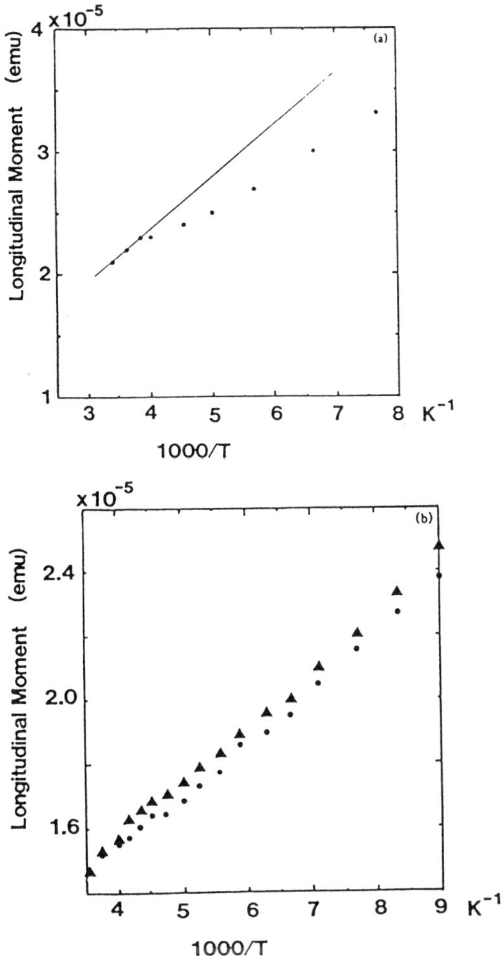

FIG. 3. Longitudinal moment vs $1000/T$ for $YBa_2Cu_3F_2O_y$. (a) Measurements made with use of 100-G field. Note the diamagnetic deviation from a Curie-Weiss law below 250 K. (b) Weak magnetic hysteresis at 20 G. Data for warming after zero-field cooling are indicated by circles and data from cooling with field applied are indicated by triangles.

the calculated average. The fact that the average resistivity of the multiphase ceramic material at a temperature of 91 K is 5 times lower than copper provides another indication of the existence of higher-temperature superconducting phases.

The magnetic measurements in Fig. 3 suggest that only a very small volume fraction of the sample is superconducting at these high temperatures. Figure 3(a) plots the magnetic moment as a function of temperature in a magnetic field of 100 G. The diamagnetic deviation from the Curie-Weiss law at temperatures below 250 K is an indication of the Meissner effect in this small volume fraction. A large diamagnetic response below 90 K is also observed but not plotted here. Weak magnetic hysteresis is observed as shown in Fig. 3(b), where the data on warming after zero-field cooling are indicated by circles and the data on cooling with field applied, by triangles. The facts that (1) the data of warming and cooling measurements coincide from room temperature until 260 K and deviate below this temperature and (2) the paramagnetism is stronger in the cooling process provide additional verification of the existence of high-temperature superconducting phases.

In summary, a zero-resistance state at 155 K has been observed in the fluorinated Y-Ba-Cu-O system and phases with even higher T_c have also been observed. The abnormally rapid decline of resistivity starting at room temperature suggests the existence of phases exhibiting superconducting onset above room temperature. The high-temperature phases are in the process of being identified. In the samples reported here, the volume fraction of the high-T_c material is very small as indicated by the magnetic data. Since a zero-resistance state requires a percolation path, the small-volume-fraction superconducting phase appears to be in a filamentary form. Work continues both to identify clearly the structure of the high-T_c phase or phases and to optimize materials synthesis to increase their yield.

We wish to thank Dr. S. J. Hudgens, Professor J. T. Chen, Dr. E. Teller, and G. Wicker for stimulating discussions on various aspects of this work. We would also like to thank B. Chao, L. Contardi, and D. Pawlik for structure analysis and G. Fournier for technical assistance.

[1]J. G. Bednorz and K. A. Müller, Z. Phys. B **64**, 189 (1986).

[2]C. Michel and B. Raveau, J. Solid State Chem. **43**, 73 (1982); J. Provost, F. Studer, C. Michel, and B. Raveau, Synth. Met. **4**, 157 (1981).

[3]M. K. Wu, J. R. Ashburn, C. J. Tong, P. H. Hor, R. L. Wong, L. Gao, Z. J. Huang, Y. Q. Wang, and C. W. Chu, Phys. Rev. Lett. **58**, 908 (1987); P. H. Hor, L. Gao, R. L. Meng, Z. J. Huang, Y. O. Wang, K. Forster, J. Vassiliow, and C. W. Chu, Phys. Rev. Lett. **58**, 911 (1987).

[4]A. M. Saxena, J. E. Crow, and M. Strongin, Solid State Commun. **14**, 799 (1974).

[5]H. Sadate-Akhavi, J. T. Chen, A. M. Kadin, J. E. Keem, and S. R. Ovshinsky, Solid State Commun. **50**, 975 (1984).

[6]J. T. Chen, L. E. Wenger, C. J. McEwan, and E. M. Logothetis, Phys. Rev. Lett. **58**, 1972 (1987).

[7]See, for example, A. R. Moodenbaugh, M. Suenaga, T. Asano, R. N. Shelton, H. C. Ku, R. W. McCallum, and P. Klavins, Phys. Rev. Lett. **58**, 1885 (1987); P. H. Hor, R. L. Meng, Y. Q. Wang, L. Gao, Z. J. Huang, J. Bechtold, K. Forster, and C. W. Chu, Phys. Rev. Lett. **58**, 1891 (1987).

[8]S. R. Ovshinsky, in *Physical Properties of Amorphous Materials*, edited by D. Adler, B. B. Schwartz, and M. C. Steele (Plenum, New York, 1985), and Rev. Roum. Phys. **26**, 893 (1981), and J. Phys. (Paris), Colloq. **42**, C4-1095–C4-1104 (1981), and J. Non-Cryst. Solids **32**, 17 (1979).

[9]S. R. Ovshinsky, to be published.

[10]Data taken from G. K. White and S. B. Woods, Phil. Trans. Roy. Soc. London, Ser. A **251**, 272 (1959).

Rapid Commun. in High T_c

Modern Physics Letters B Vol. 1, Nos. 7 & 8 (1987) 275–288

A STRUCTURAL CHEMICAL MODEL FOR HIGH T_C CERAMIC SUPERCONDUCTORS

S. R. OVSHINSKY, S. J. HUDGENS, R. L. LINTVEDT,[†] and D. B. RORABACHER[†]
Energy Conversion Devices, Inc., 1675 West Maple Road, Troy, Michigan 48084, U.S.A.

Received 15 December 1987.

A structual chemical model for high T_c ceramic superconductors is proposed in which carriers pairs, coupled through a superexchange process, undergo Bose condensation. In the Y-Ba-Cu-O system, Cu atoms on "chain" sites and "sheet" sites are initially assigned formal oxidation states of +3 and +2, respectively, and "sheet"/"chain" interactions are then introduced to bring about the electrically conductive mixed oxidation state. This model explains the presence of labile oxygen in the "chain" structures, describes the structure of deoxygenated phases exhibiting ordered oxygen vacancies in the "chains", and allows interpretation of the effects of oxygen removal on magnetic and electronic properties of the material.

Introduction

During the past year a large number of models have been proposed to explain high T_c superconductivity in Cu-O ceramic compounds. A nearly equal number of novel charge carrier pairing mechanisms have been invoked in these models as alternatives to the conventional phonon mediated process described in the BCS theory. Some of these models have recently been reviewed by Rice.[1]

In the current paper we propose a model for high T_c superconductivity with a carrier pair mechanism based on antiferromagnetic spin coupling which explicitly focuses on the role of chemical bonding and the oxidation states of constituent copper atoms in the reported crystal structure of the Y-Ba-Cu-O ceramic superconductors. Although electron-pairing mechanisms which involve antiferromagnetic spin coupling have previously been discussed,[2–6] the current proposal uniquely interprets the manner in which this mechanism is manifested in the Cu-O ceramic superconductors. Specific attention is directed to $YBa_2Cu_3O_{7-X}$ as a model system. However, the proposed model has greater generality as will be discussed.

In $YBa_2Cu_3O_{7-X}$, each element plays a role—in some cases several roles—in

[†] Dept. of Chemistry, Wayne State University, Detroit, Michigan 48202, U.S.A.

creating conditions required for achieving high T_c superconductivity. For yttrium this role is not unique as experiments have shown that it can be substituted completely by most of the trivalent lanthanides with essentially unchanged superconducting properties.[7–10] As exceptions, substitution of Pr, Tb or Ce results in changes in T_c. However, it has been argued[11] that these elements, alone among the lanthanides, have stable tetravalent states, and that these states are likely to be present in materials prepared at high temperature in oxidizing environments. Thus, substitution of these elements in Y-Ba-Cu-O system, unlike the other lanthanides, necessitates oxidation state changes elsewhere in the compound, similar to those produced by oxygen deficiency. The partial substitution of divalent Ba by Sr or Ca has also been shown to have little effect on T_c.[8] The small effects which have been observed could be attributed to an alteration in structural equilibrium, but might involve chemical or electronic effects as well.[11]

By contrast, substitution for Cu[12] and changes in O content have profound consequences. In particular, reduction of O content, through the formation of vacancies, is found[13–17] to have a significant effect on both the transition temperature and the crystal structure of the Y-Ba-Cu-O materials. Therefore, our consideration of the chemical basis for the existence of high T_c superconductivity in these materials will focus primarily on the Cu-O chemistry, structure, and magnetic properties. Generalization to other elements which might play analogous roles in new materials will then be considered.

Proposed structural chemical model

Assignment of oxidation states

Figure 1 presents a schematic diagram of the arrangement of copper and oxygen atoms in orthorhombic $YBa_2Cu_3O_7$ as revealed by crystallographic studies.[18] As illustrated, the copper atoms are arranged in two distinct patterns: (i) upper and lower, two-dimensional dimpled "sheets" in which the copper atoms are coordinated to five oxygen atoms of which four are nearly coplanar with the copper while the fifth is in an apical position resulting in an overall square pyramidal coordination sphere; (ii) central copper one-dimensional "chains" (represented in boldface in Fig. 1) in which the copper atoms are coordinated to four oxygens in a flat planar array. These differences in chemical bonding and the apparent formal oxidation states of the two types of Cu form the basis for the proposed high T_c superconducting model.

We construct a model for high T_c superconductivity by first conceptually treating the "sheets" and the "chains" as isolated systems. Interactions between these structures, which play a key role in the phenomenon, will be introduced later. We additionally simplify the initial description by postulating that, in each of the two structures, the Cu atoms are all in the same oxidation state. We therefore start with $YBa_2Cu_3O_7$ by proposing that the 5-coordinate copper atoms in the upper and lower "sheets" are formally in the +2 oxidation state. Although

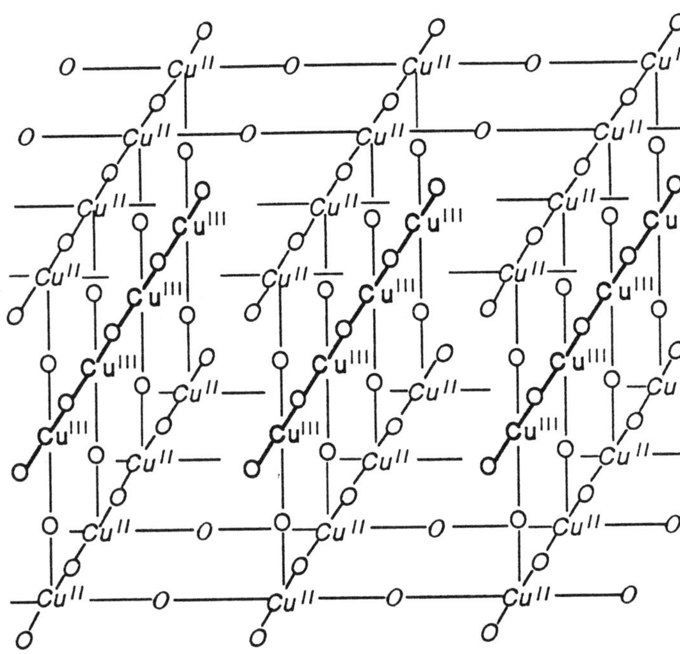

Fig. 1. Schematic representation of the Cu-O skeletal network in the $YBa_2Cu_3O_7$ superconductor. The 4-coordinate copper atoms in the "chains" (shown in boldface) are assigned a formal oxidation number of +3. The 5-coordinate copper atoms in the upper and lower "sheets" (shown in italics) are assigned a formal oxidation number of +2. (Note that some of the cross-linking oxygen atoms in the "sheets" are omitted for clarity.)

Cu^{II} commonly exhibits a coordination number of 6, with the two donor atoms along one axis being elongated as a result of Jahn-Teller distortion, 5-coordinate Cu^{II} is well known.[19] If not otherwise constrained, 5-coordinate Cu^{II} is square pyramidal with the copper being displaced approximately 0.1–0.3 Å above the basal plane,[20] this distance increasing as the strength of the apically coordinated donor atom increases. The Cu atoms in the "sheets" are, in fact, displaced from the basal oxygen plane by approximately 0.27 Å,[18] creating the "dimples" referred to earlier. (Note that no attempt has been made to represent this latter feature in Fig. 1.)

We further propose that the copper atoms in the central "chains" are formally in the +3 oxidation state. Although reported occurrences of Cu^{III} are still relatively rare, there are a number of examples of this oxidation state in the literature. In particular, Margerum has shown [21] that Cu^{III} is readily obtained when copper is coordinated to "hard" donor atoms such as deprotonated amide nitrogens (or, presumably, oxide ions). In the one crystal structure of a Cu^{III} complex which has been reported to date,[22] Cu^{III} is shown to exhibit planar 4-coordination as expected for an internally spin-paired d^8 system.

These assignments for the copper atom valencies are consistent with the overall stoichiometry of $YBa_2Cu_3O_7$. Assuming the elements Y, Ba, and O have their expected valencies of +3, +2, and −2, respectively, the average oxidation state for Cu must be 2.33 which implies that two-thirds of the copper atoms are in the +2 oxidation state while the remaining one-third are +3. Moreover, recent iodometric titrimetric analyses by Appelman *et al.*[23] of $YBa_2Cu_3O_{6.9}$ have established the presence of Cu^{III} in quantities which are consistent with the preceding oxidation state assignments.

Spin state assignments

The foregoing oxidation state assignments allow us to formulate a hypothesis regarding the mechanism for superconductivity in $YBa_2Cu_3O_7$ materials which can account for the observed properties at various stages of deoxygenation. Furthermore, the proposed mechanism permits conjectures to be made regarding modifications in the crystal lattice which might lead to higher T_c values.

The Cu^{II} atoms in the upper and lower "sheets" have a d^9 electronic configuration and, thus, posses an unpaired spin. However, since we know from magnetic susceptibility measurements that carefully prepared samples of $YBa_2Cu_3O_7$ exhibit a zero Curie-Weiss term,[24] we must conclude that the majority of Cu^{II} spins in the "sheets" are coupled antiferromagnetically, resulting in no local magnetic moment. Such antiferromagnetic coupling comes about through an O intermediary in the superexchange process. The anti-parallel spin alignment utilizes the *p*-orbitals of the oxygen atom to form the coordinate bond and is favored by the large Cu-O-Cu bond angle of approximately 165°.[18] This coupled spin state in the "sheets" need not result in a spatially extensive, time-independent, two-dimensional Nе̊el antiferromagnet but may, in fact, manifest itself in the form of long range instantaneous antiferromagnetic spin correlations such as the "Quantum Spin Fluid" state found by Shirane *et al.*[25] to exist in the Cu-O "sheets" of La_2CuO_4.

In the "chain" structure, we note that the O-Cu-O bond angle along the "chain" direction is exactly 180°[18] with the Cu atoms sitting at the center of a square planar array of O atoms. Here again our identification of the oxidation state of these atoms as Cu^{III}, which are internally spin paired with a d^8 electronic configuration, is consistent with the observed coordination geometry and bond angles and is dictated by our previous assignment of Cu^{II} for the Cu atoms in the "sheets". The oxygen atoms perpendicular to the "chain" direction are, in fact, the apically-coordinated oxygen atoms of the Cu^{II} atoms in the upper and lower "sheets". Thus they provide a direct communication between the Cu atoms in the "chains" and those in the "sheets" and form the basis in our model for charge propagation in these materials as outlined later.

Structural effects of deoxygenation

Experiments have shown oxygen is readily removed from $YBa_2Cu_3O_7$. At relatively low temperature (400–500°C), the careful removal of oxygen produces a range of compositions of the general formulation $YBa_2Cu_3O_{7-X}$. A number of remarkable properties of these oxygen deficient compounds have been reported which must be addressed by any model for this overall system. First, neutron scattering experiments[17,26] indicate that, over the range of oxygen deficiency $0 \leq x \leq 1$, oxygen is preferentially removed from "chain" sites, producing at $x = 1$ "chains" which are totally free of bridging oxygen. Second, over this range of x, room temperature resistivity, magnetic susceptibility, and, particularly, superconductivity transition temperature reveal the occurrence of two clearly demarcated regions with distinctly different T_c.[24,27]

Based on our assignment of the oxidation states of the copper atoms in $YBa_2Cu_3O_7$, the removal of a single oxygen from a "chain" would result in the reduction of the two adjacent Cu^{III} atoms to Cu^{II} each of which will then be 3-coordinate (Fig. 2). Such a coordination number for Cu^{II} is virtually unknown and, presumably, unstable. As a result, these Cu^{II} atoms should readily undergo the loss of the other adjacent oxygen atom in the "chain" resulting in their further reduction to Cu^{I} with a coordination number of 2 (Fig. 2). (Such linearly coordinated Cu^{I} atoms are well known in other compounds[28] and are presumably stable.) At the same time, however, two new 3-coordinate Cu^{II} atoms are produced at the edge of the deoxygenated region. This process will result in a cascade in which a Cu-O-Cu "chain" will begin to "unzip" once a single bridging

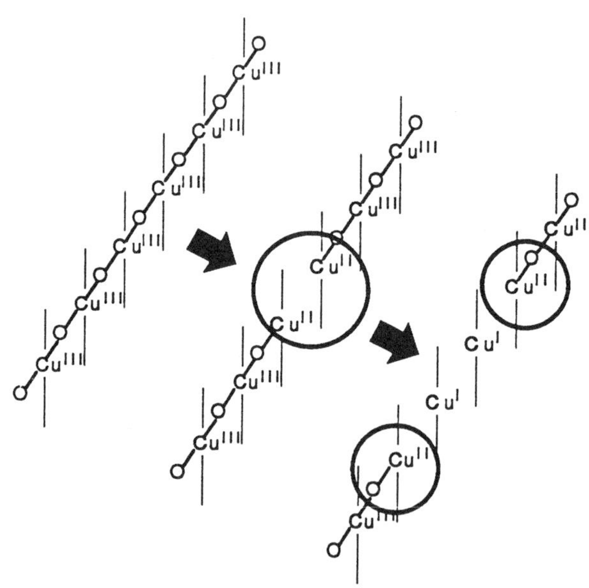

Fig. 2. Schematic representation of the proposed process of deoxygenation in a single "chain" (left) in which the loss of a single bridging oxygen (center) creates two Cu^{II} atoms (circled) adjacent to the vacated site. Such 3-coordinate Cu^{II} atoms are thermodynamically unstable, thereby destabilizing the remaining adjacent bridging oxygen atoms and leading to further oxygen loss to create stable 2-coordinate Cu^{I} (right) and two new unstable Cu^{II} atoms (circled) at the edge of the deoxygenated region. This process continues in a single chain until the entire chain is converted to disconnected Cu^{I} atoms.

oxygen atom is removed. As a result, deoxygenation should continue in any single "chain" until all Cu^{III} atoms have been reduced to linearly coordinate Cu^{I} atoms. This process accounts for both the preferential removal of the oxygen atoms in the "chains" and for the absence of local magnetic moments in partially deoxygenated material.[24]

When all copper "chains" have been fully reduced, the overall stoichiometry conforms to $YBa_2Cu_3O_6$ (i.e., $x = 1$) and the former "chains" will consist entirely of disconnected Cu^{I} atoms, a proposition which has been previously expressed by Cava *et al.*[27] However, according to the deoxygenation mechanism described above, partially deoxygenated systems ($0<x<1$) should consist of a mixture of Cu^{III}-O-Cu^{III} "chains" and oxygen-free linear arrays of Cu^{I} atoms as illustrated in Fig. 3.

Electrical properties

The Y-Ba-Cu-O system, as described above, is an insulator. To obtain the

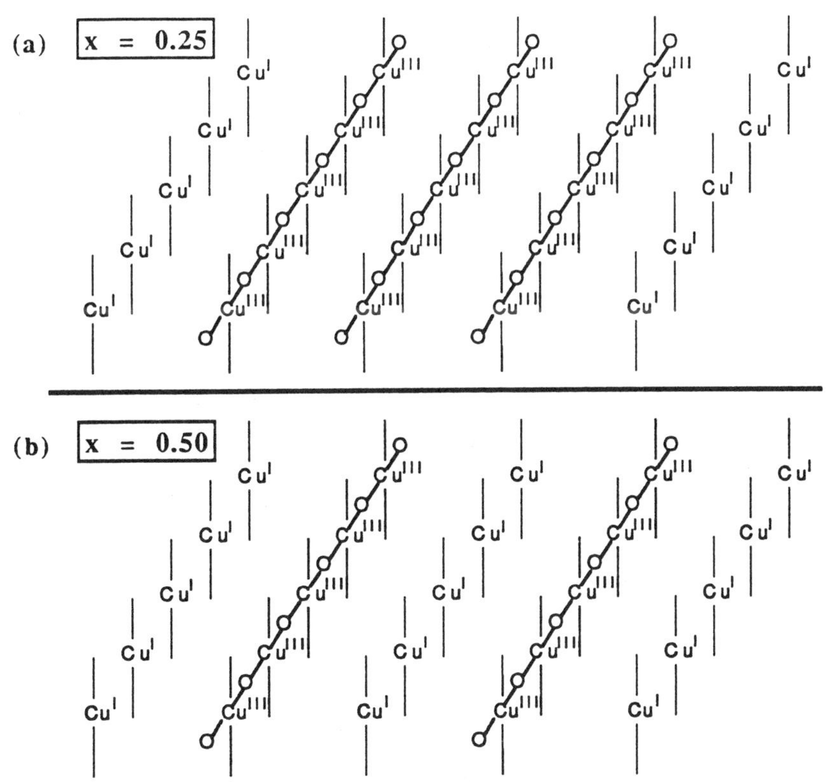

Fig. 3. Schematic representation of the predicted patterns of alternating "chains" of Cu^{III}-O-Cu^{III} and disconnected linear arrays of Cu^{I} atoms postulated to predominate at various stages of deoxygenation. At $x = 0.25$ (Fig. 3a), three adjacent Cu^{III}-O-Cu^{III} "chains" exist between linear arrays of disconnected Cu^{I} atoms. At $x = 0.5$ (Fig. 3b), each Cu^{III}-O-Cu^{III} "chain" is sandwiched between linear arrays of Cu^{I}, thereby eliminating cooperativity between the "chains."

observed electrical properties of the system, both above and below T_c, we must now consider charge transfer interactions between the "sheets" and the "chains". "Sheets"/"chain" interaction allows the "sheets" to act as electron reservoirs which can transfer carriers to the "chains", providing the basis for electrical conduction and a means of establishing conditions to give high T_c superconductivity. The O atoms lying at the apex of the square pyramidal Cu atoms in the "sheet" provide an accessible pathway for such "sheet"/"chain" coupling. Each electron promoted from the "sheet" in effect leaves behind a Cu^{III} atom and converts one of the "original" Cu^{III} atoms in the "chain" to Cu^{II}. As electrons are transferred up and down from the adjacent "sheets" to the Cu-O-Cu "chains", a mixed Cu^{II}/Cu^{III} system will be formed in both the "sheets" and the "chains" as illustrated conceptually in Fig. 4.

This process does not, of course, produce localized Cu^{II} atoms on the "chains" nor does it leave Cu^{II} atoms with unpaired spins on the "sheets" since this would produce local magnetic moments. Rather the free carriers present in this mixed oxidation state are delocalized, resulting in the observed Pauli-Landau temperature-independent magnetic susceptibility and the metallic type electrical conductivity. In fact, the carrier delocalization and increase in entropy, which result from the "sheet" to "chain" charge transfer, provides the thermodynamic driving force to produce the mixed oxidation state. Although the holes and electrons

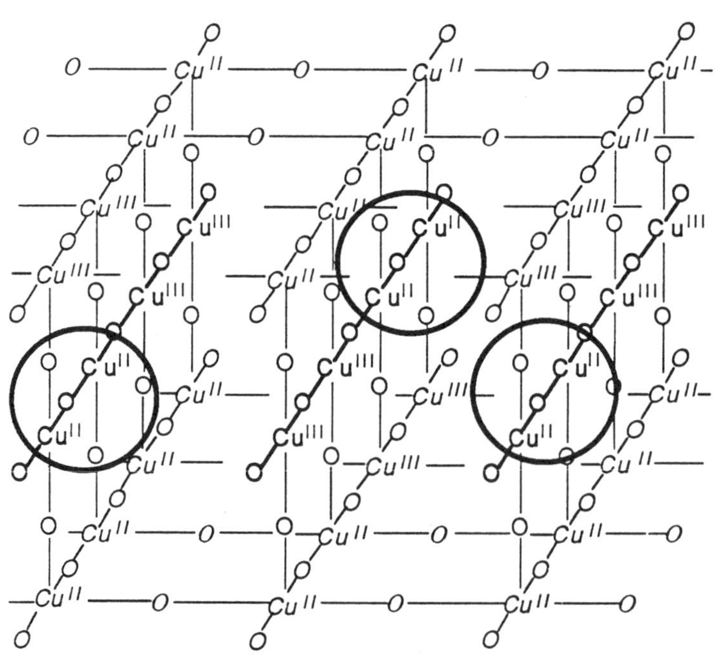

Fig. 4. Conceptualized representation showing the instantaneous generation of spin-paired Cu^{II} atoms (circled) in the "chains" and the corresponding generation of paired Cu^{III} "holes" in either the upper (left and right) or lower (center) "sheets". In reality, the spin-paired Cu^{II} atoms are delocalized.

created in this manner must exhibit a coulombic interaction, bound excitons are not formed because the screening length which will be present in this metallic system is shorter than the "sheet" to "chain" distance.

With about ⅔ of the Cu^{III} "chain" atoms converted to Cu^{II} and, correspondingly, ⅓ of the Cu^{II} atoms on each of the "sheets" converted to Cu^{III}, there will be a density of electrons and holes of 6×10^{21} cm^{-3} — a value approximately equal to the measured free carrier density above T_c. If we assume that the holes created on the "sheets" by this process have higher mobility than the electrons transferred to the "chains", we can also explain the *p*-type thermopower measured[29] for this system. The delocalized carriers on the "sheets" and "chains" also interact themselves through the O atom intermediaries. Because of delocalization, however, the coupling for this interaction is weaker than the previously described antiferromagnetic coupling which exists between the localized spins on the Cu^{II} atoms in the "sheets". Despite the reduced strength of this free carrier antiferromagnetic coupling, at sufficiently low temperatures two spins on alternate sides of a bridging O atom can be in a favorable position, because of the large bond angle, to couple through the superexchange process to produce an antiparallel spin pair.

The superexchange coupled spin pairs both on the "chains" and on the "sheets" are mobile, strongly bound, spinless composite particles which will obey Bose statistics. At any given concentration of these spin pairs, therefore, there will exist a Bose condensation temperature at which a transition to a superconducting ground state will occur. Thus, there are three important temperatures in the problem. The first two are the spin-pairing temperatures for the holes on the "sheets" and electrons in the "chains", respectively, and the third is the Bose condensation temperature.

Based on the foregoing concepts, the superconducting state can be achieved in three different ways. First, if both spin pairing temperatures are lower than the Bose condensation temperature, superconductivity will occur, in principle, at a temperature just below the highest spin pairing temperature. This situation is analogous to that found in conventional low temperature metallic superconductors where T_c is determined by the strength of the carrier pairing interaction.

In the second possibility, one spin pairing temperature could lie above the Bose condensation temperature and one below. In our model, the higher spin pairing temperature would likely be that of the "chain". This would give rise to a situation in which, above T_c, normal conduction would occur through both ordinary Fermi particles and uncondensed spin pairs. Since Hall Effect and thermopower measurements in Y-Ba-Cu-O show normal state *p*-type conductivity, this could be explained, as previously mentioned, by postulating a higher mobility for the holes on the "sheets," allowing them to dominate the electrical properties of the material, thus "hiding" the electrical consequences of the uncondensed spin pairs on the "chains" above T_c.

The third possibility is that the lowest temperature is the Bose condensation temperature. This novel situation would have dramatic consequences for electrical properties of the normal state, to say the least, resulting in the occurrence of charge transport exclusively through uncondensed spin pairs! It is not clear whether one could successfully explain all of the measured transport and magnetic properties of these materials in terms of such peculiar charge carriers.

To consider these possibilities, we obtain a simple estimate of the Bose condensation temperature for the model system in the following way. The volume occupied by the wavefunction of a Bose particle can be approximated as a cube with dimension equal to the particle's de Broglie wavelength. The de Broglie wavelength, in turn, is determined by the momentum of the particle and, therefore, by its thermal energy. This temperature dependent Bose particle interaction volume increases with decreasing temperature. When the interaction volume grows to become equal to the volume available per Bose particle in the system, the Bose particles interact so as to bring about condensation. The condensation temperature, T_0, can be written as

$$T_0 = 3.31 \frac{h^2}{(2\pi)^2 k_B m} n^{2/3}$$

where h is Planck's constant; k_B is Boltzmann's constant; m is the effective mass of the Bose particle; and n is the number density. For $YBa_2Cu_3O_7$ the greatest density of Bose particles (Cu^{II}-O-Cu^{II} spin pairs) will occur when $\frac{2}{3}$ of the "chain" Cu atoms are in the +2 oxidation state. If we use this density and take $m = 2m_e$, where m_e is the electron mass, we obtain for the Bose condensation temperature, $T_0 \approx 3000$ K. We can obtain another estimate which takes into account the two-dimensionality of the "sheets" by equating the "sheet" area available per spin pair to the square of the de Broglie wavelength, but this also gives a value T_0 equal to a few thousand Kelvin.

These results could be overestimated to some extent as a result of using too small an effective mass. However, an effective mass of $m > 20m_e$ would be required to obtain $T_0 = T_c$. Therefore, we should conclude that in these high T_c systems, as in the low temperature metallic superconductors, the transition to the normal state is likely to be determined by pair breaking and not by the Bose condensation.

Although it is difficult to obtain an estimate of the spin pairing temperatures in terms of this simple model, it is clear that, because of the increased orbital overlap between Cu d-electrons and O p-electrons which occur in the 180° bond angles along the "chains", we would expect spin pairs to be more strongly bound through superexchange on these structures than in the "sheets." Support for this is obtained from recent nuclear spin lattice relaxation experiments on the Y-Ba-Cu-O system reported by Warren *et al.*[30] which show a clear indication of the presence of two distinct pairing energies, with substantially larger energies for quasiparticle formation (pair breaking) on the "chains" than on the "sheets." Their data indicate spin pairing temperatures of approximately 60K and 200K when one assumes the standard BCS weak-coupling gap ratio of $2\Delta/kT_c \approx 4$.

Superconductivity changes upon deoxygenation

According to the recent studies of Cava *et al.*[28], as illustrated in Fig. 5, partial deoxygenation of $YBa_2Cu_3O_{7-x}$ causes only minor changes in T_c up to $x \approx 0.25$ when the value of T_c suddenly drops from ~90 K to ~60 K. Upon further removal of oxygen this latter T_c value is maintained until $x \approx 0.5$ when the T_c value decreases again by over 30K.

In accordance with the model which we have proposed, $x = 0.25$ can be viewed as corresponding to the point at which one fourth of all "chains" have been reduced to oxygen-free linear arrays of Cu^I atoms resulting in an average of three adjacent Cu^{III}-O-Cu^{III} "chains" between Cu^I linear arrays (Fig. 3a). The sudden change in properties at this point (Fig. 5) suggests that the higher T_c value derives from extended three-dimensional communication between the coupled spin pairs in adjacent "chains." When $x = 0.5$, half of the "chains" have been reduced and, on the average, all Cu^{III}-O-Cu^{III} "chains" are sandwiched between linear arrays of Cu^I atoms (Fig. 3b). Since the Cu^I atoms are incapable of participating in the mixed valence charge propagation process, three-dimensional interaction between adjacent intact "chains" is virtually ruled out and a further significant reduction in T_c results (Fig. 5).

In this same range of the oxygen deficiency parameter, x, we see in Fig. 5 that an increase in room temperature resistivity occurs in the region of $x < 0.25$. Presumably this is due to elimination of conductive Cu-O-Cu mixed valence "chain" structures resulting in an inhibition in the formation of mixed oxidation states on the "sheets" and "chains". This reduction in free carrier density is also accompanied by slight decrease in magnetic susceptibility as observed.

Once the transition to the $T_c = 60$K phase has occurred, where $0.25 \leq x \leq 0.5$, the room temperature resistivity dips, and the magnetic susceptibility shows a dramatic increase (Fig. 5). These data suggest the formation of an ordered phase as has been pointed out by Cava *et al.*[28] Indeed, these authors have reported evidence, from electron diffraction measurements, which suggest vacancy ordering of some type, although the lattice still retains orthorhombic symmetry. In terms of our model, this ordering is predicted to be an arrangement of intact Cu^{III}-O-Cu^{III} "chains" and completely oxygen deficient linear Cu^I arrays. This stands in contrast to the ordered arrangement of empty and occupied oxygen sites on single "chains" as proposed by the previous authors. The increase in free carrier magnetic susceptibility in this range is remarkable and is much greater than would be associated with the slight increase in conductivity observed, suggesting a change in effective mass, which may be due to details of band structure related to the ordered "chains."

Two ordered arrangements of "chains" can occur in this range of x. At $x = 0.25$ groups of three intact "chains" can be separated by a single oxygen-free array of

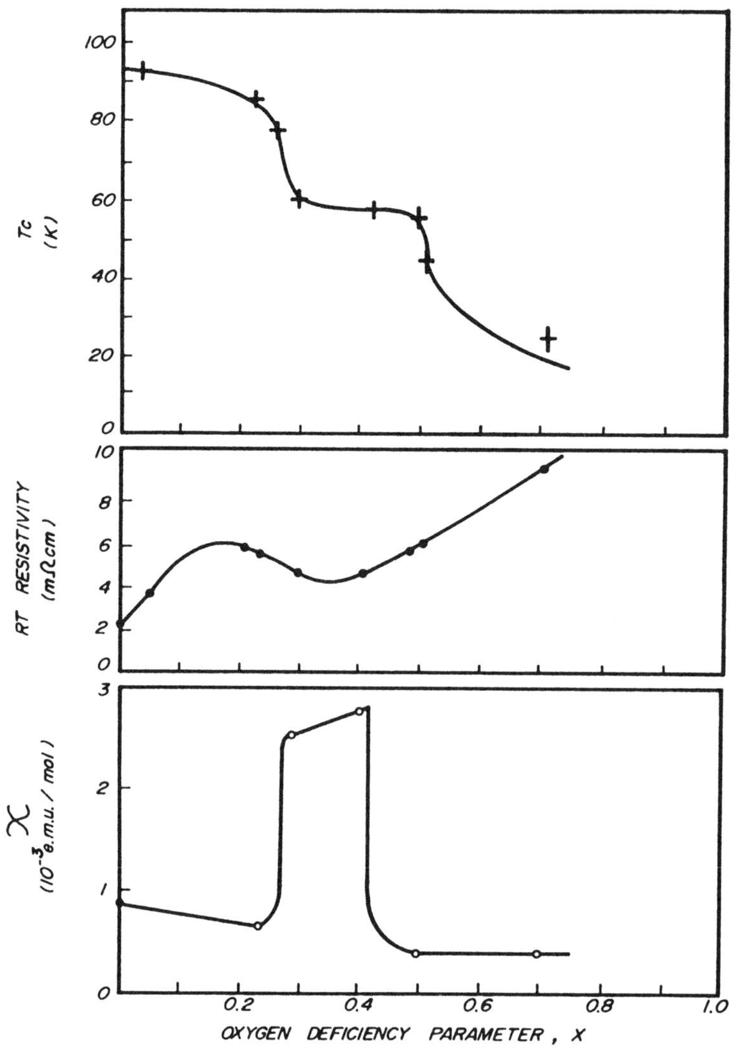

Fig. 5. Effects of partial deoxygenation in $YBa_2Cu_3O_{7-x}$ on the superconducting transition temperature, T_c (top), the room temperature resistivity (center), and the room temperature magnetic susceptibility, X (bottom). Data are from references 24 and 27.

Cu^I atoms (Fig. 3a), while at $x = 0.33$ pairs of intact "chains" can be separated by an oxygen-free array. When $x = 0.5$, half of the "chains" have been reduced and, on the average, all intact Cu^{III}-O-Cu^{III} "chains" are sandwiched between oxygen-free Cu^I arrays (Fig. 3b). It is interesting to note that the T_c value reached for $x > 0.5$ is typical of the "sheet" type ceramic superconductors La_2CuO_4 and $La_{2-x}(Ba,Sr)_xCuO_4$, suggesting that, with the loss of all "chain"/"chain" interaction, the effective dimensionality of the system has become "sheet-like".

As $x \rightarrow 1.0$, the remaining Cu^{III}-O-Cu^{III} "chains" are reduced to Cu^{I} and at $x = 1$ all superconductivity is lost. This loss in superconductivity can be attributed to the decreasing possibility of creating the required mixed oxidation state either in the "sheets" or in the "chains" and, ultimately, to the transition from an orthorhombic to a tetragonal structure.

Clearly, a more thorough analysis is required to understand details both of the antiferromagnetic coupling between delocalized carriers, and the proposed "chain"/"chain" coupling. The origin of the free carrier magnetic susceptibility enhancement in the ordered oxygen deficient phase will also require closer attention. Additionally, we must develop a better understanding of how the observed T_c in $YBa_2Cu_3O_7$ is arrived at through interaction of the "chain" and "sheet" systems. The small spatial separation of these structures and the existence of interactions, including superconducting proximity effects, should result in both structures participating in the superconducting state.

Application of this simple model to the La-Cu-O system is also possible. In these structures, although the same antiferromagnetic spin pairing mechanism is presumably operative, no "chains" exist and the mixed valence Cu^{II}/Cu^{III} system in established on the "sheets" either by doping with Sr or Ba or through to the incorporation of excess O. As has been mentioned, the La-Cu-O systems exhibit transition temperatures which are near those found for the Cu-O "sheet" systems which remain after deoxygenation of Y-Ba-Cu-O. These systems, of course, will not exhibit the complex behavior with O removal seen in the Y-Ba-Cu-O system, nor will they ever likely exhibit the high T_c behavior, both as a result of the limited dimensionality of their structure.

Predicted requirements for achieving enhanced T_c

The foregoing interpretation of the behavior of partially deoxygenated Y-Ba-Cu-O systems suggests that optimal T_c values are obtained when electron pairs in the "chains" interact in three-dimensions involving the adjacent "chains" as well as the adjacent "sheets." This suggests that T_c might be further enhanced by synthesizing new materials in which layers of "sheets" and "chains" are fully interleaved in an extended three-dimensional array. This would have the additional benefit of converting Cu atoms in the "sheets" to a 6-coordinate geometry, thereby removing "dimples" from the "sheets". By changing the 165° O-Cu-O bond angles to 180°, the spin pairing temperature on the sheets would be increased. In addition, the spin pairing temperature in the "chains" might be increased by shortening the distance between the Cu^{III} atoms by the substitution of more tightly bound bridging anions, thus increasing the strength of the antiferromagnetic coupling.

Such anion substitutions might require altering the charge on the other cations (i.e., making appropriate substitution for Y and Ba) to achieve balance of the overall charges. Moreover, as some of the bond distance change in the overall crystal lattice, care needs to be taken to maintain optimal Cu-X-Cu bond angles, both for promoting charge propagation and to optimize antiferromagnetic coupling. We believe that such substitutions can be made in a fruitful manner as, for example, may be indicated in preparations of multi-phase samples of partically fluorinated $YBa_2Cu_3O_{7-x}F_y$ which initially shown by Ovshinsky *et al.*,[31] and subsequently confirmed by others,[32] to contain some regions with $T_c = 159K$. The effect of fluorine on oxygen diffusion and microcrystal orientation have been discussed elsewhere.[33]

Finally, one can consider the possibility of synthesizing new mixed oxidation state systems in which other transition metal atoms with appropriate multiple oxidation states replace Cu. Very recent indications of high T_c superconductivity in La-Sr-Nb-O [34] and Y-Ba-Ag-O[35] ceramic films support our belief that the Cu based systems may not be unique but rather only members of a larger family of antiferromagnetically spin coupled mixed oxidation state systems.

Acknowledgement

The authors would like to acknowledge stimulating discussions with Prof. H. Fritzsche.

References

1. T. M. Rice, *Z. Physik* **B67** (1987) 141.
2. J. E. Hirsch, *Phys. Rev. Lett.* **59** (1987) 228.
3. L. Pauling, *Phys. Rev. Lett.* **59** (1987) 225.
4. P. W. Anderson, *Science* **235** (1987) 1196.
5. V. J. Emery, *Phys. Rev. Lett.* **57** (1987) 2794.
6. H. Kamimura, *Jpn. J. Appl. Phys.* **26** (1987) L627.
7. P. H. Hor, R. L. Meng, Y. Q. Wang, L. Gao, Z. J. Huang, J. Bechtold, K. Forster, C. W. Chu, *Phys. Rev. Lett.* **58** (1987) 1891.
8. E. M. Engler, V. Y. Lee, A. I. Nazzal, R. B. Beyers, G. Lim, P. M. Grant, S. S. P. Parkin, M. L. Ramirez, J. E. Vasquez, R. J. Savoy, *J. Am. Chem. Soc.* **109** (1987) 2848.
9. L. Soderholm, K. Zhang, D. G. Hinks, M. A. Beno, J. D. Jorgensen, C. U. Segre, I. K. Schuller, *Nature* **328** (1987) 604
10. B. W. Veal, W. K. Kwok, A. Umezawa, G. W. Crabtree, J. D. Jorgensen, J. W. Downey, L. J. Nowicki, A. W. Mitchell, A. P.Paulikas, C. H. Sowers, *Appl. Phys. Lett.* **51** (1987) 279
11. S. R. Ovshinsky. In *Festschrift in Honor of Heinz Henisch*; S. R. Ovshinsky, R. W. Pryon, B. B. Schwartz, Eds.; Institute for Amorphous Studies Series (Plenum, New York, 1988) in press.
12. Y. Maeno, T. Tomita, M. Kyogoku, S. Awaji, Y. Aoki, K. Hoshino, A. Minami, T. Fujita, *Nature* **328** (1987) 512.
13. P. K. Gallagher, H. M. O'Bryan, S. A. Sunshine D. W. Murphy, *Mater. Res. Bull.* **22** (1987) 995
14. A. Santors, S. Miraglia, F. Beech, S. A. Sunshine, D. W. Murphy, L. F. Schneemeyer, J. V. Waszczak, *Mater. Res. Bull.* **22** (1987) 1007.
15. I. K. Schuller, D. G. Hinks, M. A. Beno, D. W. Capone, II, L. Soderholm, J. P. Locquet, Y. Bruynseraede, C. V. Segre, K. Zhang, *Solid State Commun.* **66** (1987) 385.
16. J. D. Jorgensen, B. W. Veal, W. K. Kwok, G. W. Crabtree, A. Umezawa, L. S. Nowicki, A. P. Paulikas, *Phys. Rev.* **B36** (1987) 5719.

17. R. Beyers, G. Lim, E. M. Engler, R. J. Savoy, T. M. Shaw, T. R. Dinger, W. T. Gallager, R. L. Sandstrom, *Appl. Phys. Lett.* **50** (1987) 1918.
18. M. A. Beno, L. Soderholm, D. W. Capone, II, D. G. Hinks, J. D. Jorgensen, J. D. Grace, I. K. Schuller, C. U. Segre, K. Zhang, *Appl. Phys. Lett.* **51** (1987) 57.
19. (a) B. J. Hathaway. *Coord. Chem. Rev.* **35** (1981) 211. (b) B. J. Hathaway, D. E. Billing, *Coord. Chem. Rev.* **5** (1970) 143.
20. See, e.g., (a) P. W. R. Corfield, C. Ceccarelli, M. D. Glick, I. W.-Y. Moy, L. A. Ochrymowycz, D. B. Rorabacher, *J. Am. Chem. Soc.* **107** (1985) 2399. (b)V. B. Pett, L. L. Diaddario, Jr., E. R. Dockal, P. W. R. Corfield, C. Ceccarelli, M. D. Glick, L. A. Ochrymowycz, D. B. Rorabacher, *Inorg. Chem.* **22** (1983) 3661, and references therein.
21. D. W. Margerum, *Pure Appl. Chem.* **55** (1983) 23.
22. L. L. Diaddario, W. R. Robinson, D. W. Margerum, *Inorg. Chem.* **22** (1983) 1021.
23. E. H. Appelman, L. R. Morss, A. M. Kini, V. Geiser, A. Umezawa, G. W. Crabtree, D. K. Carlson, *Inorg. Chem.* **26** (1987) 3237.
24. R. J. Cava, B. Batlogg, C. H. Chen, E. A. Rietman, S. M. Zahurak, D. Werder, *Nature* **329** (1987) 423.
25. G. Shirane, Y. Endoh, R. J. Birgeneau, M. Kastner, Y. Hidaka, M. Oda, M. Suzuki, T. Murakami, *Phys. Rev. Lett.* **59** (1987) 1613.
26. F. Beech, S. Miraglia, A. Santors, R. S. Roth, *Phys. Rev.* **B36** (1987) 8778.
27. R. J. Cava, B. Batlogg, C. H. Chen, E. A. Rietman, S. M. Zahurak, D. Werder, *Phys. Rev.* **B36** (1987) 5719.
28. H. Hope, P. P. Power, *Inorg. Chem.* **23** (1984) 936.
29. N. Mitra, J. Trefny, M. Young, B. Yarar, *Phys. Rev.* **B36** (1987) 5581.
30. W. W. Warren, Jr., R. E. Walstedt, G. F. Brennert, G. P. Espinosa, J. P. Remeika, *Phys. Rev. Lett.* **59** (1987) 1860.
31. S. R. Ovshinsky, R. T. Young, D. D. Allred, G. DeMaggio, G. A. Van der Leeden, *Phys. Rev. Lett.* **58** (1987) 2579.
32. R. N. Bharagawa, S. P. Heako, W. N. Osborne, *Phys. Rev. Lett.* **59** (1987) 1468.
33. S. R. Ovshinsky, R. T. Young, B. S. Chao, G. Fournier, D. A. Pawlik, *Proc. Int'l. Conf. on High Temperature Superconductivity,* Drexel University, (World Scientific, Singapore, 1988) to be published.
34. T. Ogushi, Y. Hakuraku, Y. Honjo, G. N. Suresha, S. Higo, Y. Ozono, I. Kawano, T. Numata, *Low Temp. Phys.* (Jan. 1988), to be published.
35. K. K. Pan, H. Mathias, C. M. Rey, W. G. Moulton, H. K. Ng, L. R. Testardi, Y. L. Wang, *Phys. Lett.* **A125** (1987) 147.

Volume 195, number 4 CHEMICAL PHYSICS LETTERS 24 July 1992

The origin of pairing in high-T_c superconductors

S.R. Ovshinsky

Energy Conversion Devices, Inc., Troy, MI 48084, USA

Received 1 April 1992

Imbedded in the structural chemistry of the high-temperature cupric superconductors lies the new physics which can answer the unsolved fundamental problem of pairing. A mechanism is offered which uniquely shows how pairing can occur by virtue of intimate Cu^{III} valence relationships that are generated in the mixed valence copper-oxygen planes. This mechanism leads to condensation to the lowest free energy superconducting state.

1. Introduction

Despite the very large number of papers published (for a recent review, see ref. [1]) during the past six years on the subject of copper-oxide-based high-temperature ceramic superconductors, the origin of the carrier pairing interaction that produces the superconducting spin-zero state remains an unresolved problem [2,3], that is crucial for understanding the nature of high-T_c superconductivity.

A major problem with many proposed pairing mechanisms is that they are quite general and have not specifically addressed the observation that all of the known high-T_c materials belong to a specific chemical structural class of solids. This point has been made by others and is exemplified by Jorgensen's statement: Questions concerning direct relationships between structure and superconductivity – for example, the possibility of structure itself playing a role in the superconducting mechanism – remain to be answered [4].

2. Discussion

In the present model, we address the origin of pairing in high-T_c superconducting copper-oxide ceramics by proposing an explicit stereochemical mechanism. It is widely accepted that, at high temperatures, charge carriers in these materials are holes that reside in the copper oxide planes. These holes are charge compensated in the rest of the material either by chemical substitution of certain elements or by oxygen vacancies. The presence of holes in the copper-oxide planes converts Cu^{II} atoms to Cu^{III} atoms producing a mixed valence state [5,6]. Numerous experiments [1] have confirmed that the hole is shared by the d orbital of the Cu^{III} and the lone pair p orbitals of the adjacent O atoms by hybridization.

In this paper we propose a pairing process which may be related to the general class of bipolaronic mechanisms [7], but which differs in a fundamental way by invoking lattice deformations due to specific orbital configurations of the O–Cu–O chemical bonds that occur in the mixed valence planar superconducting ceramic systems. Instead of describing polaron formation in terms of a general polarization of a deformable continuum, the present model notes that the configurational and bonding environments of Cu atoms in the copper-oxide planes depend, fundamentally, on their charge state (valency), and this environment is changed in a specific way with a change in charge state. Cu^{II} configurations in the copper-oxide planes result in puckered O–Cu–O bond angles of 165°; Cu^{III} configurations result in flat O–Cu–O 180° bond angles. As a consequence, the bond angles in the copper-oxide planes of the ceramic high-T_c superconductors exhibit reversible transforma-

Correspondence to: S.R. Ovshinsky, Energy Conversion Devices, Inc., Troy, MI 48048, USA.

tions in a Cu^{II}/Cu^{III} conducting, mixed valence system.

In copper-oxide planes with no holes present, all of the copper atoms will be Cu^{II} and the bond angles will be 165°. Since Cu^{III} bond angles represent the lowest energy configurations of the two valence states, we suppose that the bond angle environment of an isolated hole or Cu^{III} atom imbedded in a mixed valence copper-oxygen plane will always relax toward the flat configuration. When two holes are in close proximity on the copper-oxygen plane, i.e. two neighbouring Cu^{III} atoms, we propose that the energy gain of the system due to the required bond angle transformations around the neighboring pair will exceed the energy gain from lattice relaxation around two isolated Cu^{III} atoms. The difference between the energy gained from the bond-angle transformation for the intimate pair of Cu^{III} atoms and energy gained from bond-angle transformations associated with an isolated pair constitutes a carrier pairing energy which we estimate [8] to be at least of the order of 10–30 meV. Since this carrier pairing energy is comparable to or greater than kT_c, one expects that these bound pairs will contribute to, and possibly dominate normal state conductivity in the copper-oxide superconductor materials.

The bound pair forms a spin-zero boson by preserving the anti-alignment of the surrounding Cu^{II} spins that originates with superexchange coupling in the copper-oxide plane. Thus, in our model, the carrier pairing energy results from the shared lattice strain associated with a neighboring pair of Cu^{III} atoms, and the spin-zero state of the boson comes about, naturally, from short range antiferromagnetic interactions between lone pair orbitals on the oxygen atoms with the Cu d orbitals in the Cu^{II} valence state.

A sufficiently large concentration of these bound pairs will result in their delocalization giving mobile, spin-zero bosons. The superconducting transition of these carriers can then be regarded as a Bose condensation, at T_c, of pre-existing mobile, spin-zero bosons.

3. Summary

The mixed valency and chemical bonding of the copper-oxide planes is essential for the occurrence of high-T_c superconductivity in the present model. The mixed valency not only provides a sufficient concentration of charge carriers and equivalent sites for their motion, but in the copper-oxide ceramics, it also provides a basis for the pairing mechanism itself.

Acknowledgement

I wish to express my deep appreciation to Hellmut Fritzsche and to Stephen Hudgens for their constructive discussions.

References

[1] B. Batlogg, Phys. Today 44 (1991) 44.
[2] R. Schrieffer, Science 251 (1991) 1005.
[3] P.W. Anderson, Science 251 (1991) 1005.
[4] J.D. Jorgensen, Physics Today 34 (1991) 40.
[5] S.R. Ovshinsky, S.J. Hudgens, R.L. Lintvedt and D.B. Rorabacher, Modern Physics Letters B 1 (1987) 275.
[6] S.R. Ovshinsky, Revue Roumaine de Physique 36 (1991) 761.
[7] D. Emin, in: Physics and materials science of high temperature superconductors, Vol. 2, eds. R. Kossowsky, N. Raveau and S. Patapis (Kluwer, Dordrecht, 1992).
[8] H. Fritzsche, private communication.

Physica C 200 (1992) 437–441
North-Holland

PHYSICA C

High quality epitaxial YBCO(F) films directly deposited on sapphire

R.T. Young, K.H. Young, M.D. Muller and S.R. Ovshinsky
Energy Conversion Devices, Inc., 1675 West Maple Road, Troy, MI 48084, USA

J.D. Budai and C.W. White
Solid State Division, Oak Ridge National Laboratory, Oak Ridge, TN 37831, USA

J.S. Martens
Sandia National Laboratories, Albuquerque, NM 87185, USA

Received 29 June 1992

High-quality epitaxial $YBa_2Cu_3O_{7-\delta}(F)$ films were grown directly on r-plane $(1\bar{1}02)$ sapphire by a pulsed laser deposition technique using a multiphase fluorinated target. X-ray diffraction data showed that the film is highly *c*-axis oriented and in-plane aligned. The *c*-axis and in-plane mosaic spread are 1.1° and 2.4°, respectively. The microwave surface resistance measured by a con-focal resonator configuration is 56 mΩ at 94.1 GHz at 77 K which is the best reported value for a YBCO film on sapphire.

The improved film quality is attributed to the fluorine which is transported from the target to the growing surface of the film where it acts as an etching and/or catalytic agent to preferentially remove defects and second phase impurities and to promote better in-plane epitaxy.

One of the most important and immediate commercial applications of high-T_c superconductors is in microwave devices. At microwave frequencies, the performance of the device is critically dependent on the quality of the superconducting film and the dielectric loss of the substrate. Sapphire, because of its excellent dielectric properties (i.e. small dielectric constant and very low loss tangent), good thermal and mechanical properties and the availability of large wafers, is considered the most desirable substrate. However, the growth of high-quality $Y_1Ba_2Cu_3O_{7-\delta}$ (YBCO) superconducting films directly on sapphire with good microwave properties has not yet been demonstrated. The reasons are: (1) sapphire chemically reacts with YBCO at the temperature required for high-quality film growth; (2) sapphire has poor lattice match with YBCO. Furthermore, because of the extremely short coherence length of high-T_c films, any grain boundaries and/or crystalline disorder will give rise to residual surface impedance [1]. Therefore, very stringent criteria are imposed on the quality of the films for microwave applications.

The best reported microwave surface resistance (R_s) of a YBCO film directly on sapphire at 10 GHz is 1 mΩ measured at 4 K by Char et al. [2]. The relatively high residual resistance of the film was attributed to the fact that, in order to minimize the chemical reaction, the film was deposited at 670°C instead of the 750°C required for optimal epitaxial growth. This resulted in a film with poor in-plane epitaxy and a high surface resistance. Epitaxial YBCO film on sapphire has also been reported by Varlamov et al. [3]; however, no R_s data were presented.

To prevent the interface reaction and to be able to grow films at the optimal deposition temperature, numerous efforts have been directed toward growing an epitaxial buffer layer such as $SrTiO_3$, $CaTiO_3$, CeO_2, ZrO_2 or MgO, on sapphire [4–7]. To our knowledge, the best published result was obtained using a thin (500 Å) layer of $SrTiO_3$ as the buffer layer [4]. The films show improved in-plane epitaxy

0921-4534/92/$05.00

and lower surface resistance. For example, the *c*-axis and in-plane mosaic spreads from X-ray diffraction were reported as 2.4° and 3.8°, respectively. At 10 GHz, R_s is given as 850 μΩ at 77 K which is considerably better than films deposited directly on sapphire. However, due to the extremely poor dielectric properties, even a thin buffer layer of $SrTiO_3$ is detrimental to microwave performance.

In this work, we describe a unique approach using fluorine at the film growing surface to achieve a high quality epitaxial YBCO(F) film growth *directly* on sapphire without a buffer layer. Some advantages of adding fluorine in the high-T_c films have been reported previously [8,9]. Our film, from X-ray diffraction measurement, has a *c*-axis and in-plane mosaic spread of 1.1° and 2.4°, respectively. At 77 K, for example, the R_s is 56 mΩ at 94.1 GHz. Using the frequency-squared scaled to 10 GHz the R_s is 630 μΩ.

The YBCO(F) films were deposited by a pulsed laser deposition technique using a 50 W Questek excimer laser at 248 nm. The unique feature of our technique is that a fluorinated multi-phase target is used instead of a conventional single-phase YBCO target. The target has a nominal composition of $YBa_{2.6}Cu_{3.0}O_{4.5}F_{5.2}$. The films were deposited with 50 mTorr of oxygen partial pressure at a substrate holder temperature of 680–700°C. The substrate surface temperature was measured to be 50°C to 70°C lower than the temperature of the holder. At this temperature, interface reaction does not appear to be a serious problem. Even though there is a high concentration of F (~32 at.%) in the target, this resulted in only a very small amount (~0.5 at.%) of F being incorporated into the film. However, the considerable amount of fluorine transferred to the growing surface of the film provides etching and some catalytic effects which promote better in-plane epitaxy at relatively low surface temperature. The YBCO(F) film on sapphire has a $T_c(R=O)$ of 89 K and a ratio of $R(300\ K)/R(100\ K) \approx 3.0$. These are typical values for a "true" epitaxial film on $SrTiO_3$, $LaAlO_3$, etc.

The quality of the film was characterized using a four circle X-ray diffraction and Rutherford backscattering-ion channeling measurements. The surface morphology was studied by scanning electron microscopy (SEM). The microwave surface resistances were measured by a con-focal resonator configuration [10] in the frequency range of 30 to 100 GHz.

An X-ray diffraction θ–2θ scan near the surface normal reveals that the YBCO(F) sample is a predominantly *c*-axis film containing a small amount of *a*-axis grains (of order 2%). Very weak peaks from an unidentified impurity phase are also observed at *d*-spacings of 3.09 Å and 2.80 Å. A rocking curve (θ-scan) through the YBCO (006) reflection shows a mosaic full width at half maximum (FWHM) of ~1.1° in this direction. Interestingly, the average position of the YBCO (00*l*) axis is not exactly aligned with either the sample surface normal or the normal to the Al_2O_3 $(1\bar{1}02)$ planes. Instead, we find that the substrate surface is miscut 0.55° away from the $(1\bar{1}02)$ planes and that the average position of the YBCO (00*l*) planes lies approximately halfway between these two orientations. Char et al. [2] similarly report a misorientation between the YBCO and Al_2O_3 axes of about 1°, and these observations lead us to speculate that the best epitaxy (smallest mosaic) should occur on substrates with zero miscut angle.

X-ray scans obtained away from the surface normal were used to establish the in-plane epitaxial relation between the film and the substrate. Figure 1(a) is a low-resolution ϕ scan of the YBCO (225) peak. Consistent with previous reports [2], the sharp peaks observed every 90° indicate that the in-plane YBCO (110) directions are aligned with the orthogonal Al_2O_3 $(1\bar{1}01)$ and $(1\bar{1}20)$ directions. The FWHM of the in-plane mosaic spread is ~2.4°. The results clearly demonstrate that high quality epitaxial YBCO(F) films can be grown on sapphire without a buffer layer. Figure 1(b) shows a high-resolution θ–2θ scan through the (205) and (025) peaks. The presence of two separate peaks indicates that the YBCO(F) film has long-range orthorhombic rather than tetragonal order. However, the relatively broad reflections indicate some in-plane disorder. The positions of X-ray peaks yield lattice parameters of $a=3.83$ Å, $b=3.89$ Å, and $c=11.67$ Å. The *c*-lattice parameter is slightly smaller than is typically observed in bulk YBCO samples and suggests that the film is under a slight tensile stress, presumably due to differential thermal contraction.

Rutherford backscattering-ion channeling mea-

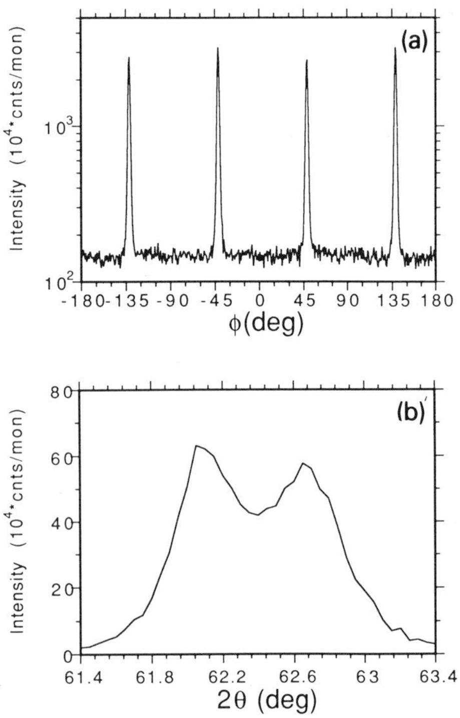

Fig. 1. X-ray diffraction measurement of a YBCO(F) film directly deposited on (1102) sapphire. (a) ϕ scan of the (225) peak, (b) θ–2θ scan of the (205) and (025) peaks.

surements were made along the (001) axis of the film using a beam of 5.0 MeV He^+ ions at a scattering angle of 160°. At these energies, scattering from the individual constituents Ba, Y, and Cu was completely separated, allowing the atom ratios of these constituents to be determined. Figure 2 shows random and aligned backscattering spectra from that part of the spectrum which contains the scattering from these atoms on the surface. From the random spectra, a film thickness of ~260 nm is obtained, and the atom ratios are determined to be $N(\mathrm{Ba})/N(\mathrm{Y}) = 1.6$ and $N(\mathrm{Cu})/N(\mathrm{Y}) = 2.6$. In fig. 2, the aligned spectrum measured from each constituent has a minimum yield χ_{min} (defined as χ_{min} = yield in the aligned direction/yield in the random direction) at the surface of ~50%, increasing to ~75% at the interface. The fact that channeling is observed at all demonstrates that the film is crystalline and epitaxial with the substrates. These values for χ_{min} are not unreasonable considering the mosaic spread of the film measured by X-ray diffraction (~1°) and the poor lattice match. Qualitatively, the same behavior in the aligned spectra is observed for epitaxial silicon-on-sapphire films where the high defect density near the interface gives rise to a high value for χ_{min} in the silicon lattice depth; the defect density (and the value for χ_{min}) decrease as a function of distance

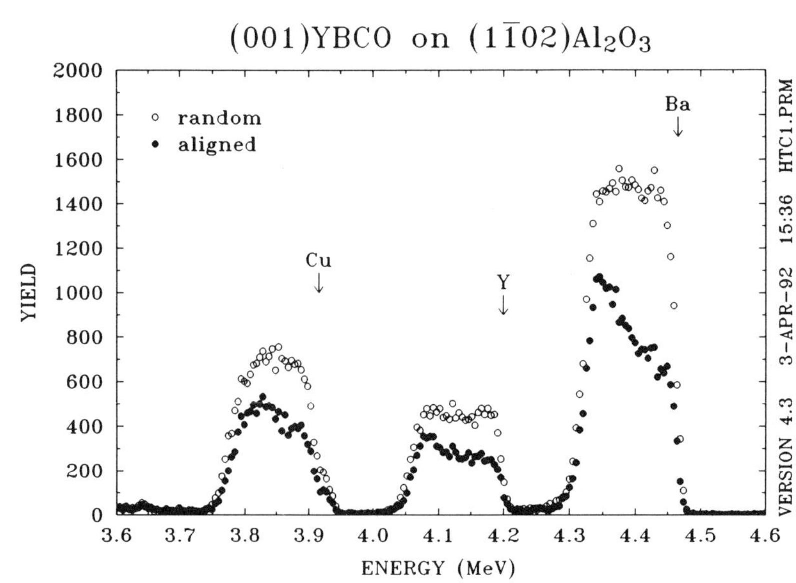

Fig. 2. The random and aligned backscattering spectra of a YBCO(F) film on sapphire.

away from the silicon/sapphire interface.

For comparison, YBCO films have also been deposited on sapphire from a high-density single-phase commercial target without fluorine. The films were deposited using the same deposition parameters as the fluorinated films. The film quality was found to be similar to that reported by Char et al. [2]. However, cracks are frequently observed on the YBCO film. Figure 3 shows SEM micrographs that provide a comparison of films deposited from the two different targets. Figures 3(a) and (c) were taken under the backscattering mode to study the cracks in the films. Figures 3(b) and (d) were taken using the secondary electron mode to reveal details of the surface morphology. Both films show smooth surfaces. The cracks in the YBCO film were identified to be primarily along the *a*- and *b*-axes as shown in fig. 3(a). Because of the poor lattice match, the YBCO film would be highly stressed during growth and cracks will develop, most likely, in the highly disordered boundaries. The cracks in the YBCO film suggest the presence of the high in-plane disorder. No surface cracks were observed in the YBCO(F) films. It is evident from these results that fluorine at the film growing surface can help to reduce the in-plane disorder and/or release the stress so that a high quality epitaxial film of YBCO(F) can be grown directly on sapphire.

The surface resistance (R_s) has been considered the most important figure-of-merit for microwave devices. The R_s of the YBCO(F) was measured by a con-focal resonator configuration in the frequency range of 30 to 100 GHz. Figure 4 is a plot of R_s at different frequencies (f), which indicates that R_s fol-

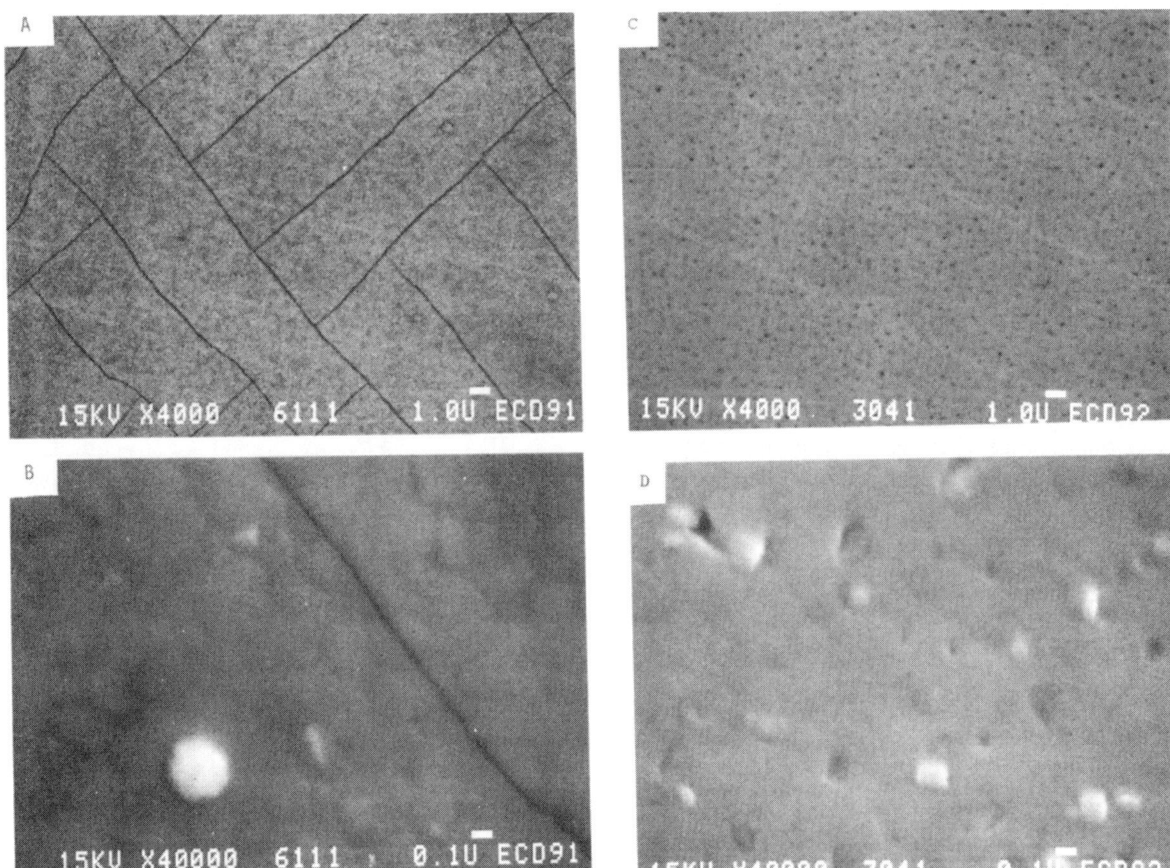

Fig. 3. SEM micrographs showing a comparison of films deposited from a conventional YBCO target, (a) and (b), and a multiphase YBCO(F) target, (c) and (d).

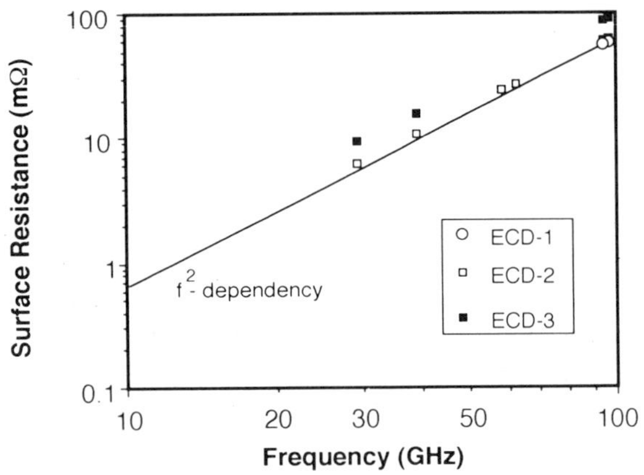

Fig. 4. Surface resistance R_s vs. frequency plot of three different YBCO(F) films directly deposited on (1102) sapphire.

lows the typical $R_s \propto f^2$ relation. The R_s for example, of ECD-1 was measured to be 56 mΩ at 94.1 GHz at 77 K. Using the f^2 scaled to 10 GHz, R_s is 630 μΩ. This is the best reported value of YBCO film on sapphire.

In summary, we have demonstrated for the first time that high quality epitaxial YBCO(F) films with low surface resistance can be grown directly on sapphire. By using fluorine, we have minimized and circumvented the problems of large lattice mismatch and chemical reaction between the YBCO and the sapphire. This is attributed to the possibility that fluorine plays an important role in removing defects and/or impurities at the growing surface and promoting better epitaxial growth. The ability to grow device-quality high-T_c films on sapphire has enabled the fabrication of high-performance microwave devices operating over a much wider frequency range.

Acknowledgements

We would like to thank Ben Chao and R.F. Wood for helpful discussions and critical reading of the manuscript. The authors from Energy Conversion Devices, Inc. are grateful to Inland Steel Industries, Inc. for support, especially to Florence Metz for the stimulation and encouragement of this work. The research at Oak Ridge National Laboratory is supported by the Division of Materials Sciences, US Department of Energy under contract DE-AC05-84OR21400 with Martin Marietta Energy Systems, Inc.

References

[1] T.L. Hylton, A. Kapitulnik, M.R. Beasley, J.P. Carini, L. Drabeck and G. Gruner, Appl. Phys. Lett. 53 (1988) 1343.

[2] K. Char, D.K. Fork, T.H. Geballe, S.S. Laderman, R.C. Taber, R.D. Jacowitz, F. Bridges, G.A.N. Connell and J.B. Boyce, Appl. Phys. Lett. 56 (1990) 785.

[3] Yu.D. Varlamov, M.R. Predtechenskii and A.N. Smal, Supercond. Phys. Chem. Technol. 3 (1990) 1970.

[4] K. Char, N. Newman, S.M. Garrison, R.W. Barton, R.C. Taber, S.S. Laderman and R.D. Jacowitz, Appl. Phys. Lett. 57 (1990) 409.

[5] M. Schieber, Y. Ariel, M. Schwartz, M. Levinsky, Sh. Shukran, M. Maharizy, B.L. Zhou and S.C. Han, Supercond. Sci. Technol. 4 (1991) 268.

[6] J. Gao, B.B.G. Klopman, W.A.M. Aarnink, A.E. Reitsma and G.J. Gerritsma, J. Appl. Phys. 71 (1992) 2333.

[7] R.E. Muenchausen, D.W. Cooke, S.R. Foltyn, X.D. Wu and N.S. Nogar, Physica C 190 (1991) 46.

[8] S.R. Ovshinsky and R.T. Young, SPIE vol. 1324 (1990) p. 32.

[9] R.T. Young, S.R. Ovshinsky, B.S. Chao, G. Fournier and D.A. Pawlik, Mat. Res. Soc. Symp. Proc., vol. 99 (1988) p. 549.

[10] J.S. Martens, V.M. Hietala, D.S. Ginley, T.E. Zipperian and G.K.G. Hohenwarter, Appl. Phys. Lett. 58 (1991) 2543.

Applied Superconductivity Vol.1, Nos 3–6, pp. 263 – 267, 1993
Printed in Great Britain.

0964-1807/93 $6.00 + 0.00

A MECHANISM FOR HIGH TEMPERATURE SUPERCONDUCTIVITY

Stanford R. Ovshinsky
Energy Conversion Devices, Inc.
1675 West Maple Road, Troy, Michigan 48084 U.S.A.

ABSTRACT

We describe a pairing mechanism that is based on changes in the O-Cu-O bonds and orbital configurations that occur in the mixed valence planar superconducting ceramics. We discuss the mobility of these spin-zero bosons and their distinctive differences from other proposed real-space bosons.

INTRODUCTION

The origin of the pairing interaction that produces the superconducting spin-zero state remains a challenging and crucial problem for understanding high-T_c superconductivity.[1,2]

In contrast to BCS-type superconductors which become normal metals above the superconducting transition temperature T_c, the copper-oxide-based high-T_c ceramic superconductors remain highly anomalous even in their normal state. This suggests that the pairing mechanism that yields high-T_c superconductivity manifests itself already above T_c. One problem with a number of proposed pairing mechanisms is that they commence at the superconducting transition temperature; another problem is that they fail to associate the pairing mechanism with the specific chemical structural class of the known high-T_c materials. This latter point was emphasized by Jorgensen's statement:[3] "Questions concerning direct relationships between structure and superconductivity - for example, the possibility of structure itself playing a role in the superconducting mechanism - remain to be answered." We address these questions in this paper.

THE PAIRING MODEL

In the following, we propose a pairing mechanism which may be a crucial component for the occurrence of high-T_c superconductivity. It is intimately related to the stereochemical nature of the copper-oxide planes that are vital to high-T_c superconductivity in the copper-oxide-based ceramics.

The charge carriers in these materials are known to be holes that reside in the copper-oxide planes. These holes are charge compensated in the material between the copper-oxide planes either by chemical substitution or by oxygen vacancies. The presence of holes in the copper-oxide planes converts Cu^{II} atoms into Cu^{III} atoms producing a mixed valence state.[4] Experiments have confirmed that the hole is shared by the d-orbitals of the Cu^{III} and the lone pair p-orbital of the neighboring oxygen atom by hybridization. The configurational and bonding environment of an atom depends generally on its charge state (valency) and is altered with a change in charge which, in effect, is a change in valency. The degree of change is largest for covalent bonds and diminishes as the bonds become more ionic. We believe that the lone pair bonding configuration is pertinent in this regard since it is not as strong as a covalent structural bond and has greater flexibility.[5]

Despite the strong ionic contribution to the O-Cu-O bonds, the bond angles in the copper-oxide planes of the YBCO ceramic high-T_c superconductors, for example, inherently change with the charge state: Cu^{II} configurations in the copper-oxide planes result in puckered O-Cu-O bond angles of 165° while Cu^{III} configurations result in flat O-Cu-O bond angles of 180°.[4] Since these bond angles represent the lowest energy configurations, we suppose that the Cu^{III} environment of a hole in a plane of mixed valency Cu^{II}/Cu^{III} will relax toward a flat 180° bond angle configuration. When two holes are in close proximity in the copper-oxide plane, i.e. two neighboring Cu^{III} atoms, we propose[6] that the energy gain of the system from the configurational relaxation of the neighboring pair will exceed the energy gain of two isolated Cu^{III} atoms. The difference between the energy gained from the bond relaxation of an intimate pair of Cu^{III}

atoms and the energy gained from the relaxation of two isolated Cu^{III} atoms constitutes a pair bonding energy which we estimate to be of the order of 10-30 meV.[7]

This bound pair forms a spin-zero boson by preserving the anti-alignment of the surrounding Cu^{III} spins that originates from the superexchange coupling in the copper-oxide planes. The lone pair configuration may also be relevant here.[5]

Since the carrier pairing energy is comparable to or larger than T_c, one expects that these bound pairs exist above T_c in the normal state of these materials. A crucial question that needs exploring is whether these real-space bosons are sufficiently mobile to initiate a Bose condensation to a superconducting quantum state.

COHERENT TUNNELING OF REAL-SPACE BOSONS

For understanding the mobility of the real-space bosons formed by the configurational relaxation mechanism just described, it is important to realize that they are distinctly different from bipolarons that are bound together by large phonon displacements. One usually distinguishes small and large bipolarons in which two carriers are self-trapped in a common potential well.[8] Small bipolarons localize two charges in a region of a few interatomic distances. Their large strain and binding energies of a few eV make them essentially immobile. Large bipolarons extend over considerably larger distances such that the relevant coulomb energies are greatly reduced by dielectric screening. This enhances their mobility even though the charge motion must involve the displacements of a large number of surrounding atoms.[8]

In contrast, the real-space bosons resulting from our pairing mechanism are small, extending only a few interatomic distances, and of low energy. This makes them distinctly different from bipolarons either large or small. They are not bipolarons but a new class of bosons whose existence depends upon reversible changes of chemical structure. They therefore have implications for the development of interesting new materials and devices.

In order to yield Bose condensation and the superconducting quantum state, our real-space bosons must coherently tunnel. This tunneling probability is governed exponentially by the ratio of the configurational relaxation energy and a typical phonon energy. The small size as well as the low energy of our real-space bosons exponentially favor their mobility which, in turn, allows Bose condensation at T_c.

We conclude with a few remarks about the role of our real-space bosons in the normal state properties above T_c.

Even though the observed temperature dependence of the resistivity and other properties can be explained with real-space bosons,[4,5,9] there is a variety of normal state properties that suggest the presence of fermions.[10] This fact may not constitute a contradiction[5] as fermions and bosons can coexist and dynamically interchange in equilibrium, yet, the superconducting quantum state is governed by spin-zero bosons.

SUMMARY

We pointed out that configurational relaxation associated with the mixed valence state provides a pairing mechanism in copper-oxide-based high-T_c superconductors. The resulting real-space bosons are therefore unique in that they are small and of low energy and, hence, have sufficient coherent tunneling mobility to promote the superconducting quantum state.

ACKNOWLEDGEMENTS

I wish to express my appreciation and thanks to Hellmut Fritzsche for his contributions and to Stephen Hudgens for his helpful discussions.

REFERENCES

1. Schrieffer, J. R., Science 251, 1005 (1991).

2. Anderson, P. W., Abrahams, E., and Laughlin, R., Science 251, 1005 (1991).

3. Jorgensen, J. D., "Defects and Superconductivity in the Copper Oxides," Physics Today 44, 40 (1991).

4. Ovshinsky, S. R., Hudgens, S.J., Lintvedt, R.L., and Rorabacher, D.B., "A Structural Chemical Model for High T_c Ceramic Superconductors," Mod. Phys. Lett. B 1, 275 (1987).

5. Ovshinsky, S.R., "The Chemical Basis of High Temperature Superconductivity," Revue Roumaine de Physique, 36, 761 (1991).

6. Ovshinsky, S. R., "The Origin of Pairing in High-T_c Superconductors", Chem. Phys. Letters 195, 455 (1992).

7. Fritzsche, H., Personal communication.

8. Emin, D., "Large Polarons and Superconductivity," Lattice Effects in High Temperature Superconductors, Y. Bar-Yam, editor (World Scientific, New York, 1992).

9. Mott, N. F., "Real-Space Bosons in the Bismuth Oxide and Copper Oxide Superconductors," Superconductor Science and Technology 4, S 59 (1991).

10. Levin, K., Kim, J.H., Lu, J.P., and Si, Q., "Normal State Properties in the Cuprates and their Fermi Liquid Based Interpretation", Physica C 175, 449 (1991).

Superconductivity Publications

Correlation Between the Superconducting and Normal State Properties of Amorphous Molybdenum – Silicon Alloys (with A.S. Edelstein, H. Sadate-Akhavi and J. Wood), Solid State Comm. 41 (1982) 139.

Asymmetric Flux-Flow Behavior in Superconducting Multi-layered Composites (with A.M. Kadin, R.W. Burkhardt, J.T. Chen and J.E. Keem), Proc. 17th Intl. Conf. on Low Temperature Physics, (Elsevier Science Publishers, 1984).

Superconducting Properties of Amorphous Multilayer Metal-Semiconductor Composites (with A.M. Kadin, R.W. Burkhardt, J.T. Chen and J.E. Keem), in "Layered Structures Epitaxy and Interfaces," edited by J. M. Gibon and L. R. Dawson; Mat. Res. Soc. Symp. Proc. 37 (1985) 530.

Superconducting Properties of Sputtered Mo-C Films and Columnar Microstructure (with J. Wood, J.E. Keem, J.T. Chen, A.M. Kadin and R.W. Burkhardt), IEEE Trans. on Magnetics MAG-21 (1985) 842.

Superconductivity at 155K (with R.T. Young, D.D. Allred, G. DeMaggio and G.A. Van der Leeden), Phys. Rev. Lett. 58 (1987) 2579.

A Structural Chemical Model for High Tc Ceramic Superconductors (with S.J. Hudgens, R.L. Lintvedt and D.B. Borabacher) Modern Physics Lett. B1 (1987) 275.

A Simplified Summary of the ECD Model Explaining the Mechanism of High Temperature Superconductivity in "Topics in Non-Crystalline Semiconductors – In Memory of David Adler 1937 – 1987," edited by Hellmut Fritzsche and Ai-Lien Jung, Beijing University of Aeronautics and Astronautics, (1987), p.186.

Superconductivity in Fluorinated Copper Oxide Ceramics (with R.T. Young, B.S. Chao, G. Fournier and D.A. Pawlik), Reviews of Solid State Science 1 (1987) 209.

Superconductivity in the Fluorinated YBaCuO (with R.T. Young, B.S. Chao, G. Fournier and D.A. Pawlik), Mat. Res. Soc. Proc. (1987).

This Week's Citation Classic [S.R. Ovshinsky, R.T. Young, D.D. Allred, G. DeMaggio and G.A. Van der Leeden, Superconductivity at 155K, Phys. Rev. Lett. 58, 2579 (1987)], Current Contents 30 (1990) 20.

Unusual Fluorination Effects of Superconducting Films (with R.T. Young), SPIE Proc.1324 (1990) 32.

An Approach to the Puzzle of High Temperature Superconductivity – A Letter to David Adler, Epilogue to "Disordered Materials: Science and Technology – Selected Papers by Stanford R. Ovshinsky," 2nd Edition, edited by David Adler, Brian B. Schwartz and Marvin Silver (Plenum Press, New York, 1991) p.375.

The Chemical Basis of High Temperature Superconductivity: Structure and Function, Revue Roumaine De Physique 36 (1991) 761.

High Quality Epitaxial YBCO (F) Films Directly Deposited on Sapphire (with R. Young, K. Young and M. Muller), Physica C 200 (1992) 437.

The Origin of Pairing in High-Tc Superconductors, Chem. Phys. Lett. 195 (1992) 455.

A Mechanism for High Temperature Superconductivity, Applied Superconductivity 1 (1993) 263.

US patents - superconductivity

Superconducting films and devices exhibiting ac to dc conversion
4608296 08/26/1986

Parametrically modified superconductor material
5004725 04/02/1991

Method of controllably introducing a parametric modifier into a ceramic oxide which includes at least one superconducting phase
5102860 04/07/1992

Aligned superconducting film and epitaxial-like method of growing same
5124310 06/23/1992

Superconducting structure and method of fabricating same
5198414 03/30/1993

Parametrically modified superconductor material
5227362 07/13/1993

Aligned superconducting film and epitaxial-like method of growing same
5426092 06/20/1995

Method of aligning grains of a multi-grained superconducting material
5520953 05/28/1996

Chapter VIII: Other Topics of Interest

After reading, absorbing and marking books with colorful stickers, Ovshinsky often bought extra copies of the best to send to his friends. A brief sample:

The 3 volume set “Twentieth Century Physics” (AIP Press, 1995)
“The New Physics”, Ed. P. Davies (Cambridge Univ. Press, 1989)
“Niels Bohr’s Times in Physics, Philosophy and Politics” by Abraham Pais (Oxford, 1991)
“My Universe” by Ya. B. Zeldovich (Harwood Academic Publ., 1992)
“Genesis of the Big Bang” by R.A. Alpher and R. Herman (Oxford 2001)
“Rabi, Scientist and Citizen” by J.C. Rigden (Basic Books 1987)
“Quarks, the Stuff of Matter” by Harald Fritzsch (Basic Books 1983)
“Dreams of a Final Theory” by Steve Weinberg (Pantheon Books 1992)
“Big Bang, the Origin of the Universe” by Simon Singh (Harper Collins Publ. 2004)
“Facing Up, Science and its Cultural Adversaries” by Steven Weinberg (Harvard U. Press 2001)
“Warped Passages, Unraveling the Mysteries of the Universe’s Hidden Dimensions” by Lisa Randall (Harper Collins Pub. 2005).

Coming home after a long day of meetings dealing with technical problems, new experiments, patent and financial issues and negotiations with business partners, Ovshinsky liked to relax with Iris and friends discussing or reading about fundamental questions of physics, cosmology, neuroscience, history or problems facing our world and humanity. Talks and lectures for numerous international congresses and scientific meetings had to be prepared. Ovshinsky rarely turned down an opportunity to respond to the curiosity of young people. The memories of these encounters live on and have made their mark. The formal papers in this book focused on science and technology fail to capture the full range of Ovshinsky’s curiosity or the depth of his personality. His poetry opens but a small window to him as a human being, his warm heart, filled with tenderness and love of humanity. Looking at gorgeous fall colors he wrote:

While the world is in horror and dismay
Nature is neutral—Beauty is on display

It is more than a wish, it is more than hope
It is the strength of our desire and will
To change the world—the inhumanity

So that like the leaves, the earth
Will represent all shades of color and subtleties
In a harmonious palette
With which we can paint our goal—
A future of peace and harmony that must be.

Not in despair but in dedication,
Humanity shall overcome.

What follows are reprints of a few publications that evolved from the relaxing evening hours.

Comment on "Vacuum catastrophe: An elementary exposition of the cosmological constant problem," by Ronald J. Adler, Brendan Casey, and Ovid C. Jacob [Am. J. Phys. 63 (7), 620–626 (1995)]

Stanford R. Ovshinsky
Energy Conversion Devices, Inc., 1675 West Maple Road, Troy, Michigan 48084

Hellmut Fritzsche
Department of Physics, University of Chicago, Chicago, Illinois 60637

(Received 9 February 1997; accepted 14 March 1997)

The article by Ronald J. Adler *et al.*[1] is a pedagogically beautiful exposition of the vacuum catastrophe caused by the huge density of the zero-point energies of the electromagnetic and other field quanta. As much as we enjoyed that this article explained in fascinating clarity the fundamental discrepancy between the present theory of gravity and quantum field theory, we believe that one of their arguments is erroneous and needs correction.

The authors state correctly that a high energy density does not cause a problem for energy considerations because only energy differences matter in nature so that the zero of energy can be set arbitrarily. Of concern, however, is the gravitational force of such energy density, which cannot be dismissed. The authors proceed to calculate the gravitational force of a homogeneous and uniform vacuum mass density ρ, caused by the energy density of the zero-point energies, on the planets in our solar system. This force [their Eq. (25)] would greatly disturb the planetary motion. Since it does not do so with a high degree of accuracy, the authors conclude that ρ must be infinitesimally smaller than the expected value.

We suggest that this argument ignores a crucial piece of physics. The gravitational force of a uniform energy density must be zero at any point of a homogeneous and isotropic universe because there is no preferred direction in which it can point. The fact that Newton's gravitational theory yields contradictory and nonunique solutions for an infinite universe of uniform mass density was noted quite early by Seeliger[2] and Einstein.[3] More recently, this issue has been discussed both historically and from a modern point of view by Lemons.[4] Newton's symmetry principle obviously requires the gravitational force to be zero at any point. This holds true also for the enormous energy density of the vacuum fluctuations of all quantum particles which have finite rest mass m_0 provided the vacuum is homogeneous and isotropic. With vacuum fluctuations we mean the brief particle–antiparticle pair creation events allowed by Heisenberg's uncertainty principle. During the short time $\tau \approx \hbar/E$ these vacuum fluctuations act gravitationally like a mass $m = E/c^2$. A homogeneous, uniform, and isotropic vacuum is gravitationally unobservable, and there seems to be no contradiction between the presence of the vacuum fluctuations or zero-point energies and the observed planetary motions.

There are, on the other hand, observations of the motion of stars in galaxies, which cannot be explained by their visible mass. One hypothesizes the presence of a galaxial halo of dark matter which provides the missing gravitational effects. We alternatively suggest in the following that the real vacuum is not uniform and homogeneous and that the deviations from homogeneity act gravitationally. The presence of matter such as stars, galaxies, and black holes causes curvatures of spacetime in their vicinity. We suggest that the rate of the vacuum fluctuations of the elementary particles depends on the spacetime curvature. We specifically suggest that the rate is increased with increasing curvature. This increase in fluctuation rate need, of course, not be the same for particles having different rest mass m_0. Such increase in fluctuation rates of different particles with spacetime curvature introduces the nonuniformity of the vacuum that acts gravitationally. Moreover, the time dilation associated with spacetime curvature affects both the rate and the uncertainty time τ of the fluctuations. These effects link quantum mechanics with general relativity theory.

The gravitational effects of the zero-point energies also become noticeable only at inhomogeneities of the fields produced, for example, by the boundaries of conductors which give rise to the Casimir force between closely spaced conducting plates, as explained by Adler *et al.*[1]

If our suggestion of the relation between the vacuum fluctuation rates and duration and the spacetime curvature proves to be correct, it will influence estimates of the matter in the universe which is believed to significantly exceed that of total baryon matter.

[1] Ronald J. Adler, Brendan Casey, and Ovid C. Jacobs, "Vacuum Catastrophe: An elementary exposition of the cosmological constant problem," Am. J. Phys. **63**, 620–626 (1995).
[2] H. Seeliger, "Über das Newton'sche Gravitationsgesetz," Astron. Nachr. **137**, 129–135 (1895).
[3] A. Einstein, "Kosmologische Betrachtungen zur allgemeinen Relativitätstheorie," Sitzungsber. Preuss. Akad. Wiss. **32**, 142–152 (1917).
[4] Don S. Lemons, "A Newtonian cosmology Newton would understand," Am. J. Phys. **56**, 502–504 (1988).

Mott's Room

Stan Ovshinsky

Iris and I met and spent quite a bit of time with Nevill in San Francisco at the 1967 meeting of the American Physical Society. I recall an earlier brief introduction at a Gordon Conference I attended with Hellmut Fritzsche.

Several months earlier, in Leningrad, I had been invited to give a talk at a small gathering of primarily Russian and East European researchers, no more than thirty people. I was due to give one of the first scientific disclosures of my switching work. Kolomiets in his summary of the meeting drew an almost flat line on the board to show the progress in amorphous materials until this meeting and then said, 'With the work of Ovshinsky the field will now go like this,' and drew an almost vertical line.

That acceptance in Leningrad encouraged me to discuss our work with Nevill, Adler* and several others in San Francisco, and a group of us had a very nice dinner at a French restaurant where Hellmut and I carried on an intensive conversation with Nevill about the Ovonic memory, the Ovonic threshold switch and amorphous materials in general. He became very interested, albeit a bit skeptical that there could be such unusual effects. We therefore invited him to come and see for himself and arranged for him to visit our laboratory on his way back to England.

When he saw the devices actually working, he said that this work was very important, became quite enthusiastic and asked what he could do to help. This started a very warm and constructive working and personal relationship that continued until his death. He often said that our meeting led to his serious attention and contributions to the field of amorphous and disordered materials. He strongly supported my 1968 publication on switching in *Physical Review Letters* when it came under attack. He focused his powerful creative mind on the problems and possibilities of the amorphous and disordered field, made many important contributions, and working together with Ted Davis and others, he became its acknowledged leader.

We missed him at the international meeting on amorphous materials that fall in Bucharest; we were told he was unable to come as his father was very ill. He spoke about our work at various meetings and invited us to discuss it

* I had arranged to meet Dave Adler at the meeting because of my work in the 1950s with amorphous mixed-valence transition metal oxides which resulted in switching and memory devices. In 1960 I developed amorphous chalcogenides to demonstrate unique switching and memory devices, and they are the ones that we have commercialized.

and demonstrate our devices at the Cavendish Laboratory at the next international amorphous meeting that was held in Cambridge in 1969. This was a memorable occasion for two reasons. It was a wonderful experience to show our work at the Cavendish Laboratory. Our friendship became much deeper as he shared with me his personal feelings and the scientific and political frustrations that he had. His advice and encouragement were heartwarming. This closeness continued until the very end.

One of our great disappointments is that we were not able to set up a plant near the Cavendish, as Nevill had suggested, so that we could work more closely together. He introduced me to several university officials and we discussed plans of how an industrial park could be set up. One did come into being, but, unfortunately, we were not able to participate due our not having the financial wherewithall at the time. One of Mott's characteristics I admired was that he considered the technology that ensued from scientific discovery and activity should be put to everyday use to solve important problems in society. He wanted assurance that photovoltaics could be a realistic approach in Great Britain and was very interested in science policy.

The mechanism for the Ovonic threshold switch always intrigued him. Indeed, he was still discussing this with me in his last letter, shortly before he died. He had always believed it to be electronic in nature. He wrote several papers on it and was very pleased that experiments by various groups had proven this. The mechanism that I suggested was based on the fact that the chalcogenides, as pointed out by Kastner, were lone-pair materials. I suggested that the lone pairs were the basis for the unique electronic activity since they are a source of nonbonded electrons and, as such, they can respond reversibly to a suitable electric field without affecting the bonding electrons, which are much deeper in energy and responsible for the structural integrity of the material. The excitation process permitted injection from the electrodes and produced exceedingly fast switching. As long as the highly dense conducting plasma of carriers was sustained, the switch would remain in the On condition. Recombination of the carriers took place when the holding voltage was reduced. The Ovonic memory switches were designed through their bonding energies to permit a reversible phase change in response to the switching mechanism I have described. Nevill was delighted that the optical version of the phase-change switch (PD) had gained widespread commercial application.

We had ongoing scientific discussions about superconductivity. He had developed a polaronic model and I argued for a new class of bosons whose existence arose from changes of chemical structure in response to changes in local charge. It was very gratifying to see his great mind always at work with the deepest problems. He was particularly interested that we could apply the principles of disorder across a wide spectrum of usage: batteries for electric cars, photovoltaics, information systems, etc.

Whenever he came to the United States, which in the earlier days was once or twice a year, he would visit us and stay at our home in what we affectionately called Mott's room. When he stayed with us at the Ann Arbor meeting in 1971, he made breakfast for me, since I had never learned to cook and Iris suddenly had to go to the hospital.

When he visited our working group – including Hellmut Fritzsche, Dave Adler, Artie Bienenstock, Marc Kastner, Heinz Henisch and Morrell Cohen – all of us would meet and there would be much discussion and debate. On our many trips to Europe, we often visited him in Cambridge for a few days.

One of our most memorable scientific meetings was in 1976, where the Kastner–Adler–Fritzsche (KAF) model had its inception.

He came for the inauguration of our Institute for Amorphous Studies, as a member of the international advisory committee, and he gave one of the early lectures there. That year was a good one for octogenarians at our Institute, three lecturers in a row: Mott, Linus Pauling and Doc Edgerton of MIT! On behalf of the Institute, Hellmut Fritzsche, David Adler and I presented him in Cambridge with a three-volume festschrift for his eightieth birthday.

He was greatly interested in our work in photovoltaics and urged me to submit a paper to *Nature*. Published in 1978, it showed how the physical properties and the density of states in amorphous silicon alloys could be minimized and controlled.

He kept up a continuous correspondence with me to the end. His handwritten letters came in envelopes similar to wartime V-mail and besides his personal comments about what was going on in the various fields of science, they always concentrated on the scientific problems at hand, which later

A meeting held at Energy Conversion Devices: Stan Ovshinsky is seated on Nevill's right.

David Adler, Hellmut Fritzsche, Stan Ovshinsky, Nevill Mott and Iris Ovshinsky.

included high temperature superconductivity. In his letter writing, he was of the old English school, always informative and stimulating, and his mind never flagged at attacking the most difficult problems. The richness and depth of his scientific thinking was always apparent.

We celebrated many of his birthdays with him and Lady Mott. Iris and I were deeply appreciative of what he said at our last meeting, his ninetieth birthday celebration, where he acknowledged with great generosity our influence on him in the scientific sphere. We have lost a wonderful friend who was a model of how one could age and yet be ageless in one's interests and important contributions. He was truly a great man.

***MR S. R. OVSHINSKY** is President of Energy Conversion Devices (ECD), a company he founded to develop and exploit the unique properties of amorphous and disordered materials. His wife, Iris, works with him and shares his scientific interests. Together they met Nevill on numerous occasions in Cambridge, at ECD and at many locations (often conferences venues) around the world, maintaining a close friendship for thirty years.*

CREATIVITY AND INTUITION
A Physicist Looks at East and West

By Hideki Yukawa
Translated by John Bester

206 pp., published by Kodansha International Ltd.,
Tokyo, New York, and San Francisco;
distributed by Harper and Row, New York, 1973

Reviewed by Sanford R. Ovshinsky

Hideki Yukawa, the Japanese theoretical physicist and Nobelist, has written a provocative book of essays including personal memoirs and thoughts on peace. He unselfconsciously discusses his inspiration for doing science with all of the beauty and contradictions inherent in the subject. Straight away let us understand that Yukawa states that he achieved his inspiration and insight from reading the early Chinese philosophers from Confucius to Lao Tzu and Chuangtse. It is "well known" that these philosophers are in fact anti-scientific. He speaks of romance, imagination, the world of fancy, the value of defeat, and the utilization of fables. Do these terms apply to science? That they do is an experimental fact as witness Yukawa's ability to utilize them in conceiving his pioneering work in elementary particles--his prediction of the meson. All new ideas must be proven and Yukawa had to wait 12 years for final confirmation. To understand the value of the process of elucidating a new idea, in Yukawa's case, several other particles were discovered in the investigation of his concept. It is in this rigorous process that many scientists can participate and which greatly adds to existing knowledge.

The precepts implicit in Yukawa's thinking show intuition to be dependent upon the ability to draw analogies. How or whence one draws them depends upon the individual. Yukawa can use the famous Japanese novel, "The Tale of the Genji," and the parables of Taoism of Chuangtse because they inspire his imagination. Great philosophies are vague enough so that the individual brings to them his own interpretations. Chinese philosophers see Taoism as mysticism, but Needham, the famous historian, can read into it exemplary early class-consciousness with proto-anarchistic aims. Such ambiguities can be understood when one considers that Darwin's inspiration was Malthus.

This is not another book about two cultures. It is about two cultures in science, and inferentially in society, about people who have "an excessive conservatism in the realm of ideas" as against those who have "a spirit of adventure."

Intuition has been a little-understood but powerful means of advancing science in quantum steps, yet somehow not considered legitimate. There are a few honorable exceptions to this thinking. Holton, explaining Einstein, emphasizes that major advances in science owe much to intuition. Szent-Gyorgyi's, Watson's, Medawar's, Tribus's and Ermenc's ideas can be at least indicated by Szent-Gyorgyi's statement, "A discovery must be, by definition, at variance with existing knowledge." Yukawa would heartily agree. This is confusing to many scientists since they consider that science <u>is</u> existing knowledge. Imagine the confusion to the rest of the world. Most scientists are uncomfortable with the use of such terms as creativity and intuition, and if they practice them, it is done in Marrano-like secrecy.

Is there then some utopia where creativity can find a home? Nowhere, for its home is in the mind of the individual. It is the "seed beneath the snow," most often beneath the concrete. Newton could express it, isolated on a farm by bubonic plague; Einstein, unknown and alienated in a Swiss patent office, without friends or scientific dialogue, could articulate it. Yukawa writes of his own early isolation. Each was alone, not part of the mass. Each was self-motivated and persisted in the face of apparent contradictions in his work.

At most a few people can be convinced of the soundness of a new idea. Japan, which has also produced Yukawa, has become sensitive to its lack of innovators. The answer is its dependence on authority in all strata of society and its reliance on consensus which, although an effective means of mobilizing mass effort, puts peer pressure on individual thought.

Under the best of conditions, it will always be difficult for new ideas to be introduced, for skepticism is a necessary and first reaction to them. History is replete with great minds, such as Einstein's, resisting the intuition of others, e.g., his refusal to accept quantum theory.

The United States in the past, because of its multiple and conflicting groups in government, industry and the university, allowed some options so that in the search for institutional support for new concepts, there was a possibility for innovation. This was the crack in the concrete, but is rapidly disappearing. American society is becoming more corporate and conformist, resembling the homogeneous and monolithic societies that do not have readily available institutional means for the expression of innovation. While Japan has, through its culture, achieved psychological conformity, China and the USSR are attempting to create such a consensus but must enforce it. Channeling free ideas into science and technology without overlap into the more sensitive areas of art, culture, and politics is virtually impossible.

Mankind normally shares its inhumanity. A humanist such as Yukawa transcends the specialities and national barriers which he deplores. The universal intuitive process, whether expressed in art, literature, music or science, is the ability to "see" connections between facts or concepts which to others are unrelated. Creativity links insight in such a way that a meaningful pattern leaps out of interlocking steps and becomes a bridge or pathway.

Many people who are merely imaginative are not insightful or creative, for the path has to lead somewhere and the bridge must be a means of fording a stream. Intuition, the basis of science, is therefore not an exotic tool but the most utilitarian of arts and its practitioners are the craftsmen of imagination.

Stanford R. Ovshinsky

A Nation of Fliers: German Aviation and the Popular Imagination

by Peter Fritzsche, Harvard University Press, 1992. 282 pp. $27.95

reviewed by Stanford R. Ovshinsky, ***DISSENT***, Summer 1993

Manipulation of Minds

A Nation of Fliers examines the rise of aviation in Germany, from its beginnings through the eve of World War II, to illuminate the relationship between technology and culture during the Nazi period. It is not only an interesting book but a profound one.

We take it for granted that technology is somehow the base of modern society, yet insufficient attention to how technology shapes our society and culture. How, for example, have technological advances shaped popular attitudes toward war?

Radicals have never felt quite comfortable going beyond their tradition's standard explanations as to the causes of war. But the accepted wisdom – that capitalism and imperialism inherently lead to war – sheds little light on how dominant groups foster ideological hatred within their societies.

Peter Fritzsche, a historian at the University of Illinois, presents a case study in the manipulation of popular culture. The Nazis prepared a nation for war in part by uniting the classes around symbols of aggression. The soaring, magical field of aviation offered images to which everyone in Germany could respond – socialists as well as Nazis. But it was the Nazis who seized the moment to forever change the character of mass destruction.

Tennyson in 1842 presciently posed the dilemma between the possibilities of science and technology and their potential for destruction:

Saw the Heavens fill with commerce,
Argosies of magic sails
Pilots of the purple twilight, dropping
Down with costly bales;
Heard the heavens fill with shouting,
And there reigned a ghastly dew
From the nation's airy navies grappling
In the central blue.

Aviation, of course, was only one of the areas of popular culture on which the Nazis put their stamp. But in Fritzsche's hands it becomes a sort of magnifying glass on the Nazi manipulation of the German mind.

Fritzsche shows how the idealistic and creative impulses of the German youth, intertwined with nationalist sentiment from the beginning, were twisted further in that direction by the Nazis. It is well known that German youth across the spectrum embraced mystical and romantic beliefs in the period between the wars. One might have thought, however, that the movement that embraced technology would have had a primarily leftist base – after all, the idealization of the industrial worker, the smokestack, and the tractor is considered to be a hallmark of the Old Left. What is often forgotten is that the Italian futurists, with their mechanical orientation, were also predecessors of the Fascist movement, and that Mussolini preached a corrupted version of syndicalism in order to appeal to the Italian working class.

The appropriation of idealistic and noble aspirations by Fascism need not come as a surprise. After all, the cross had been turned into a bloody hatchet throughout history in religious wars.

Each Fascist country, whether black, red, or brown, used aviation as a symbol of its new world. The Nazis were the most accomplished in this, because they tapped into the deepest spring, and because their appeal reached a wide constituency across class lines. The beginnings of aviation in this century brought not only the airplane and the glider but the distinctly German giant dirigible, which captured the imagination of Germany. Having looked up as a child in Akron, Ohio and seen giant dirigibles such as the Akron and the Macon floating in the sky, it is not difficult for me to understand how such airships could capture the public imagination.

In the 1920s, in Europe as in America, pulp magazines were filled with the derring-do of the World War I aces, those cavaliers of the sky who, with daring, skill, and chivalry, shot down their enemies. This mythology – a modern version of the hand-to-hand combat of the old warrior class – provided a new romantic hero who could transcend the agonies of trench warfare. This new icon was nowhere more celebrated than in Germany, where a monster like Goering, who was a popular aviation war hero, could become one of the leading figures of Nazism.

One could not romanticize a tank, but the soaring gliders and airplanes in the heavens could engage a spectrum of feelings – foremost among the, nationalistic pride – and serve as a most effective form of weaponry at the same time.

Many idealistic, hobby-driven young men of post-World War I Germany, cut off from the use of airplanes by Allied restrictions, turned to the art of gliding to keep aviation alive in that battered country. In its early days, gliding combined individualistic strivings with the camaraderie of the youth nature movement then popular in Germany; at the same time its appeal was strongly nationalistic in that it was a symbol of how, despite great difficulties, Germany would rise again to its rightful place in the world. Its appeal was not limited to the right: socialist working-class youth also played an important role. There were even Jews and women prominent in the gliding movement. But, as Fritzsche shows, this pastime – which drew on individualism, personal initiative, and small-group cooperation – became the basis of the new Luftwaffe when the Nazis, upon their seizure of power, channeled these initial strivings into rigid, centralized, totalitarian organizations.

Fritzsche's discussion of the glider movement does much to explain how Nazism captured the minds of young Germans. Conventional historical emphasis has dealt extensively with the effects of the economic crisis of the Weimar Republic and the impact of totalitarian philosophy on working- and middle-class Germans. But, for the Nazis to succeed, they had to go beyond the appeal of blood and soil, the demonization of the Jews, and the call to avenge their defeat in

World War I. They did this in part by showing that technology could help Germany reassert its leadership in the world.

The Nazis also pointed to the new technology to evoke fears among the population – fears that technology could then allay. Through skillful propaganda, the Nazis persuaded the Germans that they were vulnerable to air attack from all sides – including, though it is hard to believe, Czechoslovakia. The horror of air warfare was hammered into the German psyche through mass drills and propaganda campaigns until it became a self-fulfilling prophecy when the Nazis began to "preempt" these dangers by destroying any cities within their reach.

In several ways the Nazis' use of aviation paralleled their use of radio. Could Hitler have succeeded without his mastery over the radio waves? Just as Germans could hear the message of their superiority broadcast constantly over the radio, so too could they look to the heavens for signs of German greatness and the reaffirmation of German power.

By tracing the history of German aviation from its innocent beginnings in the work of brave and creative inventors to the death and destruction that rained from the skies, Fritzsche has given us an important, thought-provoking study.

book reviews

The road to decarbonized energy

Speeding towards a hydrogen economy — and the obstacles along the way.

Powering the Future: The Ballard Fuel Cell and the Race to Change the World
by Tom Koppel
Wiley: 1999. 276 pp. £16.99, $27.95

Prospects for Sustainable Energy: A Critical Assessment
by Edward S. Cassedy
Cambridge University Press: 2000. 284 pp. £45, $69.95

Stanford R. Ovshinsky

AP/FORD MOTOR COMPANY

Upping the HP and PR: Ford's vice-president Bill Powers extols the virtues of the fuel-cell-powered car.

Ever since humans first struck flint to ignite fire, energy and technology have been inextricably intertwined. The seeds of science and technology lay in intuitive insight and experiment: these led to new materials that linked energy servo-mechanistically to invention. Yet while materials (and the innovative products made from them) have changed since the Stone, Bronze and Iron ages, the hydrocarbon age is still with us.

The interaction between materials and the products they have spawned through the ages has been greatly affected by the energy source. The burning of coal made railways possible: steam became king and helped usher in the age of electricity. While historians could conclude that electronics transformed the world economy, one could also show that the petroleum-fuelled vehicle industry was the other fundamental base of progress.

Starting with the Industrial Revolution, a basic change has clearly been taking place: energy was being decarbonized from wood to coal to oil and gas, and so on. The curve of energy decarbonization directly overlays that of the growth of modern society. Extrapolating from that curve, it was clear that hydrogen, the ultimate energy source, would not begin its dominance for another 50 years. There are still reserves of oil to be used (although they are depletable), but, as has been pointed out, "the Stone Age did not end because of lack of stone".

Societal problems jarringly intervene in what appears to be orderly progress. We are familiar with the problems caused by fossil fuels: pollution, climate change, and a strategic dependence on oil that results in global conflict, inflation and the economic burdens of developing countries.

This litany would be an exercise in futility if there were no solutions. But there is a solution that can dramatically shorten the 50-year scenario. Based on hydrogen, it requires a complete systems approach to make it realistic and achievable in the short term. Short term in that we had already entered the hydrogen age some years ago when nickel metal hydride batteries became the enabling technology for electric and hybrid vehicles — electric vehicles can achieve over 200 miles on a single charge and hybrids get 80 miles to the gallon of petrol.

However, there is more than one road to Mecca, because hydrogen can be used as a transition fuel in internal-combustion engines that emit no climate-change gases and in fuel cells that convert chemical into electrical energy. Hydride batteries shuttle ions back and forth and are charged by electricity while fuel cells use up the hydrogen in generating electricity.

Fuel cells have been the centre of attention, but although these are necessary they are not sufficient in the energy equation. Not only must hydrogen be produced in various ways, but it must also be stored and transported safely, economically and practically, without the need for drastic changes in the energy infrastructure. The hydrogen economy must be systems-based with a simple universal means of seamlessly utilizing it. The demonstrably realistic approach that will open a new age is the use of thin-film, multijunction photovoltaics to obtain hydrogen by breaking up water.

Powering the Future and *Prospects for Sustainable Energy* address the subject from different perspectives. The first is an anecdotal journalistic account of how the proton-exchange membrane (PEM) fuel cell — invented by Sir William Grove in 1839 and developed by General Electric in the 1950s — was redeveloped by a small Canadian team of committed, creative individuals under the inspired leadership of Geoffrey Ballard. As described in the book, their success required excellent engineering and packaging rather than scientific breakthroughs. It is a gritty, fascinating story of how entrepreneurial and leadership skills combined with engineering and management talents made Ballard Power Systems the leader in fuel cells.

The book has clear limitations, since it accepts without question that the PEM fuel cell is fuel-neutral. It does not put the problems of storage into proper perspective, except to quote an executive: "But think about what would happen here if decent hydrogen storage systems are invented for the car". Equally important is that the fuel providers for the automotive industry are not likely to provide the methanol infrastructure so blithely discussed, since they can only afford to change their infrastructure once. It is clear to them, and to most of those knowledgeable in fuel cells, that pure hydrogen is required and that this will be very difficult to produce on board the vehicle. The use of gaseous and liquid hydrogen presents problems that limit these two forms to a transient place in the hydrogen economy.

A more feasible approach that is gaining support is the onboard solid storage of hydrogen as a light-weight hydride capable of providing a vehicular range of over 300 miles. This also offers a solution to the important problems of infrastructure, as hydrogen in hydride form is at its lowest free energy and can be transported by conventional means.

Technological progress is based on new materials. The use of hydrogen in a solid is

book reviews

made practical by atomic engineering, as various chemical, electronic and topological functions can be integrated in one material, through such means as catalysis, hydrogen-diffusion paths and acceptor sites.

Edward Cassedy's book is a well-meaning compendium dealing with "the prospects for technological change in the energy industry". It includes a great deal of interesting background information on various forms of energy conversion. Its views are flawed by not being up to date, but it does have value in providing historical reference.

The consequences of the hydrogen economy are of great import, requiring many new books on the subject. Social scientists will seek to understand the dynamics of petroleum companies becoming energy companies: the most progressive of these see themselves as agents of change addressing societal needs. The huge transportation industry will be irrevocably changed as it attempts to respond to societal pressures. New industries based on new science will be developed, and society will benefit. ■

Stanford R. Ovshinsky is at Energy Conversion Devices, 1675 West Maple Road, Troy, Michigan 48084, USA.

Technology's tortoise and hare

The sociological dynamics are now right for the electric car to eclipse its rival.

The Electric Vehicle and the Burden of History
by David A. Kirsch
Rutgers University Press: 2000. 256 pp. $20

Stanford Ovshinsky

History is not a linear progression of events. The history of science and technology is packed with unexpected basic advances that are not recognized until they have reached a critical mass and then surged forward to change the way we live, work and understand the world around us.

We live in an age when the foundations of the global economy — energy, transport and information systems — are being transformed. In the United States Jack Smith, chairman of General Motors, and Bill Ford, chairman of the Ford Motor Company, agree that no car manufacturer will go far in the new century with the internal-combustion engine (ICE). It is hardly necessary to survey the serious societal problems underlying these opinions. They are emphasized by the current global oil crisis, with its profound consequences for the world economy.

The transport system based on the ICE is undergoing a transition to the use of electric, hybrid and fuel-cell vehicles. The dependence of these vehicles on hydrogen — in metal-hydride batteries for electric and hybrid vehicles, or as the fuel for fuel cells — is initiating a new age: the age of the hydrogen economy.

How can we understand the emerging electric-vehicle industry? The simplest approach might appear to be to study the events of 100 years ago, when electric vehicles were overtaken by the incredible growth of the automotive industry, as the success of the ICE with an electric starter-motor made cars practical and affordable.

However, what if such a history turned out not to be truly relevant to what is unfolding now? David Kirsch's book is an informative history, full of excellent case studies of the various attempts to build a transport system based on battery-driven vehicles. Its strength lies in the fact that it takes a systems approach, combining social, environmental and business perspectives to provide a well-researched analysis of the early battle between the ICE and battery-powered vehicles. As history, it is an excellent, insightful book.

However, as the book reaches modern times, it lacks both technological depth and political understanding. Kirsch does not appear to be up to date with California's struggle to enforce the mandate set by its Air Resources Board that, by 2003, 10% of vehicles sold in the state should produce no emissions — despite heated opposition from various groups. The mandate will be enforced by the Air Resources Board from 2003. It is expected that car companies will comply rather than face possibly severe monetary penalties.

The problem now is not, as Kirsch believes, whether a breakthrough in battery technology has provided a dramatic increase in range, nor whether there is a market for electric vehicles, nor whether costs can compete with those of internal-combustion vehicles. The crucial hearings of the California Air Resources Board this September answered all of these in the affirmative so definitively as to result in a unanimous vote to continue the mandate.

Past is not prelude. Just as many generals fight the previous war, so it is a mistake to think that the present situation reflects the sociological dynamics of the turn of the last century. Heraclitus was right: it is not possible to step twice into the same river.

What is involved is the task of changing a huge, powerful, entrenched global industry, with enormous resources as well as financial and psychological investments, from a petrol-engine base to an electric one. It is a dynamic process of competing strategies and, for the automotive industry, a very traumatic event in a vital historical drama. The industry's attitude can be likened to that of St Augustine: "Make me chaste, Lord, but not yet."

A recent report from the US Department of Energy provides data showing that

A burden undeserved

Far from being the encumbrance and sign of retribution portrayed in Samuel Coleridge's *Rime of the Ancient Mariner*, an albatross is a welcome and impressive sight — and one that Coleridge himself never experienced. The so-called shy albatross shown above belongs to this diverse grouping of seabirds, which inhabits the Antarctic and Pacific oceans. *Albatrosses* by W. L. N. Tickell (Yale University Press, $60), from which the picture is taken, is a comprehensive account of the 13 groups of albatrosses, and compares their natural history, breeding, flight and ecology. Measures to reduce the numbers that in the past have been inadvertent victims of the fishing industry are also described. The book's final chapter looks at some poems in which these birds have figured; it is called "The *Mariner* syndrome".

electric vehicles propelled by nickel metal-hydride batteries based on new scientific principles can achieve a range of well over 200 miles, answering the question of viability of pure electric vehicles and hybrids. The electric vehicle's superiority in city driving was demonstrated in 1998 during a mileage comparison test between two Geo Metro vehicles, one converted to run on nickel metal-hydride batteries, the other a conventional petrol version. In heavy New York city driving conditions, the electric version demonstrated a projected range of 362 kilometres, as against 192 kilometres for its ICE counterpart.

The Institute of Electrical and Electronics Engineers monitored a 347-kilometre trip from Boston to New York city by a four-passenger Solectria using similar batteries. It completed the journey on a single charge, with 15% energy remaining — having used the energy equivalent of less than one gallon of petrol.

Impressive as these results are, they are not the whole story. For, as the recurring problems of availability and affordability of petrol have shown, a city must still function. Global pollution and climate change have made the electric vehicle a necessity as a delivery vehicle, taxi or bus. In addition, hybrids, which can do 100 kilometres on 3.5 litres or less (80 miles or more to the gallon), using a small petrol-based engine and similar nickel metal hydride batteries, are becoming part of the accepted vehicle mix so necessary in establishing an automotive industry. Even now, the ICE automobile needs greater battery power to fuel the increasing number of electrical components it contains.

The question of affordability has been answered by Robert Stempel, former head of General Motors, the world's largest corporation. Stempel, a pioneer whom many consider to be the best automotive engineer in the industry, is the father of the first commercial electric vehicle, EV1. He gave testimony to the California Air Resources Board that an electric vehicle made in the same numbers as conventional cars today would be cheaper to produce than its petrol-based counterpart. Until such volumes are achieved, the industry should use the time-honoured mechanism of 'forward pricing' — introducing a new product at an affordable price so that it can achieve sufficient volume to become profitable.

It is clear that events have overtaken academic history in this field. The new electric vehicle does not carry the burden of history. On the contrary, in its various manifestations it is spurring new scientific, technological and social inventions and is the vehicle for the kind of change that is desperately needed in the new millennium. ■

Stanford Ovshinsky is at Energy Conversion Devices, 1675 West Maple Road, Troy, Michigan 48084, USA.

Reviews of Solid State Science, Vol. 1, No. 2 (1987) 207–219

SUPERCONDUCTIVITY IN FLUORINATED COPPER OXIDE CERAMICS

S. R. Ovshinsky, R. T. Young, B. S. Chao,
G. Fournier, D. A. Pawlik

Energy Conversion Devices, Inc.
1675 West Maple Road
Troy, Michigan 48084

ABSTRACT

Our report of 155-168K zero resistance superconductivity in fluorinated copper oxide ceramics has recently been confirmed by several other groups. Fluorine substitution for some oxygen in the high T_c superconducting ceramic oxides is both fundamentally and technologically interesting. In addition to our zero resistance measurements, we have observed diamagnetic signals and flux trapping at temperatures as high as 305K.

The early samples were made by solid state reaction in pressed pellet form. The BaF_2 formation and the small volume fraction created difficulties in identifying the specific high T_c phases. In the course of elucidating and optimizing the fluorination effect, we carried out fluorination by other low temperature means. The observation of zero resistance in the same temperature range (154K) was repeated utilizing different preparation methods, e.g., a plasma process.

In addition to the observation of these T_c results, we showed that dopant amounts of fluorine promote oriented crystal growth, a much desired attribute. We also proved that the replacement of weakly bonded oxygen by fluorine can greatly enhance the material's stability by eliminating the serious problem of oxygen diffusion.

I. Background

It is difficult to do justice to the scientific and technological impact of the recent discoveries in high T_c superconducting ceramic oxides. It is now clear that the oxides that were the basis of the lanthanum copper oxide ceramics which moved the superconducting temperature from 23 to 30 to 40K [1] have been around for some time.[2,3]

Substitution of elements which alter the stereochemistry of these materials was shown by Wu, Chu and associates [4] to have an important effect on zero resistance superconductivity, moving the range to approximately 95K.

We previously reported that through the use of fluorine [5], we have been able to dramatically increase the zero resistance temperature by 60-70K--by far the highest increase--to 155-168K. (See Fig. 1.)

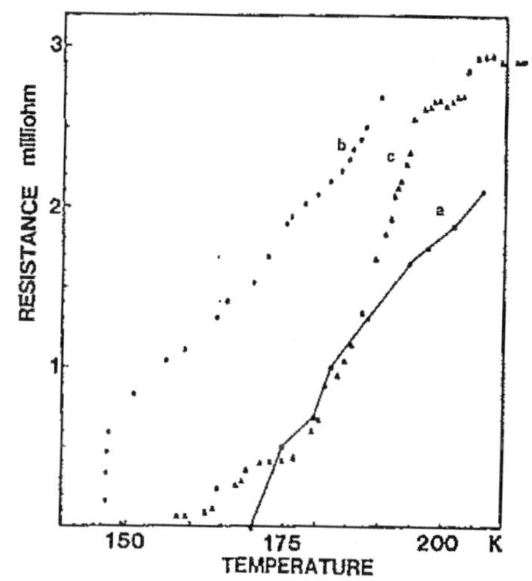

FIG. 1 Resistance vs temperature measured at constant current of 1 mA for a sample with nominal composition $YBa_2Cu_3F_2O_y$. Curve a shows the resistance upon initial cooling (line drawn to guide the eyes); curve b, data obtained upon second cooling; and curve c, data from warming after second cooling.

This result has been confirmed by several groups, for example, in China,[6] Taiwan [7] and Sweden.[8] The most recent confirmation was by a group at North American Philips in Physical Review Letters [9]. The Philips group carefully confirmed our fluorinated findings in various samples. They were able to increase T_c from 95K to a maximum of 110K by thermal cycle treatment of the conventional materials.

Only with the addition of fluorine were they able to go up to 159K. The paper also confirms our results that the increase in T_c occurs systematically and monotonically with increasing fluorine addition. The 159K temperature was achieved by exactly following the published ECD recipe with two fluorines substituted for oxygen. In fact, they duplicated the solid state reaction synthesis utilizing BaF_2 as reported by us.

We also have been able to show (Fig. 2) an abnormally rapid decline of resistivity starting at room temperature which indicates the existence of a phase exhibiting superconducting onset above room temperature. It is also important to point out that the resistivity of this sample above T_c is four times lower than that of copper.

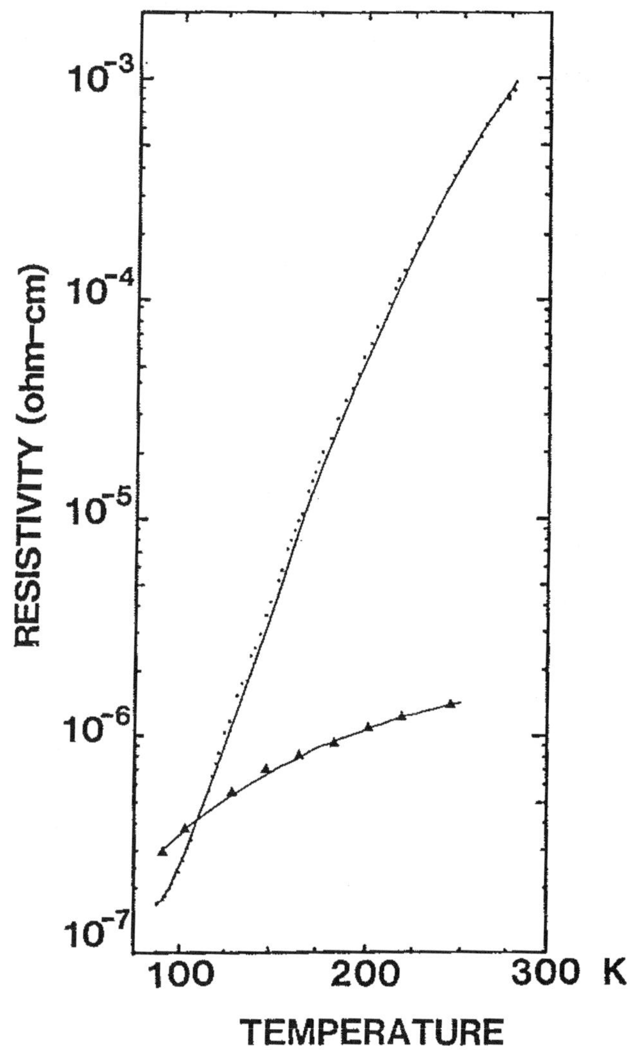

FIG. 2 The logarithm of the average of a YBCOF sample. Average resistivity was calculated on the assumption of uniform current density. Resistivity was found to follow T^n, where $n \approx 8.3$. The ideal resistivity of pure copper is also plotted (triangles).

Since there is a small volume fraction of such high T_c material allowing for a conducting percolation path, we have utilized magnetic measurements to probe the high temperature phases that are dispersed through the material and have observed the diamagnetic signal as well as flux trapping below 280K as shown in Fig. 3. Indeed, we have observed similar phenomena at temperatures as high as 305K.

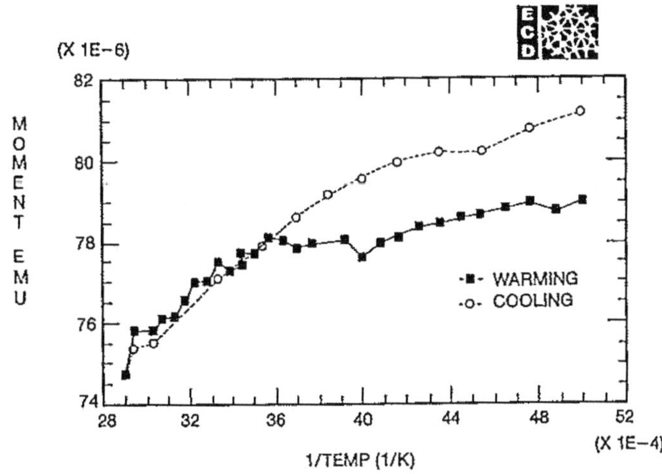

FIG. 3 Magnetic moment vs 1/T for a YBCOF sample. Measurements made with use of 40G field. Data for warming after zero-field cooling are indicated by "■" and data from cooling with field applied are indicated by "o".

The early samples were made in a typical solid state reaction in pressed pellet form. In order to get better control of the fluorination reaction we have since utilized the plasma technique described in the next section.

X-ray diffraction and electron microprobe analyses indicate that the samples prepared from solid state reaction are multiphasic, containing at least four phases. Three of them, i.e., $Y_2Ba_1Cu_1O_y$:F, BaF_2, and CuO, are nonsuperconducting. The fluorine content of the superconducting $Y_1Ba_2Cu_3O_{7-x}$ (YBCO) phase is very small (≈ 0.2 at.%).

The volume fraction of the high-temperature phase estimated from the strength of the diamagnetic signal was only about 1 part in 10^4. Because of the multiphases and because of such small volume fraction, we have not yet been able to identify the specific high T_c phases.

Various fluorides have been used in the starting reagent including BaF_2, CuF_2, NH_4HF_2. We found that regardless of what initial fluoride was used, the BaF_2 phase was always observed. Due to the strong affinity of barium to fluorine, barium will preferentially react with fluorine to form BaF_2 before the formation of YBCO. The amount of YBCO phase in the material is therefore critically dependent upon the amount of fluorine additive in the mixture. Since the majority of the fluorine reacted with the barium, the fluorine incorporation into the YBCO phase became very minimal. High fluorine content in the YBCO may only exist at the interface between YBCO and BaF_2. The role that fluorination plays with respect to surfaces, twin boundaries and/or grain boundaries still must be clarified.

In this paper, we report two low-temperature fluorination techniques with the aim of controlling the amount of fluorine incorporated into the YBCO structure without the formation of BaF_2.

II. Gas Phase Reaction

In this approach, NH_4HF_2 was used as the fluorine source. At room temperature, NH_4HF_2 is in a solid form. It melts at 120°C and subsequently decomposes to NH_3 and 2HF.

Single phase fully reacted YBCO powder was first mixed with appropriate amounts of NH_4HF_2 to a nominal composition of $F_{0.2}$, $F_{0.5}$, and $F_{1.0}$. These mixed powders then reacted in the alumina boats in air at a temperature range of from 300 to 900°C. The X-ray powder data of the $F_{0.5}$ and $F_{1.0}$ samples revealed that BaF_2 forms at temperatures as low as $300^{o}C$ and its peak intensity increases linearly with increasing temperatures. The amount of fluorine incorporated into the YBCO powder is very low (< 1 at.%). These powders were then pressed into pellets and sintered under O_2 at $950^{o}C$ followed by slow cooling. All samples show typical 90K superconducting transition. However, magnetic measurements showed diamagnetic signals at high temperatures (250-300K) indicating other high temperature phases.

Since single crystal materials have little practical use, in order for technological progress to be made, reliance upon epitaxial high temperature crystalline formation must be overcome. Through the use of fluorine, we have been able to change the random crystalline structure of the conventional material and have been able to get preferred orientation.

With our approach, we observed a strong preferred orientation effect on these pellets, especially in the pellets with $F_{0.2}$. Figure 4 shows the comparison of the X-ray diffraction pattern of a standard YBCO pellet and a slightly fluorinated YBCO pellet. The much stronger [00ℓ] peaks in the fluorinated pellet clearly indicate such orientation effect. From the intensity ratio, we estimated that more than 90% of the crystalline grains are aligned with c-axis perpendicular to the sample surface. A dopant amount of fluorine, or perhaps HF, plays an important role in controlling the nucleation and promoting the preferred crystal growth. We anticipate that this same approach should be suitable to the growth of oriented polycrystalline films. Due to the very anisotropic nature of these materials, producing oriented microcrystalline films would be an essential step in achieving the high critical current needed for practical applications.

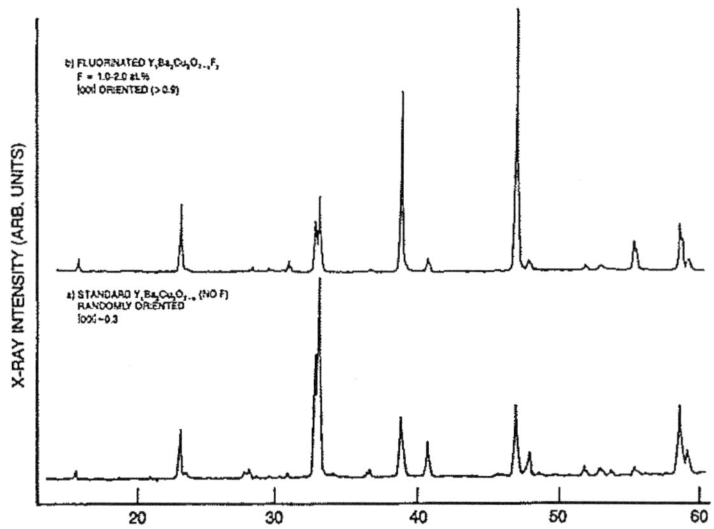

FIG. 4 X-ray diffraction showing the preferred orientation in the fluorinated material.

III. Microwave Plasma Treatment

The objective of this treatment is to use microwave to generate F^+ free radicals and subsequently to replace a controlled amount of oxygen in the YBCO at a temperature below the BaF_2 formation. The microwave plasma experiment is performed in a tubular reactor at a power level of 50-100W with a frequency of 2.45 GHz. The gas mixtures are Ar and NF_3. The operating pressure is ≈20mτ. The NF_3 flow rate, sample temperature, reaction time, and microwave power level are important controlling parameters. Both orthorhombic $YBCO_7$ and tetragonal $YBCO_6$ [10] were fluorinated under similar conditions.

Figure 5 is a cross-sectional SEM micrograph of $YBCO_7$ pellet after being treated with fluorine at 400°C for 1 hour. Several important features can be seen from this figure:

(1) Fluorine diffuses rather fast in this material--at a given time and temperature, a penetration depth of ≈50 um is obtained.

(2) The microprobe analysis indicates that (i) fluorine primarily substitutes for oxygen; (ii) the metal ratio remains unchanged; and (iii) no BaF_2 is observed. However, like any diffusion-controlled process, the F concentration gradient is seen.

3) A layer type of structure is developed in the fluorinated region. Detailed microprobe analysis indicates that fluorine concentration is higher in the darker and closely layered region. We speculated that these layers are associated with twin boundaries. The fluorine may preferentially diffuse through the twin boundaries and then laterally diffuse to the grains. However, a correlation with TEM is needed before the model can be confirmed.

Region 1		Region 2		Region 3	
Y	7.1	Y	7.5	Y	6.6
Ba	14.3	Ba	15.3	Ba	15.9
Cu	21.1	Cu	22.6	Cu	22.7
O	34.6	O	46.9	O	51.9
F	22.9	F	7.7	F	2.9

FIG. 5 SEM cross-sectional micrograph (backscattered mode).

We have applied similar plasma treatment to treat the $YBCO_6$ pellets. The results are shown in Figure 6. The starting pellets are orthorhombic $YBCuO_{6.8}$. The sample is then converted to tetragonal with an expanded c-axis after annealing at 600^oC for three hours in UHV (curve b). The microprobe analysis indicated thatthe composition becomes $YBCO_{6.0}$. The lower Cu valence resulting from the reduction of oxygen is believed to be the cause of the expansion of the c-axis. After fluorination, the structure is converted into orthorhombic with the (006) peak shifting back to the initial position (curve c). This result suggests that the fluorine is not only incorporated into the structure but also is located at the ordered Cu linear chain. Furthermore, the fluorine substitution restored the Cu valence. The microprobe data shows that after fluorine treatment, the F and O contents are ≈15 at.% and 41 at.%, respectively, which resulted in a nominal composition of $Y_1Ba_2Cu_3(OF)_{7.4}$. The 0.4 extra fluorine may occupy the ordered vacancy site. This may be the reason for the modified orthorhombic structure. The fluorinated sample was then reannealed in UHV at 600^oC for 4 hours. There was no detection of fluorine evolution from the sample monitored from a quadropole mass spectometer. The X-ray diffraction data also confirm that the structure remains as the modified orthorhombic structure with no shift of the (006) peak. Furthermore, there is no indication of BaF_2 formation.

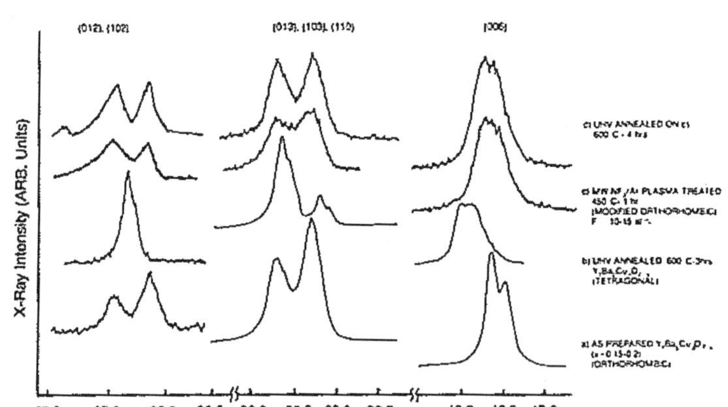

FIG. 6 X-ray difraction showing the orth-teta-orth phase transformation.

We should emphasize that the process has not yet been optimized. However, these data strongly indicate that once fluorine is bonded into the structure, it forms much stronger bonds and is much more stable than oxygen. The results also show that fluorine can be incorporated into the structure without reacting with Ba to form BaF_2. This finding is of scientific and technological importance. There has been a great fear that the oxygen diffusion problem might be the Achilles heel of high T_c superconductivity. We have shown that oxygen can be replaced by fluorine at which point neither the remaining oxygen nor the fluorine diffuses. The oxygen retention and stability problem is in principle solved. The fluorinated material is therefore qualitatively different from the conventional YBCO material.

The electrical properties of the microwave fluorinated samples can generally be classified into three categories. The over-fluorinated samples usually exhibit semiconductor behavior which are either superconducting at lower temperature or not superconducting at all. (Fig. 7)

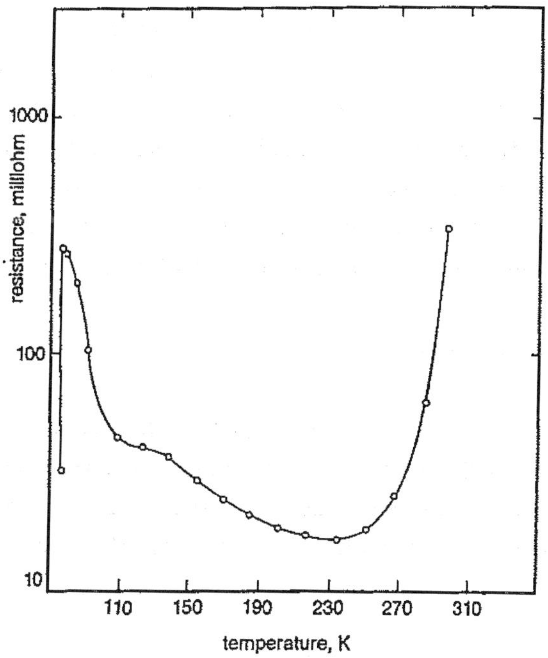

FIG. 7 Resistance vs temperature plot of an over-flourinated sample.

The lightly-fluorinated sample normally shows typical 90K transition. The samples which show zero resistance transition in the 155-168K range are in a delicate balance of fluorination (Fig. 8). That such a balance can be achieved by the crude means of solid phase reactions and the more sophisticated means of plasma is of great importance.

FIG. 8 Resistance vs temperature plot showing 154K zero resistance transition of a microwave treated YBaCuOF sample.

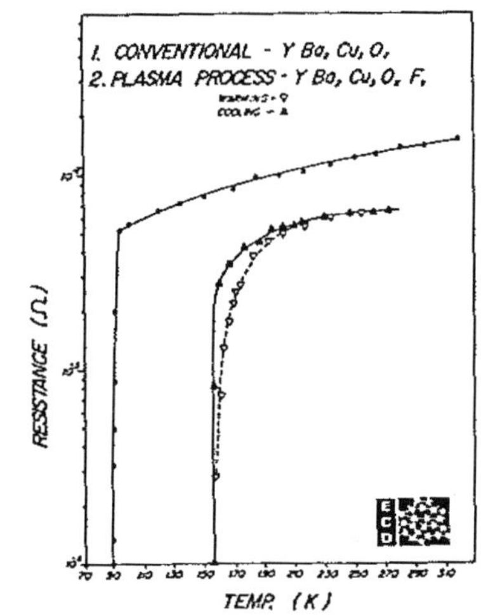

Since microwave plasma reaction is a surface treatment, the fluorine incorporation is primarily through a diffusion controlled process. Therefore, the concentration gradient is present. Furthermore the fluorine incorporation is very sensitive to local power fluctuation, gas depletion, temperature variation as a result of microwave heating, etc. Therefore, other methods are being developed so that the narrow window of reproducibility can be widened.

IV. Summary

In summary, the superconducting zero resistance transitions at 155-168K have been observed in multiphase solid-state synthesized samples as well as low-temperature fluorine-treated samples. The amount of fluorine and its direct location have been generally but not specifically established. Due to the extremely reactive nature of fluorine, methods that will enhance reproducibility still require attention.

In addition to the very high temperature achieved, we showed that dopant amounts of fluorine promote the much desired oriented crystal growth We also found that the replacement of weakly-bonded oxygen by fluorine offers a solution to the material stability problem.

High T_c materials are critically dependent on the Cu-O chain, oxygen stoichiometry, ordered oxygen vacancies and the stereochemistry associated with mixed valency control.[11] The amount of fluorine substitution and its location affect all of these parameters and therefore are critical in the determination of T_c.

REFERENCES

1. J. G. Bednorz and K. A. Muller, 2 Phys. B64, 289 (1986).
2. C. Michel and B. Raveau, J. Solid State Chem. 43 73, (1982); J. Provost, F. Studer, C. Michel and B. Raveau. Synth. Met. 4, 157, (1981).
3. I. S. Shaplygin, B. G. Kakhan and V. B. Lazarev, Russian journal of Inorganic Chemistry 24, 1478 (1979).
4. M. K. Wu, J. R. Ashburn, C. J. Tong, P. H. Hor, R. L. Wong, L. Gao, Z. J. Huang, Y. Q. Wang, and C. W. Chu, Phys. Rev. Lett. 58, 908 (1987); P. H. Hor, L. Gao, R. L. Meng, Z. J. Huang, Y. Q. Wang, K. Forster, J. Vassiliow, and C. W. Chu, Phys. Rev. Lett. 58, 911 (1987).
5. S. R. Ovshinsky, R. T. Young, D. D. Allred, G. DeMaggio and G. A. Van der Leeden, Phys. Rev. Lett. 58, 2579 (1987).

6. Y. R. Meng, Y. R. Ren, M. Z. Lin, Q. Y. Tu, Z. J. Lin, L. H. Sang, W. Q. Bing, Proc. Int. Sym. on High T_c Superconductors, July 1, (1987) Beijing, China (to be published).

7. J. H. Kung. Proc. 1987 Sym on Low Temp. Phys. Sept. 7-8, (1987) Hsin-Chu, Taiwan.

8. C. Krontiras, Private Communication.

9. R. N. Bhargava, S. P. Herko and W. N. Osborne, Phys. Rev. Lett. 59, 1468 (1987).

10. J. S. Swirnea, H. Steinfink, J. Mat. Res. 2, 424, (1987)

11. S. R. Ovshinsky, to be published.

Stan and Iris Ovshinsky

Happiness can be everywhere
Where two people share love

Life is so much easier on a porch
On a beautiful summer day

It is a pity that a beautiful summer day
Is not reflected in a peaceful beautiful world

Maybe an answer for our earth and its strife
Is a global porch and people sharing their love

Poem by Stan
Photo by Dr. Takeo Ohta